Black Holes, White Dwarfs, and Neutron Stars

The Physics of Compact Objects

Black Holes, White Dwarfs, and Neutron Stars

THE PHYSICS OF COMPACT OBJECTS

Stuart L. Shapiro

Saul A. Teukolsky

Cornell University, Ithaca, New York

A Wiley-Interscience Publication

JOHN WILEY & SONS

New York • Chichester • Brisbane • Toronto • Singapore

Library of Congress Cataloging in Publication Data:

Shapiro, Stuart L. (Stuart Louis), 1947–
    Black holes, white dwarfs, and neutron stars.

  "A Wiley-Interscience publication."
  Includes bibliographical references and index.
    1. Black holes (Astronomy)   2. White dwarfs.
  3. Neutron stars.   I. Teukolsky, Saul A. (Saul Arno),
  1947–   . II. Title.
  QB843.B55S5   1983   521′.5   82-20112

ISBN 0-471-87317-9
ISBN 0-471-87316-0 (pbk.)

Printed in the United States of America

10 9 8 7 6 5 4 3 2

To Our Families

Preface

This textbook is the outgrowth of a course on the physics of compact objects, which we have taught at Cornell University since 1975. As a class, compact stars consist of white dwarfs, neutron stars, and black holes. As the endpoint states of normal stellar evolution, they represent fundamental constituents of the physical Universe.

This book, like the course itself, is a product of the burst of scientific activity commencing in the 1960s which centered on compact objects. During this period, pulsars and binary X-ray sources were discovered in our Galaxy. These discoveries proved to be milestones in the development of the field. They furnished definitive proof of the existence of neutron stars, which had previously existed only in the minds of a few theorists. They made plausible the possibility of black holes and even pointed to a few promising candidates in the night sky. More important, perhaps, these discoveries triggered new theoretical studies and observational programs designed to explore the physical nature of compact stars. A whole generation of experimental and theoretical physicists and astronomers has been trained to participate in this exciting, ongoing investigation.

An area of active current research and great popular interest, the study of compact objects is far from complete. Not all—nor even most—of the basic questions concerning their structure and evolution are fully resolved. Nevertheless, answers to these questions seem to be within our eventual reach. The field is now firmly established both as an *observational* and as a rigorous *theoretical* branch of physics. New information and new insights are generated constantly. Moreover, as compact objects undergo interactions involving all four of the fundamental forces of Nature, some of these insights should have an important impact on other branches of physics. Who could foresee, for example, that the question of whether or not nuclear matter makes a phase transition to a quark state at high density might be settled by observations made from an X-ray satellite!

Our book is intended for beginning graduate students or advanced undergraduates in physics and astronomy. *No* prior knowledge of astrophysics or general relativity is assumed. Instead, we introduce the necessary concepts and mathematical tools in these areas as they are required. We *do* assume that the reader has familiarity with electromagnetism, statistical mechanics and thermodynamics, classical and quantum mechanics, and special relativity at the advanced undergraduate or first-year graduate level.

Since we develop only enough general relativity for our purposes, advanced students may wish to consult one of the excellent recent texts on general relativity for reference or further reading. We recommend *Gravitation* by C. W. Misner, K. S. Thorne, and J. A. Wheeler or *Gravitation and Cosmology* by S. Weinberg. Students wishing to explore the physics of "normal" nuclear-burning stars, prior to their gravitational collapse to compact objects, are encouraged to read *Principles of Stellar Evolution* by D. Clayton or *Principles of Stellar Structure*, Vols. I and II, by J. P. Cox and R. T. Giuli. We reemphasize, however, that we have made a serious attempt to keep our book entirely self-contained.

We present the material in its natural order: For each type of compact object (white dwarf, neutron star, or black hole) we first analyze the physical properties of the star in its "ground" state. For example, we initially focus on spherically symmetric, nonrotating, zero-temperature configurations. Next, we analyze the stars when they are subjected to "perturbations," for example, rotation, magnetic fields, transient thermal effects, accretion, and so on. As in most physical systems, the structure of compact stars is best revealed when they are perturbed in some manner. (Indeed, unperturbed compact stars in deep space are unobservable!) We invoke observations to motivate and elucidate the theoretical discussion whenever possible.

We try to provide simple (e.g., "one-dimensional") analytic model calculations for very complicated results or somewhat inaccessible numerical calculations. Such analytic models serve to highlight underlying physical principles, even though they may not be terribly precise. When these back-of-envelope calculations are presented in lieu of more exact computations, we are always careful to clearly state the exact results—if they exist.

In order to keep the book to a manageable length, we have been forced to be selective in our choice of topics. Some of our choices were arbitrary and based on personal preference. In other cases, we have deliberately chosen to emphasize certain topics over others so that the book will not become rapidly outdated. For example, an understanding of polytropic stellar models or the equation of state of an ideal Fermi gas is likely to be useful to the student forever. Similarly, while an exact treatment of neutron star cooling lies still in the future, it is clear already what the important physical principles are and how the calculation will be done. We have accordingly included a detailed "prototype" calculation; the numbers may change, but the ideas will remain the same. On the other hand, we are still ignorant of the detailed mechanism of pulsar emission. It is not even clear which aspects of the physics underlying present models will survive. We have accordingly given a briefer treatment to this topic. In 10 years or so, the reader will have the pleasure of seeing which of our emphases were justified!

To make the book suitable for use as a course textbook, we have included over 250 exercises to be worked by the student. These exercises are sprinkled liberally throughout the main text. Some involve the completion of derivations begun or sketched in the text; others are of a more challenging variety. Answers to many exercises are given. Because most of the results contained in the exercises form an integral part of the discussion and are referred to frequently, we suggest that the student at least stop to *read* those exercises he or she does not intend to solve. Of course, as is true in all branches of physics, true mastery of the concepts only emerges after one *works* in the field: in this case, work means the solving of problems. To make this activity more interesting, we have included a number of "computer exercises" in the book. These are somewhat longer, numerical exercises that can be solved on a programmable hand calculator or on any small computer. They serve not only to illustrate a physical point, but also to introduce the student to numerical problem solving on the computer.

There exist a number of outstanding books and review articles that discuss many aspects of the subject matter developed here. We have referred to them frequently throughout the volume. In addition to the texts cited above, *Relativistic Astrophysics*, Vol. 1, by Ya. B. Zel'dovich and I. D. Novikov may serve as a particularly useful reference volume for interested students.

Not surprisingly, a great many people at numerous institutions contributed to the preparation of this volume. Indeed, it is virtually impossible for us to recall all of the occasions when students and colleagues gave us invaluable criticism, suggestions, and instruction. However, we are particularly grateful to several colleagues for carefully reading portions of an earlier draft of the book and providing us with crucial feedback. For their enthusiasm, unselfish investment of time, and numerous contributions we thank C. Alcock, J. Arons, J. N. Bahcall, J. M. Bardeen, H. A. Bethe, R. D. Blandford, S. Chandrasekhar, J. M. Cordes, T. Gold, K. Gottfried, P. C. Joss, D. Q. Lamb, F. K. Lamb, A. P. Lightman, C. W. Misner, J. P. Ostriker, F. Pacini, D. Pines, S. A. Rappaport, E. E. Salpeter, S. W. Stahler, J. H. Taylor, Y. Terzian, K. S. Thorne, H. M. Van Horn, R. V. Wagoner, and I. Wasserman. In addition, quite a few colleagues offered needed encouragement and support during the preparation of this volume. Among others, we are indebted to W. D. Arnett, G. A. Baym, G. W. Clark, D. M. Eardley, W. A. Fowler, R. Giacconi, J. B. Hartle, S. W. Hawking, M. Milgrom, C. J. Pethick, W. H. Press, R. H. Price, M. J. Rees, M. A. Ruderman, D. N. Schramm, B. F. Schutz, D. W. Sciama, P. R. Shapiro, L. L. Smarr, S. Weinberg, and J. A. Wheeler. For meticulously scrutinizing a revised draft of the manuscript, including the exercises, for errors, we acknowledge with gratitude R. C. Duncan, P. J. Schinder, H. A. Scott, and J. Wang. Finally, we will be forever grateful to D. E. Stewart and G. L. Whitacre for typing the manuscript and the countless revisions that preceded the final draft.

For assistance in the research that went into this book we thank the National Science Foundation for grants awarded to Cornell University; the Alfred P. Sloan

Foundation for a fellowship awarded to one of us (S. L. S.), and the John Simon Guggenheim Memorial Foundation for a fellowship awarded to the other (S. A. T.).

STUART L. SHAPIRO
SAUL A. TEUKOLSKY

Ithaca, New York
January 1983

Suggestions for Using the Book

In our effort to be reasonably complete and self-contained, we have included more material in the book than can be covered comfortably in a one-semester survey course. Therefore, we have prepared the accompanying table to assist instructors in choosing a manageable amount of "essential" subject matter for use in such a course. General readers with similar time constraints may also wish to use the table as a rough guide for independent study.

The basic prerequisites for each section in the table are listed by chapter in the second column. Key chapter sections that can be adequately developed in lectures are listed in the third column. Additional reading assignments for the students may be chosen from the last column. In general, only those selections in the text previously suggested for lectures or further reading constitute the prerequisite material of a cited chapter.

The order of presentation of topics is somewhat flexible. However, we recommend the order presented in the table to capture the faint threads of a "storyline" running throughout the book.

Instructors may feel it necessary to cut their coverage of the text by an additional 10 to 20% in order to develop the material in sufficient depth. On the other hand, individual readers may wish to read those sections omitted from the table which focus on their particular interests. For example, the omitted sections of Chapters 3 and 4 may be of interest to students in solid-state physics; the omitted sections of Chapters 8 and 11 may be of interest to students of nuclear and particle physics, and so on. In such instances, a glance at the section heading or a quick skimming of the introductory paragraphs will usually be sufficient to surmise the content and personal relevance of a particular section.

The sections of the book omitted from the table, together with the appendixes, can be more than adequately covered in a two-semester course. In this circumstance, the instructor may occasionally want to supplement the general discussion in the text with additional, more advanced material contained in some of the references.

		One-Semester Course	
Chapter	Prerequisites	Lectures	Reading
1	None	All	Appendix A
2	None	2.1–2.6, Box 2.1	2.7
3	2	3.2–3.6	3.1
4	2, 3	4.1, 4.2, 4.5, 4.6	4.3, 4.4
5	None	All	
6	None	6.1, Box 6.1, 6.9[a], 6.10	
7	6	7.1, 7.4[a]	7.3
8	2	8.1, 8.5, 8.6, Box 8.1	8.2, 8.4, 8.10, 8.12, 8.14
9	8	9.2–9.4	9.1
10	None	10.2, 10.3, 10.5, 10.8	10.1, 10.7, 10.9–10.11
11	None	11.1–11.4, 11.8–11.9	11.5–11.7
12	5	12.1, 12.3, 12.4, 12.6	12.2, 12.5, 12.7, 12.8
13	None	13.2, 13.3, 13.5, 13.7, Box 13.1	13.1, 13.4, 13.6
14	13	14.3[a]	14.1, Appendix J[a], 14.5[a]
15	13	15.1, Fig. 15.1	15.2
16	5	16.3, 16.5, Box 16.1	16.1, 16.2, 16.4, 16.7
17	5, 6	Omit	Omit
18	2, 8, 11	18.1, 18.5, 18.7	18.2, 18.4, 18.6

[a]Summarize only

Contents

Black Holes, White Dwarfs, and Neutron Stars

The Physics of Compact Objects

1

Star Deaths and the Formation of Compact Objects

1.1 What Are Compact Objects?

A book on compact objects logically begins where a book on normal stellar evolution leaves off. Compact objects—white dwarfs, neutron stars, and black holes—are "born" when normal stars "die," that is, when most of their nuclear fuel has been consumed.

All three species of compact object differ from normal stars in two fundamental ways. First, since they do not burn nuclear fuel, they cannot support themselves against gravitational collapse by generating thermal pressure. Instead, white dwarfs are supported by the pressure of degenerate electrons, while neutron stars are supported largely by the pressure of degenerate neutrons. Black holes, on the other hand, are completely collapsed stars—that is, stars that could not find *any* means to hold back the inward pull of gravity and therefore collapsed to singularities. With the exception of the spontaneously radiating "mini" black holes with masses M less than 10^{15} g and radii smaller than a fermi, all three compact objects are essentially static over the lifetime of the Universe. They represent the final stage of stellar evolution.

The second characteristic distinguishing compact objects from normal stars is their exceedingly small size. Relative to normal stars of comparable mass, compact objects have much smaller radii and hence, much stronger surface gravitational fields. This fact is dramatically illustrated in Table 1.1 and Figure 1.1.

Because of the enormous density range spanned by compact objects, their analysis requires a deep physical understanding of the structure of matter and the nature of interparticle forces over a vast range of parameter space. All four fundamental interactions (the strong and weak nuclear forces, electromagnetism,

1

Table 1.1

Distinguishing Traits of Compact Objects

Object	Mass[a] (M)	Radius[b] (R)	Mean Density (g cm^{-3})	Surface Potential (GM/Rc^2)
Sun	$M_\odot$	$R_\odot$	1	10^{-6}
White dwarf	$\lesssim M_\odot$	$\sim 10^{-2} R_\odot$	$\lesssim 10^7$	$\sim 10^{-4}$
Neutron star	$\sim 1\text{--}3 M_\odot$	$\sim 10^{-5} R_\odot$	$\lesssim 10^{15}$	$\sim 10^{-1}$
Black hole	Arbitrary	$2GM/c^2$	$\sim M/R^3$	~ 1

$^a M_\odot = 1.989 \times 10^{33}$ g
$^b R_\odot = 6.9599 \times 10^{10}$ cm

and gravitation) play a role in compact objects. Particularly noteworthy are the large surface potentials encountered in compact objects, which imply that general relativity is important in determining their structure. Even for white dwarfs, where Newtonian gravitation is adequate to describe their equilibrium structure, general relativity turns out to be important for a proper understanding of their stability.

Because of their small radii, luminous white dwarfs, radiating away their residual thermal energy, are characterized by much higher effective temperatures than normal stars even though they have lower luminosities. (Recall that for a blackbody of temperature T and radius R, the flux varies as T^4, so the luminosity varies as $R^2 T^4$). In other words, white dwarfs are much "whiter" than normal stars, hence their name.

Conversely, no light (or anything else, for that matter), can escape from a black hole. Thus isolated black holes will appear "black" to any observer. (This statement has to be modified slightly when certain quantum mechanical effects are taken into account, as we will discuss in Chapter 12.)

Neutron stars derive their name from the predominance of neutrons in their interior, following the mutual elimination of electrons and protons by inverse β-decay. Because their densities are comparable to nuclear values, neutron stars are essentially "giant nuclei" (10^{57} baryons!), held together by self-gravity.

White dwarfs can be observed directly in optical telescopes during their long cooling epoch. Neutron stars can be observed directly as pulsating radio sources ("pulsars") and indirectly as gas-accreting, periodic X-ray sources ("X-ray pulsars"). Black holes can only be observed indirectly through the influence they exert on their environment. For example, they could be observed as gas-accreting, aperiodic X-ray sources under appropriate circumstances. We will discuss these as well as other observable phenomena involving compact stars in subsequent chapters.

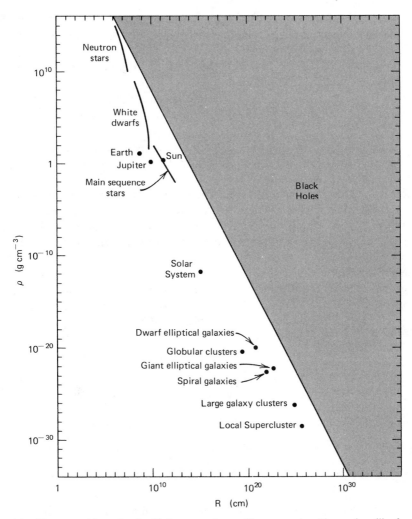

Figure 1.1 Compact objects in the Universe at large. The mean densities and radii of various astronomical bodies are shown.

1.2 The Formation of Compact Objects

Compact objects are the end products of stellar evolution. The primary factor determining whether a star ends up as a white dwarf, neutron star, or black hole is thought to be the star's mass.

White dwarfs are believed to originate from light stars with masses $M \lesssim 4M_\odot$. As we shall see in Chapter 3, there is a maximum allowed mass for white dwarfs, which is around $1.4 M_\odot$. White dwarf progenitor stars probably undergo relatively gentle mass ejection (forming "planetary nebulae") at the end of their evolutionary lifetimes before becoming white dwarfs.

Neutron stars and black holes are believed to originate from more massive stars. However, the dividing line between those stars that form neutron stars and those that form black holes is very uncertain because the final stages of evolution of massive stars are poorly understood. Neutron stars also have a maximum mass (in the range of 1.4–$3 M_\odot$), but numerical calculations that attempt to deal with steady mass loss from stars, or catastrophic mass ejection and supernova explosions, are in a very primitive stage. Thus the fate of a star with a mass $M \gtrsim 4 M_\odot$ is not clear at the present time.

Table 1.2 gauges our current level of ignorance regarding the fate of evolved stars. A further uncertainty is that in the calculations underlying Table 1.2, one generally assumes that factors other than mass (e.g., magnetic fields, rotation, binary star effects) are less important than mass in determining the ultimate fate of a star.

Total gravitational collapse leading to a black hole can, in principle, occur by routes other than by the direct collapse of an evolved, massive star. For example, since there is a definite maximum mass above which a white dwarf or a neutron star can no longer support itself against collapse, the accretion of gas by either of these objects (e.g., in a binary system) can lead to black hole formation.

Table 1.2
Outcome of Stellar Evolution vs. Mass[a]

Range of Mass	Result Expected
$\lesssim M_\odot$	Lifetime longer than age of universe
$1 \lesssim M/M_\odot \lesssim (3 \text{ to } 6)$	White dwarf + planetary nebula, Mass loss

$(3 \text{ to } 6) \lesssim M/M_\odot \lesssim (5 \text{ to } 8)$	(a) Degenerate ignition of $^{12}C + {}^{12}C$
	  (1) "fizzle" $\rightarrow$ core collapse, or
	  (2) detonation and dispersal of core, or
	  (3) deflagration and (??)
	(b) Pulsationally driven mass loss to white dwarf

$(5 \text{ to } 8) \lesssim M/M_\odot \lesssim (60 \text{ to } 100)$	Core collapse + supernova $\rightarrow$ neutron star, sometimes black hole (?)
$(60 \text{ to } 100) \lesssim M/M_\odot$	Instability

[a]After Arnett (1979), in *Sources of Gravitational Radiation*, edited by L. Smarr, Cambridge University Press.

At least two additional black hole formation processes have been proposed by theoreticians (although they have not yet been confirmed by observers!). The first involves the collapse of a hypothetical "supermassive star," which leads to the formation of a "supermassive black hole." We will discuss this process in more detail in Chapter 17, but for now simply note that such supermassive stars are known to be unstable when they reach a certain critical density, depending on mass. Accordingly, when a supermassive star has evolved to this density by radiative cooling and contraction, it may undergo catastrophic collapse to a black hole. This may be the origin of supermassive holes with $M/M_\odot \sim 10^6$–10^9 proposed to explain the violent activity observed in quasars and energetic galactic nuclei.

The second mechanism involves the formation of primordial black holes in the early Universe, due to perturbations in the homogeneous background density field.[1] Since all "mini-black holes" with $M \lesssim 10^{15}$ g will have radiated away their mass by the Hawking process (see Section 12.8) in a time less than the age of the Universe, only primordial black holes with $M \gtrsim 10^{15}$ g could exist today.

Astronomical observation can, in principle, corroborate the idea that compact objects are the end products of stellar evolution. By counting the number of stellar deaths in the Galaxy since the initiation of star formation, we can presumably estimate the number (and number density) of compact objects now present in the Galaxy. We can then compare this number with observations.

The calculation will be most reliable in the case of white dwarfs. There exists a wealth of observational data on white dwarfs and planetary nebulae, so that any estimate of their number density in space based upon stellar death rates can be checked.

A similar estimate in the case of neutron stars or black holes is much more uncertain. Besides the greater uncertainty in the mass range of the progenitor stars, another complication is that neutron stars and black holes can only be observed during astronomically brief epochs when they are active as pulsars or compact X-ray sources.

Despite the uncertainties, the data already allow us to draw some interesting conclusions. As we shall see, compact objects are apparently as common as other stars in the Galaxy. The observed mass density in white dwarfs is a significant fraction of the mass density in normal stars. Many white dwarfs and neutron stars (i.e., pulsars) have been observed, and at least one good black hole candidate has been identified (Cygnus X-1).

In the rest of this chapter, we shall discuss how the number density of compact objects in the solar neighborhood can be determined from the statistics of stellar birth and death rates. Although the numbers we quote are uncertain, once a

---

[1] Zel'dovich and Novikov (1966); Hawking (1971).

detailed understanding of the late stages of stellar evolution emerges, more reliable estimates will be possible. Fortunately, most of the properties of compact objects do not depend on the uncertain histories of their progenitors. In the next chapter, we shall begin to analyze the physics on which their properties do depend.[2]

1.3   The Statistics of Stellar Births and Deaths

All quantitative determinations of stellar birth rates and death rates proceed along the following lines[3]:

Define

$$\phi(M_v) \equiv \text{luminosity function of field stars,} \qquad (1.3.1)$$

that is, the number of stars of *all* types (not just main-sequence stars) per unit absolute visual magnitude and per *cubic* parsec in the plane of the Galaxy that are not in star clusters. [See Appendix A for a discussion of the astronomical units "absolute magnitude" (power) and "parsec" (distance), and for a brief summary of stellar evolution and a description of the main sequence.]

Next, define

$$\phi_{\text{MS}}(\log M) \equiv \begin{array}{l} \text{present-day mass function (PDMF) of} \\ \textit{main-sequence} \text{ field stars in the solar} \\ \text{neighborhood,} \end{array} \qquad (1.3.2)$$

that is, the number of main-sequence stars per unit logarithmic mass interval per *square* parsec. Note that all masses in this section [such as M in Eq. (1.3.2)] are in units of solar masses, and that all logarithms are to *base 10*. The quantities ϕ_{MS} and $\phi(M_v)$ are related by

$$\phi_{\text{MS}}(\log M) = \phi(M_v)\left|\frac{dM_v}{d\log M}\right| 2H(M_v) f_{\text{MS}}(M_v). \qquad (1.3.3)$$

---

[2] Many readers may wish to omit the technical astronomical details in the rest of this chapter on a first reading. They should nevertheless glance at Table 1.4 before moving on, bearing in mind the large uncertainties in its entries.

[3] For example, Ostriker, Richstone, and Thuan, (1974); Audouze and Tinsley, (1976); Miller and Scalo, (1979); Bahcall and Soneira (1980).

Here the derivative term converts a luminosity function to a mass function. The factor $2H(M_v)$ arises from integrating the luminosity function over distance z measured perpendicular to the Galactic plane, assuming that stars are distributed as $\exp(-|z|/H)$, where $H(M_v)$ is the scale height. The factor $f_{MS}(M_v)$ gives the fraction of stars at a given magnitude that are on the main sequence.

The basic observational ingredient in Eq. (1.3.3) is $\phi(M_v)$. The results of numerous determinations of this function are in excellent agreement,[4] and the values adopted by Bahcall and Soneira are given in Table 1.3. Also shown in the table is the main sequence mass–luminosity relation, M_v vs. $\log M$, which is reasonably well-determined both observationally and theoretically. The variation of scale height H with M_v (and hence M) is not as well-determined except for the brightest stars. It is clear from Table 1.3, however, that high-mass, high-luminosity stars are more strongly concentrated towards the plane of the disk than low-mass stars. The correction factor f_{MS} deviates from unity because of the presence of evolved stars that are no longer burning only hydrogen in their interiors. Typically, f_{MS} satisfies $f_{MS} \sim 1$ for faint stars with $M_v \gtrsim 3$ ($M \lesssim 1.4 M_\odot$), but falls to about $\frac{1}{2}$ for bright stars with $M_v \lesssim 0$ ($M \gtrsim 3.5 M_\odot$). The correction f_{MS} is not well determined for bright stars. The resulting PDMF determined by Eq. (1.3.3) is also given in Table 1.3. The uncertainty for low-mass stars is largely due to $H(M_v)$; for high-mass stars it is largely due to the M-M_v relation and to $\phi(M_v)$.

Now define the field star initial mass function (IMF):

$$\xi(\log M) = \text{ total number of field stars that have } \textit{ever} \text{ formed, per unit area per unit logarithmic mass interval.} \qquad (1.3.4)$$

Assuming a constant birthrate,[5] the rate of formation of field stars per unit area per $\log M$ is simply $\xi(\log M)/T_0$. Here T_0 is the age of the Galaxy (essentially equal to the age of the Universe determined from Big Bang cosmology), which we take to be 12×10^9 years.

Now we can relate ϕ_{MS} to ξ by introducing T_{MS}, the main-sequence lifetime. Most massive stars born since star formation began have long since evolved off the main sequence ($T_{MS} < T_0$) and so do not contribute to ϕ_{MS}. Thus

$$\phi_{MS}(\log M) = \xi(\log M)\frac{T_{MS}}{T_0}, \qquad T_{MS} < T_0. \qquad (1.3.5)$$

---

[4] See, for example, McCuskey (1966); Miller and Scalo (1979); Bahcall and Soneira (1980).
[5] Miller and Scalo (1979) discuss the evidence that this is a reasonable approximation.

Table 1.3

Relationships for the PDMF

M_v	$\phi(M_v)^a$ (stars pc^{-3} mag^{-1})	log $M/M_\odot$ ^{b}	$-\dfrac{dM_v^b}{d\log M}$	$2H^c$ (pc)	log T_{MS}(yr)^d	f_{MS}^e	ϕ_{MS}(log M) (stars pc^{-2} log M^{-1})
−6	1.49(−8)	2.07	3.7	180	6.42	0.40	3.97(−6)
−5	7.67(−8)	1.80	3.7	180	6.50	0.40	2.04(−5)
−4	3.82(−7)	1.53	3.7	180	6.58	0.41	1.04(−4)
−3	1.80(−6)	1.26	3.7	180	6.84	0.42	5.03(−4)
−2	7.86(−6)	0.99	3.7	180	7.19	0.43	2.25(−3)
−1	3.07(−5)	0.72	3.7	180	7.68	0.46	9.41(−3)
0	1.04(−4)	0.45	10.8	180	8.36	0.50	1.01(−1)
1	2.95(−4)	0.36	10.8	180	8.62	0.56	3.21(−1)
2	6.94(−4)	0.26	10.8	180	8.93	0.64	8.63(−1)
3	1.36(−3)	0.17	10.8	300	9.24	0.78	3.44(+0)
4	2.26(−3)	0.08	10.8	465	9.60	0.98	1.11(+1)
5	3.31(−3)	−0.02	10.8	630	9.83	1.00	2.25(+1)
6	4.41(−3)	−0.11	10.8	650	10.28	1.00	3.10(+1)
7	5.48(−3)	−0.20	10.8	650	—	1.00	3.85(+1)
8	6.52(−3)	−0.29	10.8	650	—	1.00	4.58(+1)
9	7.53(−3)	−0.39	10.8	650	—	1.00	5.29(+1)
10	8.52(−3)	−0.48	10.8	650	—	1.00	5.98(+1)
11	9.54(−3)	−0.57	10.8	650	—	1.00	6.70(+1)
12	1.06(−2)	−0.67	10.8	650	—	1.00	7.44(+1)
13	1.17(−2)	−0.76	10.8	650	—	1.00	8.21(+1)
14	1.29(−2)	−0.85	10.8	650	—	1.00	9.06(+1)
15	1.41(−2)	−0.94	10.8	650	—	1.00	9.90(+1)
16	1.41(−2)	−1.04	10.8	650	—	1.00	9.90(+1)

^{a}From Bahcall and Soneira (1980), Eq. (1).
^{b}From Bahcall and Soneira (1980), Eq. (17).
^{c}From Bahcall and Soneira (1980), Figure 2.
^{d}From Miller and Scalo (1979). We have interpolated their results to agree with the Bahcall–Soneira values for the same M, not M_v, since theoretical calculations generally give T_{MS} in terms of M.
^{e}From Bahcall and Soneira (1981), Eq. (1).

Less massive stars are still on the main sequence, and so

$$\phi_{MS}(\log M) = \xi(\log M), \qquad T_{MS} \geqslant T_0. \qquad (1.3.6)$$

The quantity T_{MS} is given as a function of M in Table 1.3; a rough analytic expression is

$$T_{MS} = \frac{\Delta X_{MS} M E^*}{L}, \qquad (1.3.7)$$

where ΔX_{MS} is the mass fraction of hydrogen burned during the main-sequence phase, ~ 0.13, and E^* is the energy released per gram when H reacts to form He via nuclear fusion—that is, $E^* \simeq 0.007c^2 \simeq 6.4 \times 10^{18}$ erg g^{-1}. From stellar structure theory we know that, very crudely,

$$\frac{L}{L_\odot} \simeq \left(\frac{M}{M_\odot}\right)^{3.5}, \qquad M \lesssim 10M_\odot \qquad (1.3.8)$$

(the main-sequence "mass–luminosity law"). Equations (1.3.7) and (1.3.8) then give

$$T_{MS} \simeq 13 \times 10^9 \left(\frac{M}{M_\odot}\right)^{-2.5} \text{yr}, \qquad M \lesssim 10M_\odot. \qquad (1.3.9)$$

Exercise 1.1 Determine the reliability of Eqs. (1.3.8) and (1.3.9) by comparing them with more accurate values of L and T_{MS} vs. $M/M_\odot$ obtained from Table 1.3 and Eq. (A.6).

Using Eqs. (1.3.5) and (1.3.6) and taking ϕ_{MS} from Table 1.3, we have computed $\xi(\log M)$. The results can be summarized by the analytic fit

$$\log \xi(\log M) = A_0 + A_1 \log M + A_2(\log M)^2,$$

$$A_0 = 1.41, \qquad A_1 = -0.90, \qquad A_2 = -0.28, \qquad \log M > -1.$$

$$(1.3.10)$$

Here M is measured in units of $M_\odot$. The slope of the IMF is approximately

$$\frac{d\log \xi}{d\log M} \simeq -(0.9 + 0.6\log M). \qquad (1.3.11)$$

These values are not too different from those of Miller and Scalo (1979), bearing in mind the uncertainties.

It is customary to compare modern determinations of ξ with Salpeter's (1955) birth rate function. In his pioneering study of this problem, Salpeter considered star densities in the Galactic disk per unit volume, neglecting the factor $H(M_v)$, and used "old" values for T_{MS}. The Salpeter birth rate function is

$$\psi_s d\left(\frac{M}{M_\odot}\right) = 2 \times 10^{-12} \left(\frac{M}{M_\odot}\right)^{-2.35} d\left(\frac{M}{M_\odot}\right) \text{ stars pc}^{-3} \text{ yr}^{-1} \quad (1.3.12)$$

over the range $0.4 \leqslant M/M_\odot \leqslant 10$.

Exercise   1.2   Show that the IMF, ξ_s, corresponding to ψ_s, is given by

$$\xi_s \, d\log\left(\frac{M}{M_\odot}\right) = 2H(M)T_0\psi_s \, d\left(\frac{M}{M_\odot}\right). \qquad (1.3.13)$$

From Eqs. (1.3.12) and (1.3.13), we get

$$\frac{d\log\xi_s}{d\log M} \simeq -1.35, \qquad (1.3.14)$$

neglecting the variation of H with M. This is in reasonable agreement with the more recent determination (1.3.11) for the range 2–$10M_\odot$. Above $10M_\odot$, the IMF is steeper than this, while below $2M_\odot$ it is flatter. The differences are due to the neglect of $H(M)$ and to the updated values of T_{MS}.

A consistency check on the PDMF, $\phi_{\mathrm{MS}}(M)$, is provided by a comparison with the *Oort limit*. From a study of the dynamical motions of stars in the solar neighborhood of the Galaxy, Oort (1960) derived a value for the *total* amount of matter responsible for the observed accelerations. A recent determination of the Oort limit gives $(0.14 \pm .03)M_\odot$ pc^{-3} (Krisciunas, 1977). Using the PDMF, one can compute the total mass of main-sequence stars in the solar neighborhood. Bahcall and Soneira (1980) obtain $0.040M_\odot$ pc^{-3}. In what form is the remaining matter? Interstellar gas accounts for $0.045M_\odot$ pc^{-3} (Spitzer, 1978). Observed white dwarfs account for another $0.005M_\odot$ pc^{-3}.[6] This leaves a "missing mass" of around $0.05M_\odot$ pc^{-3} in the solar neighborhood. There are numerous speculations concerning the nature of this "missing mass" (rocks, planets, low velocity M dwarfs, black dwarfs, black holes, etc.). We do not know the answer at present.

We can use the Oort limit to conclude that roughly half of the mass of the Galaxy has already been processed by stars that have completed their evolution. The main-sequence lifetime, T_{MS}, is shorter than the age of the Galaxy, $T_0 = 12 \times 10^9$ yr, for $M \gtrsim 0.9M_\odot$. (This value is derived by interpolation in Table 1.3.) Using the Salpeter birth rate function as an approximation, we compute

$$\int_{0.9}^{\infty}\psi_s M \, d\left(\frac{M}{M_\odot}\right) = 5 \times 10^{-12}M_\odot \text{ pc}^{-3} \text{ yr}^{-1}. \qquad (1.3.15)$$

Multiplying by T_0 gives the total mass consumed by bright stars; we obtain $0.06M_\odot$ pc^{-3}, roughly half the Oort limit.

---

[6]We quote the value of Bahcall and Soneira (1980). They assume that the sharp drop off in the number of very faint white dwarfs reported by Liebert et al. (1979) is real, and so determine the number density of white dwarfs in the solar neighborhood to be 0.008 pc^{-3}. Multiplying by an average "observed" white dwarf mass of $0.65M_\odot$ (see footnote 14 in Chapter 4) gives $0.005M_\odot$ pc^{-3}. This is a factor of 4 smaller than Weidemann's (1977) value.

We can now compute the death rate of massive stars, and hence the birth rate of compact objects. For stars with masses $\gtrsim 0.9 M_\odot$, we have $T_{MS} < T_0$ so that we may assume the stellar population is stationary—the average death rate is in approximate balance with the birth rate. For a simple treatment we can use the analytic Salpeter birth rate function, which is not greatly in error for the mass range we are interested in. (See Exercise 1.7 for a "modern" calculation.)

For illustrative purposes, we shall take $1-4 M_\odot$ as the mass of white dwarf progenitors (cf. Table 1.2). The mass range for neutron star progenitors is even more uncertain, but let us use $4-10 M_\odot$ and assume that all stars with masses greater than $10 M_\odot$ end up as black holes.

The Salpeter birth rate function ignores the variation with height H above the Galactic plane. We shall convert rates per unit volume to rates by multiplying by the effective volume of the Galaxy,

$$V_{\text{disk}} = \pi r^2 (2H) = 1.3 \times 10^{11} \text{ pc}^3, \qquad (1.3.16)$$

where we have taken $2H = 180$ pc, appropriate for stars more massive than $2 M_\odot$ (cf. Table 1.3) and $r = 15$ kpc, the characteristic radius of the Galactic disk.[7]

The birth rate ($=$ death rate) for stars in the range M_1 to M_2 is

$$V_{\text{disk}} \int_{M_1/M_\odot}^{M_2/M_\odot} \psi_s \, d\left(\frac{M}{M_\odot}\right) = 0.19 \left(\frac{M}{M_\odot}\right)^{-1.35} \Bigg|_{M_2/M_\odot}^{M_1/M_\odot} \text{ yr}^{-1}. \qquad (1.3.17)$$

The number density n of compact objects formed from parent stars with masses in the range M_1 to M_2 is

$$n = T_0 \int_{M_1/M_\odot}^{M_2/M_\odot} \psi_s \, d\left(\frac{M}{M_\odot}\right) = 0.018 \left(\frac{M}{M_\odot}\right)^{-1.35} \Bigg|_{M_2/M_\odot}^{M_1/M_\odot} \text{ pc}^{-3}. \qquad (1.3.18)$$

To determine the mass density ρ of these compact objects, we cannot simply integrate the product $M\psi_s$. This would be correct *if* there were no mass loss and all the mass of the progenitor star ended up in the compact object. For white dwarfs and neutron stars, which have maximum masses of around $1.4 M_\odot$ and $2-3 M_\odot$, respectively, we know that this would be wrong. Instead, we determine ρ

---

[7]This procedure is very crude because we are assuming that rates determined for the solar neighborhood apply throughout the Galaxy. Also, the "characteristic" radius of the disk is not really a well-defined quantity. (The sun is located ≈ 10 kpc from the Galactic center.)

by multiplying n by an average mass $\langle M \rangle$, where we take $\langle M \rangle_{\text{wd}} = 0.65 M_\odot$ and $\langle M \rangle_{\text{ns}} = 1.4 M_\odot$ for illustrative purposes. For black holes we are even less sure what to do, so we simply neglect mass loss. Forming the ratio of ρ to the total mass density $\rho_T = 0.14 M_\odot \text{ pc}^{-3}$ (Oort limit), we have

$$\frac{\rho}{\rho_T} = \frac{n \langle M \rangle}{\rho_T} \quad \text{(white dwarf or neutron star)}, \qquad (1.3.19)$$

while for black holes,

$$\frac{\rho}{\rho_T} = \frac{T_0}{\rho_T} \int_{M_1/M_\odot}^{M_2/M_\odot} M \psi_s \, d\left(\frac{M}{M_\odot}\right) = 0.49 \left(\frac{M}{M_\odot}\right)^{-0.35} \Bigg|_{M_2/M_\odot}^{M_1/M_\odot}. \qquad (1.3.20)$$

The average separation of each kind of compact object in the solar neighborhood is

$$\langle d \rangle = \frac{1}{(4\pi n/3)^{1/3}}. \qquad (1.3.21)$$

Using the above formulas, we can fill in the entries in Table 1.4.

Several entries in this table can be checked with observations. Bahcall and Soneira find the local white dwarf number density to be 0.008 pc^{-3} (see footnote 6), within a factor of 2 of our theoretical estimate. Dividing by T_0 gives an estimated white dwarf birth rate function of about 10^{-12} pc^{-3} yr^{-1}.

Table 1.4

Compact Objects in the Solar Neighborhood[a]

Object	Mass Range of Parent Star ($M_\odot$)	Integrated Galactic Birth Rate (yr^{-1})	Number Density (pc^{-3})	$\dfrac{\rho}{\rho_T}$	$\langle d \rangle$ (pc)
White dwarfs	1–4	0.16	1.5×10^{-2}	0.070	2.5
Neutron stars	4–10	0.021	2.0×10^{-3}	0.020	4.9
Black holes	> 10	0.0085	8.0×10^{-4}	0.22	6.7

[a]These values are obtained from Eqs. (1.3.17)–(1.3.21).
Note: Nearest known white dwarf: Sirius B, 2.7 pc. Nearest known neutron star: PSR 1929 + 10, 50 pc. Nearest known black hole candidate: Cygnus X-1, ~2 kpc.

Uncertainties in the planetary nebula distance scale make it difficult to derive a birth rate function, but most estimates[8] are consistent with the white dwarf rate.

These results seem to support the hypothesis that stars in the mass range $\sim 1-4 M_{\odot}$ go through the planetary nebula stage at the end of their lifetime and that all white dwarfs are produced by this process.

Data for neutron stars are somewhat more uncertain. Arnett (1979) estimates the rate of pulsar formation in the Galaxy as

$$R_{\mathrm{PSR}} \sim \frac{1}{35 \times 10^{\pm 1}\ \mathrm{yr}}, \tag{1.3.22}$$

$$\dot{\sigma}_{\mathrm{PSR}} \sim 4 \times 10^{-11 \pm 1}\ \mathrm{pc}^{-2}\ \mathrm{yr}^{-1}. \tag{1.3.23}$$

Taylor and Manchester (1977) quote a range

$$\dot{\sigma}_{\mathrm{PSR}} \sim (3-10) \times 10^{-11}\ \mathrm{pc}^{-2}\ \mathrm{yr}^{-1}, \tag{1.3.24}$$

whereas Gunn and Ostriker's (1970) original value is

$$\dot{\sigma}_{\mathrm{PSR}} = 5 \times 10^{-11}\ \mathrm{pc}^{-2}\ \mathrm{yr}^{-1}. \tag{1.3.25}$$

These estimates are sensitive to the distance scale for pulsars, which is determined from the dispersion measure (cf. Section 10.4).

A rough comparison of the above pulsar data with our Table 1.4 is provided by dividing Eq. (1.3.24) by $2H = 180$ pc and multiplying by $T_0 = 12 \times 10^9$ yr to get

$$n_{\mathrm{PSR}} \sim (2-6) \times 10^{-3}\ \mathrm{pc}^{-3}, \tag{1.3.26}$$

in good agreement. A more careful analysis is given by Shipman and Green (1980).

It is of interest to compare the pulsar birth rate with the rate of supernovae. Theorists believe that most, if not all, pulsars are the remnants of supernova explosions. (The Crab supernova of 1054 A.D. certainly did give rise to a pulsar.) Tammann (1978) has estimated the historical supernova rate for the Galaxy as

$$R_{\mathrm{SN,hist}} = \frac{N}{f \Delta t} = \frac{1}{28\ \mathrm{yr}}, \tag{1.3.27}$$

where the number of historical supernovae is $N = 6$ in a time interval $\Delta t = 10^3$ yr. The quantity $f = 60°/360°$ is the fraction of the Galactic disk in which the

---

[8] Osterbrock (1973); Alloin, Cruz-González, and Peimbert (1976); Cahn and Wyatt (1976).

supernovae have been observed; presumably supernovae in other directions from the sun are invisible because of Galactic absorption. This rate gives [dividing by the disk area $\pi \times (15 \text{ kpc})^2$]

$$\dot{\sigma}_{SN} \sim 5 \times 10^{-11} \text{ pc}^{-2} \text{ yr}^{-1}. \tag{1.3.28}$$

The typical rate of supernovae observed in other galaxies is (Tammann, 1978)

$$R_{SN} \sim \frac{1}{300 \text{ yr}} \quad \text{per galaxy.} \tag{1.3.29}$$

If the historical rate for our Galaxy is typical, then presumably most extragalactic supernovae are not seen by observers. Tammann believes that Eq. (1.3.27) is a good estimate for the actual rate in our Galaxy, while Van den Bergh (1978) has obtained a rate of

$$R_{SN} \sim \frac{1}{60 \text{ yr}} \quad \text{for our Galaxy,} \tag{1.3.30}$$

a factor of 2 lower than Tammann's rate.

Note that $\dot{\sigma}_{SN} \sim \dot{\sigma}_{PSR}$, in support of our theoretical ideas.

A comparison of the theoretical birth rates of different compact objects with the statistics of Galactic X-ray sources would, in principle, be quite illuminating. However, since these sources are likely to be compact objects in *binary* systems (see Chapter 13), fundamental uncertainties regarding their evolution (e.g., angular momentum and mass loss) make it difficult to perform a reliable comparison at present.

Exercise   1.3   Using the Salpeter birth rate function and assuming that all stars with masses $> 10 M_\odot$ form black holes, estimate the average mass of black holes formed by star collapse. Neglect mass loss.

Exercise   1.4   How are the entries for neutron stars and black holes in Table 1.4 affected if $T_0 = 18 \times 10^9$ yr? If $T_0 = 9 \times 10^9$ yr?

Exercise   1.5   The main-sequence of the Pleiades cluster consists of stars with masses $M \lesssim 6 M_\odot$; the more massive stars have already evolved off the main-sequence (see Section A.2). The discovery that this cluster may contain a white dwarf implies that stars with masses as large as $6 M_\odot$ eventually form white dwarfs (why?), and not only those up to $4 M_\odot$, as assumed in constructing Table 1.4. Use the Salpeter birth rate function to redetermine the entries in Table 1.4 in light of this result. Compare your theoretical predictions with the observed stellar statistics for white dwarfs, planetary nebulae, supernovae, and pulsars. [Note: Romanishin and Angel (1980) have studied four other star clusters containing white dwarfs and have tentatively concluded that stars as large as $7 M_\odot$ form white dwarfs.]

Exercise   1.6   (based on Ostriker, Richstone, and Thuan 1974) Suppose that high-mass stars ($M > 8M_\odot$) produce pulsars, while moderate-mass stars (4–$8M_\odot$) explode completely (Arnett and Schramm, 1973), each explosion producing from the processed core about $1.4M_\odot$ of elements close to iron in atomic number.

(a)   Use the Salpeter birth rate function and $T_0 = 12 \times 10^9$ yr to compute the total density of iron returned to the interstellar medium by such explosions.

Answer: $2 \times 10^{-3} M_\odot$ pc^{-3}.

(b)   Use the Oort limit to predict the abundance of iron (fraction by mass) in the Galaxy disk, assuming most of the iron originates from such explosions. Compare with the observed abundance of 1.4×10^{-3} (Withbroe, 1971).

Answer: Predict 1.7×10^{-2}.

Exercise   1.7   Recompute the entries in Table 1.4 using the birth rate function based on the Bahcall–Soneira PDMF. Note that integrals involving Eq. (1.3.10) can be expressed in terms of error functions. Alternatively, one can use the following analytic approximation:

$$\xi(\log M) = D_0 M^{D_1},$$

$$D_0 = 33, \qquad D_1 = -0.5, \qquad 0.1 \leqslant M \leqslant 1,$$

$$D_0 = 35, \qquad D_1 = -1.5, \qquad 1 \leqslant M \leqslant 10,$$

$$D_0 = 163, \qquad D_1 = -1.9, \qquad 10 \leqslant M,$$

where M is in units of $M_\odot$. In this case, note that $d\log M = dM/(M \ln 10)$.

At this stage it is appropriate to inject a cautionary note. Besides all the observational uncertainties, we have made a number of theoretical assumptions that may not be valid. In addition to those mentioned earlier, we have also neglected the effects of a possible *steady* mass loss during the late stages of evolution of massive stars. In P Cygni supergiants, for example, the mass-loss rate is close to $\dot{M} \sim 10^{-5} M_\odot$ yr^{-1}; for (rotating) Be-type stars, $\dot{M} \sim 10^{-6}$–$10^{-10} M_\odot$ yr^{-1}. No one knows at present which stars pass through mass ejection phases or how long these phases last. There are three binary systems containing white dwarfs in orbit around normal stars where accurate mass determinations have been possible (see Section 3.6). In all three cases the mass of the white dwarf is *less* than the mass of the companion, even though it has evolved more rapidly [cf. Eq. (1.3.9)]. Since the white dwarf has reached the endpoint of thermonuclear evolution, significant mass loss from the white dwarf progenitor must have occurred. Since the separation of the stars in these systems is quite large, it is conceivable that the mass loss does not depend on the binary nature of these systems.

Exercise   1.8   A crude estimate of the number of extragalactic, stellar collapse events in a given volume centered at the Earth can be made as follows (Arnett, 1979). The luminosity of the Galaxy is $L_{\text{Gal}} \simeq 10^{10.5} L_\odot$. Assume the Galactic rate is the historical supernova rate, $R_{\text{Gal}} \simeq (30 \text{ yr})^{-1}$.

(a)   Use these values and the assumption that $R \propto L$ inside any large volume centered on the Earth to determine the extragalactic event rate R as a function of L, the volume luminosity. Motivate the linearity assumption.

(b)   Estimate a "cosmic emissivity" ε in units of $L_\odot \text{ Mpc}^{-3}$ from L_{Gal} and the mean space density of galaxies $n_G \simeq 0.02 \text{ Mpc}^{-3}$. Determine $R(D)$, where D is the radius in Mpc of the volume surveyed. Compute R for the Virgo cluster ($D \sim 10 \text{ Mpc}$).

(c)   If the typical apparent magnitude of an observed supernova (SN) is 14 and the absolute magnitude is -18 (near maximum), determine a typical value of D_{obs}. Compute $R(D_{\text{obs}})$ and compare with the *observed* extragalactic rate of SN events of $R_{\text{obs}} \sim (300 \text{ yr})^{-1}$ per spiral galaxy.

Exercise   1.9   (based on Bahcall, 1978b, 1981)

(a)   A typical supernova event should emit $\sim 0.1 M_\odot c^2$ in the form of neutrinos with an average energy $E \sim 10$ MeV (see Chapter 18). If a supernova occurred at a distance R from the Earth, it might induce transitions in Davis' ^{37}Cl solar neutrino experiment. The time between successive collections of the products of neutrino capture in Davis' apparatus is about one month. What is the effective flux ϕ_{eff} of neutrinos in terms of R over this period (neutrino s^{-1} cm^{-2})?

(b)   Adopt as a detection criterion $\phi_{\text{eff}} \sigma(E) \gtrsim 3\langle \phi\sigma \rangle$. Here $\sigma(E)$ is the capture cross section ($= 2.7 \times 10^{-42}$ cm^2 per ^{37}Cl atom at 10 MeV), while $\langle \phi\sigma \rangle$ is the average capture rate for solar neutrinos which the experiment is designed to detect. For this we may use the measured value of $\langle \phi\sigma \rangle$, which is ~ 2 SNU (1 SNU $= 10^{-36}$ captures per atom per second). Out to what value of R is the experiment able to detect a supernova?

Answer: $R \sim 4$ kpc

Note: In all of Davis' runs to date, only one had a capture rate as high as 6 SNU. We conclude that there has been at most one stellar collapse in the last 10 years within the closest 30% of the Galaxy.

2

Cold Equation of State Below Neutron Drip

The theory of compact objects involves two distinct categories of input physics. These categories may be loosely distinguished as the "global" and "local" properties of matter. Global properties of matter describe the large-scale, dynamical response of matter to gravity, electromagnetic fields, rotation, and so on. These global properties are governed by the equations of motion for the matter. In addition to terms describing gravity, electromagnetic fields, and so forth, the equations of motion include the effects of internal stresses, such as pressure gradients, and energy dissipation, due to viscosity or radiation, for example. Quantities like the pressure, viscosity, or emissivity are generally local properties of the matter, determined by the local thermodynamic state of an individual matter element.

We shall develop both the local and global properties of compact objects in subsequent chapters. We shall begin by treating the microphysics necessary to discuss white dwarfs—namely, the equation of state (the pressure-density relation) below the regime of "neutron drip" (density less than $\sim 4 \times 10^{11}$ g cm^{-3}). With appropriate modifications, some of this discussion will carry over to neutron stars.

First we review some key thermodynamic relations we employ frequently in our subsequent discussion.

2.1 Thermodynamic Preliminaries

Conventionally, thermodynamic quantities refer to a number of particles N in a volume V. The relativistic invariance of thermodynamics is more transparent if all quantities describe measurements made in a local inertial frame *comoving* with the fluid. We thus imagine a local Lorentz frame moving with the same velocity as some fluid element. Let n be the number density of baryons as measured in this reference frame, and let ε be the total energy density (including rest-mass energy). Then ε/n is the energy per baryon. It is convenient to define quantities on a per baryon basis as baryon number is a conserved quantity.[1] The first law of thermodynamics takes the general form

$$dQ = d\left(\frac{\varepsilon}{n}\right) + P d\left(\frac{1}{n}\right), \tag{2.1.1}$$

where dQ is the heat gained per baryon, P is the pressure, and $1/n$ is the volume per baryon. The slash notation on dQ is a reminder that it is not a perfect differential.

If a process occurs in a fluid element that is in equilibrium at all times, then

$$dQ = T\,ds, \tag{2.1.2}$$

where s is the entropy per baryon and T is the temperature. Combining Eqs. (2.1.1) and (2.1.2) gives *in equilibrium*

$$d\left(\frac{\varepsilon}{n}\right) = -P\,d\left(\frac{1}{n}\right) + T\,ds. \tag{2.1.3}$$

In writing Eq. (2.1.3), we are tacitly assuming that ε/n is a function of n and s only, that is, $\varepsilon = \varepsilon(n, s)$. In general, the energy density of a system containing different species of particles depends on the relative amounts of each species, as well as on the volume $(1/n)$ and the adiabat (value of s). If we define the concentration of the ith species of particle by

$$Y_i \equiv \frac{n_i}{n}, \tag{2.1.4}$$

where n_i is the number density of particle i, then

$$\varepsilon = \varepsilon(n, s, Y_i). \tag{2.1.5}$$

We therefore should write in general

$$d\left(\frac{\varepsilon}{n}\right) = -P\,d\left(\frac{1}{n}\right) + T\,ds + \sum_i \mu_i\,dY_i, \tag{2.1.6}$$

---

[1] We shall ignore the possibility of baryon and lepton nonconserving reactions that may occur at super-high energies ($> 10^{15}$ GeV) in some "grand unified" gauge theories (see, e.g., Weinberg, 1979). In this case, it would be necessary to introduce explicitly the volume of a given fluid element.

where

$$P \equiv \frac{-\partial(\varepsilon/n)}{\partial(1/n)} = n^2 \frac{\partial(\varepsilon/n)}{\partial n}, \tag{2.1.7}$$

$$T \equiv \frac{\partial(\varepsilon/n)}{\partial s}, \tag{2.1.8}$$

and

$$\mu_i \equiv \frac{\partial(\varepsilon/n)}{\partial Y_i} = \frac{\partial \varepsilon}{\partial n_i}. \tag{2.1.9}$$

The quantity μ_i is called the *chemical potential* of species i. It has the interpretation of the change in energy density for a unit change in number density of species i, while the volume, entropy, and all other number densities are kept constant. Note that since ε is defined to include the rest-mass energy, so is μ_i.

In equilibrium, reactions between the particles produce a state of detailed balance where each reaction is balanced by its inverse and the concentration of each species remains constant. Thus in equilibrium, the concentrations Y_i are not all independent of the other thermodynamic quantities. We can determine the equilibrium relations as follows:

Consider first the special case where the system is infinitesimally close to equilibrium. Reactions are allowed to proceed to bring the system to equilibrium, but the system is kept thermally isolated ($dQ = 0$) and at constant volume ($dn = 0$) so that no work is done on the system. In this case, Eq. (2.1.1) gives $d(\varepsilon/n) = 0$; that is, the energy of the system remains constant. Entropy is generated by the reactions, but because entropy is a maximum in equilibrium (from the second law of thermodynamics), $ds = 0$ to first order. Thus Eq. (2.1.6) gives in equilibrium

$$\sum_i \mu_i \, dY_i = 0. \tag{2.1.10}$$

Suppose, for example, one is considering the reaction

$$e^- + p \leftrightarrow n + \nu_e \tag{2.1.11}$$

in equilibrium. Then $dY_e = dY_p = -dY_n = -dY_{\nu_e}$ and so

$$\mu_e + \mu_p = \mu_n + \mu_{\nu_e}. \tag{2.1.12}$$

A similar relation among chemical potentials holds for each reaction producing equilibrium. If the chemical potentials are known as a function of composition at the relevant values of n and s or n and ε (e.g., from statistical mechanics), then Eq. (2.1.10) gives the equilibrium concentrations.

Even when the initial state is very far from equilibrium, if the system proceeds to equilibrium with $dQ = 0$ and $dn = 0$, then its energy still remains constant. Eventually it will be infinitesimally close to equilibrium and the above argument will apply. Thus the composition can again be determined if the chemical potentials are known at the fixed values of n and ε.

Now consider the general case where the system is not necessarily thermally isolated, and work may be done on the system. If the system were to achieve equilibrium by quasistatic reactions, then $T\,ds = dQ$. In general, however, the second law implies

$$dQ \leqslant T\,ds. \tag{2.1.13}$$

Energy conservation [Eq. (2.1.1)] therefore implies

$$d\left(\frac{\varepsilon}{n}\right) + P\,d\left(\frac{1}{n}\right) \leqslant T\,ds. \tag{2.1.14}$$

If equilibrium is achieved at constant n and s, Eq. (2.1.14) yields

$$d\varepsilon \leqslant 0. \tag{2.1.15}$$

The equilibrium state corresponds to no further changes in ε (that is, $d\varepsilon = 0$), and is clearly a *minimum* of ε at fixed n and s. Using Eq. (2.1.6), we can recover Eq. (2.1.10) from this principle.

Similarly, if T and n are kept constant, Eq. (2.1.14) yields

$$df \leqslant 0, \tag{2.1.16}$$

where

$$f \equiv \frac{\varepsilon}{n} - Ts \tag{2.1.17}$$

is the *Helmholtz free energy* per baryon.

If T and P are kept constant (the most common situation in practice), then

$$dg \leqslant 0, \tag{2.1.18}$$

where

$$g \equiv \frac{\varepsilon + P}{n} - Ts \tag{2.1.19}$$

is the *Gibbs free energy* per baryon. Equilibrium corresponds to a minimum in this free energy at constant T and P. This expression of the equilibrium condition is particularly useful when there are phase transitions accompanied by discontinuities in n, with T and P remaining continuous (see Section 2.7).

Using Eq. (2.1.6), we find from Eq. (2.1.19) that

$$dg = \frac{1}{n} dP - s \, dT + \sum_i \mu_i \, dY_i. \qquad (2.1.20)$$

Thus the requirement that g be a minimum at constant T and P again implies Eq. (2.1.10).

Quantities like energy, volume, entropy, and number of particles are *extensive* quantities: if one divides a given volume in two by means of a partition, the energy, entropy, and number of particles on each side are half of their values for the whole volume. Quantities like pressure and temperature which remain the same are called *intensive* quantities. The requirement that all the extensive quantities of the system scale with volume in the same way leads to the relation[2]

$$g = \sum_i \mu_i Y_i. \qquad (2.1.21)$$

Let us be a little more precise about the number of independent thermodynamic quantities required to specify an equilibrium state. Consider, by way of illustration, an interacting mixture of baryons (including, e.g., neutrons and protons) and leptons (including electrons and muons and their associated neutrinos). All reactions in a given volume will conserve baryon number density n, electron lepton number density,[3] n_{Le}, muon lepton number density, $n_{L\mu}$, and electric charge density n_Q. Choose four basic chemical potentials corresponding to these four conserved quantities. For example, choose these to be μ_p (associated with n), μ_n (associated with n_{Le}), μ_μ (associated with $n_{L\mu}$), and μ_e (associated with n_Q). Then in equilibrium all other chemical potentials will be linear combinations of these four. Thus for example Eq. (2.1.12) gives μ_{ν_e} from reaction (2.1.11).

Now in equilibrium all thermodynamic quantities associated with species i are functions only of T and μ_i. (In the next section, we will see this explicitly for ideal gases.) Thus in general one needs to specify T and four of the μ_i's for a complete description of the equilibrium state. Equivalently, one needs to specify any five

---

[2] See, e.g., Reif (1965), p. 314, transcribed to our notation, for a derivation.

[3] Here and throughout the book we assume that neutrinos are *massless*, spin $\frac{1}{2}$ fermions of definite helicity, with corresponding antineutrinos being of opposite helicity. If neutrinos are not massless, then the possibility of neutrino oscillations exists, whereby an observed neutrino will be a mixed state of electron, mu and tau neutrinos. In this case electron, muon, and tau lepton numbers will *not* be separately conserved.

independent thermodynamic quantities. Usually $n_Q = 0$, so only four quantities are required.

Later in this chapter we shall consider an ideal gas at $T = 0$, where all neutrinos are allowed to escape from the system. This is equivalent to choosing n_{Le} and $n_{L\mu}$ so that the neutrino chemical potentials are all zero. Since T, n_Q, n_{Le}, and $n_{L\mu}$ have all been specified, we will find that all thermodynamic properties of the system depend on only *one* quantity, for example, the baryon number density.

In some situations the concept of a *limited* equilibrium applies. This means that certain reactions necessary to achieve complete equilibrium are too slow on the timescale of interest. This implies that there will be *more* than four conserved quantities, and one has to specify more n_i's to describe the system. For example, the dynamical timescale of a star is generally much shorter than the timescale to change its composition by nuclear reactions. To determine the pressure, internal energy, and so on in the star, one has to specify the concentration of H, He, and so on explicitly, and not simply the baryon density n. A similar situation prevails for chemical reactions in a terrestrial laboratory.

2.2 Results from Kinetic Theory

In kinetic theory, the number density in phase space for each species of particle, $d\mathfrak{N}/d^3x\,d^3p$, provides a complete description of the system. Equivalently, one can specify the dimensionless *distribution function* in phase space, $f(\mathbf{x}, \mathbf{p}, t)$, defined by

$$\frac{d\mathfrak{N}}{d^3x\,d^3p} = \frac{g}{h^3}f. \tag{2.2.1}$$

Here h^3 is the volume of a cell in phase space ($h = $ Planck's constant), and g is the statistical weight—that is, the number of states of a particle with a given value of momentum $\mathbf{p}$. For massive particles, $g = 2S + 1$ ($S = $ spin), for photons $g = 2$, for neutrinos $g = 1$. The function f gives the average occupation number of a cell in phase space.[4]

Exercise 2.1 Show that $d^3x\,d^3p$ is a Lorentz invariant (i.e., scalar under Lorentz transformations) and hence conclude that f is also.

The number density of each species of particle is given by

$$n = \int \frac{d\mathfrak{N}}{d^3x\,d^3p}\,d^3p, \tag{2.2.2}$$

---

[4] The reader will not find it too difficult to distinguish from the context whenever the symbol f or g is used for a free energy, as in Section 2.1.

where the integral extends over all momenta. The energy density is given by

$$\varepsilon = \int E \frac{d\mathfrak{N}}{d^3x\,d^3p}\,d^3p, \tag{2.2.3}$$

where

$$E = \left(p^2c^2 + m^2c^4\right)^{1/2}, \quad m = \text{particle rest mass.} \tag{2.2.4}$$

Note that ε *includes* the rest-mass energy of the particles. The pressure of a system with an isotropic distribution of momenta is given by

$$P = \frac{1}{3}\int pv\frac{d\mathfrak{N}}{d^3x\,d^3p}\,d^3p, \tag{2.2.5}$$

where the velocity is $v = pc^2/E$. This relation merely states that pressure is a momentum flux, and the factor of $\frac{1}{3}$ comes from isotropy.

For an ideal gas in equilibrium, f has the simple form

$$f(E) = \frac{1}{\exp[(E-\mu)/kT] \pm 1}, \tag{2.2.6}$$

where the upper sign refers to fermions (Fermi–Dirac statistics) and the lower sign to bosons (Bose–Einstein statistics). Here k is Boltzmann's constant and μ is the chemical potential.

For sufficiently low particle densities and high temperatures, $f(E)$ reduces to

$$f(E) \approx \exp\left(\frac{\mu - E}{kT}\right), \tag{2.2.7}$$

the Maxwell–Boltzmann distribution. In this case $f(E) \ll 1$.

For completely degenerate fermions ($T \to 0$, i.e., $\mu/kT \to \infty$), μ is called the *Fermi energy*, E_F, and

$$f(E) = \begin{cases} 1, & E \leqslant E_F \\ 0, & E > E_F. \end{cases} \tag{2.2.8}$$

2.3 Equation of State of a Completely Degenerate, Ideal Fermi Gas

An isolated white dwarf or neutron star ultimately cools to zero temperature, and it is the pressure associated with matter at $T = 0$ that supports these stars against gravitational collapse. The simplest cold, degenerate equation of state is that due to a single species of ideal (noninteracting) fermions. We proceed to review this case immediately below.[5] In subsequent sections we shall consider more realistic, but also more complex, equations of state appropriate for degenerate matter in compact stars.

For definiteness, let us take the gas to be one of electrons at zero temperature. The gas can be treated as ideal if we ignore all electrostatic interactions.

If we define the *Fermi momentum* p_F by

$$E_F \equiv \left(p_F^2 c^2 + m_e^2 c^4 \right)^{1/2}, \tag{2.3.1}$$

then Eqs. (2.2.2) and (2.2.8) give

$$n_e = \frac{2}{h^3} \int_0^{p_F} 4\pi p^2 \, dp = \frac{8\pi}{3h^3} p_F^3. \tag{2.3.2}$$

It is convenient to define a dimensionless Fermi momentum, or "relativity parameter," x, by

$$x = \frac{p_F}{m_e c}. \tag{2.3.3}$$

Then

$$n_e = \frac{1}{3\pi^2 \lambda_e^3} x^3, \tag{2.3.4}$$

where $\lambda_e = \hbar/m_e c$ is the electron Compton wavelength.

The pressure is given by Eq. (2.2.5),

$$
\begin{aligned}
P_e &= \frac{1}{3} \frac{2}{h^3} \int_0^{p_F} \frac{p^2 c^2}{\left(p^2 c^2 + m_e^2 c^4 \right)^{1/2}} 4\pi p^2 \, dp \\
&= \frac{8\pi m_e^4 c^5}{3h^3} \int_0^x \frac{x^4 \, dx}{\left(1 + x^2\right)^{1/2}} \\
&= \frac{m_e c^2}{\lambda_e^3} \phi(x) \\
&= 1.42180 \times 10^{25} \phi(x) \text{ dyne cm}^{-2}, \tag{2.3.5}
\end{aligned}
$$

[5] For further discussion see Chandrasekhar (1939) or Clayton (1968).

where

$$\phi(x) = \frac{1}{8\pi^2}\left\{x(1 + x^2)^{1/2}(2x^2/3 - 1) + \ln\left[x + (1 + x^2)^{1/2}\right]\right\}. \quad (2.3.6)$$

Similarly, Eq. (2.2.3) gives

$$\varepsilon_e = \frac{2}{h^3}\int_0^{p_F}(p^2c^2 + m_e^2c^4)^{1/2}4\pi p^2\, dp$$

$$= \frac{m_ec^2}{\lambda_e^3}\chi(x), \quad (2.3.7)$$

where

$$\chi(x) = \frac{1}{8\pi^2}\left\{x(1 + x^2)^{1/2}(1 + 2x^2) - \ln\left[x + (1 + x^2)^{1/2}\right]\right\}. \quad (2.3.8)$$

Even in applications where degenerate electrons contribute most of the pressure, the density is usually dominated by the rest-mass of the ions. This density is

$$\rho_0 = \sum_i n_i m_i, \quad (2.3.9)$$

where m_i is the mass of ion species i. If we define the mean baryon rest mass as

$$m_B \equiv \frac{1}{n}\sum_i n_i m_i = \frac{\sum_i n_i m_i}{\sum_i n_i A_i}, \quad (2.3.10)$$

where A_i is the baryon number (integer atomic weight) of the ith species, then

$$\rho_0 = nm_B = \frac{n_e m_B}{Y_e}. \quad (2.3.11)$$

Here Y_e is the mean number of electrons per baryon, as in Eq. (2.1.4). For example, for completely ionized, pure ^{12}C, $m_B \equiv m_u \equiv 1.66057 \times 10^{-24}$ g (atomic mass unit) and $Y_e = Z/A = 0.5$, so

$$\rho_0 = 1.9479 \times 10^6 x^3 \text{ g cm}^{-3}. \quad (2.3.12)$$

Sometimes the quantity

$$\mu_e = \frac{m_B}{m_u Y_e} \quad (2.3.13)$$

(mean molecular weight per electron) is used, so that

$$\rho_0 = \mu_e m_u n_e = 0.97395 \times 10^6 \mu_e x^3 \text{ g cm}^{-3}, \tag{2.3.14}$$

or

$$x = 1.0088 \times 10^{-2} \left(\frac{\rho_0}{\mu_e} \right)^{1/3}, \tag{2.3.15}$$

where ρ_0 is in g cm^{-3}. Often the distinction between m_B and m_u in Eq. (2.3.13) can be ignored. For example, for a completely ionized element of atomic weight A and number Z, $\mu_e = A/Z$ to about 1 part in 10^4.

Similarly, a quantity μ (the mean molecular weight) is sometimes used to specify the density. It is defined by

$$\rho_0 = \left(n_e + \sum_i n_i \right) \mu m_u, \tag{2.3.16}$$

and so by Eq. (2.3.11) we have

$$\frac{1}{\mu} = \left(Y_e + \sum_i Y_i \right) \frac{m_u}{m_B}. \tag{2.3.17}$$

Except for very accurate work, m_u/m_B can again be taken equal to unity. The mean molecular weight is particularly useful in the nondegenerate limit, when the pressure is given by the perfect gas law

$$P = \left(n_e + \sum_i n_i \right) kT$$

$$= \frac{\rho_0}{\mu m_u} kT. \tag{2.3.18}$$

Equations (2.3.5) and (2.3.14) give the ideal degenerate equation of state $P = P(\rho_0)$ parametrically in terms of x. (Note that $\rho = \rho_0 + \varepsilon_e/c^2$ is the total density and usually the term ε_e/c^2 is negligible.)

Exercise   2.2   Show that in the limit $x \ll 1$ (nonrelativistic electrons) we have

$$\phi(x) \rightarrow \frac{1}{15\pi^2} \left(x^5 - \frac{5}{14} x^7 + \frac{5}{24} x^9 \cdots \right),$$

$$\chi(x) \rightarrow \frac{1}{3\pi^2} \left(x^3 + \frac{3}{10} x^5 - \frac{3}{56} x^7 \cdots \right), \tag{2.3.19}$$

while for $x \gg 1$ (relativistic electrons),

$$\phi(x) \rightarrow \frac{1}{12\pi^2}\left(x^4 - x^2 + \frac{3}{2}\ln 2x \cdots\right),$$

$$\chi(x) \rightarrow \frac{1}{4\pi^2}\left(x^4 + x^2 - \frac{1}{2}\ln 2x \cdots\right). \tag{2.3.20}$$

The equation of state can be written in the *polytropic* form

$$P = K\rho_0^\Gamma, \tag{2.3.21}$$

where K and Γ are constants, in two limiting cases:

1. Nonrelativistic electrons, $\rho_0 \ll 10^6$ g cm^{-3}, $x \ll 1$, $\phi(x) \rightarrow x^5/15\pi^2$,

$$\Gamma = \frac{5}{3}, \quad K = \frac{3^{2/3}\pi^{4/3}}{5}\frac{\hbar^2}{m_e m_u^{5/3}\mu_e^{5/3}} = \frac{1.0036 \times 10^{13}}{\mu_e^{5/3}} \text{ cgs.} \tag{2.3.22}$$

2. Extremely relativistic electrons, $\rho \gg 10^6$ g cm^{-3}, $x \gg 1$, $\phi(x) \rightarrow x^4/12\pi^2$,

$$\Gamma = \frac{4}{3}, \quad K = \frac{3^{1/3}\pi^{2/3}}{4}\frac{\hbar c}{m_u^{4/3}\mu_e^{4/3}} = \frac{1.2435 \times 10^{15}}{\mu_e^{4/3}} \text{ cgs.} \tag{2.3.23}$$

The above results may be scaled trivially with particle mass m_i and statistical weight g_i to determine the equation of state of an arbitrary species i of ideal fermions. For example, for pure neutrons, Eqs. (2.3.7), (2.3.14), and (2.3.21)–(2.3.23) become, respectively,

$$\varepsilon_n = \frac{m_n c^2}{\lambda_n^3}\chi(x_n) \quad \left(x_n = \frac{p_F}{m_n c}\right)$$

$$= 1.6250 \times 10^{38}\chi(x_n) \text{ erg cm}^{-3}, \tag{2.3.24}$$

$$\rho_0 = m_n n_n$$

$$= \frac{m_n}{\lambda_n^3}\frac{1}{3\pi^2}x_n^3$$

$$= 6.1067 \times 10^{15}x_n^3 \text{ g cm}^{-3}, \tag{2.3.25}$$

and

$$P = K\rho_0^\Gamma, \tag{2.3.26}$$

where the two limiting regimes are:

1. Nonrelativistic neutrons, $\rho_0 \ll 6 \times 10^{15}$ g cm^{-3},

$$\Gamma = \frac{5}{3}, \qquad K = \frac{3^{2/3}\pi^{4/3}}{5}\frac{\hbar^2}{m_n^{8/3}} = 5.3802 \times 10^9 \text{ cgs.} \qquad (2.3.27)$$

2. Extremely relativistic neutrons, $\rho_0 \gg 6 \times 10^{15}$ g cm^{-3},

$$\Gamma = \frac{4}{3}, \qquad K = \frac{3^{1/3}\pi^{2/3}}{4}\frac{\hbar c}{m_n^{4/3}} = 1.2293 \times 10^{15} \text{ cgs.} \qquad (2.3.28)$$

In this case the mass density $\rho = \varepsilon_n/c^2$ is due entirely to neutrons and greatly exceeds ρ_0 whenever the neutrons are extremely relativistic ($\rho_0 \gg 6 \times 10^{15}$ g cm^{-3}).

Exercise 2.3 In equilibrium, one can get the pressure from the energy density by means of the relation (2.1.7). By assuming that only electrons are present (i.e., $\varepsilon \to \varepsilon_e$, $n \to n_e$), show that Eq. (2.3.5) can be recovered from Eqs. (2.3.4) and (2.3.7).

Exercise 2.4 Show that

$$\frac{\varepsilon_e + P_e}{n_e} = E_F, \qquad (2.3.29)$$

and compare with Eq. (2.1.21).

Exercise 2.5 Consider completely ionized matter consisting of hydrogen, helium, and heavier atomic species $i > 2$. Let X and Y denote the fractions by *mass* of hydrogen and helium, respectively. Show that

$$\mu_e \simeq \frac{2}{1 + X}. \qquad (2.3.30)$$

Approximate m_i by $m_i = A_i m_u$ for all i, and take $Z_i/A_i \approx \frac{1}{2}$ for $i \geqslant 2$.

Exercise 2.6 Show that the mean kinetic energy of an electron in a degenerate gas is $\frac{3}{5}E_F'$ in the nonrelativistic limit and $\frac{3}{4}E_F$ in the relativistic limit. Here $E_F' \equiv E_F - m_e c^2 \simeq p_F^2/2m_e$.

Exercise 2.7
(a) Show from Eq. (2.2.7) that for a nonrelativistic Maxwell–Boltzmann gas,

$$n = g\left(\frac{mkT}{2\pi\hbar^2}\right)^{3/2}\exp\left(\frac{\mu - mc^2}{kT}\right) \qquad (2.3.31)$$

$$P = nkT, \qquad (2.3.32)$$

$$\varepsilon = nmc^2 + \tfrac{3}{2}nkT. \qquad (2.3.33)$$

(b)   Use Eq. (2.1.21) (for a single species) to show that

$$\frac{s}{k} = \frac{5}{2} + \ln\left[\frac{g}{n}\left(\frac{mkT}{2\pi\hbar^2}\right)^{3/2}\right]. \tag{2.3.34}$$

Exercise   2.8   Suppose that the gas particles in Exercise 2.7 have internal degrees of freedom (e.g., the excited states of an atom or nucleus). Then one can write for their energy $E = E_{\text{c.m.}} + E_j$, where the center-of-mass energy $E_{\text{c.m.}}$ is given by Eq. (2.2.4) and where we assume E_j is independent of p and chosen to be zero in the ground state. Show that Eq. (2.3.31) is modified by the replacement.

$$g \rightarrow \sum_j g_j e^{-E_j/kT}, \tag{2.3.35}$$

where g_j is the degeneracy of the jth excited state. How are the expressions for P, ε, and s changed?

Exercise   2.9   Show by a suitable integration by parts on Eq. (2.2.2) that the expression $P = nkT$ is generally valid for a Maxwell–Boltzmann gas, whether relativistic or not.

2.4   Electrostatic Corrections to the Equation of State

The ideal Fermi gas equation of state of the previous section was employed by Chandrasekhar (1931a, b) in his pioneering analysis of equilibrium white dwarfs (see Chapter 3). There are two important corrections to this equation of state in practice. One of them, *inverse β-decay*, will be discussed in Sections 2.5–2.7. Corrections due to *electrostatic interactions* among the ions and electrons are the subject of this section.

The principal electrostatic correction arises because the positive charges are not uniformly distributed in the gas but are concentrated in individual nuclei of charge Z. This decreases the energy and pressure of the ambient electrons: the repelling electrons are, on the average, further apart than the mean distance between nuclei and electrons, so repulsion is weaker than attraction.

In a nondegenerate gas, Coulomb effects become more important as the density increases: the ratio of Coulomb energy to thermal energy is approximately

$$\frac{E_c}{kT} = \frac{Ze^2/\langle r \rangle}{kT} \simeq \frac{Ze^2 n_e^{1/3}}{kT}, \tag{2.4.1}$$

which increases with n_e. Here $\langle r \rangle \sim n_e^{-1/3}$ is the characteristic electron–ion separation. In contrast, for a degenerate gas we have

$$\frac{E_c}{E_F'} = \frac{Ze^2/\langle r \rangle}{p_F^2/2m_e}. \qquad (2.4.2)$$

Using Eq. (2.3.2) for p_F, this becomes

$$\frac{E_c}{E_F'} = 2\left(\frac{1}{3\pi^2}\right)^{2/3} \frac{Z}{a_0} \frac{1}{n_e^{1/3}} = \left(\frac{n_e}{Z^3 \times 6 \times 10^{22} \text{ cm}^{-3}}\right)^{-1/3}, \qquad (2.4.3)$$

where $a_0 = \hbar^2/m_e e^2$ is the Bohr radius. Thus $E_c \ll E_F'$ in most astrophysical degenerate gases.

We can derive an approximate expression for the correction to the ideal degenerate equation of state using the fact [implied by Eq. (2.4.3)] that n_e is to first approximation uniform.

As $T \to 0$, the ions are located in a lattice that maximizes the inter-ion separation. Consider a "spherical" cell of the lattice of volume $4\pi r_0^3/3 = 1/n_N$, where n_N = number density of nuclei. In this *Wigner–Seitz approximation*[6] the gas is imagined to be divided into neutral spheres of radius r_0 about each nucleus, which contain the Z electrons closest to the nucleus.

The total electrostatic energy of any one sphere is the sum of the potential energies due to electron–electron (e-e) interactions and electron–ion (e-i) interactions. To assemble a uniform sphere of Z electrons requires energy

$$E_{e\text{-}e} = \int_0^{r_0} \frac{q \, dq}{r}, \qquad (2.4.4)$$

where

$$q = -Ze\frac{r^3}{r_0^3} \qquad (2.4.5)$$

is the charge inside radius r. Carrying out the integration gives

$$E_{e\text{-}e} = \frac{3}{5}\frac{Z^2 e^2}{r_0}. \qquad (2.4.6)$$

---

[6]The Wigner–Seitz approximation is much better for white dwarfs than for laboratory solids, where n_e is much more nonuniform.

To assemble the electron sphere about the central nucleus of charge Ze requires energy

$$E_{e\text{-}i} = Ze \int_0^{r_0} \frac{dq}{r}$$

$$= -\frac{3}{2}\frac{Z^2 e^2}{r_0}.$$ (2.4.7)

Thus the total Coulomb energy of the cell is

$$E_c = E_{e\text{-}e} + E_{e\text{-}i} = -\frac{9}{10}\frac{Z^2 e^2}{r_0}.$$ (2.4.8)

Since the cells are neutral, interactions between electrons and nuclei of different cells can be ignored.

The electrostatic energy per electron is

$$\frac{E_c}{Z} = -\frac{9}{10}\left(\frac{4\pi}{3}\right)^{1/3} Z^{2/3} e^2 n_e^{1/3},$$ (2.4.9)

where we have used

$$n_e = \frac{Z}{4\pi r_0^3/3}.$$ (2.4.10)

The numerical coefficient in Eq. (2.4.9) is 1.45079, very close to the exact value of 1.44423 for a body-centered cubic lattice.[7] The corresponding pressure is negative and given by Eq. (2.1.7):

$$P_c = n_e^2 \frac{d(E_c/Z)}{dn_e}$$

$$= -\frac{3}{10}\left(\frac{4\pi}{3}\right)^{1/3} Z^{2/3} e^2 n_e^{4/3}.$$ (2.4.11)

Let us first examine this result in the extreme relativistic limit. There the ideal Chandrasekhar result is

$$P_0 \rightarrow \hbar c (3\pi^2)^{1/3} \frac{n_e^{4/3}}{4}$$ (2.4.12)

[7]Coldwell–Horsfall and Maradudin (1960).

[cf. Eqs. (2.3.5), (2.3.20), and (2.3.4)], so

$$\frac{P}{P_0} = \frac{P_0 + P_c}{P_0}$$

$$= 1 - \frac{2^{5/3}}{5}\left(\frac{3}{\pi}\right)^{1/3}\alpha Z^{2/3}, \tag{2.4.13}$$

where $\alpha = e^2/\hbar c = 1/137$ is the fine structure constant. The next term in this series of powers of $\alpha Z^{2/3}$ results from the deviations of the electron distribution from uniformity (Thomas–Fermi corrections; cf. below and Salpeter, 1961).

Although the above Coulomb correction is relatively small, we shall find it is nevertheless important for high-density white dwarfs and low-density neutron stars.

In the nonrelativistic limit,

$$P_0 \rightarrow \hbar^2(3\pi^2)^{2/3}\frac{n_e^{5/3}}{5m_e} \tag{2.4.14}$$

[cf. Eqs. (2.3.5), (2.3.19), and (2.3.4)], so

$$\frac{P}{P_0} = 1 - \frac{Z^{2/3}}{2^{1/3}\pi a_0 n_e^{1/3}}. \tag{2.4.15}$$

This predicts that $P = 0$ when

$$n_e = \frac{Z^2}{2\pi^3 a_0^3}, \tag{2.4.16}$$

corresponding to a density

$$\rho_0 \simeq 0.4Z^2 \text{ g cm}^{-3} \tag{2.4.17}$$

by Eq. (2.3.11), with $A \sim 2Z$. Equation (2.4.17) gives $\rho_0 \approx 250$ g cm^{-3} for iron, instead of the laboratory value of 7.86 g cm^{-3}. The reason for the discrepancy is that at low densities it is no longer a good approximation to treat the electron gas as uniform [cf. Eq. (2.4.3)].

An accurate equation of state at laboratory densities is very complicated to derive, because electron shell effects mask the simpler statistical effects. Above a few times laboratory densities, however, a statistical approach to the equation of state works well and is sufficient to treat the low-mass range of white dwarfs (and even the gross structure of large planets).

The simplest statistical treatment of atomic structure is the *Thomas–Fermi* method. One assumes that within each Wigner–Seitz cell, the electrons move in a slowly varying spherically symmetric potential $V(r)$. Because the potential is approximately constant at any point, we can use free-particle Fermi–Dirac statistics. This means we are taking the interaction energy between the electrons to be much less than the kinetic or potential energies of an individual electron. Thus at any r, all states up to $E = E_F$ are occupied. The energy E_F is independent of r, otherwise electrons would migrate to a region of smaller E_F. Thus

$$E_F = -eV(r) + \frac{p_F^2(r)}{2m_e} = \text{constant,} \qquad (2.4.18)$$

where p_F is the maximum momentum of electrons at r. By choosing the arbitrary constant in $V(r)$ appropriately, we can choose E_F to have any convenient value, but we shall not make use of this freedom here.

Exercise   2.10   Show that the constancy of E_F may be derived from the condition that the electron cloud support itself in hydrostatic equilibrium via ideal Fermi gas pressure.

Analogously to Eq. (2.3.2), we have

$$n_e = \frac{8\pi}{3h^3} p_F^3 = \frac{8\pi}{3h^3} \{2m_e[E_F + eV(r)]\}^{3/2}. \qquad (2.4.19)$$

The potential $V(r)$ is determined by Poisson's equation:

$$\nabla^2 V = 4\pi e n_e + \text{nuclear contribution.} \qquad (2.4.20)$$

The nuclear contribution is a delta function about the origin, so we can omit it for $r > 0$ and impose the boundary condition

$$\lim_{r \to 0} rV(r) = Ze. \qquad (2.4.21)$$

The boundary condition at the cell boundary r_0 is that the electric field vanish (neutral cell):

$$\left. \frac{dV}{dr} \right|_{r_0} = 0. \qquad (2.4.22)$$

Equations (2.4.19) and (2.4.20) give the relation

$$\frac{1}{r} \frac{d^2}{dr^2}(rV) = \frac{32\pi^2 e}{3h^3} [2m_e(E_F + eV)]^{3/2}, \qquad (2.4.23)$$

to be solved subject to the boundary conditions (2.4.21) and (2.4.22).

It is convenient to change to nondimensional units by putting

$$r = \mu x, \tag{2.4.24}$$

$$E_F + eV(r) = \frac{Ze^2\phi(x)}{r}, \tag{2.4.25}$$

where

$$\mu = \left(\frac{9\pi^2}{128Z}\right)^{1/3} a_0. \tag{2.4.26}$$

After some simplification, Eq. (2.4.23) becomes the Thomas–Fermi equation,

$$\frac{d^2\phi}{dx^2} = \frac{\phi^{3/2}}{x^{1/2}}, \tag{2.4.27}$$

subject to the boundary conditions

$$\phi(0) = 1, \tag{2.4.28}$$

$$\phi'(x_0) = \frac{\phi(x_0)}{x_0}. \tag{2.4.29}$$

Exercise 2.11   Check that Eq. (2.4.29) can also be obtained by imposing the charge neutrality requirement that

$$Z = \int_0^{r_0} 4\pi r^2 n_e \, dr. \tag{2.4.30}$$

Equation (2.4.27) is nonlinear and has to be solved numerically; such an integration has been performed by Feynman, Metropolis, and Teller (1949). Starting at the origin with $\phi(0) = 1$, there is a special value[8] of $\phi'(0)$ (= -1.5880710) for which the solution is asymptotic to the x axis at large x and Eq. (2.4.29) is satisfied by having $\phi'(x_0) \to 0$, $\phi(x_0) \to 0$, as $x_0 \to \infty$ (see Fig. 2.1). This, as we shall see below, is the zero pressure case, which corresponds thus to zero density and infinite radius ($x_0 \to \infty$). This defect of the Thomas–Fermi model—namely, that free atoms have infinite radius—is remedied by incorporating exchange effects (the "Thomas–Fermi–Dirac model").

---

[8]Kobayashi et al. (1955).

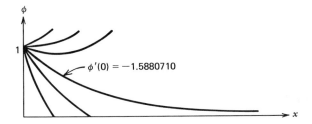

Figure 2.1 Behavior of solutions to the Thomas–Fermi equation, Eq. (2.4.27).

For $\phi'(0) > -1.5880710$, $\phi(x)$ does not vanish anywhere and diverges for $x \to \infty$. Equation (2.4.29) is satisfied at some finite value x_0, which determines the cell radius and corresponds to neutral atoms under pressure. We shall not be concerned with the case $\phi'(0) < -1.5880710$, for which $\phi(x) = 0$ at finite x_0 corresponding to free positive ions.

We now compute the pressure at the cell boundary using the free particle expression:

$$P = \frac{2}{h^3}\frac{1}{3}\int_0^{p_F}\frac{p^2}{m_e}4\pi p^2\,dp$$

$$= \frac{8\pi}{15h^3 m_e}p_F^5(r_0). \tag{2.4.31}$$

Note that since $dV/dr = 0$ at the boundary, no force is exerted on a cell by neighboring cells. Using Eqs. (2.4.19) and (2.4.24)–(2.4.26), Eq. (2.4.31) becomes

$$P = \frac{1}{10\pi}\frac{Z^2e^2}{\mu^4}\left[\frac{\phi(x_0)}{x_0}\right]^{5/2}. \tag{2.4.32}$$

The density is simply given by the total rest mass inside the cell,

$$\rho_0 = \frac{Am_B}{4\pi\mu^3 x_0^3/3}. \tag{2.4.33}$$

Equations (2.4.32) and (2.4.33) give the equation of state $P = P(\rho_0)$ in terms of the parameter x_0. Even though the *local* pressure in the cell is everywhere given by the ideal Fermi relation for nonrelativistic electrons, the deviations of $n_e(r)$ from uniformity result in different behavior for the "cell averaged" equation of state, $P(\rho_0)$.

For low densities, $x_0 \to \infty$. The asymptotic solution of Eq. (2.4.27) in this case is

$$\phi(x) \sim \frac{144}{x^3}, \quad x \to \infty. \qquad (2.4.34)$$

Exercise 2.12   Show that Eq. (2.4.34) solves Eq. (2.4.27) but not Eq. (2.4.28).

Thus

$$P \sim x_0^{-10} \sim \rho_0^{10/3}. \qquad (2.4.35)$$

This relation shows the characteristic stiffening of the equation of state at low densities[9] with an adiabatic index ~ 3.3. The consequences of this result for low-mass white dwarf structure will be discussed in Chapter 3.

In the high-density limit, the Thomas–Fermi result (2.4.32) reduces to the result (2.4.15) as one might expect (cf. Exercise 2.18).

A more complete treatment of corrections to the equation of state in the regime explored in this section is given by Salpeter (1961) and Salpeter and Zapolsky (1967). For most applications, the results of Feynmann, Metropolis, and Teller for the Thomas–Fermi–Dirac model are adequate for the low-density regime up to 10^4 g cm^{-3}. For higher densities, one uses the Chandrasekhar ideal gas result with the Coulomb lattice correction (2.4.11).

Exercise 2.13

(a)   Show that the kinetic energy of the electrons in the Thomas–Fermi model is

$$E_{\text{K.E.}} = \int_0^{r_0} 4\pi r^2 \, dr \int_0^{P_F(r)} \frac{p^2}{2m} \frac{2}{h^3} 4\pi p^2 \, dp$$

$$= \frac{3}{5} \frac{Z^2 e^2}{\mu} \int_0^{x_0} \phi^{5/2} x^{-1/2} \, dx. \qquad (2.4.36)$$

(b)   Show that the potential energy of the electrons in the field of the nucleus is

$$E_{e\text{-}n} = -Ze^2 \int_0^{r_0} 4\pi r^2 \, dr \frac{n_e}{r}$$

$$= -\frac{Z^2 e^2}{\mu} \int_0^{x_0} \phi^{3/2} x^{-1/2} \, dx. \qquad (2.4.37)$$

---

[9]Zel'dovich and Novikov (1971) give a simple but heuristic derivation of the exponent $10/3$ appearing in Eq. (2.4.35).

(c) Show that the potential energy of the electrons due to their mutual interaction in the spherical cell $\mathcal{V}$ is

$$E_{e-e} = \frac{1}{2}e^2 \int_{\mathcal{V}} d^3r\, n_e(\mathbf{r}) \int_{\mathcal{V}} d^3r'\, n_e(\mathbf{r}')|\mathbf{r}-\mathbf{r}'|^{-1} \qquad (2.4.38)$$

$$= \frac{1}{2}\frac{Z^2 e^2}{\mu} \int_0^{x_0} dx\, \phi^{3/2}(x)x^{1/2}$$

$$\times \left[\frac{1}{x}\int_0^x dx'\, \phi^{3/2}(x')(x')^{1/2} \right.$$

$$\left. + \int_x^{x_0} dx'\, \phi^{3/2}(x')(x')^{-1/2} \right]. \qquad (2.4.39)$$

Exercise 2.14 Derive the Thomas–Fermi equation (2.4.27) by minimizing the total energy E subject to the constraint (2.4.30). The variation of ϕ is to be performed keeping the endpoint x_0 fixed.

Hints: (i) Implement the constraint by varying

$$E + \lambda \int_0^{x_0} \phi^{3/2}x^{1/2}\, dx,$$

where λ = constant is a Lagrange multiplier.

(ii) By the symmetry of Eq. (2.4.38), the variation of E_{e-e} is twice the value obtained by varying ϕ in the first integrand in Eq. (2.4.39).

Exercise 2.15
(a) Show that

$$\frac{dE}{dx_0} = -\frac{2}{5}\frac{\phi^{5/2}(x_0)}{x_0^{1/2}}\frac{Z^2 e^2}{\mu}. \qquad (2.4.40)$$

Note that $\phi(x)$ depends implicitly on x_0 through the boundary condition (2.4.29), and hence must be differentiated with respect to x_0 inside the integrals. The derivative of the constraint (2.4.30) with respect to x_0 will give a useful relation.

(b) Rederive Eq. (2.4.32) for P from the first law of thermodynamics.

Exercise 2.16
(a) Show from Eq. (2.4.36) that

$$E_{K.E.} = \frac{3}{7}\frac{Z^2 e^2}{\mu}\left[\tfrac{4}{5}x_0^{1/2}\phi^{5/2}(x_0) - \phi'(0)\right]. \qquad (2.4.41)$$

Hint[10]:   The integral $I = \int_0^{x_0} \phi^{5/2} x^{-1/2} \, dx$ can be evaluated by writing $\phi^{5/2} = \phi^{3/2} \phi$, substituting for $\phi^{3/2}$ from Eq. (2.4.27), integrating by parts twice, substituting for $\phi^{3/2}$ once more, and integrating by parts again.

(b)   Evaluate the inner integrals in Eq. (2.4.39) for E_{e-e} by substituting for $\phi^{3/2}$ from Eq. (2.4.27). Hence show that

$$E_{\text{P.E.}} \equiv E_{e-n} + E_{e-e}$$

$$= -\frac{6}{7} \frac{Z^2 e^2}{\mu} \left[\frac{1}{3} x_0^{1/2} \phi(x_0)^{5/2} - \phi'(0) \right]. \qquad (2.4.42)$$

Exercise   2.17   Verify that the Thomas–Fermi approximation satisfies the virial equation

$$E_{\text{K.E.}} + \tfrac{1}{2} E_{\text{P.E.}} = \tfrac{3}{2} P \mathcal{V}$$

where $\mathcal{V} = \tfrac{4}{3} \pi r_0^3$.

Exercise   2.18   (based on Salpeter and Zapolsky 1967)   Solve the Thomas–Fermi equation in the high-density limit ($x_0 \to 0$) as follows: Put $y = x/x_0$ and

$$\phi(y) = \frac{1}{x_0} f_0(y) + f_1(y) + x_0 f_2(y) + \cdots .$$

Show that the boundary condition (2.4.29) can be satisfied by setting

$$f_0'(1) = f_0(1), \qquad f_1'(1) = f_1(1), \text{ and so on,}$$

while the boundary condition (2.4.28) can be satisfied by setting

$$f_1(0) = 1, \qquad f_2(0) = f_3(0) = \cdots = 0.$$

Show that one recovers the earlier result (2.4.15); that is,

$$P = P_0 + P_c.$$

[Why can one not simply set $x = x_0 = 0$ and use Eq. (2.4.28) in Eq. (2.4.32) together with Eq. (2.4.33) to get $P(\rho_0)$ at high density?]

---

[10] The manipulations described here are originally due to Milne (1927), who knew the analogous calculation of the gravitational potential energy of a polytrope due to Emden (1907); see Chapter 3.

2.5 Inverse *β*-decay: The Ideal, Cold *n-p-e* Gas

At high densities, the most important correction to the equation of state is due to inverse *β*-decay[11]:

$$e^- + p \rightarrow n + \nu. \qquad (2.5.1)$$

The proton and neutron are generally bound in nuclei. In this section, however, we will get an overview of the effects of inverse *β*-decay by examining the case of a gas of *free* electrons, protons, and neutrons. We assume that the neutrinos generated in reaction (2.5.1) escape from the system. In Sections 2.6 and 2.7 we will treat the more interesting case of *bound* nuclei.

The reaction (2.5.1) can proceed whenever the electron acquires enough energy to balance the mass difference between proton and neutron, $(m_n - m_p)c^2 = 1.29$ MeV. This process is effective for transforming protons into neutrons if *β*-decay, that is,

$$n \rightarrow p + e + \bar{\nu} \qquad (2.5.2)$$

does not occur. Reaction (2.5.2) is blocked if the density is high enough that all electron energy levels in the Fermi sea are occupied up to the one that the emitted electron would fill. Thus there is a critical density for the onset of reaction (2.5.1).

We can compute the properties of such a mixture of electrons, protons, and neutrons by assuming that they are in equilibrium. Then Eq. (2.1.12) gives

$$\mu_e + \mu_p = \mu_n, \qquad (2.5.3)$$

where we have set the chemical potential of the neutrinos to zero; that is, we have taken their number density to be zero. Define

$$x_e = \frac{p_F^e}{m_e c}, \qquad x_n = \frac{p_F^n}{m_n c}, \qquad x_p = \frac{p_F^p}{m_p c}, \qquad (2.5.4)$$

analogously to Eq. (2.3.3). Then since $\mu_e = [(p_F^e c)^2 + m_e^2 c^4]^{1/2}$ and so on, Eq. (2.5.3) becomes

$$m_e\left(1 + x_e^2\right)^{1/2} + m_p\left(1 + x_p^2\right)^{1/2} = m_n\left(1 + x_n^2\right)^{1/2}. \qquad (2.5.5)$$

Charge neutrality $n_e = n_p$ implies that

$$\frac{1}{3\pi^2 \lambda_e^3} x_e^3 = \frac{1}{3\pi^2 \lambda_p^3} x_p^3 \qquad (2.5.6)$$

---

[11]Whenever it is clear from the context that only electron-type neutrinos are present, we shall employ the notation ν instead of ν_e.

[cf. Eq. (2.3.4)]; that is,

$$m_e x_e = m_p x_p. \tag{2.5.7}$$

The equation of state can now be determined, for example, in terms of the parameter x_e: Choose a value of x_e. Then Eq. (2.5.7) gives x_p, and Eq. (2.5.5) gives x_n. Then

$$P = \frac{m_e c^2}{\lambda_e^3} \phi(x_e) + \frac{m_p c^2}{\lambda_p^3} \phi(x_p) + \frac{m_n c^2}{\lambda_n^3} \phi(x_n), \tag{2.5.8}$$

$$\varepsilon = \frac{m_e c^2}{\lambda_e^3} \chi(x_e) + \frac{m_p c^2}{\lambda_p^3} \chi(x_p) + \frac{m_n c^2}{\lambda_n^3} \chi(x_n), \tag{2.5.9}$$

$$n = \frac{1}{3\pi^2 \lambda_p^3} x_p^3 + \frac{1}{3\pi^2 \lambda_n^3} x_n^3. \tag{2.5.10}$$

Exercise 2.19 Derive Eq. (2.5.5) by minimizing ε at fixed n and imposing charge neutrality.

The minimum density at which neutrons appear can be found by setting $x_n = 0$ in Eq. (2.5.5). Since the protons at this density will turn out to be nonrelativistic, we have $x_p \ll 1$, so from Eq. (2.5.5),

$$m_e \left(1 + x_e^2\right)^{1/2} = Q, \tag{2.5.11}$$

where $Q = m_n - m_p$. Solving for x_e, we find from Eqs. (2.5.6) and (2.5.10) that

$$n = \frac{1}{3\pi^2 \lambda_e^3} \left[\left(\frac{Q}{m_e} \right)^2 - 1 \right]^{3/2}$$

$$= 7.37 \times 10^{30} \text{ cm}^{-3}, \tag{2.5.12}$$

and hence

$$\rho_0 \approx n m_p \approx 1.2 \times 10^7 \text{ g cm}^{-3}. \tag{2.5.13}$$

Exercise 2.20 Is Eq. (2.5.5) satisfied for $\rho_0 < 1.2 \times 10^7$ g cm^{-3}?

The equilibrium composition consists of an increasing proportion of neutrons above this density. The composition can be determined by substituting Eq. (2.5.7) in Eq. (2.5.5):

$$\left(m_e^2 + m_p^2 x_p^2\right)^{1/2} + m_p\left(1 + x_p^2\right)^{1/2} = m_n\left(1 + x_n^2\right)^{1/2}. \qquad (2.5.14)$$

Squaring Eq. (2.5.14) twice and simplifying gives

$$4m_n^2 m_p^2 x_p^2\left(1 + x_n^2\right) = \left(Q^2 - m_e^2\right)\left[\left(m_n + m_p\right)^2 - m_e^2\right]$$

$$+ 2m_n^2 x_n^2\left(m_n^2 - m_p^2 - m_e^2\right) + m_n^4 x_n^4. \qquad (2.5.15)$$

Thus

$$\frac{n_p}{n_n} = \left(\frac{m_p x_p}{m_n x_n}\right)^3$$

$$= \frac{1}{8}\left\{\frac{1 + \dfrac{2\left(m_n^2 - m_p^2 - m_e^2\right)}{m_n^2 x_n^2} + \dfrac{\left(Q^2 - m_e^2\right)\left[\left(m_n + m_p\right)^2 - m_e^2\right]}{m_n^4 x_n^4}}{1 + \dfrac{1}{x_n^2}}\right\}^{3/2}.$$

$$(2.5.16)$$

Now Q and m_e are both $\ll m_n$, so

$$\frac{n_p}{n_n} \simeq \frac{1}{8}\left\{\frac{1 + 4Q/m_n x_n^2 + 4\left(Q^2 - m_e^2\right)/m_n^2 x_n^4}{1 + 1/x_n^2}\right\}^{3/2}. \qquad (2.5.17)$$

Thus the proton–neutron ratio initially decreases as x_n increases—that is, as the density increases. It reaches a minimum value of

$$\left(\frac{n_p}{n_n}\right)_{\min} = \left[\frac{Q}{m_n} + \frac{\left(Q^2 - m_e^2\right)^{1/2}}{m_n}\right]^{3/2} = 0.0026 \qquad (2.5.18)$$

at

$$n_n = \frac{2^{3/2}}{3\pi^2 \lambda_n^3}\left(\frac{Q^2 - m_e^2}{m_n^2}\right)^{3/4},$$

$$\rho_0 \approx m_n n_n \approx 7.8 \times 10^{11} \text{ g cm}^{-3}, \qquad (2.5.19)$$

and then rises monotonically to $\frac{1}{8}$ for $x_n \to \infty$ or $\rho_0 \to \infty$.

Note that the equilibrium determined here is stable because it corresponds to a minimum of ε.

Exercise 2.21 Verify Eqs. (2.5.18) and (2.5.19).

Exercise 2.22 Show that the result $n_e : n_p : n_n = 1 : 1 : 8$ in the limit of very large densities is a trivial consequence of charge neutrality, β-equilibrium, and extreme relativistic degeneracy.

Exercise 2.23 Compute the maximum momentum of an electron emitted in the reaction (2.5.2). Show that p_F^e is greater than this value for all densities above that determined by Eq. (2.5.12), thus verifying that the equilibrium is stable.

The results of this section are not, strictly speaking, applicable to the case of gravitational collapse with escaping neutrinos, even when the collapse occurs quasistatically and at zero temperature. Thermodynamic equilibrium is not achieved in an open system, and the *n-p-e* composition must be determined by solving appropriate rate equations for the various reactions.

The above equations do precisely describe an equilibrium system of fixed charge (zero), baryon number, and lepton number. The lepton number in this system has been chosen to have its minimum possible value; that is, we have taken a limit where $n_\nu \to 0$ ($\mu_\nu \to 0$) while preserving detailed balance.

2.6 Beta-Equilibrium Between Relativistic Electrons and Nuclei: The Harrison–Wheeler Equation of State

We now turn to a quantitative treatment of inverse beta decay in an attempt to determine more precisely the correct equation of state in the density range $10^7 \leqslant \rho \leqslant 4 \times 10^{11}$ g cm^{-3}. We wish to determine the lowest energy state of a system of $A \sim 10^{57}$ baryons (mass $\sim 1 M_\odot$) consisting of separated nuclei in beta-equilibrium with a relativistic electron gas. We must determine (1) which nuclei are present—that is, the values of A and Z that minimize the energy, and (2) the corresponding pressure.

We assume in this section that given enough time following nuclear burning, cold catalyzed material will achieve complete thermodynamic equilibrium. The matter composition and equation of state will then be determined by the lowest possible energy state of the matter. The ability of degenerate matter in dwarf stars in Nature to actually reach this minimum energy state will be discussed in Chapter 3.

It is well known that for assemblages of baryons with $A \lesssim 90$ the state of lowest energy consists of a *single* nucleus, with $^{56}_{26}$Fe the nucleus with the tightest binding. For $A \gtrsim 90$ the state of lowest energy corresponds to *more* than one nucleus; the *tightest* binding is obtained for A values that are integral multiples of 56. Thus, as A becomes larger and larger, it becomes more and more appropriate to treat the composition of matter of minimum energy content as pure $^{56}_{26}$Fe. The story changes, however, when A increases above $A \sim 10^{57}$ and the self-gravitational attraction becomes important. For baryons in hydrostatic equilibrium, densities exceed $\rho \sim 10^7$ gm cm^{-3}. Consequently, the electrons are relativistic and combine with bound nuclear protons to form neutrons (inverse β-decay), altering the equilibrium nuclear composition step-by-step away from $^{56}_{26}$Fe to more neutron-rich material.

The overall physical picture can be understood as follows. If nuclear forces alone determined the equilibrium nuclear structure, nucleons would accumulate into nuclei of unlimited size. However, the Coulomb repulsive forces become so great that such large nuclei undergo fission. For low densities, these two opposing effects strike a compromise at $A = 56$. However, when relativistic electrons enter the picture, the balance shifts. The nucleus contains a larger proportion of neutrons to protons (because of inverse β-decay), and Coulomb forces play a smaller role. Hence there is a greater tendency for larger nuclei to form. When the density increases to $\sim 4 \times 10^{11}$ g cm^{-3}, the ratio n/p reaches a critical level. Any further increase in the density leads to "neutron drip"—that is, a two-phase system in which electrons, nuclei, and *free neutrons* co-exist and together determine the state of lowest energy. Increasing the density above 4×10^{11} g cm^{-3} leads to higher n/p ratios and more and more free neutrons. Finally, when the density exceeds about 4×10^{12} g cm^{-3}, more pressure is provided by neutrons than by electrons. The neutron gas so controls the situation that one can describe the medium as one vast nucleus with lower-than-normal nuclear density!

For a quantitative treatment, we start by writing the energy density of a mixture of nuclei, free electrons, and free neutrons in the form

$$\varepsilon = n_N M(A, Z) + \varepsilon'_e(n_e) + \varepsilon_n(n_n). \tag{2.6.1}$$

Here $M(A, Z)$ is the energy of a nucleus (A, Z), including the rest mass of the nucleons. It is conventional in nuclear physics also to include the electron rest mass in $M(A, Z)$. We therefore must subtract $n_e m_e c^2$ from the expression (2.3.7) for ε_e and denote the remainder by ε'_e in Eq. (2.6.1). The quantity n_N is the number density of nuclei, while n_n is the number density of free neutrons. The baryon density n and electron density n_e are then given by

$$n = n_N A + n_n, \qquad n_e = n_N Z, \tag{2.6.2}$$

that is,

$$1 = A Y_N + Y_n, \qquad Y_e = Y_N Z. \tag{2.6.3}$$

Thus, instead of regarding ε at $T = 0$ as a function of (n, Y_N, Y_e, Y_n), we can equally well take it to be a function of (n, A, Z, Y_n). The equilibrium composition and resulting equation of state are determined by minimizing ε with respect to A, Z, and Y_n at fixed n.

Note that while we are nominally only interested in the equation of state at densities below neutron drip in this chapter, we must allow for the presence of free neutrons in Eq. (2.6.1) to find the onset of neutron drip. It will in fact turn out to be very easy to continue at least the Harrison–Wheeler equation of state beyond neutron drip, and hence we shall do so in this section.

The quantity $M(A, Z)$ is not known experimentally for the very neutron-rich nuclei produced above 10^{10} g cm^{-3} and must be inferred theoretically. This is usually done by means of a *semiempirical mass formula*, at least up to the point of neutron drip.

A fairly simple treatment of the equation of state was given by Harrison and Wheeler in 1958 (hereafter HW).[12] They used the semiempirical mass formula of Green (1955),

$$M(Z, A) = \left[(A - Z)m_n c^2 + Z(m_p + m_e)c^2 - A\bar{E}_b \right]$$

$$= m_u c^2 \left[b_1 A + b_2 A^{2/3} - b_3 Z + b_4 A \left(\frac{1}{2} - \frac{Z}{A} \right)^2 + \frac{b_5 Z^2}{A^{1/3}} \right], \quad (2.6.4)$$

where $\bar{E}_b$ is the mean binding energy per baryon and where[13]

$$b_1 = 0.991749, \qquad b_3 = 0.000840, \qquad b_5 = 0.000763,$$

$$b_2 = 0.01911, \qquad b_4 = 0.10175. \qquad (2.6.5)$$

This expression is based on the "liquid-drop" nuclear model, and the terms have the following interpretation: The dominant contribution to $\bar{E}_b$ is proportional to the volume of the nucleus, as in the case of a liquid drop. The difference of b_1 from unity is due largely to this volume binding energy (nuclear radii are roughly proportional to $A^{1/3}$). The b_2 term represents a surface energy term, b_4 is the symmetry term (favors nuclei with equal numbers of protons and neutrons), while b_5 gives the Coulomb energy. The quantity b_3 is just $(m_n - m_p - m_e)/m_u$. Pairing and shell effects have been neglected.[14]

---

[12] Harrison and Wheeler (1958). See also Harrison et al. (1965), Chapter 10, for a more detailed discussion and convenient analytic fits to the equation of state.

[13] We have multiplied b_1 through b_5 as given by Green by 0.999682 to convert from the ^{16}O mass scale to the modern ^{12}C scale.

[14] See any nuclear physics textbook for a discussion of these effects.

Exercise   2.24   Assume the nucleus consists of two ideal, nonrelativistic Fermi gases, one of Z protons and the other of $N = A - Z$ neutrons. Show that the energy of the nucleus (excluding rest energy) is

$$E = \frac{3}{5} E'_F \left[\left(\frac{2N}{A} \right)^{2/3} N + \left(\frac{2Z}{A} \right)^{2/3} Z \right].$$

Here E'_F (the Fermi energy excluding rest energy) is defined by putting A nucleons in a volume V with four nucleons per momentum state. Determine E'_F from the measured density of nuclei, $n = 1.72 \times 10^{38}$ particles/cm³. By considering deviations from the case $N = Z$ (minimum of E), derive the symmetry term and show that about $\frac{1}{2}$ the value of b_4 comes from this simple Pauli Principle picture.

Exercise   2.25   Calculate the nuclear surface energy in the Fermi gas model as follows: The number density in phase space, excluding spin effects, is

$$\frac{d\mathfrak{N}}{d^3x \, d^3p} = \frac{1}{h^3}, \qquad E < E_F.$$

For a box of volume V, this gives

$$d\mathfrak{N} = \frac{V 4\pi p^2 \, dp}{h^3}$$

for the number of states between p and $p + dp$. One can also obtain this expression by solving the free-particle Schrödinger equation inside a cubical box. In this solution, however, one should not count states for which the components p_x or p_y or p_z are zero, as these correspond to a vanishing wave function. Show that this refinement leads to the expression

$$d\mathfrak{N} = \frac{V 4\pi p^2}{h^3} \left(1 - \frac{h}{4} \frac{S}{Vp} \right) dp,$$

where S is the area of the box. Use this expression to expand the mean binding energy per particle in the form

$$\frac{E}{A} = a_0 + a_1 \frac{S}{V} + \cdots$$

where $a_0 = \frac{3}{5} E'_F$. Use the relations $r_N = r_0 A^{1/3}$, $n = A/V = 3/4\pi r_0^3$ for nuclear radius and density to relate a_1 to b_2, and show that the model is very close to the observed value.

Exercise   2.26   Derive the Coulomb term in Eq. (2.6.4) approximately by suitably reinterpreting Eq. (2.4.6) and using $r_N = 1.5 \times 10^{-13} A^{1/3}$ cm.

Exercise   2.27   Assume equal numbers of neutrons and protons and plot $\bar{E}_b$ in MeV as a function of A from Eq. (2.6.4) for stable nuclei in the region below $A = 130$. Locate

A_{max}, the nucleus of maximum binding energy. What is $\bar{E}_b|_{max}$? For what $A > A_{max}$ is it energetically favorable to fission into two nuclei?

Equation (2.6.1) now becomes

$$\varepsilon = n(1 - Y_n)\frac{M(A, Z)}{A} + \varepsilon'_e(n_e) + \varepsilon_n(n_n), \qquad (2.6.6)$$

where

$$n_e = n(1 - Y_n)\frac{Z}{A},$$

$$n_n = nY_n. \qquad (2.6.7)$$

Note that

$$\frac{d\varepsilon'_e}{dn_e} = \frac{\varepsilon'_e + p_e}{n_e} = E_{F_e} - m_e c^2,$$

$$\frac{d\varepsilon_n}{dn_n} = \frac{\varepsilon_n + p_n}{n_n} = E_{F_n}, \qquad (2.6.8)$$

[cf. Eqs. (2.1.7) and (2.3.29)]. Harrison and Wheeler made the approximation of treating Z and A as continuous variables in the spirit of the semiempirical mass formula. Thus $\partial\varepsilon/\partial Z = 0$ implies

$$\frac{\partial M}{\partial Z} = -\left(E_{F_e} - m_e c^2\right). \qquad (2.6.9)$$

This is simply the continuum limit of the β-stability condition: $M(Z - 1, A)$ in equilibrium with $M(Z, A)$, the free electron being at the top of the Fermi sea.
    Similarly, $\partial\varepsilon/\partial A = 0$ implies

$$A^2\frac{\partial}{\partial A}\left(\frac{M}{A}\right) = Z\left(E_{F_e} - m_e c^2\right). \qquad (2.6.10)$$

This is the continuum limit of $(A - 1)$ atoms of type (Z, A) in equilibrium with A atoms of type $(Z, A - 1)$. An additional Z free electrons must be created with energies at the top of the Fermi sea as the nuclear charge increases from $(A - 1)Z$ to AZ. Equations (2.6.9) and (2.6.10) can be combined to give

$$Z\frac{\partial M}{\partial Z} + A\frac{\partial M}{\partial A} - M = 0. \qquad (2.6.11)$$

The condition $\partial \varepsilon / \partial Y_n = 0$ gives, using Eq. (2.6.10),

$$\frac{\partial M}{\partial A} = E_{F_n}, \tag{2.6.12}$$

which is the continuum version of the condition for $M(Z, A)$ to be in equilibrium with $M(Z, A - 1)$ and a free neutron.

Equation (2.6.9) gives

$$b_3 + b_4 \left(1 - \frac{2Z}{A}\right) - 2b_5 \frac{Z}{A^{1/3}} = \left[(1 + x_e^2)^{1/2} - 1\right] \frac{m_e}{m_u}, \tag{2.6.13}$$

where x_e is defined as in Eq. (2.5.4). Equation (2.6.11) gives

$$Z = \left(\frac{b_2}{2b_5}\right)^{1/2} A^{1/2} = 3.54 A^{1/2}, \tag{2.6.14}$$

while Eq. (2.6.12) gives

$$b_1 + \frac{2b_2 A^{-1/3}}{3} + b_4 \left(\frac{1}{4} - \frac{Z^2}{A^2}\right) - \frac{b_5 Z^2}{3A^{4/3}} = (1 + x_n^2)^{1/2} \frac{m_n}{m_u}, \tag{2.6.15}$$

where x_n is defined as in Eq. (2.5.4). Note that while Z increases with A, the ratio Z/A decreases as $A^{-1/2}$ under these high-density conditions.

To construct the equation of state from the above relations, first pick a value of $A > 56$. Equation (2.6.14) gives Z. Test whether neutron drip has been reached—that is, whether Eq. (2.6.15) gives a value of $x_n > 0$. If so compute ε_n, P_n, and n_n from Eqs. (2.3.4), (2.3.5), and (2.3.7) with m_n replacing m_e; otherwise set these quantities to zero. Equation (2.6.13) gives x_e and hence ε_e', P_e, and n_e. Then

$$\rho = \frac{\varepsilon}{c^2} = \frac{n_e M(A, Z)/Z + \varepsilon_e' + \varepsilon_n}{c^2}, \tag{2.6.16}$$

$$P = P_e + P_n, \tag{2.6.17}$$

$$n = n_e \frac{A}{Z} + n_n. \tag{2.6.18}$$

The resulting equation of state $P = P(\rho)$ is shown in Figure 2.2. Deviations from the ideal electron gas-pure $^{56}_{26}\text{Fe}$ equation of state become apparent for $\rho \gtrsim 10^7$ g cm^{-3}. The onset of neutron drip occurs at $\rho \sim 3.18 \times 10^{11}$ g cm^{-3},

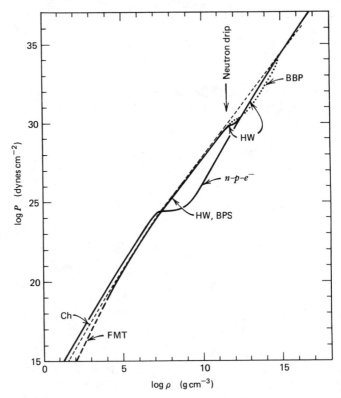

Figure 2.2   Representative equations of state below neutron drip. The letters labeling different curves are defined in Table 2.2. The Ch equation of state is shown for $\mu_e = \frac{56}{26}$. The HW and BPS equations of state match smoothly onto the FMT equation at $\rho = 10^4$ g cm^{-3}. Note that above neutron drip, the HW equation of state ($\rho_{\text{drip}} \approx 3.2 \times 10^{11}$ g cm^{-3}) merges smoothly onto the ideal n-p-e^- equation of state, while the BPS equation of state ($\rho_{\text{drip}} \approx 4.3 \times 10^{11}$ g cm^{-3}) is continued via BBP (see Chapter 8).

where $(A, Z) \sim (122, 39.1)$—that is, near the element yttrium— and $E_{F_e} \sim 23.6$ MeV. Above this density, the free neutrons begin to supply a larger and larger fraction of the total pressure and density. At $\rho \sim 4.54 \times 10^{12}$ g cm^{-3}, where $(A, Z) \sim (187, 48.4)$, the neutrons provide 60% of the pressure and density—the nuclei are rapidly becoming unimportant. Above this density, Harrison and Wheeler simply used the ideal n-p-e equation of state of Section 2.5, which matches onto their results quite smoothly both in P and $dP/d\rho$. We shall return to improvements in the equation of state above neutron drip in Chapter 8. In the next section, we discuss improvements in the regime from 10^7 g cm^{-3} to neutron drip.

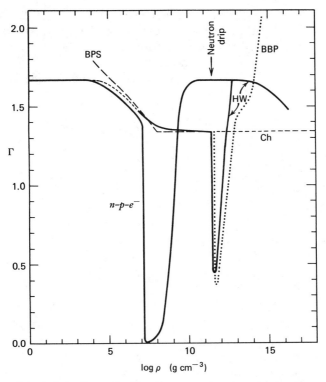

Figure 2.3   The adiabatic index $\Gamma = (d\ln P)/(d\ln \rho)$ as a function of ρ for the equations of state shown in Figure 2.2.

2.7   The Baym–Pethick–Sutherland Equation of State

The HW equation of state has the advantage that $M(A, Z)$ is represented by a fairly simple function. However, real nuclei have discrete values of A and Z, and shell effects play an important role in determining the binding energy of a given nucleus. Salpeter (1961) calculated the composition and equation of state from 10^7 g cm^{-3} to 3.4×10^{11} g cm^{-3} (onset of neutron drip) taking these effects into account.

Baym, Pethick, and Sutherland (1971b) (hereafter BPS) improved on Salpeter's treatment by using a better semiempirical mass formula[15] and also by noting that the lattice energy, Eq. (2.4.9), was important in determining the equilibrium *composition*, even though it was only a small correction to the electron *pressure*.

[15] Myers and Swiatecki (1966).

The reason is that the equilibrium nuclide is determined largely by a competition between the nuclear surface energy and the Coulomb energy. At a density of 10^{11} g cm^{-3}, the lattice energy reduces the positive nuclear Coulomb energy by $\sim 15\%$ and this has an effect on the composition.

To derive the BPS equation of state, we start by adding the lattice energy to Eq. (2.6.6):

$$\varepsilon = n(1 - Y_n)\frac{M(A, Z)}{A} + \varepsilon'_e(n_e) + \varepsilon_n(n_n) + \varepsilon_L, \qquad (2.7.1)$$

where, from Eq. (2.4.9), assuming a bcc lattice structure,

$$\varepsilon_L = -1.444 Z^{2/3} e^2 n_e^{4/3}. \qquad (2.7.2)$$

The condition $\partial \varepsilon / \partial Y_n = 0$ now gives [using Eq. (2.6.7)]

$$E_{F_n} = \frac{M(A, Z) + Z\left(E_{F_e} - m_e c^2\right) + 4Z\varepsilon_L/3n_e}{A}. \qquad (2.7.3)$$

Thus neutron drip occurs when the right-hand side of Eq. (2.7.3) equals $m_n c^2$. (Strictly speaking, one should allow for the interaction energy of the liberated neutron with the nuclei, but this is a small effect.)

We shall assume that we are below the density for neutron drip given by Eq. (2.7.3) and set $\varepsilon_n = 0$ in the following discussion.

One can determine the equilibrium composition as follows: Fix a value of n. Now try a pair of values (A, Z). Since $M(A, Z)$ is tabulated and since $n_N = n/A$, $n_e = Zn/A$, one can compute ε. Try all possible values of (A, Z). The value that minimizes ε gives the equilibrium nucleus. The pressure is given by

$$P = n^2 \left.\frac{\partial(\varepsilon/n)}{\partial n}\right|_{A, Z}$$

$$= P_e + P_L, \qquad (2.7.4)$$

where P_e is the ideal Fermi expression (2.3.5) and

$$P_L = \tfrac{1}{3}\varepsilon_L \qquad (2.7.5)$$

[cf. Eq. (2.4.11)].

Whenever there is a phase transition from one stable nuclide to another, it is accompanied by a discontinuity in n and $\rho = \varepsilon/c^2$. This occurs because the pressure must be a continuous function of radius inside a star. We can estimate

Table 2.1
Equilibrium Nuclei Below Neutron Drip

Nucleus	ρ_{max} (g cm^{-3})	
	(a) BPS	(b) JGK^a
^{56}Fe	8.1×10^6	8.1×10^6
^{62}Ni	2.7×10^8	2.8×10^8
^{64}Ni	1.2×10^9	1.3×10^9
^{66}Ni	—	1.5×10^9
^{86}Kr	—	3.1×10^9
^{84}Se	8.2×10^9	7.6×10^9
^{82}Ge	2.2×10^{10}	2.6×10^{10}
^{80}Zn	4.8×10^{10}	6.0×10^{10}
^{78}Ni	1.6×10^{11}	8.4×10^{10}
^{76}Fe	1.8×10^{11}	—
^{126}Ru	—	1.2×10^{11}
^{124}Mo	1.9×10^{11}	1.7×10^{11}
^{122}Zr	2.7×10^{11}	2.5×10^{11}
^{120}Sr	3.7×10^{11}	3.6×10^{11}
^{122}Sr	—	3.8×10^{11}
^{118}Kr	4.3×10^{11}	4.4×10^{11}

^{a}The first six nuclear masses are known experimentally. The remainder are from the Jänecke–Garvey–Kelson mass formula (cf. Wapstra and Bos, 1976).

the size of this discontinuity by noting that, since $P_L \ll P_e$ and P_e depends only on n_e, the quantity n_e is essentially continuous at the phase boundary. But $n_e = nZ/A$, so

$$\frac{\Delta\rho}{\rho} \approx \frac{\Delta n}{n} \approx -\frac{\Delta(Z/A)}{Z/A}. \tag{2.7.6}$$

For example, a phase transition from ^{56}Fe ($Z/A = 0.4643$) to ^{62}Ni ($Z/A = 0.4516$) implies a 2.9% increase in density.

Because of these discontinuities, one must make a Maxwell equal-area construction to determine the pressure at which the transition actually takes place.[16] This is a drawback of the above procedure. It is more convenient instead to minimize the Gibbs free energy g at constant P [cf. Eq. (2.1.18)]. At $T = 0$, Eq.

---

[16]See, e.g., Reif (1965), Section 8.6 for a discussion.

(2.1.19) gives

$$g = \frac{\varepsilon + P}{n}$$

$$= \frac{M(A, Z) + ZE_{F_e} + \frac{4}{3}Z\varepsilon_L/n_e}{A}. \tag{2.7.7}$$

Now fix a value of P. Take trial values of A and Z and solve the transcendental Eq. (2.7.4) for $n_e = x^3/3\pi^2\lambda_e^3$. Then compute g from Eq. (2.7.7). Repeat this procedure until the minimum of g with respect to (A, Z) is found. The resulting sequence of nuclides as found by BPS is given in Table 2.1 and the equation of state in Figure 2.2.

One can see here the importance of shell effects: the nuclides from ^{84}Se to ^{76}Fe have a closed shell of 50 neutrons; those from ^{124}Mo to ^{118}Kr a closed shell of 80 neutrons. Neutron drip occurs at $\rho \simeq 4.3 \times 10^{11}$ g cm^{-3}, where Z/A for ^{118}Kr is 0.3051.

We have repeated the BPS calculations using the experimental values tabulated by Wapstra and Bos (1977) for $M(A, Z)$ in the regime that they can be measured, and the theoretical extrapolation of Jänecke, Garvey, and Kelson as tabulated in Wapstra and Bos (1976) for the neutron-rich regime. The resulting sequence of nuclides is given in column (b) of Table 2.1. The equation of state is indistinguishable from that of BPS except right in the neighborhood of a phase transition, where P may differ by $\sim 5\%$ for a given value of ρ. We conclude that the equation of state is relatively well-established up to the onset of neutron drip.

Computer Exercise   2.28   Repeat the BPS procedure employing a different formula for $M(A, Z)$ as described, for example, in Wapstra and Bos (1976). How does your equation of state compare with BPS?

Computer Exercise   2.29   Show that in fact only *one* nuclear species is present at any given pressure under the assumptions of the BPS equation of state. Rewrite Eq. (2.7.1) allowing for a fraction f of species (A_1, Z_1) and $1 - f$ of species (A_2, Z_2). Show that $f = 0$ or 1 except at a density discontinuity.

Box 2.1

Cold Equation of State Below Neutron Drip: Summary

1. The equation of state for zero-temperature matter is well understood in the density regime below "neutron drip," where $\rho_{drip} \approx 4 \times 10^{11}$ g cm^{-3}. The pressure is dominated by degenerate electrons that become fully relativistic above $\sim 10^7$ g cm^{-3}. Positive charge is concentrated in separated nuclei, which form a regular Coulomb lattice embedded in the electron gas.

2. If the matter has settled into its ground state, it may be assumed to be in nuclear equilibrium; that is, its energy cannot be lowered by changing its composition via strong, weak, or electromagnetic interactions. The equilibrium nuclide can be determined as a function of density. Below $\sim 10^7$ g cm^{-3} the ground state corresponds to $^{56}_{26}$Fe nuclei. At higher densities the equilibrium nuclide becomes increasingly neutron-rich. The nuclei are stabilized against β-decay by the filled Fermi sea of electrons.

3. The equation of state below neutron drip determines the structure of planets and stable white dwarfs. In white dwarfs, matter has probably not achieved complete equilibrium. The Chandrasekhar equation of state with Coulomb corrections (Salpeter, 1961) therefore applies. The composition depends on the evolutionary history of the star and determines the mean molecular weight per electron, μ_e (cf. Chapter 3). In neutron stars, matter in this density regime is in complete equilibrium and an equilibrium equation of state (e.g., BPS) is applicable.

4. Models for the cold equation of state below neutron drip are summarized in Figures 2.2 and 2.3 and in Table 2.2.

Table 2.2
Representative Equations of State Below Neutron Drip

Equation of State	Density Regime (g cm⁻³)	Composition	Theory
Chandrasekhar (1931a, b; Ch): ideal electron gas	$0 \leq \rho \leq \infty$	e^- (nuclei specified by μ_e)	Noninteracting electrons
Ideal n-p-e^- gas	$0 \leq \rho \leq 1.2 \times 10^7$ $1.2 \times 10^7 < \rho \leq \infty$	e^-, p $n, p,$ and e^-	Equilibrium matter
Feynman–Metropolis–Teller (1949; FMT)	$7.9 \leq \rho \leq 10^4$	e^- and $^{56}_{26}$Fe	Thomas–Fermi–Dirac atomic model
Harrison–Wheeler (1958; HW)	$7.9 \leq \rho \leq 10^4$	Same as FMT	Same as FMT
	$10^4 < \rho \leq 10^7$	e^- and $^{56}_{26}$Fe	Noninteracting electrons
	$10^7 < \rho \leq 3 \times 10^{11}$	e^- and equilibrium nuclide	Semiempirical mass formula; equilibrium matter
	Above "Neutron-drip" $\begin{cases} 3 \times 10^{11} < \rho < 4 \times 10^{12} \\ 4.5 \times 10^{12} < \rho \leq \infty \end{cases}$	$e^-, n,$ and equilibrium nuclide	Same as ideal n-p-e^-
Baym–Pethick–Sutherland (1971b; BPS)	$7.9 \leq \rho \leq 10^4$	Same as FMT	Same as FMT
	$10^4 < \rho \leq 8 \times 10^6$	e^- and $^{56}_{26}$Fe	Ideal electrons with Coulomb lattice corrections
	$8 \times 10^6 < \rho \leq 4.3 \times 10^{11}$	e^- and equilibrium nuclide	Laboratory nuclear energies (with extrapolations); Coulomb lattice energy; equilibrium matter

3
White Dwarfs

3.1 History of the Theory of White Dwarfs

White dwarfs are stars of about one solar mass with characteristic radii of about 5000 km and mean densities of around 10^6 g cm^{-3}. These stars no longer burn nuclear fuel. Instead, they are slowly cooling as they radiate away their residual thermal energy.

We know today that white dwarfs support themselves against gravity by the pressure of degenerate electrons. This fact was not always clear to astronomers, although the compact nature of white dwarfs was readily apparent from early observations. For example, the mass of Sirius B, the binary companion to Sirius and the best known white dwarf, was determined by applying Kepler's Third Law to the binary star orbit. Early estimates placed its mass M in the range from $0.75M_\odot$ to $0.95M_\odot$. Its luminosity L was estimated from the observed flux and known distance to be about $\frac{1}{360}$ of that of the sun. In 1914 W. S. Adams (1915) made the surprising discovery that the spectrum of Sirius B was that of a "white" star, not very different from its normal companion, Sirius. By assigning an effective temperature of 8000 K to Sirius B from these spectral measurements and using the equation for blackbody emission, $L = 4\pi R^2 \sigma T_{\text{eff}}^4$, a radius R of 18,800 km could be inferred for the star. (This is about four times bigger than the modern value.)

Referring to Sirius B in his book *The Internal Constitution of the Stars*, the great astrophysicist Sir Arthur Eddington (1926) concluded that "we have a star of mass about equal to the sun and of radius much less than Uranus." He also reported in his book the extraordinary new measurements by W. S. Adams (1925) of the gravitational redshifts of several spectral lines emitted from the surface of Sirius B. By applying the theory of general relativity, the ratio M/R could be inferred from the measured redshifts. As the mass was already known from the binary orbit, the radius of Sirius B could be determined. The redshifts obtained by Adams, though crude, confirmed the previous estimates of R and the compact nature of the white dwarf. Eddington (1926) thus wrote in his book that "Prof.

Adams has killed two birds with one stone; he has carried out a new test of Einstein's general theory of relativity and he has confirmed our suspicion that matter 2000 times denser than platinum is not only possible, but is actually present in the Universe."[1]

Eddington (1926) went on to argue that although only three white dwarfs could be firmly established at that time, white dwarfs are probably very abundant in space, since the known ones were all very close to the sun. But regarding the means by which white dwarfs supported themselves against collapse, Eddington could declare only that "it seems likely that the ordinary failure of the gas laws due to finite sizes of molecules will occur at these high densities, and I do not suppose that the white dwarfs behave like perfect gas."

In August of 1926, Dirac (1926) formulated Fermi–Dirac statistics, building on the foundations established only months earlier by Fermi. In December 1926, R. H. Fowler (1926), in a pioneering paper on compact stars, applied Fermi–Dirac statistics to explain the puzzling nature of white dwarfs: he identified the pressure holding up the stars from gravitational collapse with *electron degeneracy pressure*.

Actual white dwarf models, taking into account special relativistic effects in the degenerate electron equation of state, were constructed in 1930 by Chandrasekhar (1931a, b). In the course of this analysis, Chandrasekhar (1931b) made the momentous discovery that white dwarfs had a *maximum* mass of $\sim 1.4 M_\odot$, the exact value depending on the composition of the matter. This maximum mass is called the *Chandrasekhar limit* in honor of its discoverer. Chandrasekhar (1934) was immediately aware of the important implication of his finding, for he wrote in 1934: "The life history of a star of small mass must be essentially different from the life history of a star of large mass. For a star of small mass the natural white-dwarf stage is an initial step towards complete extinction. A star of large mass cannot pass into the white-dwarf stage and one is left speculating on other possibilities."

In 1932, L. D. Landau (1932) presented an elementary explanation of the Chandrasekhar limit. He applied his argument several months later to neutron stars when he learned of the discovery of the neutron (Section 9.1).

The role of general relativity in modifying the mass–radius relation for massive white dwarfs above about $1 M_\odot$ was first discussed by Kaplan (1949). He concluded that general relativity probably induces a dynamical instability when the radius becomes smaller than 1.1×10^3 km. The general relativistic instability for white dwarfs was discovered independently by Chandrasekhar in 1964.[2]

---

[1]Adams' value for the redshift, and also the separate value obtained by Moore, agreed with general relativity, although using an incorrect observational value for the radius. The modern situation is described in Section 3.6.

[2]See Chandrasekhar and Tooper (1964).

Following the discovery of the neutron, it was generally realized[3] that at very high densities electrons would react with protons to form neutrons via inverse beta decay. The motivation for this work was the idea that the energy source of massive normal stars might be inverse beta decay in a neutron core. Much later, Schatzman (1956, 1958a, b) and, independently, Harrison and Wheeler (1958), incorporated inverse beta decay in the equation of state for white dwarf matter. Schatzman and Harrison, Wakano and Wheeler (1958) showed that its effect is also to induce a dynamical instability in the most massive white dwarf stars above $\sim M_\odot$, which have radii less than about 4×10^3 km. Stability would not be restored until virtually all of the electrons and protons were squeezed together. At such high densities, the gas would consist almost entirely of neutrons. At this point, the configuration would have a mass of $\sim M_\odot$, and a radius of ~ 10 km. A new class of stable, compact stars would then exist—they are the neutron stars that had been proposed in the 1930s.

We conclude our sketch of the early history of the theory of white dwarfs at this point. We shall resume our story in Section 9.1 where we will trace the development of the neutron star idea.

3.2 The Onset of Degeneracy

In Chapter 1, we discussed rather convincing evidence that the final state of a star that has exhausted its nuclear fuel is a white dwarf, provided it is not very massive. The consequences of hydrostatic equilibrium give us further insight into the fate of a star whose nuclear fuel has been used up.

For a spherically symmetric distribution of matter, the mass interior to a radius r is given by

$$m(r) = \int_0^r \rho 4\pi r^2 \, dr, \quad \text{or} \quad \frac{dm(r)}{dr} = 4\pi r^2 \rho. \qquad (3.2.1)$$

Here $\rho \simeq \rho_0$, the rest-mass density, as we are considering nonrelativistic matter. If the star is in a steady state, the gravitational force balances the pressure force at every point. To derive the hydrostatic equilibrium equation, consider an infinitesimal fluid element lying between r and $r + dr$ and having an area dA perpendicular to the radial direction. The gravitational attraction between $m(r)$ and the mass $dm = \rho \, dA \, dr$ is the same as if $m(r)$ were concentrated in a point at the center, while the mass outside exerts no force on dm. The net outward pressure

---

[3]See, for example, Gamow (1937), Landau (1938), and Oppenheimer and Serber (1938).

force on dm is $-[P(r + dr) - P(r)]\,dA$, so in equilibrium

$$-\frac{dP}{dr}\,dr\,dA = \frac{Gm(r)}{r^2}\,dm$$

or

$$\frac{dP}{dr} = -\frac{Gm(r)\rho}{r^2}. \qquad (3.2.2)$$

In general, hydrostatic equilibrium implies $\nabla P = -\rho\nabla\Phi$, where Φ is the gravitational potential (cf. Section 6.1).

A consequence of the hydrostatic equilibrium equation (3.2.2) is the *virial theorem*: The gravitational potential energy of the star is

$$W = -\int_0^R \frac{Gm(r)}{r}\rho 4\pi r^2\,dr$$

$$= \int_0^R \frac{dP}{dr}4\pi r^3\,dr \qquad \text{by Eq. (3.2.2)}$$

$$= -3\int_0^R P4\pi r^2\,dr, \qquad (3.2.3)$$

where we have integrated by parts.

Now if the gas is characterized by an adiabatic equation of state

$$P = K\rho_0^\Gamma \qquad (K, \Gamma \text{ constants}), \qquad (3.2.4)$$

the energy density of the gas (excluding rest-mass energy) is

$$\varepsilon' = \frac{P}{\Gamma - 1}. \qquad (3.2.5)$$

This result follows from the first law of thermodynamics, assuming adiabatic changes:

$$d\left(\frac{\varepsilon}{\rho_0}\right) = -Pd\left(\frac{1}{\rho_0}\right). \qquad (3.2.6)$$

Integration leads to [using Eq. (3.2.4)]

$$\varepsilon = \rho_0 c^2 + \frac{P}{\Gamma - 1}, \qquad (3.2.7)$$

which gives the desired result for $\varepsilon' \equiv \varepsilon - \rho_0 c^2$.

Equation (3.2.3) can thus be rewritten as

$$W = -3(\Gamma - 1)U, \qquad (3.2.8)$$

where

$$U = \int_0^R \varepsilon' 4\pi r^2 \, dr \qquad (3.2.9)$$

is the total internal energy of the star.

Exercise 3.1 Define E_T to be the translational kinetic energy of particles *not* including any energy associated with internal degrees of freedom (e.g., rotational or vibrational energy). Show that for an ideal Maxwell–Boltzmann gas characterized by a constant adiabatic exponent Γ, E_T is related to U by $E_T = \frac{3}{2}(\Gamma - 1)U$. Next show that the virial relation (3.2.3) can be written

$$E_T = -\tfrac{1}{2}W$$

for such a gas.

The total energy of the star, $E = W + U$, is

$$E = -\frac{3\Gamma - 4}{3(\Gamma - 1)}|W|, \qquad (3.2.10)$$

where $W \sim -GM^2/R$.

Exercise 3.2 Show that if Eq. (3.2.4) holds everywhere inside the star, then the gravitational potential energy is given by

$$W = -\frac{3(\Gamma - 1)}{5\Gamma - 6}\frac{GM^2}{R}. \qquad (3.2.11)$$

Hint: Rewrite Eq. (3.2.3) in the form

$$W = -3\int_0^M \frac{P}{\rho} \, dm(r).$$

Integrate by parts and use

$$d\left(\frac{P}{\rho}\right) = \frac{\Gamma - 1}{\Gamma} Gm(r) \, d\left(\frac{1}{r}\right).$$

Integrate by parts once more.

Without nuclear fuel, E decreases due to radiation. According to Eqs. (3.2.10) and (3.2.11), $\Delta E < 0$ implies $\Delta R < 0$ whenever $\Gamma > \frac{4}{3}$; that is, the star contracts. Can the star contract forever, extracting energy from the infinite supply of gravitational potential energy until R goes to zero (or until the star undergoes total collapse to a black hole)? The answer is no for stars with $M \sim M_\odot$, as we now demonstrate.

Suppose that the pressure during such a quasi-static collapse is given by the ideal Maxwell–Boltzmann gas law

$$P = \frac{\rho_0}{\mu m_u} kT, \tag{3.2.12}$$

where, for example, for pure ionized carbon $\mu = \frac{12}{7}$ [cf. Eq. (2.3.17)]. Then by the virial relation (3.2.3),

$$-W = 3\int_0^R P 4\pi r^2\, dr$$

$$= \frac{3k\overline{T}}{\mu m_u} \int_0^R \rho_0 4\pi r^2\, dr$$

$$= \frac{3M}{\mu m_u} k\overline{T}, \tag{3.2.13}$$

where $\overline{T}$ is an averaged temperature in the star. Thus $\overline{T} \propto M/R$; that is, $\overline{T}$ increases as R decreases. But $\bar{\rho} \propto M/R^3$, so the density increases even more rapidly. We now show that this results in the breakdown of the Maxwell–Boltzmann relation (3.2.12); the electron gas becomes degenerate and provides a source of pressure support even at zero temperature.

The typical (i.e., rms) momentum difference between electrons in a Maxwell–Boltzmann gas is

$$\Delta p_e \sim (6m_e k\overline{T})^{1/2} \sim \left(\frac{12 m_e GMm_u \mu}{7R}\right)^{1/2}, \tag{3.2.14}$$

where we have set $\Gamma = \frac{5}{3}$ and used Eqs. (3.2.13) and (3.2.11).

Exercise 3.3 Verify the first relation in Eq. (3.2.14). For two electrons

$$\Delta p_{\text{rms}} = \left\langle (\mathbf{p}_1 - \mathbf{p}_2)^2 \right\rangle^{1/2} = \left\langle 2\mathbf{p}_1^2 \right\rangle^{1/2}.$$

Note that $\overline{T}$ has a slightly different meaning in Eqs. (3.2.13) and (3.2.14).

The typical separation between electrons is

$$\Delta q_e \sim \left(\frac{\mu_e m_u}{\rho_0}\right)^{1/3} \sim \left(\frac{4\mu_e m_u R^3}{M}\right)^{1/3}. \tag{3.2.15}$$

The volume occupied by an electron in phase space is thus

$$(\Delta p_e \,\Delta q_e)^3 \sim 4\mu_e \left(\frac{12\mu}{7}\right)^{3/2}\left[(Gm_e R)^{1/2} m_u^{5/6} M^{1/6}\right]^3$$

$$\sim 40\left[1 \times 10^{-26}\left(\frac{M}{M_\odot}\right)^{1/6}\left(\frac{R}{R_\odot}\right)^{1/2} \text{g cm}^2 \text{ s}^{-1}\right]^3$$

$$\sim 180h^3\left(\frac{M}{M_\odot}\right)^{1/2}\left(\frac{R}{R_\odot}\right)^{3/2}, \tag{3.2.16}$$

so that when a $1M_\odot$ star contracts to $R \sim 3 \times 10^{-2}R_\odot$, the phase space volume occupied by an electron is $\sim h^3$. At this point the Pauli exclusion principle becomes important and Fermi–Dirac statistics must be used. As we have seen in the previous chapter, even at zero temperature such a gas is a source of pressure. We now explore the properties of equilibrium configurations supported by such an electron degeneracy pressure: white dwarfs.

3.3 Polytropes

The ideal Fermi gas equation of state reduces to the simple polytropic form (3.2.4) in the limiting case of extreme nonrelativistic $\left(\Gamma = \frac{5}{3}\right)$ and ultrarelativistic electrons $\left(\Gamma = \frac{4}{3}\right)$; compare Eqs. (2.3.21)–(2.3.23).

Equilibrium configurations with such an equation of state are called *polytropes* and are relatively simple to analyze. We shall first discuss the properties of white dwarfs as polytropes in the low density $\left(\Gamma = \frac{5}{3}\right)$ and high density $\left(\Gamma = \frac{4}{3}\right)$ limits, and then describe the intermediate regime and refinements to the polytropic picture.

The hydrostatic equilibrium equations (3.2.1) and (3.2.2) can be combined to give

$$\frac{1}{r^2}\frac{d}{dr}\left(\frac{r^2}{\rho}\frac{dP}{dr}\right) = -4\pi G\rho. \tag{3.3.1}$$

Substitute the equation of state (3.2.4), and write

$$\Gamma \equiv 1 + \frac{1}{n}, \tag{3.3.2}$$

where n is called the polytropic index. The equation can be reduced to dimension-less form by writing

$$\rho = \rho_c \theta^n, \tag{3.3.3}$$

$$r = a\xi, \tag{3.3.4}$$

$$a = \left[\frac{(n+1)K\rho_c^{(1/n-1)}}{4\pi G} \right]^{1/2}, \tag{3.3.5}$$

where $\rho_c = \rho(r=0)$ is the central density. Then

$$\frac{1}{\xi^2} \frac{d}{d\xi} \xi^2 \frac{d\theta}{d\xi} = -\theta^n. \tag{3.3.6}$$

This is the *Lane–Emden equation* for the structure of a polytrope of index n. The boundary conditions at the center of a polytropic star are

$$\theta(0) = 1, \tag{3.3.7}$$

$$\theta'(0) = 0. \tag{3.3.8}$$

The condition (3.3.7) follows directly from Eq. (3.3.3). Equation (3.3.8) follows from the fact that $m(r) \approx 4\pi\rho_c r^3/3$ near the center so that by Eq. (3.2.2) $dP(\rho)/dr = 0 = d\rho/dr$ at the center.

Equation (3.3.6) can easily be integrated numerically, starting at $\xi = 0$ with the boundary conditions (3.3.7) and (3.3.8). One finds that for $n < 5$ ($\Gamma > \frac{6}{5}$), the solutions decrease monotonically and have a zero at a finite value $\xi = \xi_1$: $\theta(\xi_1) = 0$. This point corresponds to the surface of the star, where $P = \rho = 0$. Thus the radius of the star is

$$R = a\xi_1 = \left[\frac{(n+1)K}{4\pi G} \right]^{1/2} \rho_c^{(1-n)/2n} \xi_1, \tag{3.3.9}$$

while the mass is

$$M = \int_0^R 4\pi r^2 \rho \, dr$$

$$= 4\pi a^3 \rho_c \int_0^{\xi_1} \xi^2 \theta^n \, d\xi$$

$$= -4\pi a^3 \rho_c \int_0^{\xi_1} \frac{d}{d\xi} \left(\xi^2 \frac{d\theta}{d\xi} \right) d\xi \qquad \text{by Eq. (3.3.6)}$$

$$= 4\pi a^3 \rho_c \xi_1^2 |\theta'(\xi_1)|$$

$$= 4\pi \left[\frac{(n+1)K}{4\pi G} \right]^{3/2} \rho_c^{(3-n)/2n} \xi_1^2 |\theta'(\xi_1)|. \tag{3.3.10}$$

Eliminating ρ_c between Eqs. (3.3.9) and (3.3.10) gives the mass-radius relation for polytropes:

$$M = 4\pi R^{(3-n)/(1-n)}\left[\frac{(n+1)K}{4\pi G}\right]^{n/(n-1)}\xi_1^{(3-n)/(1-n)}\xi_1^2|\theta'(\xi_1)|. \quad (3.3.11)$$

Exercise 3.4 Show that the ratio of the mean density to the central density in a polytrope is $\bar{\rho}/\rho_c = 3|\theta'(\xi_1)|/\xi_1$.

The solutions[4] we are particularly interested in are

$$\Gamma = \tfrac{5}{3}, \quad n = \tfrac{3}{2}, \quad \xi_1 = 3.65375, \quad \xi_1^2|\theta'(\xi_1)| = 2.71406,$$

$$\Gamma = \tfrac{4}{3}, \quad n = 3, \quad \xi_1 = 6.89685, \quad \xi_1^2|\theta'(\xi_1)| = 2.01824. \quad (3.3.12)$$

Thus we find for low-density white dwarfs $\left(\Gamma = \tfrac{5}{3}\right)$:

$$R = 1.122 \times 10^4\left(\frac{\rho_c}{10^6 \text{ g cm}^{-3}}\right)^{-1/6}\left(\frac{\mu_e}{2}\right)^{-5/6} \text{ km}, \quad (3.3.13)$$

$$M = 0.4964\left(\frac{\rho_c}{10^6 \text{ g cm}^{-3}}\right)^{1/2}\left(\frac{\mu_e}{2}\right)^{-5/2} M_\odot, \quad (3.3.14)$$

$$= 0.7011\left(\frac{R}{10^4 \text{ km}}\right)^{-3}\left(\frac{\mu_e}{2}\right)^{-5} M_\odot. \quad (3.3.15)$$

Exercise 3.5 Derive the result $M \sim R^{-3}$ by dimensional analysis from Eqs. (3.2.1), (3.2.2), and (3.2.4).

For the high-density case $\left(\Gamma = \tfrac{4}{3}\right)$, we get

$$R = 3.347 \times 10^4\left(\frac{\rho_c}{10^6 \text{ g cm}^{-3}}\right)^{-1/3}\left(\frac{\mu_e}{2}\right)^{-2/3} \text{ km}, \quad (3.3.16)$$

$$M = 1.457\left(\frac{2}{\mu_e}\right)^2 M_\odot. \quad (3.3.17)$$

Note that M is independent of ρ_c and hence R in the extreme relativistic limit. We

[4] For an extensive list of polytropic parameters, see Chandrasekhar (1939).

conclude that as $\rho_c \to \infty$, the electrons become more and more relativistic throughout the star, and the mass asymptotically approaches the value (3.3.17) as $R \to 0$.

The mass limit (3.3.17) is called the *Chandrasekhar limit* (often abbreviated M_{Ch}), and represents the maximum possible mass of a white dwarf. For the case of a cold, perfect gas, the dependence of M_{Ch} on composition is contained entirely in μ_e.

The results of integrations of the structure equations for a white dwarf using the exact Fermi gas equation of state are given by Chandrasekhar (1939); see Figures 3.1 and 3.2 below. They agree with the polytropic approximations in the relevant domains, as expected.

3.4 The Chandrasekhar Limit

The existence of a mass limit for a degenerate star is such an important result that we should try to understand it in as simple a way as possible. We follow the original argument of Landau (1932), which applies to both white dwarfs and neutron stars.

Suppose there are N fermions in a star of radius R, so that the number density of fermions is $n \sim N/R^3$. The volume per fermion is $\sim 1/n$ (Pauli principle), so by the Heisenberg uncertainty principle the momentum of a fermion is $\sim \hbar n^{1/3}$. Thus the Fermi energy of a gas particle in the relativistic regime is

$$E_F \sim \hbar n^{1/3} c \sim \frac{\hbar c N^{1/3}}{R}. \tag{3.4.1}$$

The gravitational energy per fermion is

$$E_G \sim -\frac{GM m_B}{R}, \tag{3.4.2}$$

where $M = N m_B$. (Note that even if the pressure comes from electrons, most of the mass is in baryons.) As we will show in detail in Chapter 6, equilibrium is achieved at a minimum of the total energy E, where here

$$E = E_F + E_G = \frac{\hbar c N^{1/3}}{R} - \frac{G N m_B^2}{R}. \tag{3.4.3}$$

Note that both terms scale as $1/R$. When the sign of E is positive (i.e., when N is small), E can be decreased by increasing R. This decreases E_F and the electrons tend to become nonrelativistic, with $E_F \sim p_F^2 \sim 1/R^2$. Thus eventually E_G

dominates over E_F with increasing R, and then E becomes negative, increasing to zero as $R \to \infty$. There will therefore be a stable equilibrium at a finite value of R.

On the other hand, when the sign of E is negative (i.e., when N is large), E can be decreased without bound by decreasing R—no equilibrium exists and gravitational collapse sets in.

The maximum baryon number for equilibrium is therefore determined by setting $E = 0$ in Eq. (3.4.3):

$$N_{max} \sim \left(\frac{\hbar c}{G m_B^2} \right)^{3/2} \sim 2 \times 10^{57}, \qquad (3.4.4)$$

$$M_{max} \sim N_{max} m_B \sim 1.5 M_\odot. \qquad (3.4.5)$$

With the exception of composition-dependent numerical factors, the maximum mass of a degenerate star thus depends only on fundamental constants.

The equilibrium radius associated with masses M approaching M_{max} is determined by the onset of relativistic degeneracy:

$$E_F \gtrsim mc^2, \qquad (3.4.6)$$

where m refers to either electrons or neutrons. Using Eqs. (3.4.1) and (3.4.4.), this condition gives

$$R \lesssim \frac{\hbar}{mc} \left(\frac{\hbar c}{G m_B^2} \right)^{1/2}$$

$$\sim \begin{cases} 5 \times 10^8 \text{ cm}, & m = m_e \\ 3 \times 10^5 \text{ cm}, & m = m_n. \end{cases} \qquad (3.4.7)$$

There are thus two distinct regimes of collapse: one for densities above white dwarf values and another for densities above nuclear densities. In both cases, $M_{max} \sim M_\odot$.

Exercise   3.6   Suppose a sequence of completely degenerate white dwarf configurations has been constructed for different central densities for a composition of pure $^{56}_{26}$Fe. Consider a new sequence of models composed of pure $^{12}_{6}$C.

(a)   How are the physical parameters $P(r)$, $\rho(r)$, $m(r)$, and r in the carbon sequence related to the corresponding parameters in the iron sequence?

(b)   Determine the ratio $M_{max}(^{12}C)/M_{max}(^{56}Fe)$. Use the unmodified Chandrasekhar equation of state.

3.5   Improvements to the Chandrasekhar White Dwarf Models

A comprehensive treatment of white dwarf models, including the corrections to the Chandrasekhar equation of state discussed in Chapter 2, was given by Hamada and Salpeter in 1961 (hereafter HS). Their results are summarized in Figures 3.1 and 3.2.

The principal effect of the electrostatic corrections is to give smaller radii and larger central densities compared with Chandrasekhar's models of the same mass. Even though the Coulomb corrections are small in the high-density regime relevant for $1 M_\odot$ white dwarfs, in this regime $\Gamma \sim \frac{4}{3}$ and so $E/(GM^2/R) \ll 1$ [cf. Eq. (3.2.10)]. Thus small changes in the pressure or energy produce relatively large changes in the radius.

An important effect included by HS is the effect of neutronization (inverse β-decay) on a white dwarf of homogeneous composition. It is quite possible that a

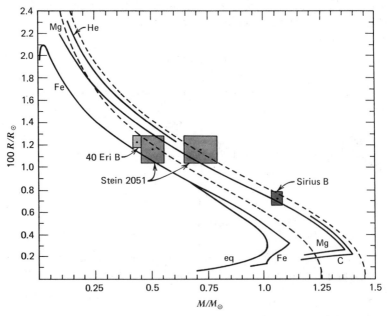

Figure 3.1   The relation between mass M and radius R, for zero-temperature stars composed of ^{4}He, ^{12}C, ^{24}Mg, and ^{56}Fe. The curve marked *eq* denotes the equilibrium composition at each density. The *dashed* curves denote the Chandrasekhar models, the upper one for $\mu_e = 2$ and the lower one for $\mu_e = 2.15$. The points inside the 1σ error boxes locate the mean masses and radii determined for the three white dwarfs in Table 3.2 (there are two solutions for Stein 2051). [After Hamada and Salpeter (1961). Reprinted courtesy of the authors and *The Astrophysical Journal*, published by the University of Chicago Press; © 1961 The American Astronomical Society.]

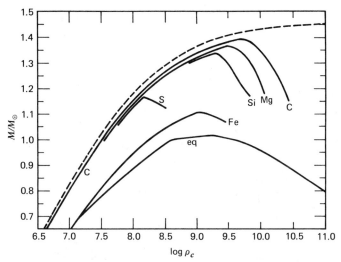

Figure 3.2   The relation between mass M and central density ρ_c (in g cm^{-3}) for zero-temperature stars composed of ^{12}C, ^{24}Mg, ^{28}Si, ^{32}S, and ^{56}Fe and for equilibrium (eq) conditions. The *dashed* curve denotes the Chandrasekhar model for $\mu_e = 2$. [After Hamada and Salpeter (1961). Reprinted courtesy of the authors and *The Astrophysical Journal*, published by the University of Chicago Press; © 1961 The American Astronomical Society.]

white dwarf in nature will never achieve the equilibrium composition of the Salpeter (1961) or BPS equations of state. First, current stellar evolution calculations of white dwarf progenitors predict that temperatures will never be high enough to burn much beyond carbon, so that massive white dwarfs are presumably largely carbon and oxygen.[5] Second, even if burning proceeded all the way to iron, followed by a gradual contraction of the white dwarf to achieve hydrostatic equilibrium, the timescales for the relevant nuclear reactions to go to completion may be too long. According to BPS, above $\rho_c = 8.1 \times 10^6$ g cm^{-3} ^{56}Fe should become ^{62}Ni. This cannot be done simply by β-decays, changing Z from 26 to 28; reactions changing A from 56 to 62 are also required and this reshuffling of nuclei from Fe to Ni may never be achieved in practice.

What happens as the density is increased above 8.1×10^6 g cm^{-3}? At a density of 1.14×10^9 g cm^{-3}, the Fermi energy of the electrons is $m_e c^2 + 3.695$ MeV. This is the threshold for the inverse β-decay reaction

$$^{56}_{26}\text{Fe} + e^- \rightarrow {}^{56}_{25}\text{Mn} + \nu.$$

[5] For a discussion of these stellar evolution calculations, see Liebert (1980). The calculations are not yet definitive, however, particularly because mass loss is important during the evolution.

The odd-even nucleus $^{56}_{25}$Mn immediately undergoes a further electron capture:

$$^{56}_{25}\text{Mn} + e^- \rightarrow {}^{56}_{24}\text{Cr} + \nu.$$

The even-even nucleus $^{56}_{24}$Cr is stable to further electron captures until a much higher density, $\sim 1.5 \times 10^{10}$ g cm^{-3}. The effect of this phase transition is to soften the equation of state: instead of compression increasing the Fermi energy of the electrons and hence the pressure, the electrons combine with Fe nuclei to make Cr. The adiabatic index, which is very close to $\frac{4}{3}$ because the electrons are highly relativistic, dips below $\frac{4}{3}$. As we shall see in Chapter 6, the M vs. R relation goes through a maximum and this signals the onset of an instability to gravitational collapse. The onset of inverse β-decay at $\rho_c = 1.14 \times 10^9$ g cm^{-3} terminates the sequence of iron white dwarfs.

In Table 3.1 we list the neutronization thresholds for the various nuclei that may be abundant in the very late stages of thermonuclear burning. We also give the corresponding densities at which the transitions occur. The threshold for ^{12}C at $\rho_0 = 3.90 \times 10^{10}$ g cm^{-3} is presumably the most relevant one, given the present predictions of stellar evolution calculations.

Exercise 3.7 Verify the values of ρ_0 in Table 3.1 corresponding to the given neutronization thresholds.

Table 3.1

Neutronization Thresholds

	Neutronization Threshold[a] (MeV)	ρ_0 (g cm^{-3})
$^1_1\text{H} \rightarrow n$	0.782	1.22×10^7
$^4_2\text{He} \rightarrow {}^3_1\text{H} + n \rightarrow 4n$	20.596	1.37×10^{11}
$^{12}_6\text{C} \rightarrow {}^{12}_5\text{B} \rightarrow {}^{12}_4\text{Be}$	13.370	3.90×10^{10}
$^{16}_8\text{O} \rightarrow {}^{16}_7\text{N} \rightarrow {}^{16}_6\text{C}$	10.419	1.90×10^{10}
$^{20}_{10}\text{Ne} \rightarrow {}^{20}_9\text{F} \rightarrow {}^{20}_8\text{O}$	7.026	6.21×10^9
$^{24}_{12}\text{Mg} \rightarrow {}^{24}_{11}\text{Na} \rightarrow {}^{24}_{10}\text{Ne}$	5.513	3.16×10^9
$^{28}_{14}\text{Si} \rightarrow {}^{28}_{13}\text{Al} \rightarrow {}^{28}_{12}\text{Mg}$	4.643	1.97×10^9
$^{32}_{16}\text{S} \rightarrow {}^{32}_{15}\text{P} \rightarrow {}^{32}_{14}\text{Si}$	1.710	1.47×10^8
$^{56}_{26}\text{Fe} \rightarrow {}^{56}_{25}\text{Mn} \rightarrow {}^{56}_{24}\text{Cr}$	3.695	1.14×10^9

[a]From Wapstra and Bos (1977); the electron rest mass-energy, $m_ec^2 = 0.511$ MeV, has been subtracted off.

Note that at $E_F \gtrsim 20$ MeV, it is energetically favorable to emit free neutrons ("neutron drip") rather than to capture electrons. Thus ${}_2^4\text{He} \rightarrow {}_1^4\text{H}$ has a threshold of ~ 22.7 MeV, compared with 20.6 MeV for ${}_2^4\text{He} \rightarrow {}_1^3\text{H} + n$.

There is one further process that must be considered before deciding on the fate of white dwarfs composed of low A nuclei at high densities—namely, *pycnonuclear* reactions. In a *thermonuclear* reaction, the thermal energy of the reacting nuclei overcomes the Coulomb repulsion between them so that the reaction can proceed. At sufficiently high densities, even at zero temperature, the zero-point energy of the nuclei in a lattice can lead to an appreciable rate of nuclear reactions. ("pyknos" is "dense" in Greek). We give an approximate treatment of these reactions in Section 3.7.

Hamada and Salpeter estimated that in 10^5 years H would be converted to ${}^4\text{He}$ via pycnonuclear reactions above a density of 5×10^4 g cm^{-3}, ${}^4\text{He}$ to ${}^{12}\text{C}$ above 8×10^8 g cm^{-3}, and ${}^{12}\text{C}$ to ${}^{24}\text{Mg}$ above 6×10^9 g cm^{-3}. These estimates were based on pycnonuclear reaction rates calculated by Cameron (1959b). Improved calculations by Salpeter and Van Horn (1969) suggest that Cameron's rates were too high. The critical density for H is about 1×10^6 g cm^{-3}, and for carbon 1×10^{10} g cm^{-3}. The densities quoted here are still quite uncertain.[6] Besides the difficulty of an accurate calculation, finite temperatures and crystal imperfections can increase the rates significantly.

The kink in the *M-R* plots of HS at low densities is produced by the stiffening of the equation of state there. As we showed in Section 2.4, $\Gamma \sim \frac{10}{3}$ (very roughly) at low densities, and so by Eq. (3.3.11) $M \sim R^{4.5}$, as opposed to the $M \sim R^{-3}$ relation for $\Gamma \sim \frac{5}{3}$. A more accurate treatment suggests that the maximum possible white dwarf radius is $R_{\max} = 3.9 \times 10^{-2} R_\odot$ at $M = 2.2 \times 10^{-3} M_\odot$ for cold carbon stars (Zapolsky and Salpeter, 1969).

3.6   Comparison With Observations: Masses and Radii

As we shall see in Chapter 4, white dwarf cooling timescales are sufficiently long that many degenerate dwarfs remain visible for quite some time due to radiation from their surfaces. This fortuitous circumstance leads to a "zero-order" test of the theory of white dwarfs—namely, their location in the Hertzsprung-Russell or HR diagram. The HR diagram is basically a log-log plot of luminosity L vs. effective temperature, T_e, where T_e is defined by

$$L = 4\pi R^2 \sigma T_e^4 \tag{3.6.1}$$

---

[6] The conversion of ${}^4\text{He}$ to ${}^{12}\text{C}$ will not go by the famous resonance in ${}^{12}\text{C}$ at zero temperature, and so requires an unlikely three-body encounter; the rate is probably negligible.

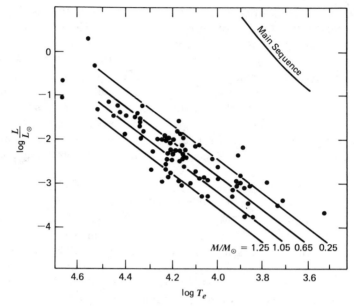

Figure 3.3 Positions of observed white dwarfs of known distances on the Hertzsprung–Russell diagram. The diagonal lines are lines of constant radius, which we have labeled by the corresponding mass from the Chandrasekhar equation of state for $\mu_e = 2$. [Data compiled by Sweeney (1976).]

and σ is the Stefan–Boltzmann constant.[7] White dwarfs of a definite mass (near $1M_\odot$) have a definite radius (near 10^9 cm), so by Eq. (3.6.1) they occupy a definite *line* in the HR diagram $L \propto T_e^4$. Since white dwarfs are expected to have masses in the vicinity of $1M_\odot$, all white dwarfs should occupy a *narrow strip* in the HR diagram well to the left and below the zero-age Main Sequence. Comparing this expectation with the observations (see Fig. 3.3) yields satisfactory agreement.

White dwarf radii are best determined by fitting model atmospheres to the observed residual radiation. The flux F_ν (ergs s^{-1} cm^{-2} Hz^{-1}) measured at the earth is

$$F_\nu = \frac{R^2 F_\nu(\text{surface})}{D^2}, \qquad (3.6.2)$$

where R is the radius and D the distance. The quantity D is determined for nearby white dwarfs by measuring their parallax. A model atmosphere, depending

[7]See Appendix A for a more complete discussion of the HR diagram.

Table 3.2

White Dwarf Masses and Radii from Optical Observations

	Mass $(M_\odot)^a$	Radius $(R_\odot)^b$	Redshift $(\text{km s}^{-1})^c$
Sirius B	$1.053 \pm .028$	$0.0074 \pm .0006$	89 ± 16
40 Eri B	$0.48 \pm .02$	$0.0124 \pm .0005$	23.9 ± 1.3
Stein 2051	$0.50 \pm .05$	$0.0115 \pm .0012$	?
	or		
	$0.72 \pm .08$		

[a]Masses: Sirius, Gatewood and Gatewood (1978); 40 Eri B, Heintz (1974) (but see Wegner, 1980); Stein 2051, Strand (1977).
[b]Radii: Shipman (1979).
[c]Redshifts: Sirius B, Greenstein et al. (1971); 40 Eri B, Wegner (1980).

on an effective temperature and a surface gravity, is constructed to reproduce the observed flux at different wavelengths, and hence R is found. An extensive compilation is given by Shipman (1979), with typical uncertainties of 5–10%.

White dwarf masses are much harder to come by, since one requires the star to be a member of a binary or triple system for a dynamical mass determination. At present, there are three white dwarfs with relatively high-precision determinations of both mass and radius. These are listed in Table 3.2 and plotted on the HS M-R diagrams in Figure 3.1. Procyon B, which used to be shown on such plots, has much larger uncertainties in M and R.

It is gratifying that the three best M-R determinations lie right on the HS curves. Indeed, the observations strongly suggest that the interiors of Sirius B and 40 Eri B are not composed of pure iron. Since the error boxes are 1σ, the observations are not inconsistent with the assertion that white dwarfs are all composed of carbon and oxygen, the currently favored outcome of stellar evolution calculations for their progenitors.

A further check on the white dwarf mass-radius relationship is provided by the Einstein gravitational redshift,[8]

$$\frac{\Delta\lambda}{\lambda} \simeq \frac{GM}{Rc^2}. \qquad (3.6.3)$$

The observed gravitational redshift is usually quoted as an equivalent Doppler

[8]We shall derive this formula in Section 5.3.

shift, $\Delta\lambda/\lambda = v/c$; that is,

$$v = 0.6362 \frac{M/M_\odot}{R/R_\odot} \text{ km s}^{-1}. \tag{3.6.4}$$

This gives 91 ± 8 km s^{-1} for Sirius B and 22 ± 1.4 km s^{-1} for 40 Eri B, in excellent agreement with the observations.

Exercise 3.8 The following redshifts have been reported in the literature: Van Maanen 2: 14 ± 18 km s^{-1} (Hershey, 1978) and 33 ± 16 km s^{-1} (Gatewood and Russell, 1974); EG 64: 131 km s^{-1}; EG 113: 52 km s^{-1}; the last two are from Greenstein and Trimble (1967), who say they may have large errors. Shipman's (1979) radii for these stars are $0.0138R_\odot$, $0.0182R_\odot$, and $0.0094R_\odot$, respectively. Are these numbers consistent with the HS M-R relation?

3.7 Pycnonuclear Reactions

Nuclear reactions can take place even at zero temperature in condensed matter. Such reactions proceed because ions fluctuating about their lattice sites with zero-point energy $E_0 \sim \hbar\omega_0$ can penetrate the Coulomb barrier of a neighboring ion. In this section we will give a very approximate calculation of the rate of such reactions. Our strategy will be to compute the transmission coefficient T for two ions to penetrate the repulsive electrostatic potential barrier keeping them apart. The quantity T depends on the detailed nature of the potential and is different for an ideal gas consisting of essentially free ions versus a crystalline solid characterized by a regular ion lattice. The former state characterizes the hot interior of a normal, main sequence star, while the latter state applies to cold, degenerate matter found in white dwarfs. Once two ions penetrate the electrostatic potential barrier and are touching, however, the probability that they interact further and undergo a nuclear reaction is basically independent of their prior state. We will exploit this fact in our analysis below.

    We first review the calculation of the rate when the two ions are essentially free particles at large distances (as in thermonuclear reactions, or reactions in a particle accelerator).[9] Recall that at low energies, the interaction cross section between two particles is proportional to $\pi\lambda^2 \sim 1/E$, where $\lambda = h/p$ is the characteristic de Broglie wavelength and E is the energy of the ion pair in the center of mass system. Recall furthermore that the probability for two ions with charges Z_1 and Z_2, moving with relative velocity v at large separation, to penetrate their repulsive Coulomb barrier is proportional to a "transmission

---

[9]See, for example, Clayton (1968), Chapter 4 for a more detailed discussion of thermonuclear reactions.

coefficient" ("Gamow penetration factor")

$$T = \exp(-2\pi\eta), \qquad \eta \equiv \frac{Z_1 Z_2 e^2}{hv}. \qquad (3.7.1)$$

The above considerations motivate writing the cross section in the form

$$\sigma(E) \equiv \frac{S(E)}{E} \exp(-2\pi\eta), \qquad E = \tfrac{1}{2}\mu v^2, \qquad (3.7.2)$$

where we expect $S(E)$ to be a *slowly* varying function of E, accounting for the contribution of the purely nuclear part of the interaction to the cross section. Here μ is the reduced mass. Laboratory measurements of $\sigma(E)$ may be used to determine the factor $S(E)$, or in some cases $S(E)$ can be determined purely theoretically. For example, for the reaction

$$p + p \rightarrow D + e^+ + \nu, \qquad (3.7.3)$$

the factor S is determined from the weak interaction rate for the conversion of a proton into a neutron in a p-p scattering, leading to positron decay and the formation of a bound n-p system (deuteron).

The Gamow factor comes from the solution of the Schrödinger equation outside the nuclei: the wave function for the relative motion of the charges Z_1 and Z_2 for $r > R_n$ (R_n = sum of nuclear radii) is

$$\psi_l(r, \theta, \phi) = \frac{\chi_l(r)}{r} Y_{lm}(\theta, \phi). \qquad (3.7.4)$$

Here we have assumed a definite angular momentum state l and neglected spin effects. The function $\chi_l(r)$ is a solution of the radial Schrödinger equation

$$\frac{d^2\chi_l}{dr^2} + \frac{2\mu}{\hbar^2}[E - V_l(r)]\chi_l = 0, \qquad (3.7.5)$$

where μ is the reduced mass and

$$V_l(r) = \frac{l(l+1)\hbar^2}{2\mu r^2} + V_c(r), \qquad r > R_n,$$

$$V_c(r) = \frac{Z_1 Z_2 e^2}{r}. \qquad (3.7.6)$$

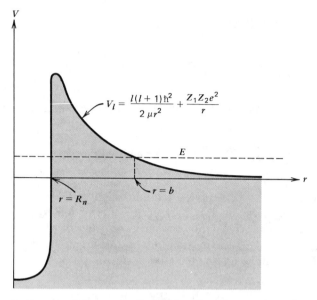

Figure 3.4   The effective potential governing the radial motion of one nucleus relative to another. For $r < R_n$ the nuclei are essentially in contact and the potential is dominated by the attractive short-range nuclear force. For $r > R_n$ the nuclear force is negligible and the potential is dominated by the Coulomb force. The classical turning point for an orbit with center-of-mass energy E is located at $r = b$.

The dominant interaction at low energies is via spherically symmetric s-wave scattering; accordingly we shall set $l = 0$. A sketch of $V(r)$ is given in Figure 3.4, showing $V_l(r)$ for $r > R_n$ and a schematic short-range, attractive nuclear potential for $r < R_n$.

The transmission coefficient can be determined without a detailed knowledge of the nuclear potential for $r < R_n$. In the one-dimensional WKB approximation, for a potential of the form sketched in Figure 3.4, the transmission coefficient is[10]

$$T \equiv \frac{|\chi_{\text{trans}}|^2 v_{\text{trans}}}{|\chi_{\text{inc}}|^2 v_{\text{inc}}} = \frac{4}{(2\theta + 1/2\theta)^2}. \qquad (3.7.7)$$

Here T is defined to be the ratio of transmitted to incident probability currents,

---

[10]Merzbacher (1970), Chapter 7.4.

and

$$\theta = \exp\left(\int_a^b |k(x)| \, dx\right),$$ (3.7.8)

$$k(x) = \left(\frac{2\mu}{\hbar^2}[E - V(x)]\right)^{1/2}.$$ (3.7.9)

For $\theta \gg 1$,

$$T \simeq \frac{1}{\theta^2}.$$ (3.7.10)

In our example, $a = R_n \to 0$, while b is defined by

$$E = \frac{Z_1 Z_2 e^2}{b}.$$ (3.7.11)

Thus

$$\int_a^b |k(x)| \, dx = \frac{(2\mu E)^{1/2}}{\hbar} \int_0^b \left(\frac{b}{x} - 1\right)^{1/2} dx$$

$$= \frac{\pi Z_1 Z_2 e^2}{\hbar} \left(\frac{\mu}{2E}\right)^{1/2}$$

$$= \pi \eta$$

and

$$T = \exp(-2\pi\eta).$$ (3.7.12)

The one-dimensional WKB approximation is able to reproduce the crucial exponential Gamow factor. However, the exact solution of the Schrodinger equation for a Coulomb potential gives

$$\frac{|\psi(0)|^2}{|\psi(\infty)|^2} = 2\pi\eta \exp(-2\pi\eta), \qquad \eta \gg 1.$$ (3.7.13)

The one-dimensional WKB approximation gives the wrong factor in front of the exponential. We shall ignore this discrepancy in our approximate treatment.

We can now relate $S(E)$ to P_n, the probability of a nuclear reaction *given* penetration to $r = R_n$. Let W be the reaction rate (probability per second) for an incident ion Z_1 to react with a nucleus (ion Z_2). Then

$$W = (\text{trans. flux at } R_n) \times 4\pi R_n^2 \times P_n$$

$$= (\text{inc. flux.}) \times \exp(-2\pi\eta) \times 4\pi R_n^2 P_n. \qquad (3.7.14)$$

But by definition of $\sigma(E)$,

$$W \equiv \sigma(E) \times (\text{inc. flux})$$

$$= \frac{S(E)}{E} \exp(-2\pi\eta) \times (\text{inc. flux}). \qquad (3.7.15)$$

Comparing Eqs. (3.7.14) and (3.7.15) gives

$$S(E) = 4\pi R_n^2 P_n E. \qquad (3.7.16)$$

The unknown probability factor P_n can thus be expressed in terms of a (potentially) measured quantity $S(E)$.

Now turn to reactions in a crystal lattice. The reaction rate per ion pair is

$$W = (\text{inc. flux}) \times T \times 4\pi R_n^2 P_n$$

$$= v|\psi_{\text{inc}}|^2 \frac{TS(E)}{E}, \qquad (3.7.17)$$

where we have to calculate $|\psi_{\text{inc}}|^2$ and T using the lattice potential for $r > R_n$. The measured nuclear factor $S(E)$ remains the same as before.

We approximate the lattice potential by considering[11] the one-dimensional motion of an ion between two fixed identical ions separated by $2R_0$. We will explore this one-dimensional ion lattice in more detail in the next chapter, but for now note that the lattice potential may be written as

$$V(x) = \frac{Z^2 e^2}{R_0 - x} + \frac{Z^2 e^2}{R_0 + x} - \frac{2Z^2 e^2}{R_0}, \qquad |x| < R_0 - R_n. \quad (3.7.18)$$

The above ionic potential, joined onto a short-range attractive nuclear potential near each fixed ion site, is illustrated in Figure 3.5. We take all the ions to have charge Ze and mass m_A, and let $x = 0$ correspond to the equilibrium position. For $x \ll R_0$, we find

$$V(x) \to \tfrac{1}{2}Kx^2, \qquad K = \frac{4Z^2 e^2}{R_0^3}. \qquad (3.7.19)$$

---

[11] The discussion presented here is patterned after Zel'dovich (1958).

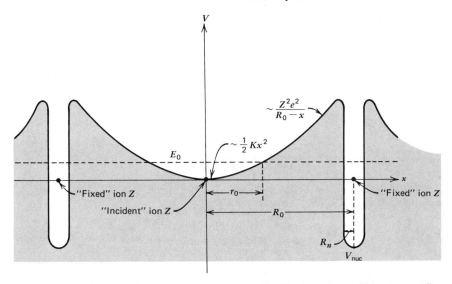

Figure 3.5 The potential governing the motion of one "incident" nucleus relative to an adjacent "fixed" nucleus in a one-dimensional ion lattice. The ions (nuclei) are separated by a distance R_0. Zero-point fluctuations (energy E_0) in the harmonic potential well near the "incident" ion lattice site can lead to Coulomb barrier penetration and nuclear reactions.

Thus the motion for small displacements about equilibrium is simple harmonic, with the zero-point energy given by

$$E_0 = \frac{1}{2}\hbar\omega_0 = \frac{Ze\hbar}{\left(\mu R_0^3\right)^{1/2}}. \tag{3.7.20}$$

Exercise 3.9 Show that ω_0 is essentially the plasma frequency for the ions, $\Omega_p \equiv (4\pi Z^2 e^2 n_A / m_A)^{1/2}$, where n_A is the mean ion density.

The classical turning point r_0 for an ion with energy $E = E_0$ is given by the relation

$$E_0 = V(r_0). \tag{3.7.21}$$

Making the harmonic approximation (3.7.19) to $V(x)$, we find

$$r_0 = \left(\frac{\hbar}{2Ze}\right)^{1/2} \left(\frac{R_0^3}{\mu}\right)^{1/4}. \tag{3.7.22}$$

Now ψ_{inc} is approximately the ground-state wave function of a simple harmonic oscillator. In three dimensions, this is

$$|\psi_{\text{SHO}}|^2 = \frac{\tau^3}{\pi^{3/2}} e^{-\tau^2 r^2}, \tag{3.7.23}$$

where

$$\tau = \left(\frac{\mu K}{\hbar^2}\right)^{1/4} = \frac{1}{r_0} \tag{3.7.24}$$

(i.e., one ion is localized in a volume $\sim r_0^3$ about each lattice site). Setting the exponential in Eq. (3.7.23) to unity, we have

$$|\psi_{\text{inc}}|^2 \approx |\psi_{\text{SHO}}|^2 \approx \frac{1}{r_0^3 \pi^{3/2}}. \tag{3.7.25}$$

The transmission coefficient T for an incident ion with energy E_0 can again be calculated in the WKB approximation:

$$T = \exp\left[-2\int_a^b |k(x)|\,dx\right], \tag{3.7.26}$$

where $k(x)$ is given by Eq. (3.7.9). Setting $u = x/R_0$, we get

$$T = \exp\left[-2\left(\frac{4\mu Z^2 e^2 R_0}{\hbar^2}\right)^{1/2} \int_{r_0/R_0}^{1-R_n/R_0} \left(\frac{1}{1-u^2} - 1 - \alpha\right)^{1/2} du\right], \tag{3.7.27}$$

where

$$\alpha = \frac{E_0}{2Z^2 e^2/R_0}. \tag{3.7.28}$$

The dimensionless integral in T can be written as

$$I = \int_\beta^{1-R_n/R_0} \left[\frac{(u-\beta)(u+\beta)}{(1-u)(1+u)}\right]^{1/2} du, \tag{3.7.29}$$

where

$$\beta \equiv \frac{r_0}{R_0} = \left(\frac{\alpha}{1+\alpha}\right)^{1/2} \simeq \alpha^{1/2}. \tag{3.7.30}$$

Letting $R_n \to 0$, we can do the integral with the aid of integral tables[12]:

$$I = (1 + \beta)E - 2\beta K, \qquad (3.7.31)$$

where E and K are complete elliptic integrals of modulus

$$k = \frac{1 - \beta}{1 + \beta}. \qquad (3.7.32)$$

Taking the limit $\beta \ll 1$, we get

$$I \approx 1 - \beta^2 \ln\left(\frac{1}{\beta}\right)^{1/2}. \qquad (3.7.33)$$

Combining Eqs. (3.7.27), (3.7.30), and (3.7.33), we get

$$T = \frac{R_0}{r_0} \exp\left(-2\frac{R_0^2}{r_0^2}\right), \qquad (3.7.34)$$

and hence, after some simplification, Eq. (3.7.17) becomes

$$W = 4\left(\frac{2}{\pi^3}\right)^{1/2} S \frac{\left(Z^2 e^2 \mu\right)^{3/4}}{\left(\hbar^2 R_0\right)^{5/4}} \exp\left[-4Ze\frac{(\mu R_0)^{1/2}}{\hbar}\right]. \qquad (3.7.35)$$

Exercise 3.10 Show that $E_0 \ll E_{\text{coul}} \sim Z^2 e^2 / R_0$ for densities $\ll 10^{10}$ g cm^{-3}; that is, a lattice does in fact exist in spite of zero-point fluctuations.

We introduce the dimensionless length parameter of Salpeter and Van Horn (1969),

$$\lambda \equiv \frac{\hbar^2}{2\mu Z^2 e^2}\left(\frac{n_A}{2}\right)^{1/3} \qquad (3.7.36)$$

$$= \frac{\hbar^2}{2\mu Z^2 e^2}\left(\frac{3}{\pi}\right)^{1/3}\frac{1}{R_0}, \qquad (3.7.37)$$

where n_A is the number density of ions, evaluated here by assigning one ion to a

[12] Byrd and Friedman (1971), Eq. (256.18).

sphere of radius $R_0/2$. The number of reactions per cm^3 per second is

$$P_0 = n_A W$$

$$= \left(\frac{\rho}{A}\right) A^2 Z^4 S \gamma \lambda^{5/4} \exp(-\varepsilon \lambda^{-1/2}), \tag{3.7.38}$$

where we have taken $\rho = A m_u n_A$. Here ρ is in g cm^{-3}, S in MeV barns (1 barn $\equiv 10^{-24}$ cm^2), and

$$\gamma = 1.1 \times 10^{44}, \qquad \varepsilon = 2.85. \tag{3.7.39}$$

We have multiplied γ by a factor of 4 because each ion in a bcc lattice has eight nearest neighbors, and W was the pairwise reaction probability.

The best calculation to date of pycnonuclear reaction rates is by Salpeter and Van Horn (1969). They used a much more realistic form for the lattice potential, including anisotropy and electron screening effects. Their result is

$$P_0 = \left(\frac{\rho}{A}\right) A^2 Z^4 S \gamma \lambda^{7/4} \exp(-\varepsilon \lambda^{-1/2}) \, \text{s}^{-1} \, \text{cm}^{-3}, \tag{3.7.40}$$

with

$$\gamma = 3.90 \times 10^{46}, \qquad \varepsilon = 2.638, \tag{3.7.41}$$

or

$$\gamma = 4.76 \times 10^{46}, \qquad \varepsilon = 2.516, \tag{3.7.42}$$

corresponding to two different approximations to the lattice potential, which roughly bracket the true potential. It is remarkable how close the naive result (3.7.38) is to Eq. (3.7.40), bearing in mind that ε appears in the exponential. We reiterate our earlier warning that lattice impurities might increase these rates considerably.

At finite temperatures, the nuclei can no longer be considered to be all in the ground state of the lattice potential. At even higher temperatures the lattice structure disappears and one enters the thermonuclear regime. Reaction rates in all these regimes are discussed by Salpeter and Van Horn (1969).

Exercise **3.11** Show that Eq. (3.7.36) can be written as

$$\lambda = \frac{1}{A Z^2} \left(\frac{1}{A} \frac{\rho}{1.36 \times 10^{11} \, \text{g cm}^{-3}}\right)^{1/3}. \tag{3.7.43}$$

Exercise 3.12 Use Eq. (3.7.40) to derive the density limits for H and C quoted in Section 3.5, by arguing that total conversion of the nuclear species occurs in a timescale given by

$$P_0 t = n_A.$$
$$(3.7.44)$$

Solve Eq. (3.7.44) for ρ, taking $t = 10^5$ yr and

$$S_{pp} = 5.38 \times 10^{-25} \text{ MeV barns,}$$

$$S_{cc} = 8.83 \times 10^{16} \text{ MeV barns.}$$

[The value for S_{cc} is from Fowler, Caughlan, and Zimmerman (1975) and is uncertain to probably at least a factor 3.]

4

Cooling of
White Dwarfs

In Chapter 3 we discussed how the mass-radius relationship for white dwarfs was subject to observational test. Another crucial observational test of the theory involves white dwarf cooling timescales. As will be described later, the test involves comparing the luminosity with the age of the white dwarf, the quantities being related by the cooling rate. Besides its astrophysical interest, the cooling theory of white dwarfs involves some elegant solid state physics in a rather novel setting.

4.1 Structure of the Surface Layers

To determine the cooling rate, we first need to find the conditions near the surface of the white dwarf.

The *interior* of a white dwarf is completely degenerate. The electrons have a large mean free path because of the filled Fermi sea, implying a high thermal conductivity and hence a uniform temperature. The isothermal interior is covered by nondegenerate surface layers which are in "radiative equilibrium": matter is essentially in thermodynamic equilibrium locally, with an energy flux carried outward by the diffusion of photons. The photon diffusion equation is

$$L = -4\pi r^2 \frac{c}{3\kappa\rho} \frac{d}{dr}(aT^4). \tag{4.1.1}$$

This equation will be derived and discussed in more detail in Appendix I. Here L is the luminosity (erg s^{-1}), aT^4 is the blackbody energy density, and κ is the *opacity* (cm^2 g^{-1}) of the stellar material. The quantity $1/\kappa\rho$ is an estimate of the photon mean free path. Equation (4.1.1) gives

$$\frac{dT}{dr} = -\frac{3}{4ac}\frac{\kappa\rho}{T^3}\frac{L}{4\pi r^2}.$$ (4.1.2)

The appropriate approximation to the opacity is Kramer's opacity

$$\kappa = \kappa_0 \rho T^{-3.5},$$ (4.1.3)

which arises from photoionization of atoms and inverse bremsstrahlung of free electrons ("bound–free" and "free–free" transitions).[1] Dividing the equation of hydrostatic equilibrium

$$\frac{dP}{dr} = -\frac{Gm(r)\rho}{r^2}$$ (4.1.4)

by Eq. (4.1.2), we get

$$\frac{dP}{dT} = \frac{4ac}{3}\frac{4\pi Gm(r)}{\kappa_0 L}\frac{T^{6.5}}{\rho}.$$ (4.1.5)

In the surface layers, which we will verify later are much thinner than the radius of the white dwarf, we can set $m(r) = M$. The density ρ can be eliminated using the equation of state of the nondegenerate matter in the surface layer as given by Eq. (2.3.18), yielding

$$P\,dP = \frac{4ac}{3}\frac{4\pi GM}{\kappa_0 L}\frac{k}{\mu m_u}T^{7.5}\,dT.$$ (4.1.6)

Equation (4.1.6) is easily integrated, with the boundary condition $P = 0$ at $T = 0$. Then P can be replaced by ρ from the equation of state, giving

$$\rho = \left(\frac{2}{8.5}\frac{4ac}{3}\frac{4\pi GM}{\kappa_0 L}\frac{\mu m_u}{k}\right)^{1/2}T^{3.25}.$$ (4.1.7)

The appropriate value of κ_0 is[2]

$$\kappa_0 = 4.34 \times 10^{24}Z(1 + X)\ \mathrm{cm^2\ g^{-1}},$$ (4.1.8)

[1] See Appendix I.
[2] Schwarzschild (1958), p. 237.

where X is the mass-fraction of hydrogen and Z the mass-fraction of "heavy" elements (i.e., all elements other than hydrogen and helium). Thus we can use Eq. (4.1.7) to give the run of ρ vs. T in the outer region of the white dwarf.

Equation (4.1.7) breaks down at some point below the surface where the electrons become degenerate. We can estimate the density, ρ_*, and temperature, T_*, at which this occurs by equating the nondegenerate expression for the electron pressure to the degenerate expression, using Eq. (2.3.22):

$$\frac{\rho_* k T_*}{\mu_e m_u} = 1.0 \times 10^{13} \left(\frac{\rho_*}{\mu_e} \right)^{5/3}. \qquad (4.1.9)$$

This gives (with T_* in degrees K)

$$\rho_* = (2.4 \times 10^{-8} \text{ g cm}^{-3}) \mu_e T_*^{3/2}. \qquad (4.1.10)$$

The temperature at this transition layer is given in terms of the luminosity by combining Eqs. (4.1.7) and (4.1.10). This gives

$$L = (5.7 \times 10^5 \text{ erg s}^{-1}) \frac{\mu}{\mu_e^2} \frac{1}{Z(1 + X)} \frac{M}{M_\odot} T_*^{3.5}. \qquad (4.1.11)$$

Thus, given L, the composition, and M, we can determine the interior temperature T_* of a white dwarf.

For illustrative purposes, set $X = 0$, $Y = 0.9$ (mass-fraction of helium), $Z = 0.1$, and $M = M_\odot$. We then find $\mu_e \approx 2$, $\mu \approx 1.4$, and so

$$L \simeq (2 \times 10^6 \text{ erg s}^{-1}) \frac{M}{M_\odot} T_*^{3.5}. \qquad (4.1.12)$$

Typically, $L \approx 10^{-2} - 10^{-5} L_\odot$, corresponding to $T_* \approx 10^6 - 10^7$ K, and hence $\rho_* \lesssim 10^3$ g cm$^{-3} \ll \rho_c$. The low density at the transition layer confirms the idea that the surface layer is relatively thin and does not alter the mass-radius relation found for cold stars. Note also that kT_* is much less than the Fermi energy of the electrons in the core of the white dwarf.

Exercise 4.1 Use Eqs. (4.1.3) and (4.1.7) to eliminate ρ from Eq. (4.1.2). Integrate to get

$$T_* = \frac{1}{4.25} \frac{\mu m_u}{k} \frac{GM}{R} \left(\frac{R}{r_*} - 1 \right), \qquad (4.1.13)$$

where r_* is the radius at which $T = T_*$. Hence show that $T_* \sim 10^6 - 10^7$ K implies

$$\frac{R - r_*}{R} \lesssim 10^{-2}. \qquad (4.1.14)$$

4.2   Elementary Treatment of White Dwarf Cooling[3]

When a star enters the white dwarf stage, the only significant source of energy to be radiated is the residual *ion* thermal energy. Very little energy can be released by further gravitational contraction since the star has already reached a degenerate state. The energy from neutrino emission is only important in the very early, high-temperature phase. The thermal energy of the electron gas cannot be readily released since most of the lower energy states are occupied in a degenerate gas. If the specific heat of the ions is c_v per ion, then the thermal energy per ion (taken to be only a function of temperature) is

$$\text{Thermal energy} = \int c_v \, dT. \tag{4.2.1}$$

Taking

$$c_v = \tfrac{3}{2}k \tag{4.2.2}$$

for a nondegenerate monatomic gas, we find for the total thermal energy of the white dwarf

$$U = \frac{3}{2}kT\frac{M}{Am_u}. \tag{4.2.3}$$

Here T is the uniform interior temperature of the white dwarf, denoted by T_* in Section 4.1. We have taken the composition to be a single ion species of baryon number A; in general $1/A$ would be replaced by $1/\mu - 1/\mu_e$. The energy store in Eq. (4.2.3) is substantial, amounting to $\sim 10^{48}$ ergs for $T_* \sim 10^7$ K, which is comparable to the energy output of some supernovae in the visual part of the spectrum. The cooling rate is given by $-dU/dt$. Equating this to the expression (4.1.12) for L, written in the form

$$L = CMT^{7/2} \tag{4.2.4}$$

with $CM_\odot \approx 2 \times 10^6$ erg s^{-1}, gives

$$-\frac{d}{dt}\left(\frac{3kT/2}{Am_u}\right) = CT^{7/2}. \tag{4.2.5}$$

[3] Mestel (1952); Schwarzschild (1958).

Integrating, we get

$$\frac{3}{5}\frac{k}{Am_u}\left(T^{-5/2} - T_0^{-5/2}\right) = C(t - t_0),\qquad(4.2.6)$$

where T_0 is the initial temperature. Taking $T_0 \gg T$, we can ignore it in Eq. (4.2.6) and write for the cooling time $\tau = t - t_0$

$$\tau = \frac{3}{5}\frac{kTM}{Am_u L}.\qquad(4.2.7)$$

From Eq. (4.2.4), we note that

$$\tau \propto \left(\frac{L}{M}\right)^{-5/7}.\qquad(4.2.8)$$

Numerically, $\tau \sim 10^9$ yr for $L \sim 10^{-3}L_\odot$, which is just what is expected—sufficiently long that many white dwarfs have not yet faded from view, but sufficiently short that their typical luminosities now are quite low. Note from Eq. (4.2.6) that most of the time is spent near the present temperature.

In the 1960s and early 1970s, the above cooling theory was in rough agreement with observations for hot, bright white dwarfs ($10^{-1} \lesssim L/L_\odot \lesssim 10^{-3}L_\odot$). For faint white dwarfs ($L \lesssim 10^{-3}L_\odot$), however, the theoretical cooling times seemed too long by more than a factor of 10. This discrepancy showed up for faint white dwarfs in star clusters, where the age of the white dwarf calculated from Eq. (4.2.7) was greater than the age of the cluster.[4] Also, since τ increases as L decreases, more and more white dwarfs should be observed at fainter luminosities. But Weidemann's (1967) luminosity function, for example, exhibited a deficiency of white dwarfs for $L \lesssim 10^{-3}L_\odot$.

We now know that this discrepancy was largely spurious. Errors in the cluster observations and age determinations were underestimated, and intensive searches have found many more noncluster white dwarfs down to $L \sim 10^{-4}L_\odot$. However, the apparent discrepancy stimulated intense theoretical attempts to "shorten" the cooling timescale.

We shall describe the most important corrections to the elementary cooling theory in the succeeding sections. We do this for two reasons. First, as the observations improve, one requires a more accurate theory to which to compare them. Second, there is strong new observational evidence[5] for a dramatic lack of

---

[4] The cluster age is essentially the main-sequence lifetime of the most luminous main-sequence star in the cluster, assuming all the stars were formed at the same time; see Appendix A.

[5] Liebert et al. (1979).

white dwarfs, but at very low luminosities satisfying $L < 10^{-4}L_{\odot}$. If this paucity is real, what physical effect might be responsible for shortening the cooling timescale?

The most important effect we have neglected is *crystallization* of the ion lattice.[6] For sufficiently low temperatures, and hence, L, the specific heat is due to *lattice vibrations* of the ions rather than free thermal motion. The critical temperature is the *Debye temperature* ($\theta_D \sim 10^7$ K typically), below which c_v falls rapidly. This leads to more rapid cooling and potentially better agreement with observations. In the succeeding sections, we will develop the basic theory of crystallization and the specific heat of an ionic lattice, and then apply the results to white dwarfs.

4.3   Crystallization and the Melting Temperature

Crystallization of an ionic lattice occurs when the dimensionless parameter

$$\Gamma \equiv \frac{(Ze)^2}{r_i kT} = \frac{\text{Coulomb energy}}{\text{Thermal energy}} \qquad (4.3.1)$$

becomes sufficiently high. Here r_i is defined by

$$\frac{n_i 4\pi r_i^3}{3} = 1, \qquad (4.3.2)$$

where n_i is the number density of ions. For $\Gamma \ll 1$, the ion plasma exhibits only small electrostatic deviations from ideal Maxwell–Boltzmann behavior. For $\Gamma \gg 1$, Coulomb forces dominate and the plasma crystallizes into a periodic lattice structure that minimizes the Coulomb energy.

We can estimate the critical value of Γ, and the corresponding melting temperature T_m, from *Lindemann's empirical rule*[7]: the ion lattice will melt when the thermally induced, mean square fluctuations in the ion position, $\langle (\delta r_i)^2 \rangle$, satisfy

$$\frac{\langle (\delta r_i)^2 \rangle}{r_i^2} \sim \frac{1}{16}. \qquad (4.3.3)$$

Exercise 4.2   Consider a small perturbation of an ion away from the center of a Wigner–Seitz cell. Treating the ambient electron cloud distribution as uniform, show that it provides a three-dimensional harmonic restoring force. Compute the spring constant K,

---

[6]Salpeter (1961); Mestel and Ruderman (1967); Van Horn (1968).

[7]Lindemann (1910).

and show that the vibration frequency is

$$\omega_0 = \left(\frac{K}{m_i}\right)^{1/2} = \frac{\Omega_p}{3^{1/2}}, \tag{4.3.4}$$

where Ω_p is the ion plasma frequency

$$\Omega_p = \left(\frac{4\pi n_i Z^2 e^2}{m_i}\right)^{1/2}. \tag{4.3.5}$$

Note that the one-dimensional model lattice used in Section 3.7 satisfies Eq. (4.3.4), too (cf. Exercise 3.9).

Exercise 4.3 (based on Salpeter, 1961) Let $r_e a_0$ be the radius of a sphere that contains one electron on average, where $a_0 = \hbar/\alpha m_e c$ is the Bohr radius and $\alpha \approx \frac{1}{137}$. Thus $r_i = Z^{1/3} r_e a_0$.

(a) Show that the total Coulomb ("Madelung") energy of the Wigner–Seitz cell, Eq. (2.4.8), may be written as

$$E_c = -\frac{9}{5}\frac{Z^{5/3}}{r_e}\, \text{Ry}$$

and that the ion zero-point energy may be written as

$$E_0 = \frac{3}{2}\hbar\omega_0 = 3\left(\frac{Zm_e}{m_i}\right)^{1/2} r_e^{-3/2}\, \text{Ry}.$$

Compare the above values to those for a bcc lattice.

Answer: The factor $\frac{9}{5}$ in E_c becomes 1.804 in a bcc lattice; the factor 3 in E_0 becomes 2.66.

(b) Show that the ratio $f \equiv |E_0/E_c|$ in a Wigner–Seitz cell satisfies

$$f \sim \frac{5}{3}\left(\frac{m_e}{m_i}\frac{x}{1.92\alpha}\right)^{1/2} Z^{-7/6} \sim 0.33\left(\frac{x}{A}\right)^{1/2} Z^{-7/6},$$

where x is the nondimensional relativity parameter introduced in Eq. (2.3.3). Clearly we require $f \ll 1$ for the ions to remain in a lattice at zero temperature (why?). How high is f for a stable ^{12}C white dwarf with density $\rho \leq 1 \times 10^{10}$ g cm^{-3}?

Answer: $f \sim 0.05$.

Now classically each harmonic degree of freedom contributes $kT/2$ to the average energy:

$$\tfrac{1}{2}K\langle(\delta r_i)^2\rangle \sim \tfrac{1}{2}kT. \tag{4.3.6}$$

Equations (4.3.4) and (4.3.6) give

$$\left\langle \left(\delta r_i\right)^2\right\rangle \sim \frac{3kT}{m_i \Omega_p^2}. \tag{4.3.7}$$

A more accurate evaluation of the numerical coefficient in Eq. (4.3.7) proceeds as follows[8]: Each normal mode of vibration of an ion in the lattice can be labeled by a wave number κ and a polarization state λ. The two transverse polarization states are $\lambda = 1, 2$, while the longitudinal mode is $\lambda = 3$. In the classical domain,

$$\left\langle \left(\delta r_i\right)^2\right\rangle_{\kappa,\lambda} = \frac{kT}{K(\kappa, \lambda)} = \frac{kT}{m_i \omega_\lambda^2(\kappa)}, \tag{4.3.8}$$

where $K(\kappa, \lambda)$ is the spring constant for that mode and $\omega_\lambda(\kappa)$ the corresponding frequency. To find $\left\langle \left(\delta r_i\right)^2\right\rangle$, we must sum Eq. (4.3.8) over the number of normal modes per ion. Recalling that the number density in phase space is $1/h^3$ per polarization state, we have for the number of modes in $d^3x\, d^3p$

$$\frac{d^3x\, d^3p}{h^3} = d^3x \frac{\kappa^2\, d\kappa}{2\pi^2}, \tag{4.3.9}$$

since $p = \hbar\kappa$. Now the volume per ion is $d^3x = 1/n_i$, so

$$\left\langle \left(\delta r_i\right)^2\right\rangle = \frac{1}{n_i} \int \frac{\kappa^2\, d\kappa}{2\pi^2} \sum_{\lambda=1}^{3} \left\langle \left(\delta r_i\right)^2\right\rangle_{\kappa,\lambda}$$

$$= \frac{kT}{m_i n_i} \int_0^{\kappa_D} \frac{\kappa^2\, d\kappa}{2\pi^2} \sum_{\lambda=1}^{3} \frac{1}{\omega_\lambda^2(\kappa)}. \tag{4.3.10}$$

The integral extends up to the Debye cut-off κ_D, given by equating the total number of normal modes of N ions in a volume V to $3N$:

$$3N = \sum_{\lambda=1}^{3} \int d^3x \frac{\kappa^2\, d\kappa}{2\pi^2}$$

$$= \frac{3V\kappa_D^3}{6\pi^2}, \tag{4.3.11}$$

---

[8] Mestel and Ruderman (1967).

or

$$\kappa_D = \left(6\pi^2 n_i\right)^{1/3}. \tag{4.3.12}$$

To proceed further with Eq. (4.3.10), we need the dispersion relation for $\omega_\lambda(\kappa)$. We shall discuss the lattice excitation spectrum in more detail in Section 4.4; for now we note that a good approximation is to take

$$\omega_{1,2} \simeq 0.7\frac{\kappa}{\kappa_D}\Omega_p,$$

$$\omega_3 \simeq 0.7\Omega_p. \tag{4.3.13}$$

Inserting these results into Eq. (4.3.10) gives

$$\left\langle \left(\delta r_i\right)^2 \right\rangle \simeq \frac{14kT}{m_i\Omega_p^2}, \tag{4.3.14}$$

to be compared with the naive result (4.3.7).

Equation (4.3.14) and Lindemann's rule (4.3.3) give for the melting point

$$\Gamma \simeq 75, \tag{4.3.15}$$

and a corresponding *melting temperature*, assuming $\mu_e \approx 2$,

$$T_m \simeq \frac{Z^2 e^2}{\Gamma k}\left(\frac{4\pi}{3}\frac{\rho}{2Zm_u}\right)^{1/3}$$

$$\simeq 2 \times 10^3 \rho^{1/3} Z^{5/3} \text{ K}. \tag{4.3.16}$$

These values are in reasonable agreement with those of Brush, Sahlin, and Teller (1966), who performed Monte Carlo calculations for a one-component Coulomb "liquid" and found $\Gamma = 126$, and with Lamb and Van Horn (1975), who allowed for ion quantum effects and found $\Gamma = 160$. Recently, Slattery et al. (1980) obtained $\Gamma = 171 \pm 3$.

We define the *Debye temperature* θ_D by

$$k\theta_D \equiv \hbar\Omega_p, \tag{4.3.17}$$

so that

$$\theta_D \simeq 4 \times 10^3 \rho^{1/2} \text{ K}. \tag{4.3.18}$$

Typically in white dwarfs with $Z > 2$, $T_m > \theta_D$ so that our classical analysis for $\langle (\delta r_i)^2 \rangle$ is valid. Otherwise we should have included zero-point fluctuations in our estimate for $\langle (\delta r_i)^2 \rangle$. It turns out that even for He, with $T_m \ll \theta_D$, Eq. (4.3.10) gives better agreement with experiment than some attempts to include zero-point effects.

When a liquid crystallizes at $T \sim T_m$, it releases an amount of latent heat per ion

$$-q \sim kT_m. \qquad (4.3.19)$$

This heat release during crystallization must be included in the total energy supply of the star, and serves to increase its thermal lifetime, as will be discussed in Section 4.5.

A third relevant temperature is T_g, the point at which the ion kinetic energy exceeds its vibrational energy. Above T_g, the lattice dissolves, yielding a dense, imperfect gas. This occurs when

$$\langle (\delta r_i)^2 \rangle \sim r_i^2, \qquad (4.3.20)$$

or $T \sim 16 T_m$ [cf. Eq. (4.3.3)]:

$$T_g \sim 3 \times 10^4 \rho^{1/3} Z^{5/3} \text{ K}. \qquad (4.3.21)$$

Note that $\theta_D < T_m < T_g$ for ^{12}C, provided $\rho \lesssim 10^6$ g cm^{-3}.

4.4   Heat Capacity of a Coulomb Lattice

For $T \gg T_g$, the ions can be treated as an ideal Maxwell–Boltzmann gas, and so the heat capacity per ion is

$$c_v \sim \tfrac{3}{2}k, \qquad T \gg T_g. \qquad (4.4.1)$$

This is the value used in the elementary discussion of white dwarf cooling in Section 4.2.

As the temperature falls below T_g, a lattice starts to form. The heat capacity *increases* by a factor of 2 because of the additional $\tfrac{1}{2}kT$ per mode from the lattice potential energy. Thus

$$c_v \sim 3k, \qquad \theta_D \ll T \ll T_g. \qquad (4.4.2)$$

Upon further cooling to $T \lesssim \theta_D$, quantum effects become important and c_v *decreases* well below the values in Eqs. (4.4.1) and (4.4.2). As $T \to 0$, $c_v \sim T^3$. Since this regime may play a crucial role in reconciling the theoretical and observational white dwarf cooling times, and thereby provide concrete evidence for white dwarf crystallization, we discuss it in some detail.

The mean energy per ion in the lattice at temperature T is given by[9]

$$\bar{\varepsilon} = \sum_{\kappa, \lambda} \hbar \omega_\lambda(\kappa) \left\{ \frac{1}{\exp[\beta \hbar \omega_\lambda(\kappa)] - 1} + \frac{1}{2} \right\}, \qquad (4.4.3)$$

where $\beta = 1/kT$. The heat capacity is given by

$$c_v = \left(\frac{\partial \bar{\varepsilon}}{\partial T} \right)_V$$

$$= k \sum_{\kappa, \lambda} \frac{\exp[\beta \hbar \omega_\lambda(\kappa)][\beta \hbar \omega_\lambda(\kappa)]^2}{\{\exp[\beta \hbar \omega_\lambda(\kappa)] - 1\}^2}. \qquad (4.4.4)$$

The sum over κ can be replaced by an integral, giving

$$c_v = \frac{k}{n_i} \int_0^{\kappa_D} \frac{\kappa^2 \, d\kappa}{2\pi^2} \sum_{\lambda=1}^{3} \frac{\exp[\beta \hbar \omega_\lambda(\kappa)][\beta \hbar \omega_\lambda(\kappa)]^2}{\{\exp[\beta \hbar \omega_\lambda(\kappa)] - 1\}^2}. \qquad (4.4.5)$$

The "usual" Debye approximation, which will not be adequate here, consists of the following[10]:

1. Assume the normal modes of vibration of the solid can be approximated by considering the propagation of *sound waves* in a *continuous* elastic medium. Hence, obtain an acoustic dispersion relation of the form

$$\omega_\lambda(\kappa) = \kappa c_{s, \lambda}, \qquad (4.4.6)$$

where the transverse sound speeds $c_{s,1} = c_{s,2}$ and the longitudinal speed $c_{s,3}$ are derivable from the elastic constants of the medium.
2. Define a "mean effective" sound velocity c_s by

$$\frac{3}{c_s^3} \equiv \frac{2}{c_{s,1}^3} + \frac{1}{c_{s,3}^3}. \qquad (4.4.7)$$

---

[9]Compare, for example, Reif (1965), Section 10.1.
[10]For example, Reif (1965), Section 10.2.

3.   The Debye cut-off frequency corresponding to Eq. (4.4.6) should be

$$\omega_{D,\lambda} = \kappa_D c_{s,\lambda}.\tag{4.4.8}$$

and thereby depends upon the polarization λ. Instead, ignore the polarization dependence and define a *common* Debye cut-off frequency for all λ,

$$\omega_D \equiv \kappa_D c_s = \left(6\pi^2 n_i\right)^{1/3} c_s,\tag{4.4.9}$$

where we have used Eq. (4.3.12)

Then Eq. (4.4.5) becomes

$$c_v = 3k f_D\!\left(\frac{\tilde{\theta}_D}{T}\right),\tag{4.4.10}$$

where the "Debye function" $f_D(y)$ is defined by

$$f_D(y) = \frac{3}{y^3}\int_0^y \frac{x^4 e^x\,dx}{\left(e^x - 1\right)^2},\tag{4.4.11}$$

and the "usual" Debye temperature $\tilde{\theta}_D$ is defined by

$$k\tilde{\theta}_D = \hbar\omega_D,\tag{4.4.12}$$

which *differs* from Eq. (4.3.17).
    Since

$$f_D(y) \to 1, \qquad y \to 0,$$

$$f_D(y) \to \frac{4\pi^4}{5}\frac{1}{y^3}, \qquad y \to \infty,\tag{4.4.13}$$

we have for the "usual" Debye heat capacity

$$c_v \to 3k, \qquad T \gg \tilde{\theta}_D,$$

$$c_v \to \frac{12\pi^4}{5}k\left(\frac{T}{\tilde{\theta}_D}\right)^3, \qquad T \ll \tilde{\theta}_D.\tag{4.4.14}$$

Unfortunately, the "usual" Debye treatment is not adequate for our purposes. Based on a crude analogy between solids and elastic media, it does not correctly describe the Coulombic nature of an ionic lattice. Most importantly, the appropriate value of the Debye cut-off frequency for an ion lattice is Ω_p, not ω_D, since the highest frequency in the normal mode spectrum results from the oscillation of a *single* ion about its equilibrium lattice site.

The correct evaluation of Eq. (4.4.5) for c_v requires a detailed normal mode analysis to determine the phonon spectrum $\omega_\lambda(\kappa)$ for an ion lattice near $T = 0$. Such an analysis has been performed for a bcc lattice by Clark (1958), Carr (1961), and others. The modes of such a lattice consist of a pair of transverse phonons and a longitudinal "plasmon"; the typical spectrum is shown in Figure 4.1.

Several features of the vibrational spectrum can be understood from an analysis of the normal modes of the one-dimensional ion lattice used in Section 3.7. Recall that we treat the ions as a row of N particles of mass m_i separated by a distance R_0 ($= 2r_i$) at equilibrium. Coulomb forces effectively serve to "connect" the ions by "springs" with spring constants

$$K = \mu\omega_0^2 = \tfrac{1}{2}m_i\omega_0^2 = \tfrac{1}{6}m_i\Omega_p^2 \qquad (4.4.15)$$

where $\mu = m_i/2$ is the reduced mass. The total length of the lattice is $(N + 1)R_0$, the endmost ions being connected to fixed points. We assume that only nearest

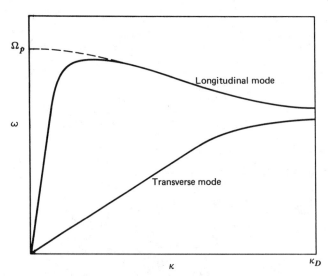

Figure 4.1 Spectrum of excitations for a body-centered-cubic lattice with Coulomb interactions and long-range screening, based upon calculations of Clark (1958) and the survey of Pines (1963). The *dashed* line is the ideal case of no screening. [From Mestel and Ruderman (1967).]

neighbors act on each other, thus ignoring *screening* due to the *long-range* character of the Coulomb field.[11]

If the displacement of the jth ion from equilibrium is q_j, the net restoring force for small displacements is

$$F_j = -K(q_j - q_{j-1}) + K(q_{j+1} - q_j). \qquad (4.4.16)$$

The equation of motion is

$$m_i \ddot{q}_j = K(q_{j-1} - 2q_j + q_{j+1}), \qquad j = 1,\ldots, N. \qquad (4.4.17)$$

We look for normal modes by substituting

$$q_j(t) = a_j e^{i\omega t}, \qquad (4.4.18)$$

so that the constants a_j satisfy

$$-Ka_{j-1} + (2K - m_i\omega^2)a_j - Ka_{j+1} = 0, \qquad j = 1,\ldots, N, \qquad (4.4.19)$$

and the boundary conditions are implemented by requiring

$$a_0 = 0, \qquad (4.4.20)$$

$$a_{N+1} = 0. \qquad (4.4.21)$$

Equation (4.4.19) is a linear second-order difference equation with constant (i.e., j-independent) coefficients. It can be solved by setting

$$a_j = a_1 e^{i(j\gamma - \delta)}, \qquad (4.4.22)$$

where it is understood that the real part of a_j is the physically meaningful solution. Then we find

$$\omega^2 = 4\frac{K}{m_i}\sin^2\frac{\gamma}{2}. \qquad (4.4.23)$$

The boundary condition (4.4.20) implies that

$$\delta = \frac{\pi}{2} \qquad (4.4.24)$$

---

[11]We will discuss screening later in this section.

so that the real part of a_0 vanishes, while Eq. (4.4.21) implies

$$\sin \gamma (N + 1) = 0, \tag{4.4.25}$$

or

$$\gamma_r = \frac{r\pi}{N + 1}, \qquad r = 1, \ldots, N. \tag{4.4.26}$$

Here r labels the N independent solutions of Eq. (4.4.19). Thus there are N independent eigenfrequencies

$$\omega_r = 2 \left(\frac{K}{m_i} \right)^{1/2} \sin \frac{r\pi}{2(N + 1)}, \tag{4.4.27}$$

and for the corresponding normal modes the displacement of the jth particle is proportional to

$$a_{jr} \sim \sin \left(j \frac{r\pi}{N + 1} \right). \tag{4.4.28}$$

We can write the dispersion relation (4.4.27) in terms of a wave number κ_r defined by

$$\kappa_r = \frac{r\pi}{R_0(N + 1)}. \tag{4.4.29}$$

Then

$$a_{jr} \sim \sin(\kappa_r x_j), \tag{4.4.30}$$

where $x_j = jR_0$, and the dispersion relation becomes

$$\omega_r = \omega_{max} \sin \frac{\kappa_r R_0}{2}, \tag{4.4.31}$$

where

$$\omega_{max} = \left(\tfrac{2}{3} \right)^{1/2} \Omega_p. \tag{4.4.32}$$

The above dispersion relation, derived for the longitudinal mode of a one-dimensional lattice, exhibits the following properties which are *universal* for large N systems:

1. At low frequencies and long wavelengths ($\omega_r \ll \omega_{max}$, $\kappa_r R_0 \ll 1$), we have the linear, acoustic result

$$\omega_r \propto \kappa_r \tag{4.4.33}$$

as in an *elastic* medium [cf. Eq. (4.4.6)].

2. At high frequencies and short wavelengths, the dispersion relation is no longer linear. The frequency ω_r reaches a maximum near Ω_p, corresponding to a wave number $\kappa_r \sim \pi/R_0$, the limit of the "Brillouin zone" of the lattice.

3. The number of independent modes is $N \times$ dimensionality of the lattice.

We observe that the *transverse* phonon spectrum plotted in Figure 4.1 for a bcc lattice and approximated by Eq. (4.3.13) is qualitatively similar to the one-dimensional spectrum given by Eq. (4.4.31).

From the result of Exercise 4.2, we would expect the longitudinal plasma spectrum to be given by $\omega_3 \sim \Omega_p$. This is the correct result at high frequencies and short wavelengths. In the low-frequency, long-wavelength limit, however, the plasmon behaves like a normal phonon, with $\omega_3 \sim \kappa$, because of Coulomb *screening* of the ions by the ambient, degenerate electron gas.

To show this, suppose that the local ionic charge density is perturbed slightly from an otherwise uniform, neutral, static background. Suppose that the mobile degenerate electron gas moves rapidly to neutralize this disturbance. We write

$$n_i = n_0 + n_1,$$

$$n_e = Zn_0 + n'_e, \tag{4.4.34}$$

where n_1 and n'_e are the small perturbations away from the static values n_0 and Zn_0. The electrostatic potential produced by the perturbation is given by

$$\nabla^2 \phi = -4\pi n_1 Ze + 4\pi en'_e. \tag{4.4.35}$$

Now in the Thomas–Fermi approximation (Section 2.4),

$$E_F = \frac{p_F^2(r)}{2m_e} - e\phi(r) = \text{constant} \tag{4.4.36}$$

and

$$n_e(r) = \frac{8\pi p_F^3}{3h^3}$$

$$= \frac{8\pi}{3h^3}(2mE_F)^{3/2}\left(1 + \frac{e\phi(r)}{E_F}\right)^{3/2}. \tag{4.4.37}$$

Expanding Eq. (4.4.37) to first order in $e\phi/E_F$ (the background is nearly uniform), we find

$$n'_e = \frac{3}{2}\frac{Zen_0\phi}{E_F}. \tag{4.4.38}$$

Thus Eq. (4.4.35) becomes

$$\left(\nabla^2 - \kappa_{sc}^2\right)\phi = -4\pi n_1 Ze, \tag{4.4.39}$$

where the screening length κ_{sc}^{-1} is defined by

$$\kappa_{sc}^2 \equiv \frac{6\pi Ze^2 n_0}{E_F}. \tag{4.4.40}$$

Exercise 4.4 Motivate the name "screening length" for κ_{sc}^{-1} by finding the electrostatic potential of a single ion at rest at the origin of a degenerate, neutral plasma in the Thomas–Fermi approximation.

Hint: The equation for ϕ is the same as Eq. (4.4.39) with $-4\pi Ze\delta(\mathbf{r})$ on the right-hand side.

Answer: $\phi(r) = Ze\exp(-\kappa_{sc}r)/r]$.

In the long-wavelength limit, we can treat the ions as a "fluid." The relevant dynamical equations for a fluid are[12] the continuity equation

$$\frac{\partial n_i}{\partial t} + \nabla \cdot (n_i \mathbf{v}) = 0, \tag{4.4.41}$$

and the momentum equation

$$\frac{\partial \mathbf{v}}{\partial t} + (\mathbf{v} \cdot \nabla)\mathbf{v} = -\frac{Ze}{m_i}\nabla\phi. \tag{4.4.42}$$

Linearizing these equations about the static, homogeneous background, where $\mathbf{v} = \phi = 0$, gives

$$\frac{\partial n_1}{\partial t} + n_0 \nabla \cdot \mathbf{v} = 0, \tag{4.4.43}$$

$$\frac{\partial \mathbf{v}}{\partial t} = -\frac{Ze}{m_i}\nabla\phi. \tag{4.4.44}$$

Consider a single Fourier component, proportional to $\exp(i\mathbf{\kappa} \cdot \mathbf{r} - i\omega t)$. Then

$$-\omega n_1 + n_0 \mathbf{\kappa} \cdot \mathbf{v} = 0, \tag{4.4.45}$$

$$-\omega \mathbf{v} = -\frac{Ze\mathbf{\kappa}\phi}{m_i}, \tag{4.4.46}$$

$$\left(\kappa^2 + \kappa_{sc}^2\right)\phi = 4\pi n_1 Ze, \tag{4.4.47}$$

[12] See Chapter 6.

where the last equation comes from Eq. (4.4.39). Dot κ into Eq. (4.4.46) and eliminate $\kappa \cdot \mathbf{v}$ from Eq. (4.4.45). Eliminate ϕ using Eq. (4.4.47). The result is

$$\omega^2 = \frac{\Omega_p^2}{1 + \kappa_{sc}^2/\kappa^2} \quad \text{(longitudinal)}, \tag{4.4.48}$$

where

$$\Omega_p^2 = \frac{4\pi Z^2 e^2 n_0}{m_i}. \tag{4.4.49}$$

Note that this is a *longitudinal* mode, since by Eq. (4.4.46) $\mathbf{v}$ is parallel to κ. We see from Eq. (4.4.48) that $\omega \sim \kappa$ for small κ and $\omega \to \Omega_p$ for large κ.

Screening does not affect the transverse modes significantly.

Exercise 4.5 Show that in the low-frequency limit, Eq. (4.4.48) reduces to the acoustic dispersion relation $\omega \to c_s \kappa$, where $c_s^2 = dP/d\rho$ and P and ρ are given by the ideal Fermi gas expressions.

We are now prepared to calculate c_v from Eq. (4.4.5) using the bcc excitation spectrum plotted in Figure 4.1. In general, the calculation must be performed numerically. We can, however, easily derive the form of c_v in two limiting cases. In the high-temperature lattice regime, we have

$$c_v \to 3k, \qquad \theta_D \ll T \ll T_g, \tag{4.4.50}$$

independent of the details of the dispersion relation, and consistent with our earlier discussion.

In the low-temperature regime, $T \ll \theta_D$, the dominant contribution to the integral in Eq. (4.4.5) comes from the low-frequency, long-wavelength region of the dispersion relation. Writing

$$\omega_\lambda \simeq \alpha_\lambda \frac{\kappa}{\kappa_D} \Omega_p \tag{4.4.51}$$

as an approximation, with α_λ constant (cf. Fig. 4.1), we find

$$c_v \simeq 2kf_D(y_1) + kf_D(y_3), \tag{4.4.52}$$

where

$$y_1 = y_2 = \frac{\alpha_1 \theta_D}{T}, \qquad y_3 = \frac{\alpha_3 \theta_D}{T}. \tag{4.4.53}$$

In the limit $T \to 0$, $y \to \infty$, we find from Eq. (4.4.13)

$$c_v \simeq \frac{4\pi^4}{5}\left(\frac{T}{\theta_D}\right)^3 k\left(\frac{2}{\alpha_1^3} + \frac{1}{\alpha_3^3}\right).$$

From Figure 4.1, we see that for small κ, $\alpha_1 \approx 0.8$, $\alpha_3 \gg 1$. Thus the longitudinal mode does not contribute and

$$c_v \simeq \frac{16\pi^4}{5}\left(\frac{T}{\theta_D}\right)^3 k, \qquad T \ll \theta_D. \qquad (4.4.54)$$

The ionic heat capacity for all T is sketched in Figure 4.2.

4.5   Refined Treatment of White Dwarf Cooling

In general, Eq. (4.2.5) for white dwarf cooling becomes

$$-\frac{d}{dt}\int c_v\, dT = CAm_u T^{7/2}, \qquad (4.5.1)$$

or

$$\frac{dt}{dT} = -\frac{c_v}{CAm_u T^{7/2}}. \qquad (4.5.2)$$

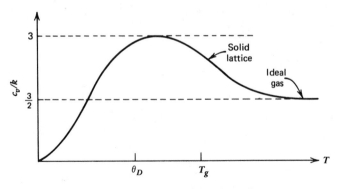

Figure 4.2 Specific heat capacity as a function of temperature (schematic drawing; ions only). For low temperatures $\theta_D < T < T_g$ lattice vibrational effects increase c_v above the ideal Maxwell–Boltzmann value of $3k/2$ as the gas crystallizes. For very low temperatures $T < \theta_D$, c_v varies as T^3.

For $\theta_D \ll T \ll T_g$, $c_v \simeq 3k$, and Eq. (4.5.2) yields

$$\tau = \frac{6}{5} \frac{kTM}{Am_u L}, \qquad \theta_D \ll T \ll T_g, \qquad (4.5.3)$$

for the cooling time. This is simply a factor of 2 bigger than Eq. (4.2.7) because of the inclusion of the lattice potential energy.

For $T \ll \theta_D$, we use Eq. (4.4.54) in Eq. (4.5.2) and obtain

$$\tau = \frac{32\pi^4}{5} \left(\frac{T}{\theta_D}\right)^3 \left[\left(\frac{T_0}{T}\right)^{1/2} - 1\right] \frac{MkT}{Am_u L}, \qquad T \ll \theta_D, \qquad (4.5.4)$$

where $T_0 \lesssim \theta_D$ is the initial temperature from which cooling begins. As first pointed out by Mestel and Ruderman (1967), this is *shorter* than the classical result (4.5.3), whenever T is less than about $0.1\theta_D$. It was hoped that this would explain the discrepancy between observation and theory mentioned in Section 4.2.

Exercise   4.6

(a)   Estimate T_c and L_c at which the transition from "classical" to Debye cooling occurs for a $1M_\odot$ white dwarf with a carbon interior but with abundances $X = 0$, $Y = 0.9$, and $Z = 0.1$ in the atmosphere. Find the approximate transition criterion by equating the expression (4.4.54) to $3k$. Obtain the density from Figure 3.2.

(b)   Estimate τ_c, the time required for the white dwarf to reach this transition.

Answers: $T_c \sim 4.8 \times 10^6$ K, $L_c \sim 1.3 \times 10^{-4} L_\odot$, $\tau_c \sim 5 \times 10^9$ yr

Exercise   4.7   Use Eq. (4.5.3) and the analytic expressions derived in Section 4.1 and Chapter 3 to estimate the dependence of τ and L_c on M for stars with a given composition.

Answers: $\tau \sim M^{-2.5}$, $L_c \sim M^{4.5}$. Other answers are possible depending on the assumptions made.

We have omitted $-q \approx kT$, the latent heat of crystallization, from the left-hand side of our analytic treatment, Eq. (4.5.1). Lamb and Van Horn (1975) carried out a detailed cooling calculation for a $1M_\odot$, pure ^{12}C white dwarf, including this effect. The result is to *increase* the classical cooling time by another factor ~ 1.6, in addition to the factor 2 for the lattice potential energy. The reason is that the energy released is

$$\frac{\delta E_{\text{latent heat}}}{E_{\text{thermal}}} \sim \frac{-q}{3kT/2} \sim \frac{2}{3}, \qquad (4.5.5)$$

so altogether

$$\frac{\delta\tau_{\text{lattice}} + \delta\tau_q}{\tau_{\text{classical}}} \sim \frac{5}{3} \Rightarrow \frac{\tau}{\tau_{\text{classical}}} \sim 3. \qquad (4.5.6)$$

However, since crystallization starts at the center of the star and gradually spreads outwards as the star cools, the contribution from q does *not* produce a peak in the luminosity function. It *cannot* provide direct evidence for white dwarf crystallization nor can it explain any sharp decrease in the observed luminosity function.

We have also omitted neutrino emission from the right-hand side of Eq. (4.5.1). Thermal neutrino emission[13] dominates photon emission if the photon luminosity $L \gtrsim 10^{-0.5} L_\odot$ and the temperature is $\gtrsim 10^{7.8}$ K. Adding L_ν to the left-hand side of Eq. (4.5.1) will cause a *decrease* in the cooling time and a corresponding *depletion* in the theoretical luminosity function compared to Mestel's cooling curve above $\log(L/L_\odot) = -0.5$. Such a depletion is visible in the Lamb and Van Horn curve shown in Figure 4.3, which will be discussed more fully below.

Finally, convection may lead to more efficient energy transfer and shorter cooling times. However, a careful analysis by Fontaine and Van Horn (1976) and Lamb and Van Horn (1975) indicates that convection is probably not important in a first approximation.

Detailed cooling calculations have also been carried out by Sweeney (1976) and Shaviv and Kovetz (1976). Shaviv and Kovetz's results for a $0.6 M_\odot$ star are often quoted because of claims that the mean white dwarf mass inferred from radius determinations is near this value.[14] From the results of Exercise 4.7 we see that the onset of Debye cooling occurs at such a low luminosity for a $0.6 M_\odot$ star that it should have no effect on the luminosity function in the observable range of interest, which is currently down to $\sim 10^{-5} L_\odot$. Thus the Shaviv–Kovetz results are much closer to the original Mestel cooling curve than the results of Lamb and Van Horn.

4.6   Comparison With Observations

A detailed comparison between theory and observation can be made in two ways: via the luminosity function of white dwarfs or via the ages of white dwarfs in star clusters. Both comparisons provide, *in principle*, rather powerful observational tests of our understanding of the solid state properties of self-gravitating, astrophysical bodies.

(a)   Comparison with Cluster Ages

Lamb and Van Horn (1975) used the observed luminosities of white dwarfs in star clusters to infer theoretical ages, assuming all the white dwarfs were similar to $1 M_\odot$, pure ^{12}C white dwarfs. They compared the theoretical ages with the cluster

---

[13] Thermal neutrino emission processes are described in Section 18.4.

[14] See, for example, Koester, Schulz, and Weidemann (1979).

ages; clearly the age of any cluster should exceed the age of any white dwarf member if, as we believe, all the stars in the cluster were formed at the same time. They find reasonable agreement (Table 4.1). A similar comparison has been carried out by Sweeney (1976).

While potentially a powerful observational test, at present this procedure is hampered by the small number of well-observed white dwarfs in clusters and by uncertainties in determining the cluster ages.

(b) Comparison with the Luminosity Function

Recall our definition of a luminosity function from Chapter 1, rewritten as

$$\phi\left(\frac{L}{L_\odot}\right) d\log\left(\frac{L}{L_\odot}\right) = \begin{array}{l}\text{space density of white dwarfs per unit}\\ \text{interval of } \log(L/L_\odot).\end{array} \quad (4.6.1)$$

If the white dwarf birth rate is uniform in space and constant in time during the age of the Galaxy, we expect ϕ for white dwarfs of a given composition and mass to satisfy

$$\phi\left(\frac{L}{L_\odot}\right) \propto \left[\frac{d\log(L/L_\odot)}{d\tau}\right]^{-1}. \quad (4.6.2)$$

This simply says that the right-hand side of Eq. (4.6.1) is proportional to $d\tau$, the time interval to cool through the unit logarithmic interval in luminosity.

Write $\tau \propto L^{-\alpha}$, where from Eqs. (4.2.4) and (4.5.3) $\alpha = \frac{5}{7}$ for $L > L_c$, and from Eq. (4.5.4) $-\frac{1}{7} \le \alpha \le 0$ for $L \le L_c$. Then

$$\log \phi = -\alpha \log\left(\frac{L}{L_\odot}\right) + \text{constant.} \quad (4.6.3)$$

Table 4.1
Comparison of Theoretical White Dwarf Ages and Cluster Ages[a]

Cluster	Luminosity Range [$\log(L/L_\odot)$]	Theoretical Ages (yr)	Cluster Ages (yr)
Hyades	-1.0 to -2.4	4×10^7 to 5×10^8	4 to 5×10^8
Praesepe	-2.3 to -2.8	4×10^8 to 9×10^8	9×10^8

[a]From Lamb and Van Horn (1975).

The empirical luminosity function derived by Liebert (1980), based on the data of Sion (1979), is shown in Figure 4.3. The data extend from $L \sim 10^{-2}L_\odot$ to $\sim 10^{-4}L_\odot$. Also shown is Green's (1977, 1980) empirical luminosity function, which extends from $\sim 10^{-1}L_\odot$ to $\sim 10^{-3}L_\odot$. Agreement is good, astrophysically speaking, with either the Mestel, Shaviv–Kovetz, or Lamb–Van Horn predictions, which are quite similar over this luminosity range.

It is doubtful whether we will ever accumulate enough statistics on hot white dwarfs ($L > 10^{-1}L_\odot$) to verify the dip in the Lamb–Van Horn calculation caused by neutrino cooling. White dwarfs spend relatively little time in this luminosity range.

The Lamb–Van Horn predictions show the onset of Debye cooling at $L \sim 10^{-4}L_\odot$, leading to a broad maximum with a width of about 2 in $\log L/L_\odot$. As mentioned earlier, this feature occurs at much lower luminosities in the Shaviv–Kovetz calculations because of the lower mass assumed.

A recent survey by Liebert et al. (1979) has concluded that there is a real deficiency of very low luminosity white dwarfs ($L \lesssim 10^{-4.5}L_\odot$). If this result remains true when even more intensive searches are mounted, it will have

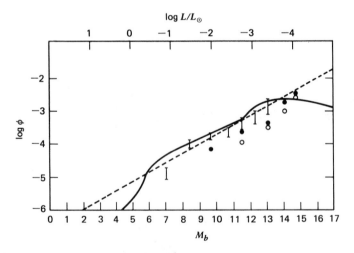

Figure 4.3 The empirical luminosity function ϕ (in units of $pc^{-3}\ M_b^{-1}$) derived by Liebert (1980), based on the data of Sion (1979), is compared with theoretical luminosity functions. *Filled* circles show values assuming relative completeness of the data to limiting magnitude $M_v \sim 13.5$ ($L \sim 10^{-3.5}L_\odot$), *open* circles to $M_v \sim 14.5$ ($L \sim 10^{-4}L_\odot$). The slope of the data is determined relatively better than the overall normalization. Green's (1977, 1980) empirical luminosity function, which has a more secure absolute scale but only goes down to $\sim 10^{-3}L_\odot$, is shown with 1σ error bars. The *solid* curve is the theoretical curve of Lamb and Van Horn (1975) for $1M_\odot$; the *dashed* curve is the Mestel cooling law [$\alpha = \frac{5}{7}$ in Eq. (4.6.3)], which is closer to the shape of the Shaviv and Kovetz (1976) $0.6M_\odot$ calculation. [After Liebert (1980). Reproduced, with permission, from the *Annual Review of Astronomy and Astrophysics*, Volume 18. © 1980 by Annual Reviews Inc.]

important consequences. The existing cooling theories still predict substantial numbers of white dwarfs at these luminosities, and no physical effect is known that can cause a *sudden* break in the luminosity function. Liebert et al. (1979) raise the possibility that Galactic disk stars that dominate this sample might have formed more recently than the currently accepted age of the Galactic halo ($\sim 10^{10}$ yr).

5
General Relativity

Modifications to the gravitational field strength due to general relativity become important when considering the stability properties of white dwarfs and the equilibrium and stability properties of neutron stars and black holes. Indeed, it is largely for this reason that compact objects are of such great theoretical interest and have so many unique and fascinating dynamical features! Although a methodical and detailed discussion of general relativity is beyond the scope of this book, we will need a brief introduction to the governing ideas and equations. Our treatment, though brief, will be entirely adequate for the applications that follow in later chapters; however, readers should not "panic" if they do not fully comprehend every concept they may be confronting here for the first time. Many aspects will become clearer in the later applications. The dedicated student can pursue a rigorous study of relativity elsewhere by reading one or more of the excellent texts on the subject.[1] It will be sufficient if, for now, the reader gains *some* familiarity with the philosophy and key equations that distinguish Newtonian theory from general relativity. In that way we can introduce relativity in subsequent chapters as required.

5.1 What is General Relativity?

General relativity is a *relativistic theory of gravitation*. To understand this statement a little better, let us ask what problems arise in trying to make Newtonian gravitation theory relativistic.

Newtonian gravitation can be described as a field theory for a scalar field Φ satisfying Poisson's equation

$$\nabla^2\Phi = 4\pi G\rho_0. \qquad (5.1.1)$$

The gravitational acceleration of any object in the field is given by $-\nabla\Phi$.

[1] For example, Misner, Thorne, and Wheeler (1973), Weinberg (1972), or, for an introduction, Landau and Lifshitz (1975).

Relativity teaches us that all forms of energy are equivalent to mass, so that a relativistic theory of gravity would presumably have all forms of energy as sources of the gravitational field, not just ρ_0. In particular, the energy density of the gravitational field itself is proportional to $(\nabla\Phi)^2$ in the Newtonian limit. Thus, if we take this term over to the left-hand side of Eq. (5.1.1), we expect a relativistic theory to involve *nonlinear* differential equations for the gravitational field. Symbolically, we may write

$$F(g) \sim GT, \tag{5.1.2}$$

where g represents the gravitational field or fields, reducing to just Φ in the weak-field limit, F is a nonlinear differential operator, reducing to ∇^2 in the weak-field limit, and T is some quantity describing all *nongravitational* forms of energy, with ρ_0 being its dominant component in the nonrelativistic limit.

One of Einstein's great insights was to make general relativity a *geometric* theory of gravitation. We shall not recount here all the motivations for this idea, but will simply start by examining the geometry of special relativity.

In special relativity, *spacetime* is the arena for physics. Spacetime consists of *events*, which require four numbers for their complete specification: three numbers to give the spatial location with respect to some chosen coordinate grid, and one number to give the time. Geometrically, spacetime is represented by a four-dimensional manifold ("surface"), each point in the manifold corresponding to an event in spacetime.

An observer in spacetime makes measurements—that is, assigns coordinates to events. Thus an observer in spacetime corresponds to some choice of coordinates on the manifold. There exists a preferred family of observers in special relativity: the inertial observers, for whom a free particle moves with uniform velocity. All inertial observers are related by Lorentz transformations. The coordinate system of an inertial observer is called an *inertial coordinate system*, or a *Lorentz frame*.

The *interval* ("distance") between two nearby events in spacetime is given by

$$ds^2 = -c^2\,dt^2 + dx^2 + dy^2 + dz^2, \tag{5.1.3}$$

where dt, dx, dy, and dz are the differences between the coordinates of the events evaluated in any Lorentz frame. The interval ds does not depend on which inertial frame is used to evaluate it—it is a *Lorentz invariant*. Writing $x^0 = ct$, $x^1 = x$, $x^2 = y$, $x^3 = z$, we can write Eq. (5.1.3) as

$$ds^2 = \eta_{\alpha\beta}\,dx^\alpha\,dx^\beta, \tag{5.1.4}$$

where

$$\eta_{\alpha\beta} = \text{diag}(-1, 1, 1, 1) \tag{5.1.5}$$

is a diagonal 4×4 matrix and there is an implied sum from 0 to 3 for the repeated indices in Eq. (5.1.4). The quantity $\eta_{\alpha\beta}$ is called the *metric tensor* of spacetime in special relativity, and completely describes spacetime geometrically. Thus, spacetime is a pseudo-Euclidean metric space—that is, Euclidean except for the minus sign in Eq. (5.1.5). This geometry is sometimes called *Minkowski space*.

One could use non-Lorentzian (or noninertial) coordinates to describe spacetime. For example, one could use polar coordinates for the spatial part of the metric, or one could use the coordinate system of an accelerated observer. If the relationship between the inertial coordinates x^α and the noninertial coordinates y^α is

$$x^\alpha = x^\alpha(y^\gamma), \tag{5.1.6}$$

then Eq. (5.1.4) becomes

$$ds^2 = g_{\alpha\beta}(y^\gamma) \, dy^\alpha \, dy^\beta, \tag{5.1.7}$$

where by the chain rule

$$g_{\alpha\beta}(y^\gamma) = \frac{\partial x^\lambda}{\partial y^\alpha} \frac{\partial x^\sigma}{\partial y^\beta} \eta_{\lambda\sigma}. \tag{5.1.8}$$

Even though the metric in the form (5.1.7) might look extremely complicated with all kinds of off-diagonal and position-dependent coefficients, spacetime is still *flat*: there exists a transformation of coordinates [the inverse of Eq. (5.1.6)] such that the metric takes the pseudo-Euclidean form (5.1.5) *everywhere*.

In general relativity, spacetime is still a four-dimensional manifold of events, but now the interval between nearby events is given by

$$ds^2 = g_{\alpha\beta}(x^\gamma) \, dx^\alpha \, dx^\beta, \tag{5.1.9}$$

where *no* choice of coordinates x^α can reduce the metric to the form (5.1.5) everywhere: spacetime is *curved*. The functions $g_{\alpha\beta}$ are used to represent the gravitational field variables; that is, the gravitational field determines the geometry. The interval ds is still an invariant so that the transformation for the components $g_{\alpha\beta}$ of the metric tensor from coordinates $\bar{x}^\alpha$ to coordinates x^α is [cf. Eq. (5.1.8)]

$$g_{\alpha\beta} = \frac{\partial \bar{x}^\lambda}{\partial x^\alpha} \frac{\partial \bar{x}^\sigma}{\partial x^\beta} \bar{g}_{\lambda\sigma}. \tag{5.1.10}$$

As in special relativity, when evaluated along the world line of a particles, ds measures the proper time interval $d\tau$ along that line: $ds^2 = -c^2\,d\tau^2$.

A metric allows one to define dot products between vectors. Equation (5.1.9) can be written as

$$ds^2 = d\vec{x} \cdot d\vec{x}, \tag{5.1.11}$$

where $d\vec{x}$ is an infinitesimal displacement vector with components dx^α. More generally, the dot product between two vectors $\vec{A}$ and $\vec{B}$ with components A^α and B^α is

$$\vec{A} \cdot \vec{B} \equiv g_{\alpha\beta} A^\alpha B^\beta. \tag{5.1.12}$$

Such vectors, with four components, are called "four-vectors" and we label them with arrows to distinguish them from three-vectors. We will sometimes write[2]

$$A_\alpha \equiv g_{\alpha\beta} A^\beta \tag{5.1.13}$$

or

$$A^\alpha = g^{\alpha\beta} A_\beta \tag{5.1.14}$$

where $\|g^{\alpha\beta}\|$ is the matrix inverse of the matrix $\|g_{\alpha\beta}\|$. Then

$$\vec{A} \cdot \vec{B} = A_\alpha B^\alpha = A^\alpha B_\alpha = A_\alpha B_\beta g^{\alpha\beta}. \tag{5.1.15}$$

In special relativity, the coordinates always correspond to the results of some physical measurement. Even noninertial coordinates can be interpreted by finding their relationship to an inertial coordinate system, where t, x, y, and z have their standard meanings in terms of measurements made by ideal clocks and rods. In general relativity, there are in general no preferred coordinate systems; in principle, any set of coordinates that smoothly labels all events in spacetime is acceptable. (Some choices may, of course, be more convenient than others!)[3] We must therefore look at the question of physical measurements more closely.

---

[2] We will not make use of the geometrical interpretation of A^α as being the contravariant components of the vector (i.e., basis vectors tangent to coordinate lines) and A_α being the covariant components (basis vectors orthogonal to coordinate surfaces), nor will we need to introduce differential forms explicitly.

[3] More precisely, general relativity does not prohibit the existence of preferred coordinate systems if some symmetry is present. For example, in the simple Big Bang cosmology the frame in which the microwave background radiation is isotropic is singled out. However, (i) the most general gravitational field imaginable has no symmetries; and (ii) even if there are symmetries, *global* inertial frames do not exist in the presence of a gravitational field.

The physical interpretation of general relativity depends on the concept of *local inertial frames*. Although in general $g_{\alpha\beta}$ in Eq. (5.1.9) cannot be diagonalized by a coordinate transformation to $\eta_{\alpha\beta}$ everywhere in spacetime, we can choose any event in spacetime to be the origin of the coordinates and then *at that point* we can diagonalize $g_{\alpha\beta}$ (it is simply a real symmetric matrix). One can in fact go further: one can find a coordinate transformation that sets the first derivatives of $g_{\alpha\beta}$ at the origin to zero too. In other words, a Taylor expansion of the metric about the chosen origin would look like

$$ds^2 = \left[\eta_{\alpha\beta} + \mathcal{O}\left(|x|^2\right)\right] dx^\alpha\, dx^\beta. \qquad (5.1.16)$$

(The reader can make this assertion plausible by simply counting degrees of freedom in the coordinate transformation. Proofs are given in standard textbooks on general relativity.) Any small coordinate patch in which the metric takes the form (5.1.16) is called a *local inertial frame*.

To see the reason for this designation, consider the local inertial frame of an observer in spacetime. An observer is represented by the worldline of all the events he or she experiences. Choose some particular event on the worldline as the origin used in Eq. (5.1.16). Erect a unit four-vector $\vec{e}_{\hat{t}}$ tangent to the t-coordinate line, and similarly construct $\vec{e}_{\hat{x}}$, $\vec{e}_{\hat{y}}$, and $\vec{e}_{\hat{z}}$. Since $g_{\hat{\alpha}\hat{\beta}} = \eta_{\alpha\beta}$, this is an *orthonormal tetrad*; that is, $\vec{e}_{\hat{t}} \cdot \vec{e}_{\hat{t}} = -1$, $\vec{e}_{\hat{x}} \cdot \vec{e}_{\hat{x}} = 1$, $\vec{e}_{\hat{t}} \cdot \vec{e}_{\hat{x}} = 0$, and so on. We use carets to denote an orthonormal tetrad. To first order in $|x|$, the geometry is the *same* as the geometry of special relativity. The observer can make measurements as in special relativity, provided the extent of his apparatus in space and time is sufficiently small. Departures from special relativity will be noticed on the scale set by the second derivatives of $g_{\alpha\beta}$—the stronger the gravitational field, the more curved spacetime is, the smaller is this scale.

An observer who carries a local orthonormal tetrad as described above is called a *local inertial* or *local Lorentz*[4] *observer*. General relativity goes beyond simply asserting that measurements in a local inertial frame are carried out as in special relativity. It further asserts that *all* the nongravitational laws of physics are the same in a local inertial frame as in special relativity. This law is called the *Principle of Equivalence*. It has its origins in the equivalence of inertial and gravitational mass, exemplified by Einstein's famous elevator thought experiment: Consider an observer performing experiments inside a closed box that is being accelerated upwards with uniform acceleration. The experimental results would be indistinguishable from those of an observer inside a closed stationary box in a

---

[4]*Any* observer can carry an orthonormal tetrad satisfying $g_{\hat{\alpha}\hat{\beta}} = \vec{e}_{\hat{\alpha}} \cdot \vec{e}_{\hat{\beta}} = \eta_{\alpha\beta}$ at an arbitrary point in spacetime. However, only for the special case of a *local inertial observer* ($\equiv$ a freely falling observer $\equiv$ an observer of zero acceleration and zero rotation) does the metric satisfy Eq. (5.1.16) to $\mathcal{O}|x|^2$; that is, $g_{\alpha\beta,\,\gamma} = 0$.

uniform gravitational field. Conversely, in a box *freely falling* in a uniform gravitational field, there would be no observable effects of the gravitational field. (Recall the pictures of astronauts in a satellite in orbit about the Earth!)

This last example gives a physical characterization of a local inertial frame: it is the reference frame of an observer freely falling in the gravitational field. A real gravitational field is not uniform, of course, and the nonuniformities spoil the inertial properties of any potential global inertial frame. However, the more "local" the frame is, the closer it is to being inertial.

Exercise   5.1   Consider two particles of mass m at distances r and $r + h$ ($h \ll r$) on the same vertical line from the center of the Earth. The particles fall freely from rest at time $t = 0$ towards the Earth's surface. Show that an observer falling with one particle will see the separation between the particles gradually increase. Translate this into a quantitative statement about the observer's local inertial frame. In particular, determine the time at which the effects of spacetime curvature will become apparent if measurements can be made to a precision of $\Delta h_{\min}$.

The Principle of Equivalence is a generalization of a principle which says that the laws of mechanics do not allow one to detect a gravitational field locally, to a principle that *no* laws of physics allow one to detect a gravitational field locally. The effects of gravitation always disappear in a freely falling (i.e., local inertial) frame.

The Principle of Equivalence tells us how to formulate nongravitational laws in the presence of a gravitational field. Start with any law of special relativity—for example, conservation of energy–momentum:

$$\nabla_\alpha T^{\alpha\beta} = 0, \tag{5.1.17}$$

$$\nabla_\alpha \equiv \frac{\partial}{\partial x^\alpha}. \tag{5.1.18}$$

Here $T^{\alpha\beta}$ is the stress–energy tensor, and Eq. (5.1.17) says that its four-divergence vanishes. By the Equivalence Principle, Eq. (5.1.17) must be valid in a local inertial frame, where the metric has the form (5.1.16). We now wish to write Eq. (5.1.17) in a form that will be valid in a general coordinate system, where the metric has the form (5.1.9). The mathematics necessary to do this is called tensor calculus, or differential geometry. We shall not need to develop this formalism in this book. It will suffice to say that one needs to define a more general derivative operator ("covariant derivative") than Eq. (5.1.18). Even in curvilinear coordinates in flat space, the reader knows that there are extra terms on the right-hand side of Eq. (5.1.18) when differentiating vectors. For example, if in spherical coordinates the vector **A** has components

$$\mathbf{A} = A^r \mathbf{e}_r + A^\theta \mathbf{e}_\theta + A^\phi \mathbf{e}_\phi, \tag{5.1.19}$$

its divergence is not simply

$$\partial_r A^r + \partial_\theta A^\theta + \partial_\phi A^\phi. \tag{5.1.20}$$

There are extra terms from the derivatives of $\mathbf{e}_r$, $\mathbf{e}_\theta$, and $\mathbf{e}_\phi$. These basis vectors are not constant in space, as is clear from the fact that the components $g_{\alpha\beta}$ are not constant in this coordinate system.

Similarly, in a curved spacetime, there are extra terms in the covariant derivative ∇_α that come from the nonconstancy of $g_{\alpha\beta}$. Now, however, these are not removable *everywhere* by a coordinate transformation, and the derivatives of $g_{\alpha\beta}$ represent the effect of the gravitational field.

This mathematical implementation of the Equivalence Principle is sometimes called the *Principle of General Covariance*: one writes down covariant equations in special relativity, and demands that they be covariant not simply under Lorentz transformations, but under general coordinate transformations.

There is no analog in general relativity of the Newtonian concept of "gravitational acceleration at a point": such local acceleration is removable by going to a freely falling coordinate system. However, the *difference* between the gravitational accelerations of two nearby test bodies is not removable in general (cf. Exercise 5.1). Thus the true gravitational field in general relativity has as its Newtonian analog the *tidal* gravitational field $\partial^2 \Phi / \partial x^i \, \partial x^j$, where Φ is the Newtonian potential. This is because the relative Newtonian gravitational acceleration of two test bodies is

$$a^i_{\text{relative}} = a^i(\mathbf{x} + \Delta\mathbf{x}) - a^i(\mathbf{x})$$

$$= \Delta x^j \frac{\partial}{\partial x^j}\left(\frac{\partial \Phi}{\partial x^i} \right). \tag{5.1.21}$$

So far we have discussed how gravitation affects all the other phenomena of physics, and how the geometry is related to physical measurements in a local inertial frame. To complete the picture we need to specify how the distribution of mass–energy determines the geometry, $g_{\alpha\beta}$, via an equation of the form (5.1.2). Although we shall not explicitly use this equation in its full generality, we write it down because it represents the crowning achievement of Einstein's theory:

$$G^{\alpha\beta} = 8\pi \frac{G}{c^4} T^{\alpha\beta}. \tag{5.1.22}$$

Here $G^{\alpha\beta}$, the Einstein tensor, is a second-order, nonlinear differential operator acting on $g_{\alpha\beta}$. The source term for Einstein's field equation is the nongravitational stress–energy tensor. This complicated equation reduces to Poisson's equation (5.1.1) in the Newtonian limit. It also guarantees the conservation of energy–momentum [cf. Eq. (5.1.17)] since $\nabla_\alpha G^{\alpha\beta} \equiv 0$.

5.2   The Motion of Test Particles

A test particle is an idealization of a material object. It is supposed to be small (does not perturb spacetime around it), uncharged (does not respond to electromagnetic forces), spherical (feels no torques), and so on. It simply moves freely in the gravitational field.

In special relativity (no gravitational field), test particles move with uniform velocity. One can get their equations of motion from a variational principle that extremizes the distance (interval) along the worldline:

$$\delta \int ds = 0. \tag{5.2.1}$$

To verify this, write the integrand in the form

$$ds = \left(-\eta_{\alpha\beta} \dot{x}^\alpha \dot{x}^\beta \right)^{1/2} d\lambda, \tag{5.2.2}$$

where

$$\dot{x}^\alpha \equiv \frac{dx^\alpha}{d\lambda}. \tag{5.2.3}$$

Here λ is any parameter along the worldline. The expression (5.2.2) for ds is invariant under a change of parameter $\lambda \to \lambda(\lambda')$. The Lagrangian for Eq. (5.2.1) is

$$L = \left(-\eta_{\alpha\beta} \dot{x}^\alpha \dot{x}^\beta \right)^{1/2}. \tag{5.2.4}$$

The Euler–Lagrange equations of motion obtained from Eq. (5.2.1) are simply

$$\frac{d}{d\lambda} \left(\frac{\partial L}{\partial \dot{x}^\alpha} \right) = \frac{\partial L}{\partial x^\alpha}. \tag{5.2.5}$$

The right-hand side is zero since L is independent of x^α. Since

$$\frac{\partial L}{\partial \dot{x}^\alpha} = -L^{-1} \eta_{\alpha\beta} \dot{x}^\beta, \tag{5.2.6}$$

we get

$$\eta_{\alpha\beta}\ddot{x}^{\beta} - \frac{1}{L}\frac{dL}{d\lambda}\eta_{\alpha\beta}\dot{x}^{\beta} = 0. \tag{5.2.7}$$

By a rescaling $\lambda \to \lambda(\lambda')$ we can make L be constant along the curve.[5] In particular, one can always parametrize the worldline of a particle by the length s along the curve—this is simply its proper time, usually denoted τ. (Actually, $s = c\tau$.) In this case $\lambda = s$, $L = 1$ along the curve, and so Eq. (5.2.7) becomes

$$\eta_{\alpha\beta}\ddot{x}^{\beta} = 0. \tag{5.2.8}$$

Multiplying by the matrix inverse of $\eta_{\alpha\beta}$, denoted $\eta^{\gamma\alpha}$, we get

$$\ddot{x}^{\gamma} = 0 = \frac{d^2x^{\gamma}}{d\tau^2}, \tag{5.2.9}$$

which is the equation for uniform velocity in a straight line.

Geometrically, curves of extremal length are called *geodesics*. The geodesics of Minkowski space (special relativity) are four-dimensional *straight lines*. The case treated above, where $ds^2 < 0$, describes *timelike* geodesics, the worldlines of free material particles. Photons, or other massless particles, travel at the speed of light so that $ds^2 = 0$; free photons are said to move along *null* geodesics. In this case the parameter λ cannot be chosen as proper time. It is convenient to choose it so that

$$\frac{dx^{\alpha}}{d\lambda} = p^{\alpha}, \tag{5.2.10}$$

the four-momentum of the photon. Since $\eta_{\alpha\beta}p^{\alpha}p^{\beta} = 0$ at all times for a photon, this is a valid choice for λ: $ds^2 = 0$. (One could also make this choice for a particle with mass m; that is, $\lambda = \tau/m$.) Equation (5.2.9) now reads

$$\frac{dp^{\alpha}}{d\lambda} = 0; \qquad \text{that is, } p^{\alpha} = \text{constant}. \tag{5.2.11}$$

One can also have *spacelike* geodesics, for which $ds^2 > 0$. These correspond, for example, to straight lines in Euclidean three-space at some instant $x^0 = $ constant. [The derivation above goes through without the minus sign in Eq. (5.2.4).]

The point of all this machinery for a simple problem in special relativity is that it immediately goes over to general relativity. By the Equivalence Principle, Eq.

---

[5]The parameter λ is then called an *affine* parameter.

(5.2.1) must be a variational principle for the motion of test particles in general relativity: free particles move along geodesics of spacetime. Now, however, Eq. (5.2.4) becomes

$$L = \left[-g_{\alpha\beta}(x^{\gamma})\dot{x}^{\alpha}\dot{x}^{\beta}\right]^{1/2}, \tag{5.2.12}$$

so Eq. (5.2.5) gives

$$g_{\alpha\beta}\ddot{x}^{\beta} + g_{\alpha\beta,\gamma}\dot{x}^{\gamma}\dot{x}^{\beta} - \tfrac{1}{2}g_{\gamma\beta,\alpha}\dot{x}^{\gamma}\dot{x}^{\beta} = 0. \tag{5.2.13}$$

Here the second term comes from

$$\frac{d}{d\lambda}g_{\alpha\beta} = \frac{\partial g_{\alpha\beta}}{\partial x^{\gamma}}\frac{dx^{\gamma}}{d\lambda}, \tag{5.2.14}$$

$$g_{\alpha\beta,\gamma} \equiv \frac{\partial g_{\alpha\beta}}{\partial x^{\gamma}}. \tag{5.2.15}$$

We have assumed an affine parametrization such that $L = $ constant. Now write

$$g_{\alpha\beta,\gamma}\dot{x}^{\beta}\dot{x}^{\gamma} = \tfrac{1}{2}(g_{\alpha\beta,\gamma} + g_{\alpha\gamma,\beta})\dot{x}^{\beta}\dot{x}^{\gamma}, \tag{5.2.16}$$

so that Eq. (5.2.13) becomes

$$g_{\alpha\beta}\ddot{x}^{\beta} + \Gamma_{\alpha\beta\gamma}\dot{x}^{\beta}\dot{x}^{\gamma} = 0, \tag{5.2.17}$$

where

$$\Gamma_{\alpha\beta\gamma} \equiv \tfrac{1}{2}(g_{\alpha\beta,\gamma} + g_{\alpha\gamma,\beta} - g_{\gamma\beta,\alpha}). \tag{5.2.18}$$

Multiplying by the matrix inverse of the metric tensor, denoted $g^{\lambda\alpha}$, and then relabeling $\lambda \leftrightarrow \alpha$, we get

$$\ddot{x}^{\alpha} + \Gamma^{\alpha}_{\beta\gamma}\dot{x}^{\beta}\dot{x}^{\gamma} = 0, \tag{5.2.19}$$

where

$$\Gamma^{\alpha}_{\beta\gamma} \equiv g^{\alpha\lambda}\Gamma_{\lambda\beta\gamma}. \tag{5.2.20}$$

The Γ-symbols have the technical name "Christoffel symbols."

Equation (5.2.19) is the final form of the geodesic equation in general relativity. Notice how the Equivalence Principle is satisfied: in a local inertial frame, one can choose coordinates so that $g_{\alpha\beta,\gamma} = 0$—that is, so that the Γ-symbols vanish. Thus the motion of a test particle in a local inertial frame is uniform velocity in a straight line. Requiring this statement to be true in any inertial frame at any point in spacetime leads to Eq. (5.2.19), where the Γ's represent the effect of the gravitational field.

Note the difference between the Principle of General Covariance employed here and the principle of Lorentz covariance in special relativity. There, one requires that a transformation from one inertial frame to another leave the form of a law of physics unchanged. The velocity of the transformation must drop out of the final equation. This requirement restricts possible laws of physics quite severely. The Principle of General Covariance does not restrict the possible laws of physics. One can postulate any law in a local inertial frame, transform it to a general coordinate system, and say that the extra terms appearing describe the effect of the gravitational field. Only experiment can then decide if the law is true.

Exercise   5.2   Show that, with λ taken to be an affine parameter, the Lagrangian

$$L = \tfrac{1}{2}g_{\alpha\beta}\dot{x}^\alpha\dot{x}^\beta \tag{5.2.21}$$

is an equivalent Lagrangian to Eq. (5.2.12) for geodesics; that is, show that the Euler–Lagrange equations for Eq. (5.2.21) give the same geodesic equations, except that now λ is no longer an arbitrary parameter and the condition $L = $ constant is built in to the variational principle.

As usual, we can define the canonically conjugate momentum to the coordinate x^α by

$$p_\alpha \equiv \frac{\partial L}{\partial \dot{x}^\alpha}. \tag{5.2.22}$$

From Eqs. (5.2.21) and (5.2.10), we see that

$$p_\alpha = g_{\alpha\beta}\dot{x}^\beta = g_{\alpha\beta}p^\beta$$

or

$$p^\alpha = g^{\alpha\beta}p_\beta, \tag{5.2.23}$$

where $g^{\alpha\beta}$ is the matrix inverse of $g_{\alpha\beta}$. Note that if L is independent of the coordinate x^1 say, then p_1 is a constant of the motion.

Exercise   5.3   The metric of two-dimensional Euclidean space is

$$ds^2 = dr^2 + r^2\,d\phi^2, \tag{5.2.24}$$

leading to the Lagrangian

$$L = \tfrac{1}{2}\left(\dot{r}^2 + r^2\dot{\phi}^2\right).\tag{5.2.25}$$

Show that the equations of motion of a test particle are

$$\ddot{r} - r\dot{\phi}^2 = 0,\tag{5.2.26}$$

$$r^2\dot{\phi} = \text{constant}.\tag{5.2.27}$$

(Taking $\lambda = t$, we recognize the usual Newtonian equations in empty space.)

The constant in Eq. (5.2.27) is p_ϕ (the angular momentum per unit mass), by Eq. (5.2.22). Since $g^{\phi\phi} = 1/g_{\phi\phi} = 1/r^2$, Eq. (5.2.23) gives $p^\phi = \dot{\phi}$, as we knew all along. The *physically measured* value of the ϕ component of the momentum is the projection of the vector $\vec{p}$ along a *unit* vector in the ϕ direction. Now the *coordinate* basis vectors $\vec{e}_r$ and $\vec{e}_\phi$ satisfy

$$\vec{e}_r \cdot \vec{e}_r = g_{rr} = 1, \qquad \vec{e}_r \cdot \vec{e}_\phi = g_{r\phi} = 0, \qquad \vec{e}_\phi \cdot \vec{e}_\phi = g_{\phi\phi} = r^2. \tag{5.2.28}$$

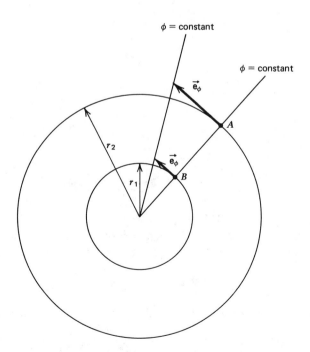

Figure 5.1   The coordinate basis vector $\vec{e}_\phi$ at A is r_2/r_1 times longer than the basis vector at B. (The lines $\phi = $ constant are supposed to be infinitesimally close.)

We can understand, for example, the last equation in (5.2.28) by recalling that $\vec{e}_\phi$ is the tangent to the ϕ coordinate line. This means (cf. Fig. 5.1) that it connects two radial lines ϕ = constant. As is clear from the figure, $\vec{e}_\phi$ at radius r_1 and $\vec{e}_\phi$ at radius r_2 have lengths in the ratio r_1/r_2—that is, $\vec{e}_\phi \cdot \vec{e}_\phi \propto r^2$, the constant of proportionality depending on the scaling of the ϕ coordinate. One always makes the simple choice $\vec{e}_\phi \cdot \vec{e}_\phi = r^2$, as in Eq. (5.2.28). In general, one has the relation

$$\vec{e}_\alpha \cdot \vec{e}_\beta = g_{\alpha\beta}. \tag{5.2.29}$$

This equation results from Eqs. (5.1.9) and (5.1.11), noting that $d\vec{x} = dx^\alpha \vec{e}_\alpha$. Special cases of Eq. (5.2.29) are

$$\mathbf{e}_i \cdot \mathbf{e}_j = \delta_{ij} \text{ (three - dimensional Cartesian)},$$

$$\vec{e}_{\hat\alpha} \cdot \vec{e}_{\hat\beta} = \eta_{\hat\alpha\hat\beta} \text{ (four - dimensional orthonormal)}. \tag{5.2.30}$$

Exercise 5.4 Show that $p_\alpha = \vec{p} \cdot \vec{e}_\alpha$.
Hint: Expand $\vec{p} = p^\beta \vec{e}_\beta$.

We can now choose an *orthonormal* set of basis vectors to be

$$\vec{e}_{\hat r} = \vec{e}_r, \tag{5.2.31}$$

$$\vec{e}_{\hat\phi} = \frac{1}{r}\vec{e}_\phi, \tag{5.2.32}$$

so that Eq. (5.2.28) gives

$$\vec{e}_{\hat\phi} \cdot \vec{e}_{\hat\phi} = \frac{1}{r^2}\vec{e}_\phi \cdot \vec{e}_\phi = \frac{1}{r^2}g_{\phi\phi} = 1, \text{ and so on.} \tag{5.2.33}$$

Then

$$p^{\hat\phi} = p_{\hat\phi} = \vec{p} \cdot \vec{e}_{\hat\phi} = \frac{1}{r}p_\phi = r\dot\phi, \tag{5.2.34}$$

a familiar result for the momentum per unit mass (i.e., velocity) along $\vec{e}_{\hat\phi}$.

We shall sometimes use carets also to denote proper infinitesimal displacements. For example, a displacement in the ϕ direction, with r = constant, is

$$d\hat\phi \equiv ds_{(r=\text{const})} = r\,d\phi. \tag{5.2.35}$$

Thus the circumference of a circle of radius r is the distance along successive ϕ displacements:

$$\oint_{r=\text{const}} ds = \oint d\hat\phi = \oint r\,d\phi = 2\pi r. \tag{5.2.36}$$

5.3   The Gravitational Redshift

The simplest example of the gravitational redshift involves an emitter and a detector of electromagnetic waves (i.e., photons) at two fixed positions in a static gravitational field. The frequency ν of the radiation at the emitter is just the inverse of the proper time between two wave crests as measured in the frame of the emitter; that is,

$$\nu_{em} = \frac{1}{d\tau_{em}}$$

$$= \frac{c}{\left(-g_{\alpha\beta}\, dx^\alpha\, dx^\beta\right)^{1/2}_{em}}. \tag{5.3.1}$$

Now $dx^1 = dx^2 = dx^3 = 0$ because the emitter stays at a fixed position while emitting. An expression similar to Eq. (5.3.1) holds for the receiver, so we have

$$\frac{\nu_{rec}}{\nu_{em}} = \frac{\left[(-g_{00})^{1/2}\, dx^0\right]_{em}}{\left[(-g_{00})^{1/2}\, dx^0\right]_{rec}}. \tag{5.3.2}$$

The *coordinate* time dx^0 between two wave crests is the same at the emitter and the receiver. This is because the gravitational field is static; that is, nothing depends on x^0. Whatever the worldline of one photon is from emitter to receiver, the next photon follows a congruent path, merely displaced by dx^0 at all points (see Fig. 5.2). Thus

$$\frac{\nu_{rec}}{\nu_{em}} = \frac{(-g_{00})^{1/2}_{em}}{(-g_{00})^{1/2}_{rec}}. \tag{5.3.3}$$

5.4   The Weak-Field Limit

One way of obtaining the weak-field limit of general relativity is to consider an alternative derivation of the redshift formula, based on energy conservation. A photon of frequency ν has an effective mass

$$m = \frac{h\nu}{c^2}. \tag{5.4.1}$$

Its total energy in a Newtonian gravitational field with potential $\Phi(\mathbf{x})$ is $h\nu +$

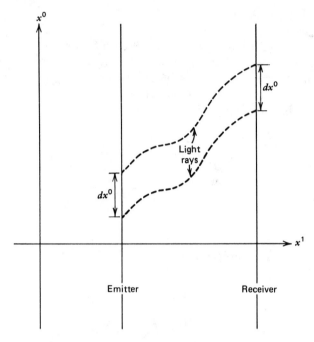

Figure 5.2 Spacetime diagram for gravitational redshift. The vertical lines are the world lines of a light-ray emitter and a receiver. The *dashed* lines are the world lines of two light rays emitted a coordinate time interval dx^0 apart. Note that the coordinates are not inertial (no global inertial frames in the presence of a gravitational field).

$m\Phi(\mathbf{x})$. Equating this energy at the emitter to a similar expression at the receiver, we obtain

$$\frac{\nu_{\text{rec}}}{\nu_{\text{em}}} = \frac{\left(1 + \Phi/c^2\right)_{\text{em}}}{\left(1 + \Phi/c^2\right)_{\text{rec}}}. \tag{5.4.2}$$

This is usually written as

$$\frac{\Delta\nu}{\nu_{\text{em}}} = -\frac{\Delta\Phi}{c^2}, \tag{5.4.3}$$

since $\Phi/c^2 \ll 1$ in a Newtonian field.

Comparing Eqs. (5.3.3) and (5.4.2), we find

$$g_{00} \simeq -\left(1 + \frac{2\Phi}{c^2}\right) \tag{5.4.4}$$

in the Newtonian limit.

An alternative way of arriving at Eq. (5.4.4) is to consider the motion of a slowly moving ($v \ll c$) test particle in a weak gravitational field ($\Phi \ll c^2$). Since gravity is weak, we can choose a coordinate system that is almost Lorentzian everywhere:

$$g_{\alpha\beta} = \eta_{\alpha\beta} + h_{\alpha\beta}, \qquad |h_{\alpha\beta}| \ll 1. \tag{5.4.5}$$

Since the velocity v^j of the particle is small compared with c, and since $g_{\alpha\beta} \simeq \eta_{\alpha\beta}$, coordinate time t is equal to the proper time τ of the particle to leading order. Thus the acceleration of the particle is

$$\frac{d^2x^i}{dt^2} \simeq \frac{d^2x^i}{d\tau^2} = -\Gamma^i_{\alpha\beta} \frac{dx^\alpha}{d\tau} \frac{dx^\beta}{d\tau}, \tag{5.4.6}$$

where we have used the geodesic equation (5.2.19). The dominant term in the sum over α and β is for $\alpha = \beta = 0$; the other terms are smaller by at least one power of v/c. Also,

$$\Gamma^i_{00} \simeq \Gamma_{i00} \quad \left(\text{since } g_{\alpha\beta} \simeq \eta_{\alpha\beta}\right)$$

$$= \tfrac{1}{2}(2h_{0i,0} - h_{00,i})$$

$$\simeq -\tfrac{1}{2}h_{00,i}, \tag{5.4.7}$$

since $h_{0i,0} \simeq h_{0i,j}v^j/c$. Thus

$$\frac{d^2x^i}{dt^2} \simeq \frac{c^2}{2}h_{00,i}. \tag{5.4.8}$$

But in the Newtonian limit,

$$\frac{d^2x^i}{dt^2} = -\Phi_{,i}. \tag{5.4.9}$$

Comparing Eqs. (5.4.8) and (5.4.9), we find

$$h_{00} = -\frac{2\Phi}{c^2},$$

$$g_{00} \simeq -\left(1 + \frac{2\Phi}{c^2}\right), \tag{5.4.10}$$

as before.[6]

---

[6] The constant of integration obtained by equating Eqs. (5.4.8) and (5.4.9) and integrating may be set equal to zero by requiring $g_{\alpha\beta} \equiv \eta_{\alpha\beta}$, $h_{\alpha\beta} = 0 = \Phi$ at $r = \infty$.

5.5   Geometrized Units

It is very convenient in general relativity to choose units so that $c = G = 1$. In other words, time is measured in cm, with $1 \text{ s} = 3 \times 10^{10}$ cm, and mass is also measured in cm, with $1 \text{ g} = 0.7425 \times 10^{-28}$ cm. The numerical factor is simply G/c^2 in cgs units. A more convenient conversion factor for astronomical objects is $M_\odot = 1.4766$ km.

Amusingly, we note that it is actually the combination $GM_\odot$ that is measured by celestial mechanics. The value of $M_\odot$ in *grams* comes from measuring the value of G via a Cavendish experiment. The quantity $GM_\odot$ is known to about a part in 10^8. The gravitational constant in only known to a part in 10^3. The mass of the sun in kilometers is more accurately known than its mass in grams!

5.6   Spherically Symmetric Gravitational Fields

A spherically symmetric metric can depend on the time and radial coordinates, denoted t and r, and also on angles, but the latter only in the combination

$$d\Omega^2 \equiv d\theta^2 + \sin^2 \theta \, d\phi^2. \tag{5.6.1}$$

Thus the most general spherically symmetric metric can be written

$$ds^2 = -A(t, r) \, dt^2 + B(t, r) \, dr^2 + 2C(t, r) \, dt \, dr + D(t, r) \, d\Omega^2. \tag{5.6.2}$$

Choose a new radial coordinate

$$r' = D^{1/2}(t, r). \tag{5.6.3}$$

Then, substituting into Eq. (5.6.2) and dropping the prime, we get

$$ds^2 = -E(t, r) \, dt^2 + F(t, r) \, dr^2 + 2G(t, r) \, dt \, dr + r^2 \, d\Omega^2, \tag{5.6.4}$$

where E, F, and G are related to A, B, and C in some way that we have not bothered to write down. We can eliminate the coefficient of $dt \, dr$ by introducing a new time coordinate, t'. The form of Eq. (5.6.4) suggests we set

$$dt' = E(t, r) \, dt - G(t, r) \, dr. \tag{5.6.5}$$

However, in general the right-hand side of Eq. (5.6.5) will not be a perfect differential. Since there are only two independent variables, t and r, we know that there always exists an integrating factor—that is, a function $H(t, r)$ such that

$$dt' = H(t, r)\left[E(t, r) \, dt - G(t, r) \, dr\right] \tag{5.6.6}$$

is a perfect differential. Making this substitution in Eq. (5.6.4) leaves only two metric functions, the coefficients of $(dt')^2$ and dr^2. Dropping the prime, we write finally

$$ds^2 = -e^{2\Phi} dt^2 + e^{2\lambda} dr^2 + r^2 d\Omega^2, \tag{5.6.7}$$

where Φ and λ are functions of t and r. We use the exponential form for the coefficients for later convenience.

An important result of Newtonian gravitation is that at any point outside a spherical mass distribution, the gravitational field depends only on the mass interior to that point. Moreover, even if the mass interior is moving spherically symmetrically, the field outside is constant in time—we have simply $\Phi = -M/r$.

This result is also true in general relativity, where it is known as *Birkhoff's Theorem*: the only vacuum, spherically symmetric gravitational field is static. It is called the *Schwarzschild metric*:

$$ds^2 = -\left(1 - \frac{2M}{r}\right) dt^2 + \left(1 - \frac{2M}{r}\right)^{-1} dr^2 + r^2 d\Omega^2. \tag{5.6.8}$$

The word "vacuum" here denotes a region of spacetime where the gravitational effect of any matter present is negligible. The constant M appearing in Eq. (5.6.8) is the mass of the source. We can see this for example by considering the weak-field limit of Eq. (5.6.8), $r \gg M$. Then Eq. (5.4.4) shows that the Newtonian potential is $-M/r$; that is, M is the mass. This mass could be measured operationally by, for example, having satellites in distant orbits about the source and using Kepler's laws as in conventional celestial mechanics. The Schwarzschild metric applies everywhere *outside* a spherical star, right up to its surface.

Because of the relatively simple form of Eq. (5.6.8), the coordinates have a direct physical interpretation. At any radius r, there is a 2-sphere of equivalent points,[7] and θ and ϕ are conventional polar coordinates on the 2-sphere. The quantity r is defined by setting the proper circumference of the 2-sphere equal to $2\pi r$, or the proper area equal to $4\pi r^2$. We see this by setting $t = $ constant, $r = $ constant in Eq. (5.6.8). Then the metric on the 2-sphere is

$$^{(2)}ds^2 = r^2 d\Omega^2. \tag{5.6.9}$$

Recalling that ds^2 gives the results of physical measurements, we see, for example, that the proper circumference of the 2-sphere is

$$\oint_{\theta = \pi/2} {}^{(2)}ds = \int_0^{2\pi} r \, d\phi = 2\pi r, \tag{5.6.10}$$

---

[7]That is, a two-dimensional spherical surface centered about the point $r = 0$.

as claimed. Note that the proper distance between two points r_1 and r_2 on a radial line is

$$\int_{r_1}^{r_2} (g_{rr})^{1/2} \, dr \neq r_2 - r_1. \qquad (5.6.11)$$

The time coordinate t has been chosen to exhibit the static nature of the solution: the field is invariant under a change $t \to t + \Delta t$. The coordinate t is normalized to be equal to the Minkowski time coordinate at $r \gg M$, where the metric (5.6.8) reduces to the metric of special relativity.

5.7   Spherical Stars

The metric (5.6.7) also describes the gravitational field inside a spherical star. For a star in hydrostatic equilibrium, we can take Φ and λ to be independent of t. We assume the stellar material can be described as a perfect fluid, and in general we would have to specify an equation of state

$$\rho = \rho(n, s). \qquad (5.7.1)$$

(Since $c = 1$, we are not distinguishing between total energy density ε and total mass density $\rho = \varepsilon/c^2$.) The pressure then follows from the first law of thermodynamics, as in Eq. (2.1.7):

$$P = P(n, s). \qquad (5.7.2)$$

Although a perfect fluid is adiabatic (the entropy s of a fluid element remains constant), it is not necessarily isentropic (s does not need to have the same value everywhere). In the case of cold white dwarfs and neutron stars, however, the temperature is essentially zero everywhere (more precisely, $kT \ll E_F'$), so $s = 0$ everywhere. Later we shall discuss supermassive stars, where s is uniform because of convection. Thus we can use an equation of state of the form

$$P = P(\rho). \qquad (5.7.3)$$

The equations of stellar structure in general relativity are derived in standard textbooks. We write them in a form that emphasizes the similarity to the Newtonian equations. First, define a new metric function $m(r)$ by

$$e^{2\lambda} \equiv \left(1 - \frac{2m}{r}\right)^{-1}. \qquad (5.7.4)$$

Then Einstein's equations give

$$\frac{dm}{dr} = 4\pi r^2 \rho, \tag{5.7.5}$$

$$\frac{dP}{dr} = -\frac{\rho m}{r^2}\left(1 + \frac{P}{\rho}\right)\left(1 + \frac{4\pi P r^3}{m}\right)\left(1 - \frac{2m}{r}\right)^{-1}, \tag{5.7.6}$$

$$\frac{d\Phi}{dr} = -\frac{1}{\rho}\frac{dP}{dr}\left(1 + \frac{P}{\rho}\right)^{-1}. \tag{5.7.7}$$

The Newtonian limit is recovered by taking $P \ll \rho$ and $m \ll r$.

Equation (5.7.6) is called the "OV equation of hydrostatic equilibrium" (OV means *Oppenheimer–Volkoff*).

The quantity $m(r)$ has the interpretation of "mass inside radius r." Equation (5.7.5) gives

$$M = \int_0^R 4\pi r^2 \rho\, dr \tag{5.7.8}$$

for the total mass of the star.[8] Note that this includes all contributions to the mass, including gravitational potential energy. This fact is somewhat obscured by the simple appearance of Eq. (5.7.8), but recall that the proper volume element is not $4\pi r^2\, dr$ here; it is

$$d\mathcal{V} = (g_{rr})^{1/2}\, dr \times 4\pi r^2 = \left(1 - \frac{2m}{r}\right)^{-1/2} 4\pi r^2\, dr. \tag{5.7.9}$$

Thus Eq. (5.7.8) does *not* simply add up $\rho\, d\mathcal{V}$, the local contributions to the total mass–energy, but also includes a global contribution from the negative gravitational potential energy of the star.

Equations (5.7.5)–(5.7.7) lend themselves to easy numerical computation of a general relativistic stellar model:

1. Pick a value of central density, ρ_c. The equation of state gives P_c, and one has a boundary condition $m(r = 0) = 0$.
2. Integrate Eqs. (5.7.5) and (5.7.6) out from $r = 0$, using the values in step 1 as initial conditions. Each time a new value of P is obtained, the equation of state gives the corresponding value of ρ.
3. The value $r = R$ at which $P = 0$ is the radius of the star, and $m(R) = M$.

[8] The quantity $m(R)$ must equal M so that the interior metric coefficient (5.7.4) will match smoothly to the exterior Schwarzschild metric (5.6.8).

4. The metric function Φ has the boundary condition

$$\Phi(r = R) = \frac{1}{2}\ln\left(1 - \frac{2M}{R}\right), \tag{5.7.10}$$

so that it will match smoothly to the Schwarzschild metric (5.6.8) at the surface. In numerical work, it is convenient to choose an arbitrary value for $\Phi(r = 0)$ and integrate Eq. (5.7.7) out from $r = 0$ simultaneously with Eqs. (5.7.5) and (5.7.6). Since Eq. (5.7.7) is linear in Φ, one can then add a constant to Φ everywhere so that Eq. (5.7.10) is satisfied.

Exercise 5.5 Show that inside a uniform density star ($\rho = $ constant),

$$\frac{P}{\rho} = \frac{\left(1 - 2Mr^2/R^3\right)^{1/2} - \left(1 - 2M/R\right)^{1/2}}{3\left(1 - 2M/R\right)^{1/2} - \left(1 - 2Mr^2/R^3\right)^{1/2}}, \tag{5.7.11}$$

$$e^\Phi = \frac{3}{2}\left(1 - \frac{2M}{R}\right)^{1/2} - \frac{1}{2}\left(1 - \frac{2Mr^2}{R^3}\right)^{1/2}. \tag{5.7.12}$$

Show that the condition $P_c < \infty$ implies

$$\frac{2M}{R} < \frac{8}{9}. \tag{5.7.13}$$

The limit given by Eq. (5.7.13) for the maximum "compaction" of a uniform density equilibrium sphere applies, in fact, to spheres of arbitrary density profile, as long as the density does not increase outwards.[9]

---

[9]For example, Weinberg (1972), Section 11.6.

6

The Equilibrium and Stability of Fluid Configurations

In this chapter we develop some basic properties of the equilibrium and stability of stars. Section 6.1 summarizes the fundamental fluid equations, and will be referred to many times later in the book. Sections 6.2–6.8 develop the Newtonian theory of the equilibrium and stability of nonrotating stars. This involves a study of perturbations of fluid configurations. The principal results of these sections are summarized in Box 6.1. The general relativistic extension of these results is carried out in Section 6.9, with a summary of the results at the beginning of the section. Finally, the results are applied to white dwarfs in Section 6.10. This latter section is a prototype of similar calculations carried out later in the book.

6.1 Basic Fluid Equations

We summarize here the equations governing the flow of a *nonrelativistic*, one-component fluid.[1]

The conservation of mass is described by the continuity equation for the density ρ and fluid velocity $\mathbf{v}$:

$$\frac{\partial \rho}{\partial t} + \nabla \cdot (\rho \mathbf{v}) = 0. \tag{6.1.1}$$

The momentum equation (Newton's law for fluids, $\mathbf{a} = \mathbf{F}/m$) is

$$\frac{d\mathbf{v}}{dt} = -\frac{1}{\rho}\nabla P - \nabla \Phi, \tag{6.1.2}$$

[1]See, for example, Landau and Lifshitz (1959).

127

where P is the pressure and Φ is the gravitational potential. Here

$$\frac{d}{dt} = \frac{\partial}{\partial t} + \mathbf{v} \cdot \nabla \qquad (6.1.3)$$

is the *total* time derivative following a fluid element (also called the convective time derivative, or *Lagrangian* time derivative), while $\partial/\partial t$ is the ordinary time derivative at fixed location (the *Eulerian* time derivative). Equation (6.1.2) is called *Euler's equation*. If dissipative terms representing the action of viscous forces are included on the right-hand side of Eq. (6.1.2), it becomes the *Navier–Stokes equation* (cf. Appendix H).

The gravitational potential is determined by Poisson's equation

$$\nabla^2 \Phi = 4\pi G\rho. \qquad (6.1.4)$$

The equation of entropy production is

$$\frac{ds}{dt} = \text{entropy production terms.} \qquad (6.1.5)$$

The right-hand side, usually expressed as a function of T and ρ, represents the production of entropy through dissipative processes such as heat conduction, viscous shearing and expansion, emission and absorption of radiation, and so on. We shall restrict ourselves in this chapter to *adiabatic* flows, where

$$\frac{ds}{dt} = 0, \qquad (6.1.6)$$

that is, s remains constant for any given fluid element. Note that in the nonrelativistic case it is conventional to express thermodynamic quantities like entropy on a per unit mass basis, rather than on a per baryon basis. The two definitions differ simply by m_B.

In the general (nonadiabatic) case, Eqs. (6.1.1), (6.1.2), and (6.1.5) provide five dynamical equations for the time evolution of ρ, $\mathbf{v}$, and s from their initial values. The potential Φ can be determined from ρ at any instant from Eq. (6.1.4), while the remaining variables P and T are found from the equation of state of the fluid. A convenient way of giving the equation of state is to specify u, the internal energy per unit mass of the fluid:

$$u = u(\rho, s). \qquad (6.1.7)$$

From the first law,

$$du = -Pd\left(\frac{1}{\rho}\right) + Tds, \qquad (6.1.8)$$

and Eq. (6.1.7), we get

$$P = \rho^2 \left. \frac{\partial u}{\partial \rho} \right|_s, \tag{6.1.9}$$

$$T = \left. \frac{\partial u}{\partial s} \right|_\rho. \tag{6.1.10}$$

This then is the framework for a complete description of the fluid flow.

One might wonder why it is justified to use the first law in the form of Eq. (6.1.8), when one is considering nonequilibrium processes in Eq. (6.1.5). After all, Eq. (6.1.8) is supposed to be valid only for quasistatic changes through a succession of equilibrium states. The reason is that we are tacitly restricting the departures from equilibrium to be small. Since the entropy is a maximum in equilibrium, the entropy production will be of second order in the departures from equilibrium. Thus the error involved in using the equilibrium relations (6.1.7), (6.1.9), and (6.1.10) is also of second order, and hence can be ignored.[2]

Exercise 6.1 The heat conduction equation

$$\mathbf{q} = -\kappa \, \nabla T, \tag{6.1.11}$$

where $\mathbf{q}$ is the heat flux and κ is the thermal conductivity, can be regarded as the first term in a series expansion about equilibrium, where $\nabla T = 0$. The entropy production is

$$\rho T \frac{ds}{dt} = -\nabla \cdot \mathbf{q}. \tag{6.1.12}$$

Derive the equation

$$\frac{\partial}{\partial t} (\rho s) + \nabla \cdot \left(\rho \mathbf{v} s + \frac{\mathbf{q}}{T} \right) = \kappa \left(\frac{\nabla T}{T} \right)^2. \tag{6.1.13}$$

By integrating over the volume of the fluid, verify that the entropy generation is second order in ∇T. What is the physical interpretation of the term $\rho \mathbf{v} s + \mathbf{q}/T$?

An important simplification occurs if s is not only a constant for each fluid element, but if all the constants have the same value. In other words, $s = $ constant everywhere; the fluid is then said to be *isentropic*. In this case one need only give a one-parameter equation of state—for example,

$$u = u(\rho), \tag{6.1.14}$$

---

[2] If one is far from equilibrium, the full Boltzmann transport equation must be used to determine the fluid behavior [see, e.g., Reif (1965), Section 14.5].

which is equivalent by Eq. (6.1.9) to

$$P = P(\rho). \tag{6.1.15}$$

Note that the hydrostatic equilibrium equation follows on setting $\mathbf{v} = 0$ in Eq. (6.1.2):

$$\nabla P + \rho \nabla \Phi = 0. \tag{6.1.16}$$

For the analysis of both the equilibrium and the stability of fluid configurations, a perturbation approach can be very powerful. We will show that equilibrium states can be found from a variational principle: the energy is an extremum in equilibrium. Oscillation frequencies about the equilibrium state can be found by considering small deviations from equilibrium. Sinusoidal oscillations correspond to a stable equilibrium (energy minimum), while exponential growth signals an instability (energy maximum). In the rest of this chapter we shall develop the formalism necessary to prove and exploit these principles. Our ultimate application of these principles will be the determination of *global* equilibrium and stability criteria for white dwarfs and neutron stars.

6.2   Lagrangian and Eulerian Perturbations

We start by distinguishing between two different descriptions of a fluid perturbation. The first is a "macroscopic" point of view: one simply considers changes in fluid variables at a particular point in space. These perturbations, written $\delta \rho$, δP, δv^i, are the Eulerian changes. More precisely, if $Q(\mathbf{x}, t)$ is any attribute of the perturbed flow and $Q_0(\mathbf{x}, t)$ is the attribute for the unperturbed flow, then

$$\delta Q \equiv Q(\mathbf{x}, t) - Q_0(\mathbf{x}, t). \tag{6.2.1}$$

In the "microscopic" approach, one defines a *Lagrangian displacement* $\boldsymbol{\xi}(\mathbf{x}, t)$, which connects fluid elements in the unperturbed state to corresponding elements in the perturbed state. The Lagrangian change ΔQ in an attribute Q is defined as

$$\Delta Q \equiv Q[\mathbf{x} + \boldsymbol{\xi}(\mathbf{x}, t), t] - Q_0(\mathbf{x}, t). \tag{6.2.2}$$

In other words, the fluid element at $\mathbf{x}$ is displaced to $\mathbf{x} + \boldsymbol{\xi}$, and we compare Q at the same fluid element. Comparing Eqs. (6.2.1) and (6.2.2), we find the operator relation

$$\Delta = \delta + \boldsymbol{\xi} \cdot \nabla \tag{6.2.3}$$

when operating on scalar quantities Q. We shall adopt the same relation (6.2.3) for vector quantities. The Lie derivative is probably a better generalization,[3] but we will not need this refinement here.

The Lagrangian change in the fluid velocity, Δv, is the velocity in the perturbed flow at $x + \xi(x, t)$ *relative* to the velocity of the same element at x in the unperturbed flow; that is,

$$\Delta v = \frac{d}{dt}(x + \xi) - \frac{dx}{dt} = \frac{d\xi}{dt}. \tag{6.2.4}$$

Exercise 6.2 Prove the following commutation relations, which will be useful in later derivations in this chapter.

(i) $\delta \dfrac{\partial}{\partial t} = \dfrac{\partial}{\partial t}\delta$ $\hspace{4cm}$ (6.2.5)

(ii) $\delta \dfrac{\partial}{\partial x^i} = \dfrac{\partial}{\partial x^i}\delta$ $\hspace{3.7cm}$ (6.2.6)

(iii) $\Delta \dfrac{\partial}{\partial t} = \dfrac{\partial}{\partial t}\Delta - \dfrac{\partial \xi}{\partial t}\cdot\nabla$ $\hspace{2.5cm}$ (6.2.7)

(iv) $\Delta \dfrac{\partial}{\partial x^i} = \dfrac{\partial}{\partial x^i}\Delta - \dfrac{\partial \xi^j}{\partial x^i}\nabla_j$ $\hspace{2cm}$ (6.2.8)

(v) $\Delta \dfrac{d}{dt} = \dfrac{d}{dt}\Delta$ $\hspace{3.8cm}$ (6.2.9)

(vi) $\delta \dfrac{d}{dt} = \dfrac{d}{dt}\Delta - (\xi\cdot\nabla)\dfrac{d}{dt}$ $\hspace{2cm}$ (6.2.10)

6.3   Perturbations of Integral Quantities

Consider the integral

$$I = \int_V Q_0(x, t)\, d^3x. \tag{6.3.1}$$

The same integral defined with respect to the perturbed flow is

$$\int_{V+\Delta V} Q(x, t)\, d^3x, \tag{6.3.2}$$

where $V + \Delta V$ is the volume derived from V by subjecting its boundary to the displacement ξ. The *first variation* of I caused by the perturbation is defined to

[3] See, for example, Friedman and Schutz (1978a).

be[4]

$$\delta I \equiv \int_{V+\Delta V} Q(\mathbf{x}, t)\, d^3x - \int_V Q_0(\mathbf{x}, t)\, d^3x. \qquad (6.3.3)$$

Make the transformation $\mathbf{x}' = \mathbf{x} - \boldsymbol{\xi}(\mathbf{x}, t)$ in the first integral. This makes the two volumes the same. The Jacobian of the transformation is, to lowest order,

$$J = \frac{\partial(\mathbf{x})}{\partial(\mathbf{x}')} = \frac{\partial(\mathbf{x}' + \boldsymbol{\xi})}{\partial(\mathbf{x}')} = 1 + \nabla' \cdot \boldsymbol{\xi} \qquad (6.3.4)$$

for $\boldsymbol{\xi}$ infinitesimal. Thus

$$\delta I = \int_V Q(\mathbf{x}' + \boldsymbol{\xi}, t)\, J\, d^3x' - \int_V Q_0(\mathbf{x}, t)\, d^3x. \qquad (6.3.5)$$

On relabeling $\mathbf{x}' \to \mathbf{x}$ in the first integral, we obtain finally

$$\delta I = \int_V (\Delta Q + Q\nabla \cdot \boldsymbol{\xi})\, d^3x. \qquad (6.3.6)$$

We can use Eq. (6.3.6) to derive an expression for $\Delta\rho$: the mass in an arbitrary volume V of fluid is conserved, so

$$\delta \int_V \rho\, d^3x = 0. \qquad (6.3.7)$$

Thus

$$\Delta\rho = -\rho\nabla \cdot \boldsymbol{\xi}, \qquad (6.3.8)$$

and hence

$$\delta\rho = -\nabla \cdot (\rho\boldsymbol{\xi}). \qquad (6.3.9)$$

Exercise 6.3 Derive Eq. (6.3.8) by perturbing the differential form of the continuity equation,

$$\frac{d\rho}{dt} + \rho\nabla \cdot \mathbf{v} = 0. \qquad (6.3.10)$$

---

[4] The notation δI is not to be confused with the notation for the Eulerian perturbation of a local fluid quantity.

Exercise   6.4   Show that

$$\delta \int_V Q\rho \, d^3x = \int_V \Delta Q \, \rho \, d^3x. \tag{6.3.11}$$

For an equation of state of the form

$$P = P(\rho, s), \tag{6.3.12}$$

we shall restrict ourselves to adiabatic perturbations

$$\Delta s = 0. \tag{6.3.13}$$

This means that

$$\frac{\Delta P}{P} = \Gamma_1 \frac{\Delta \rho}{\rho}, \tag{6.3.14}$$

where

$$\Gamma_1 = \frac{\partial \ln P}{\partial \ln \rho}\bigg|_s \tag{6.3.15}$$

is the adiabatic index *governing the perturbations*. Note that in principle Γ_1 need not equal Γ, the adiabatic index governing the equilibrium pressure–density relation. This could be, for example, because on the timescale of the perturbation certain reactions necessary to achieve complete thermodynamical equilibrium cannot proceed, or it could be simply because one has *modeled* the equilibrium pressure–density relation in some way (e.g., as a polytrope of some constant Γ), but that the local value of Γ_1 is different.

The perturbation in u, the Newtonian internal energy per unit mass of the fluid, can be found from the equation of state (6.3.12) and the first law (6.1.8):

$$\Delta u = \left(\frac{\partial u}{\partial \rho}\right)_s \Delta\rho + \left(\frac{\partial u}{\partial s}\right)_\rho \Delta s$$

$$= \frac{P}{\rho^2}\Delta\rho + T\Delta s$$

$$= \frac{P}{\rho^2}\Delta\rho \tag{6.3.16}$$

for adiabatic perturbations.

The perturbation in the gravitational potential is governed by the perturbed form of Eq. (6.1.4):

$$\nabla^2 \delta\Phi = 4\pi G \, \delta\rho. \tag{6.3.17}$$

Equations (6.3.17) and (6.3.9) imply

$$\delta\Phi = -G\int \frac{\delta\rho'}{|\mathbf{x} - \mathbf{x}'|} d^3x' \tag{6.3.18}$$

$$= G\int \frac{\nabla' \cdot (\rho'\boldsymbol{\xi}')}{|\mathbf{x} - \mathbf{x}'|} d^3x' \tag{6.3.18a}$$

$$= -G\int \rho'\boldsymbol{\xi}' \cdot \nabla' \frac{1}{|\mathbf{x} - \mathbf{x}'|} d^3x'. \tag{6.3.19}$$

We have integrated by parts to get the last line, and set the surface term to zero because $\rho = 0$ on the surface. (If $\rho \neq 0$ on the surface—e.g., for an incompressible fluid—then the last line is valid only for perturbations $\boldsymbol{\xi}$ tangential to the surface.)

Exercise   6.5   Show from Eq. (6.3.18a) that for *radial* perturbations of a spherical star,

$$\nabla_i \, \delta\Phi = -4\pi G\rho\xi^i. \tag{6.3.20}$$

Hint: The expansion of $1/|\mathbf{x} - \mathbf{x}'|$ in spherical coordinates may prove useful.

6.4   Equilibrium from an Extremum of the Energy

In this section we derive a variational principle for the equation of hydrostatic equilibrium. We show that equilibrium corresponds to an extremum of the energy of the configuration.

The total energy can be written

$$E = T + U + W, \tag{6.4.1}$$

where

$$T = \int \frac{1}{2} v^2\rho \, d^3x \tag{6.4.2}$$

is the kinetic energy,

$$U = \int u\rho \, d^3x \qquad (6.4.3)$$

is the internal energy, and

$$W = \int \frac{1}{2}\Phi\rho \, d^3x \qquad (6.4.4)$$

is the gravitational potential energy.

We restrict ourselves to static equilibria ($\mathbf{v} = 0$). Thus $\delta T = 0$ to first order for all perturbations, so we can drop it from the variational principle. Now

$$\delta U = \int \Delta u \, \rho \, d^3x$$

$$= -\int P \nabla \cdot \boldsymbol{\xi} \, d^3x, \qquad (6.4.5)$$

where we have used Eqs. (6.3.11), (6.3.16), and (6.3.8). Integration by parts gives (since $P = 0$ on the surface)

$$\delta U = \int \nabla P \cdot \boldsymbol{\xi} \, d^3x. \qquad (6.4.6)$$

The variation of W gives

$$\delta W = \frac{1}{2}\int [\delta\Phi \, \rho + \rho(\boldsymbol{\xi} \cdot \nabla)\Phi] \, d^3x, \qquad (6.4.7)$$

where we have used Eqs. (6.3.11) and (6.2.3). Now Eq. (6.3.18) gives

$$\int \delta\Phi \, \rho \, d^3x = -G\iint \frac{\delta\rho'\rho}{|\mathbf{x} - \mathbf{x}'|} d^3x' \, d^3x$$

$$= \int \delta\rho \, \Phi \, d^3x = -\int \nabla \cdot (\rho\boldsymbol{\xi})\Phi \, d^3x$$

$$= \int \rho(\boldsymbol{\xi} \cdot \nabla)\Phi \, d^3x.$$

Thus

$$\delta W = \int (\rho\nabla\Phi) \cdot \boldsymbol{\xi} \, d^3x. \qquad (6.4.8)$$

Finally, combining Eqs. (6.4.6) and (6.4.8), we obtain

$$\delta E = \int (\nabla P + \rho \nabla \Phi) \cdot \xi \, d^3x. \tag{6.4.9}$$

Thus $\delta E = 0$ implies the equation of hydrostatic equilibrium, Eq. (6.1.16).

Exercise 6.6 Show that for spherical symmetry,

$$\Phi(r) = -\frac{Gm(r)}{r} + G\int_0^r 4\pi\rho r \, dr, \tag{6.4.10}$$

where

$$m(r) = \int_0^r 4\pi\rho r^2 \, dr \tag{6.4.11}$$

is the mass inside radius r and $\Phi(0) = \Phi'(0) = 0$.

Exercise 6.7 Show that for spherical symmetry

$$W = -\int \frac{Gm}{r} \, dm + \text{constant}. \tag{6.4.12}$$

Show how the first term in Eq. (6.4.12) comes from an elementary argument, and how the value of the constant depends on the boundary value of Φ.

Exercise 6.8

(a) Show that an alternative variational principle for hydrostatic equilibrium can be based on the *Eulerian* variations of the fluid quantities: extremize E subject to the constraint that the total number of baryons N remains constant.

Hint: (i) Implement the constraint by extremizing

$$I = E - \lambda N, \tag{6.4.13}$$

where the Lagrange multiplier λ is a constant.

(ii) Use Eq. (6.3.3) in the form

$$\delta I = \int_{V+\Delta V} \delta Q \, d^3x. \tag{6.4.14}$$

Note that this is justified since Q_0 vanishes in ΔV.

(iii) Note that

$$\delta n = \frac{\delta \rho}{m_B}. \tag{6.4.15}$$

(b)   What is the physical significance of λ?

(c)   Show that extrema of E and N occur at the same point along a one-parameter sequence of equilibrium models.

Note: In the Lagrangian approach, it was unnecessary to impose explicitly the constraint that N be constant. Equation (6.3.8) automatically guarantees this.

6.5   Perturbations about Equilibrium

In this section we derive the equations governing small perturbations of a static equilibrium configuration. These equations are useful for two reasons: (i) they enable us to calculate the frequencies and normal modes of oscillation of the configuration; (ii) they enable us to settle the question of stability of the equilibrium state.

The dynamics is governed by the perturbed Euler equation

$$\Delta\left(\frac{dv^i}{dt} + \frac{1}{\rho}\nabla_i P + \nabla_i \Phi\right) = 0. \tag{6.5.1}$$

Using Eqs. (6.2.8), (6.2.9), (6.1.16), and (6.2.4), we get, on multiplying through by ρ,

$$\rho\frac{d^2\xi^i}{dt^2} - \frac{\Delta\rho}{\rho}\nabla_i P + \nabla_i \Delta P + \rho\nabla_i \Delta\Phi = 0. \tag{6.5.2}$$

Since the unperturbed configuration is static, $d/dt \to \partial/\partial t \equiv \partial_t$. Making use of Eqs. (6.3.8), (6.2.3), (6.3.14), and (6.1.16), we get

$$\rho\partial_t^2\xi^i = L_{ij}\xi^j, \tag{6.5.3}$$

where

$$L_{ij}\xi^j \equiv \nabla_i\left(\Gamma_1 P \nabla_j \xi^j\right) - \left(\nabla_j \xi^j\right)\nabla_i P + \left(\nabla_i \xi^j\right)\nabla_j P$$

$$- \rho\xi^j\nabla_j\nabla_i \Phi - \rho\nabla_i \delta\Phi. \tag{6.5.4}$$

Here all quantities are expressed in terms of ξ^i and the unperturbed variables, since $\nabla_i \delta\Phi$ can be found from Eq. (6.3.19) or Eq. (6.3.20).

Equation (6.5.3) is the dynamical equation of motion for the perturbations. If we take a time dependence $\exp(i\omega t)$ for ξ^i, then the equation becomes

$$-\omega^2\rho\xi^i = L_{ij}\xi^j. \tag{6.5.5}$$

With the imposition of suitable boundary conditions on ξ^i, Eq. (6.5.5) gives the linear eigenvalue problem for the normal modes of oscillation of the star.

Exercise 6.9 Show that the eigenvalue equation governing radial oscillations of a spherical star is

$$\frac{d}{dr}\left(\Gamma_1 P \frac{1}{r^2}\frac{d}{dr}(r^2\xi)\right) - \frac{4}{r}\frac{dP}{dr}\xi + \omega^2\rho\xi = 0, \qquad (6.5.6)$$

where ξ denotes the radial component of $\boldsymbol{\xi}$.

Hint: Use Eqs. (6.1.4) and (6.1.16) to eliminate the derivatives of Φ.

The boundary conditions for Eq. (6.5.6) are

$$\xi = 0 \qquad \text{at } r = 0, \qquad (6.5.7)$$

$$\Delta P = 0 \qquad \text{at } r = R. \qquad (6.5.8)$$

Equation (6.5.7) is obvious in spherical symmetry, and Eq. (6.5.8) says that a fluid element at the unperturbed surface is displaced to the perturbed surface. Since by Eqs. (6.3.14) and (6.3.8)

$$\Delta P = -\Gamma_1 P \frac{1}{r^2}\frac{d}{dr}(r^2\xi), \qquad (6.5.9)$$

and since P vanishes at $r = R$, it is generally sufficient to demand

$$\xi \text{ finite at } r = R. \qquad (6.5.10)$$

Equation (6.5.6), subject to the boundary conditions (6.5.7) and (6.5.10), is a *Sturm–Liouville eigenvalue equation*[5] for ω^2. The results below follow from the theory of such equations:

1. The eigenvalues ω^2 are all real.
2. The eigenvalues form an infinite discrete sequence

$$\omega_0^2 < \omega_1^2 < \omega_2^2 \cdots .$$

3. The eigenfunction ξ_0 corresponding to ω_0^2 has no nodes in the interval $0 < r < R$; more generally, ξ_n has n nodes in this interval.
4. The ξ_n are orthogonal with weight function ρr^2:

$$\int_0^R \xi_n \xi_m \rho r^2 \, dr = 0, \qquad m \neq n. \qquad (6.5.11)$$

---

[5]See, for example, Morse and Feshbach (1953), Section 6.3.

5.   The ξ_n form a complete set for the expansion of any function satisfying the boundary conditions (6.5.7) and (6.5.10).

An important consequence of item (2) is the following: if the fundamental radial mode of a star is stable ($\omega_0^2 > 0$), then all radial modes are stable. Conversely, if the star is radially unstable, the fastest growing instability will be via the fundamental mode (ω_0^2 more negative than all other ω_n^2).

As a simple example that can be solved analytically, we consider perturbations of a homogeneous star—that is, $\rho = $ constant, $\Gamma_1 = $ constant. A homogeneous star in equilibrium has an adiabatic index $\Gamma = \infty$ (incompressible gas) and polytropic index $n = 0$.

Exercise   6.10   Show that a homogeneous star in equilibrium satisfies

$$P = \frac{2\pi G\rho^2}{3}(R^2 - r^2).$$ (6.5.12)

Substituting Eq. (6.5.12) in Eq. (6.5.6), we obtain after simplifying:

$$(1 - x^2)\xi'' + \xi'\left(\frac{2}{x} - 4x\right) + \left(A - \frac{2}{x^2}\right)\xi = 0.$$ (6.5.13)

Here

$$x = \frac{r}{R}, \qquad ' = \frac{d}{dx},$$

$$A = \frac{3\omega^2}{2\pi G\rho\Gamma_1} + \frac{8}{\Gamma_1} - 2.$$ (6.5.14)

We look for a series solution in the standard way:

$$\xi = \sum_{n=0}^{\infty} a_n x^{n+s}.$$ (6.5.15)

The coefficient of x^{s-2} gives the indicial equation

$$(s + 2)(s - 1) = 0.$$ (6.5.16)

Clearly we must choose $s = 1$ to satisfy the boundary condition (6.5.7). We then find $a_1 = a_3 = a_5 = \cdots = 0$, and

$$\frac{a_{n+2}}{a_n} = \frac{n^2 + 5n + 4 - A}{n^2 + 7n + 10}, \qquad n = 0, 2, 4, \dots .$$ (6.5.17)

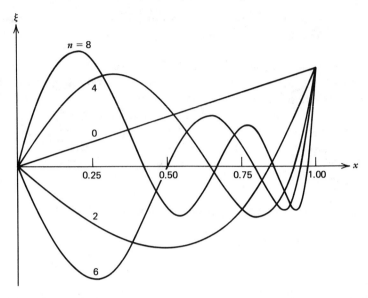

Figure 6.1 Radial oscillation amplitudes for the first five modes of the homogeneous model.

The series diverges, and hence ξ fails to satisfy boundary condition (6.5.10), unless the series terminates. This requires that

$$A = n^2 + 5n + 4, \qquad n = 0, 2, 4, \ldots . \qquad (6.5.18)$$

or, by Eq. (6.5.14),

$$\omega^2 = \frac{2\pi G\rho}{3} \left[\Gamma_1(n^2 + 5n + 6) - 8 \right], \qquad n = 0, 2, 4 \ldots . \qquad (6.5.19)$$

Note that the star is unstable if $\Gamma_1 < \frac{4}{3}$. This is a general result, as we shall discuss and exploit in Section 6.7. The timescale for stable oscillations is $\sim (G\rho)^{-1/2}$, as we could have expected on dimensional grounds. Figure 6.1 shows the oscillation amplitudes, normalized to have equal magnitude at $r = R$. Note that with $\Gamma_1 = \frac{4}{3}$ the solution for the lowest mode with $n = 0$ corresponds to $\omega^2 = 0$ and $\xi \propto r$. The general validity of this result will be shown in Section 6.7.

Computer Exercise 6.11   Find the lowest few stable pulsation frequencies of a polytrope of your favorite value of index n for $\Gamma_1 = \frac{5}{3}$.

  Hint: (i)   Nondimensionlize the eigenvalue equation.

    (ii)   Determine the Lane-Emden function by simultaneously integrating the Lane-Emden equation [Eq. (3.3.6)], or use the results conveniently provided by Service (1977).

(iii) Investigate analytically the behavior of the solution near $r = 0$ and $r = R$.

(iv) A possible numerical method is to integrate out from $r = 0$ and in from $r = R$ to some convenient r inside the star with a guessed value of ω^2, imposing the boundary conditions determined in step (iii). Since you do not know the relative magnitudes of these two solutions, start them both with unit magnitude. If you had guessed the correct value of ω^2, the Wronskian of the two solutions at the "matching radius" r would vanish. (Why?) In general, it will be nonzero. Choose another value of ω^2, integrate in from the boundaries to r, and compute the Wronskian again. Now interpolate (or extrapolate) to that value of ω^2 that zeros the Wronskian. Continue to iterate until ω^2 converges to the required number of significant digits.

6.6   Lagrangian for the Perturbations

We show in this section that the perturbation equation (6.5.3) can be derived from a variational principle. A variational principle lends itself to approximation methods for solving the perturbation equations. More importantly, in our case, the variational principle leads to stability criteria that will be useful throughout the book.

We start by showing that the operator L_{ij} of Eq. (6.5.4) is symmetric; that is,

$$\int \eta^i L_{ij} \xi^j \, d^3x = \int \xi^i L_{ij} \eta^j \, d^3x, \qquad (6.6.1)$$

where ξ^i and η^i are any Lagrangian displacements. Since

$$\eta^i \nabla_i \left(\Gamma_1 P \nabla_j \xi^j \right) = \nabla_i \left(\eta^i \Gamma_1 P \nabla_j \xi^j \right) - \Gamma_1 P \left(\nabla_i \eta^i \right) \left(\nabla_j \xi^j \right), \qquad (6.6.2)$$

$$\left(\eta^i \nabla_i \xi^j \right) \nabla_j P = \nabla_i \left(\eta^i \xi^j \nabla_j P \right) - \left(\nabla_i \eta^i \right) \xi^j \nabla_j P - \eta^i \xi^j \nabla_i \nabla_j P, \qquad (6.6.3)$$

we have from Eq. (6.5.4)

$$\int \eta^i L_{ij} \xi^j \, d^3x = - \int \left[\Gamma_1 P \left(\nabla_i \eta^i \right) \left(\nabla_j \xi^j \right) + \left(\nabla_j \xi^j \right) \eta^i \nabla_i P \right.$$

$$+ \left(\nabla_i \eta^i \right) \xi^j \nabla_j P + \eta^i \xi^j \left(\nabla_i \nabla_j P + \rho \nabla_i \nabla_j \Phi \right)$$

$$\left. + \rho \eta^i \nabla_i \, \delta\Phi \right] d^3x. \qquad (6.6.4)$$

The surface terms coming from the first terms on the right-hand sides of Eqs. (6.6.2) and (6.6.3) are zero because $P = 0$ and $\nabla_j P = 0$ on the surface. [Equation (6.1.16) implies that $\nabla_j P = 0$ if $\rho = 0$ on the surface. If $\rho \neq 0$ on the surface—e.g., for an incompressible fluid—then we require $\xi^j \nabla_j P = 0$—i.e., nonradial perturbations.] From Eq. (6.6.4), it is clear that L_{ij} is symmetric. Note that the last term on the right-hand side is, by Eq. (6.3.19),

$$\int \rho \eta^i \nabla_i \delta\Phi \, d^3x = -G \int\int \rho\rho' \eta^i \nabla_i \xi^{j'} \nabla_{j'} \frac{1}{|\mathbf{x} - \mathbf{x}'|} d^3x \, d^3x', \qquad (6.6.5)$$

which is clearly symmetric.

Now we expect a Lagrangian for the perturbation equations to be of the form

$$L = T_2 - V_2, \qquad (6.6.6)$$

where T_2 is the kinetic energy and V_2 the potential energy of the perturbations. We use the subscript 2 to denote an expression quadratic in ξ^j. The left-hand side of Eq. (6.5.3) is like a mass density times an acceleration, so the right-hand side $L_{ij}\xi^j$ is the force density. To convert the force density into a potential energy, form the dot product $\xi^i L_{ij}\xi^j$ and integrate over the volume of the star. This will give the expression (6.6.4) with η^i replaced by ξ^i. This motivates the definitions

$$T_2 \equiv \frac{1}{2} \int \rho \left(\partial_t \xi^i \right)^2 d^3x, \qquad (6.6.7)$$

$$V_2 \equiv -\frac{1}{2} \int \xi^i L_{ij} \xi^j d^3x \qquad (6.6.8)$$

$$= \frac{1}{2} \int \left[\Gamma_1 P \left(\nabla_i \xi^i \right)^2 + 2 \left(\nabla_j \xi^j \right) \xi^i \nabla_i P \right.$$

$$\left. + \xi^i \xi^j \left(\nabla_i \nabla_j P + \rho \nabla_i \nabla_j \Phi \right) + \rho \xi^i \nabla_i \delta\Phi \right] d^3x. \qquad (6.6.9)$$

We now verify that Eq. (6.6.6) is indeed the Lagrangian for Eq. (6.5.3). Since L_{ij} is symmetric, we have for the variation of the action

$$\delta S \equiv \delta \int L \, dt$$

$$= \delta \frac{1}{2} \int \left[\rho \left(\partial_t \xi^i \right)^2 + \xi^i L_{ij} \xi^j \right] d^3x \, dt$$

$$= \int \left[\rho \left(\partial_t \xi^i \right) \left(\partial_t \delta\xi^i \right) + \delta\xi^i L_{ij} \xi^j \right] d^3x \, dt$$

$$= \int \left(-\rho \partial_t^2 \xi^i + L_{ij} \xi^j \right) \delta\xi^i \, d^3x \, dt, \qquad (6.6.10)$$

where we have integrated the first term of the integrand by parts to arrive at the last line. Thus $\delta S = 0$ for arbitrary $\delta \xi^i$ implies the equation of motion, Eq. (6.5.3). Note that the symbol δ appearing in the above variational principle refers to the familiar variation of a function between neighboring paths of motion connecting the same initial and final points.[6]

Exercise 6.12 Show that the energy of an initially static configuration

$$E_2 = T_2 + V_2 \qquad (6.6.11)$$

is constant in time, using $\partial L_{ij}/\partial t = 0$. [Note: Friedman and Schutz (1978a) have shown directly that $E_2 \equiv \delta^2 E$ is the second-order variation in the energy, Eq. (6.4.1).]

Exercise 6.13 Show from Eq. (6.6.9) that for radial perturbations of a spherical star,

$$V_2 = \frac{1}{2} \int_0^R \left\{ \Gamma_1 P \left[\frac{1}{r^2} \frac{d}{dr} (r^2 \xi) \right]^2 + \frac{4}{r} \xi^2 \frac{dP}{dr} \right\} 4\pi r^2 \, dr. \qquad (6.6.12)$$

Hints: Eliminate the derivatives of Φ using Eqs. (6.1.4) and (6.1.16). Use Eq. (6.3.20) for $\nabla_i \delta \Phi$, and integrate the term involving $d^2 P/dr^2$ by parts.

Exercise 6.14 Define

$$I \equiv \frac{1}{2} \int \rho \xi^i \xi^i \, d^3 x. \qquad (6.6.13)$$

Prove the "virial theorem" for the perturbations—namely,

$$\frac{1}{2} \frac{d^2 I}{dt^2} = T_2 - V_2. \qquad (6.6.14)$$

6.7 Stability Criteria

For small deviations from equilibrium, we have

$$E = E_0 + E_2, \qquad (6.7.1)$$

where E_0 is the equilibrium energy and the first variation $\delta E \equiv E_1$ is zero because

---

[6]See, for example, Goldstein (1981), Chapter 2.

E is an extremum in equilibrium. The second variation is

$$E_2 \equiv \delta^2 E = T_2 + V_2 \equiv \delta^2 T + \delta^2 (U + W), \qquad (6.7.2)$$

where T_2 and V_2 are given by Eqs. (6.6.7) and (6.6.8).

An instability corresponds to the unbounded growth of a small initial per-turbation $(\xi^i, \partial_t \xi^i)$. [Note that both ξ^i and $\partial_t \xi^i$ form the initial data for the evolution of the perturbation, which is governed by the second-order differential equation (6.5.3).] Alternatively, we can take as the definition of an instability the unbounded growth of the kinetic energy, T_2.

Since we are considering dynamical stability (i.e., we are neglecting any effects due to dissipative forces), E_2 is a conserved quantity (Exercise 6.12). Now suppose that $V_2(t)$ is positive for all perturbations $\xi^i(\mathbf{x}, t)$. Since $E_2 = T_2 + V_2$ and T_2 is positive definite, T_2 cannot grow without bound. Thus $V_2(t)$ positive definite for all $\xi^i(\mathbf{x}, t)$ implies stability.

The converse of this statement is proved in Appendix B: if $V_2(t)$ is negative for some $\xi^i(\mathbf{x}, t)$, there exists an instability.

Since at any instant of time $t = 0$, say, $\xi^i(\mathbf{x}, t)$ gives initial data $[\xi^i(\mathbf{x}, 0),$ $\partial_t \xi^i(\mathbf{x}, 0)]$ for the subsequent evolution, the stability criterion can be formulated in terms of initial data: *A necessary and sufficient condition for stability is that the potential energy V_2 be positive definite for all initial data $\xi^i(\mathbf{x}, 0)$.*

The stability criterion can also be stated in terms of the energy. If E_2 (which is conserved and hence can be evaluated at $t = 0$, say) is positive for all initial data $(\xi^i, \partial_t \xi^i)$, then in particular it is positive for all ξ^i with $\partial_t \xi^i = 0$. Thus V_2 is positive and the system is stable. Conversely, if E_2 is negative for some $(\xi_i, \partial_t \xi^i)$, V_2 must be negative and there exists an instability. Thus $\delta^2 E \geqslant 0 \Leftrightarrow$ *stability.*

The stability criterion can also be related to a normal mode analysis. If one looks for a normal mode with time dependence

$$\xi^i(\mathbf{x}, t) = \xi^i(\mathbf{x}) e^{i\omega t}, \qquad (6.7.3)$$

instability corresponds to $\omega^2 < 0$. The equation of motion of the normal modes is Eq. (6.5.5). If we multiply that equation by $\xi^i(\mathbf{x})$ and integrate over the star, we get from Eqs. (6.6.8) and (6.6.13)

$$\omega^2 = \frac{V_2}{I}. \qquad (6.7.4)$$

Since I is positive definite, we see that V_2 positive definite $\Leftrightarrow \omega^2 \geqslant 0 \Leftrightarrow$ stability. The onset of instability occurs when $\omega^2 = 0$, which is called a *neutral mode*. This occurs when $V_2 = 0$—that is, when $E_2 = 0$.

An important result is that Eq. (6.7.4) is itself a variational principle for the normal modes:

$$I\,\delta\omega^2 = \delta V_2 - \frac{V_2}{I}\delta I$$

$$= \delta V_2 - \omega^2\,\delta I$$

$$= -\int \delta\xi^i\left(L_{ij}\xi^j + \omega^2\rho\xi^i\right) d^3x. \tag{6.7.5}$$

Thus $\delta\omega^2 = 0$ implies the normal mode equation (6.5.5).

We therefore have an alternative procedure for finding the normal modes of a star: Find displacements $\xi^i(\mathbf{x})$ that extremize Eq. (6.7.4). The eigenfrequencies are then the numerical values of Eq. (6.7.4).

An important application of these results is to radial displacements of a spherical star. Using the form of V_2 in Eq. (6.6.12), we obtain

$$\omega^2 = \frac{\int_0^R \left\{ \Gamma_1 P \frac{1}{r^2}\left[\frac{d}{dr}(r^2\xi)\right]^2 + 4r\xi^2\frac{dP}{dr} \right\} dr}{\int_0^R \rho\xi^2 r^2\,dr}. \tag{6.7.6}$$

We will now show that the star is stable if the pressure-averaged value of Γ_1, denoted by $\bar{\Gamma}_1$, is greater than $\frac{4}{3}$; unstable if $\bar{\Gamma}_1 < \frac{4}{3}$; and marginally stable ($\omega^2 = 0$) if $\bar{\Gamma}_1 = \frac{4}{3}$.

First consider a star with $\Gamma_1 = \frac{4}{3}$ everywhere. Equation (6.5.6) has a solution with no nodes (i.e., the fundamental mode) given by

$$\omega^2 = 0, \qquad \xi = \text{constant} \times r. \tag{6.7.7}$$

Exercise 6.15 Verify that Eq. (6.7.7) is a solution to Eq. (6.5.6) for $\Gamma_1 = \frac{4}{3}$.

In other words, a star with $\Gamma_1 = \frac{4}{3}$ is marginally stable to a homologous (i.e., self-similar) deformation. Now suppose Γ_1 is slightly different from $\frac{4}{3}$ at various points in the star. Then the marginally stable solution of Eq. (6.5.6) will be

$$\xi(r) = r\left[1 + \mathcal{O}\left(\Gamma_1 - \tfrac{4}{3}\right)\right]. \tag{6.7.8}$$

Thus if we use $\xi(r) = r$ as a trial function in Eq. (6.7.6), we will obtain ω^2 with error only of order $(\Gamma_1 - \frac{4}{3})^2$. [Recall that Eq. (6.7.6) is a variational principle and therefore the error in the eigenvalue ω^2 goes as the square of the error in the eigenfunction.] Thus the numerator of Eq. (6.7.6) gives

$$\omega^2 \propto 9\int_0^R \Gamma_1 Pr^2\,dr - 12\int_0^R Pr^2\,dr, \tag{6.7.9}$$

where we have integrated by parts in the second term. Therefore,

$$\omega^2 \propto 3\bar{\Gamma}_1 - 4, \tag{6.7.10}$$

where

$$\bar{\Gamma}_1 \equiv \frac{\int_0^R \Gamma_1 P r^2 \, dr}{\int_0^R P r^2 \, dr} \tag{6.7.11}$$

is the pressure-averaged adiabatic index. The transition from stability to instability as $\bar{\Gamma}_1$ decreases through $\frac{4}{3}$ is clear from Eq. (6.7.10).

It may be worthwhile to show directly that for a polytrope with $\Gamma = \frac{4}{3} = \Gamma_1$, a homologous expansion or contraction leaves it in equilibrium. Apply the homology transformation

$$P' = AP, \qquad \rho' = B\rho, \qquad r' = Cr, \tag{6.7.12}$$

to the hydrostatic equilibrium equation

$$\frac{dP}{dr} = -G\rho \frac{m}{r^2}, \tag{6.7.13}$$

with the auxiliary condition that the mass m remains constant. This gives two relations between the multiplicative constants A, B, and C:

$$AC = B, \qquad BC^3 = 1. \tag{6.7.14}$$

Eliminating C, we get

$$A = B^{4/3}, \tag{6.7.15}$$

or, by Eq. (6.7.12),

$$\frac{P'}{P} = \left(\frac{\rho'}{\rho}\right)^{4/3}, \tag{6.7.16}$$

that is, $\Gamma_1 = \frac{4}{3}$ for $\Gamma = \frac{4}{3}$.

Exercise 6.16

(a) Show that the radial oscillation frequency ω may be written in terms of the moment of inertia of the star, I, and the total gravitational potential energy, W, according to

$$\omega^2 = \frac{|W|}{I}\left(3\bar{\Gamma}_1 - 4\overline{\left(\frac{\xi}{r}\right)^2} \Big/ \overline{\left(\frac{\xi}{r}\right)^2}\right), \tag{6.7.17}$$

where

$$I \equiv \int_0^M r^2 \, dm, \qquad \overline{\left(\frac{\xi}{r}\right)^2} = \frac{1}{I} \int_0^M \xi^2 \, dm,$$

$$\widetilde{\left(\frac{\xi}{r}\right)^2} \equiv \frac{1}{|W|} \int_0^M \left(\frac{\xi}{r}\right)^2 \frac{Gm}{r} \, dm, \qquad |W| \equiv \int_0^M \frac{Gm}{r} \, dm,$$

$$\overline{\Gamma}_1 \equiv \frac{1}{9} \frac{\overline{(\nabla \cdot \xi)^2 \Gamma_1}}{\overline{(\xi/r)^2}}, \qquad \overline{(\nabla \cdot \xi)^2 \Gamma_1} \equiv \frac{\int_0^R (\nabla \cdot \xi)^2 \Gamma_1 P 4\pi r^2 \, dr}{\int_0^R P 4\pi r^2 \, dr}.$$

Hint: Recall Eq. (3.2.3).

   (b)   Show that for Γ_1 close to $\frac{4}{3}$, the two averages of $(\xi/r)^2$ are approximately equal, and hence

$$\omega^2 \simeq \frac{|W|}{I} (3\overline{\Gamma}_1 - 4). \tag{6.7.18}$$

6.8   Turning-Points and the Onset of Instability

Suppose one has constructed a one-parameter sequence of equilibrium stars with the same equation of state but different central densities. What is the significance of a maximum or minimum on a plot of E_{eq} vs. ρ_c (i.e., a "critical" point or turning point where $dE_{eq}/d\rho_c = 0$)?

   Clearly such a critical point means that there exists a neighboring equilibrium configuration such that

$$E_{eq}(\rho_c + \Delta\rho_c) = E_{eq}(\rho_c) \quad \text{(to first order)}. \tag{6.8.1}$$

The neighboring equilibrium configuration is related to the original star by some Lagrangian displacement ξ induced by $\Delta\rho_c$.[7] In general,

$$E[\xi] = E_0 + E_1[\xi] + E_2[\xi], \tag{6.8.2}$$

where $E_n[\xi]$ is the contribution of order ξ^n to the expansion of the energy E about $E_0 = E_{eq}(\rho_c)$. Since E_0 corresponds to an equilibrium configuration, $E_1[\xi] = 0$ [e.g., recall Eq. (6.4.9)]. For $E[\xi]$ to correspond to an equilibrium configuration also, E must be stationary to *second order*: $E_2[\xi] = 0$ for this ξ. But from the

---

[7]Note that since the displacement ξ takes one equilibrium configuration into another, we have in this case $\Gamma_1 = \Gamma$. The conclusions in this section depend on this equality. For a discussion of the significance of turning points when $\Gamma_1 \neq \Gamma$, see Thorne (1967).

previous section we know that this is exactly the condition for a zero-frequency mode, $\omega^2 = 0$, signaling the change in stability of the mode.

The easiest way to use this criterion in practice is to know that stars at one end of the sequence (usually the low density end) are stable. Then the critical point gives the location of the onset of instability. We shall show later in this section how to generalize this test if there are several critical points.

The critical point method can also be used on a plot of equilibrium mass vs. ρ_c. One way to see this is to find the dependence of M on ρ_c from the variational principle $\delta E = 0$, where

$$E = \int u \, dm - \int \frac{Gm}{r} \, dm. \tag{6.8.3}$$

Lumping numerical factors into constants, $\alpha_1, \alpha_2, \ldots,$ we have dimensionally that

$$E = \langle u \rangle M - \alpha_1 \frac{GM^2}{R}. \tag{6.8.4}$$

Taking $\Gamma_1 = \Gamma$ with an adiabatic equation of state

$$P = K\rho^\Gamma, \tag{6.8.5}$$

we have

$$u = \frac{K\rho^{\Gamma-1}}{\Gamma - 1}, \tag{6.8.6}$$

and so

$$\langle u \rangle = \alpha_2 K \rho_c^{\Gamma-1}. \tag{6.8.7}$$

Also

$$R = \alpha_3 \left(\frac{M}{\rho_c} \right)^{1/3}. \tag{6.8.8}$$

Thus

$$E = \alpha_2 KM\rho_c^{\Gamma-1} - \alpha_4 GM^{5/3}\rho_c^{1/3}. \tag{6.8.9}$$

Equilibrium is determined by setting $dE/d\rho_c = 0$, keeping M constant. This gives

$$M \propto \rho_c^{(\Gamma-4/3)(3/2)}. \tag{6.8.10}$$

Thus

$$\frac{dM}{d\rho_c} \propto \Gamma - \frac{4}{3}. \tag{6.8.11}$$

If there is only one critical point on the sequence and the low-density end is stable, then by Eqs. (6.7.10) and (6.8.11) equilibrium configurations with $dM/d\rho_c > 0$ are stable, while those with $dM/d\rho_c < 0$ are unstable.[8]

It is not essential to parametrize the sequence of equilibrium models with ρ_c; another convenient choice is to plot M vs. R. This choice in fact allows us to treat the case of multiple critical points. As discussed earlier, we are primarily interested in the stability of the fundamental radial mode, which has no nodes in the interior of the star. Now the eigenfunction at a critical point is simply the Lagrangian displacement ξ that carries an equilibrium configuration on the low-density side of the critical point into an equilibrium configuration on the high-density side. In such a motion, ρ_c increases and so ξ is negative near the center of the star. Since there are no nodes in ξ, it is also negative near the surface. Thus R decreases with increasing ρ_c at a critical point where the fundamental mode changes stability. (More generally, $dR/d\rho_c < 0$ for the change in stability of a mode with an even number of nodes, $dR/d\rho_c > 0$ for an odd mode.)

A schematic diagram of M vs. ρ_c with multiple critical points is shown in Figure 6.2, and M vs. R in Figure 6.3. At low density (large radius), all modes are stable. The first critical point that is reached is A, the maximum white dwarf mass. Since R is decreasing with increasing ρ_c, an even mode is changing stability. Since $\omega_0^2 < \omega_1^2 < \omega_2^2 < \cdots$, and since $\omega_0^2 > 0$ for low density, the only possibility is that ω_0^2 becomes negative. At B, an even mode is changing stability ($dR/d\rho_c < 0$). It cannot be that ω_2^2 is becoming negative, because ω_1^2 is still positive. Thus ω_0^2 must be turning positive again: B corresponds to the minimum neutron star mass. The critical point C is analogous to A and ω_0^2 becomes negative again; it corresponds to the maximum neutron star mass. At D, an *odd* mode changes stability. The only possibility is that ω_1^2 becomes negative. At E, an even mode changes stability. It is impossible for ω_0^2 to become positive while ω_1^2 remains negative, so it must be that ω_2^2 becomes negative.

Exercise   6.17   Using reasoning analogous to that above, convince yourself that the following is true for any shape of an M vs. R diagram where all modes are stable at low density: a counterclockwise bend at a critical point signals the onset of an instability with increasing ρ_c; a clockwise bend signals a change from an unstable to a stable mode.

[8]A more rigorous proof of this result is given by Tassoul (1978), p. 149.

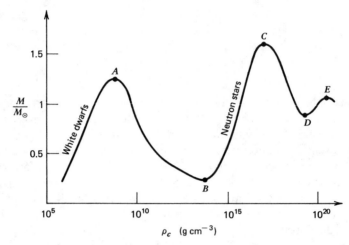

Figure 6.2 Schematic diagram showing the turning points in the mass versus central density diagram for equilibrium configurations of cold matter.

Note that the criterion $dM/d\rho_c > 0$ (< 0) for identifying stability (instability) has limited applicability: thus, the range D–E is unstable, even though $dM/d\rho_c > 0$. For typical cold equations of state, however, the lowest density at which $dM/d\rho_c = 0$ and $d^2M/d\rho_c^2 < 0$ corresponds to the maximum mass and density of a stable white dwarf (e.g., point A in Fig. 6.2). The next point at which $dM/d\rho_c = 0$ and $d^2M/d\rho_c^2 < 0$ gives the maximum mass and density of a stable

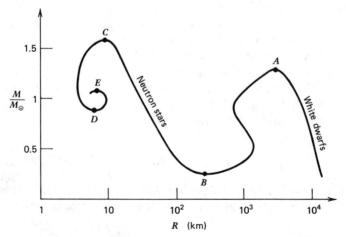

Figure 6.3 Schematic diagram showing the turning points in the mass versus radius diagram for equilibrium configurations of cold matter.

Box 6.1

Newtonian Equilibrium and Stability of Nonrotating Stars

1. Variational principle for hydrostatic equilibrium: $\delta E = 0$ [cf. Eq. (6.4.9)]. The variation is performed keeping the rest mass and entropy constant.

2. Small perturbations about equilibrium are governed by Eq. (6.5.3) or Eq. (6.5.5) (for normal modes). *Radial* modes of a spherical star are governed by the Sturm–Liouville eigenvalue equation (6.5.6).

3. The perturbations are derivable from a variational principle with Lagrangian $L = T_2 - V_2$ [Eqs. (6.6.7) and (6.6.8)].

4. Equivalent stability criteria are:
   (a) $V_2 \geqslant 0$ for all perturbations;
   (b) $\delta^2 E \equiv E_2 \geqslant 0$ for all perturbations;
   (c) $\omega^2 \geqslant 0$ for all modes;
   (d) $\bar{\Gamma}_1 \geqslant \frac{4}{3}$ [for *radial* stability only; Eq. (6.7.11)].

5. Some normal mode changes stability if $dE_{eq}/d\rho_c = 0$ or $dM/d\rho_c = 0$ (Section 6.8).

---

neutron star (e.g., point C in Fig. 6.3). We shall employ these results frequently in subsequent sections.

6.9   General Relativistic Stability Analysis

Nearly all of the results of the previous sections hold true in general relativity, provided we replace E by Mc^2, the total mass-energy of the star.[9] In particular:

1. M is an extremum in equilibrium for all configurations with the same total number of baryons N (cf. the Note at the end of Exercise 6.8 about the constancy of N). Since one can show that extrema of N and M occur at the

---

[9]See Harrison, Thorne, Wakano, and Wheeler (1965), "HTWW."

same point in a one-parameter equilibrium sequence (cf. Exercise 6.8c), one can equivalently extremize

$$E \equiv Mc^2 - m_B N. \tag{6.9.1}$$

2. Small radial deviations from equilibrium are governed by a Sturm–Liouville eigenvalue problem analogous to Eq. (6.5.6).

3. The second variation of M must be positive for stability. Equivalently, $\omega_0^2 > 0$.

4. There is a change in stability at a critical point of the M_{eq} vs. R or M_{eq} vs. ρ_c curve, which can be analyzed as at the end of Section 6.8.

5. Since the equation for ω^2 is not the same as in the Newtonian case, the criterion $\bar{\Gamma}_1 > \frac{4}{3}$ is no longer valid as a test for stability. However, if the corrections due to relativity are small (i.e., $GM/Rc^2 \ll 1$), then the revised criterion is

$$\bar{\Gamma}_1 - \frac{4}{3} > \kappa \frac{GM}{Rc^2}, \tag{6.9.2}$$

where κ is a number of order unity that depends on the structure of the star. General relativity tends to *destabilize* configurations; gravity is stronger and thus collapse is easier.

We shall spend the rest of this chapter deriving the value of κ in the important case that the structure of the star is close to being an $n = 3$ polytrope; that is, $\Gamma = \Gamma_1 \simeq \frac{4}{3}$. We also will find the corresponding stellar density at the critical point.[10] The idea is to derive an equation of the form

$$E = E_{\text{Newt.}} + \Delta E_{\text{GTR}}, \tag{6.9.3}$$

where $E_{\text{Newt.}}$ is the Newtonian energy of the star and ΔE_{GTR} is the correction due to general relativity. Minimizing the energy will give the equilibrium configuration, and the second derivative of E will give stability information.

The result for ΔE_{GTR} is given in Eq. (6.9.32); the reader who is willing to accept this on faith may skip directly to this equation. We set $c = G = 1$ in the remainder of this section.

For a spherical configuration of matter, instantaneously at rest, the total mass energy is [cf. Eq. (5.7.8)]

$$M = \int_0^R \rho 4\pi r^2 \, dr. \tag{6.9.4}$$

---

[10] Our treatment is patterned after Zel'dovich and Novikov (1971).

Here

$$\rho = \rho_0(1 + u).$$ (6.9.5)

The total number of baryons in the star is

$$N = \int_0^R n \, d\mathcal{V},$$ (6.9.6)

where

$$d\mathcal{V} = \left(1 - \frac{2m}{r}\right)^{-1/2} 4\pi r^2 \, dr$$ (6.9.7)

is the proper volume element in the Schwarzschild geometry [cf. Eq. (5.7.9)]. The energy of the star, excluding rest-mass energy, is by Eq. (6.9.1),

$$E = \int_0^R \left[\rho\left(1 - \frac{2m}{r}\right)^{1/2} - \rho_0\right] d\mathcal{V},$$ (6.9.8)

where we have used $\rho_0 = m_B n$. Substituting Eq. (6.9.5), and treating u and m/r as small quantities, we find to second order

$$E = \int_0^R \rho_0\left[u - \frac{m}{r} - u\frac{m}{r} - \frac{1}{2}\left(\frac{m}{r}\right)^2\right] d\mathcal{V}.$$ (6.9.9)

Note that $\rho_0 \, d\mathcal{V}$ is an invariant and is not expanded. The Newtonian energy is

$$E_{\text{Newt.}} = \int_0^R \rho_0 u \, d\mathcal{V} - \int_0^M \frac{M m'}{r'} \, dm',$$ (6.9.10)

where

$$dm' = \rho_0 \, d\mathcal{V},$$ (6.9.11)

$$r' = \left(\frac{3\mathcal{V}}{4\pi}\right)^{1/3}.$$ (6.9.12)

Note that the functions $m'(\mathcal{V})$ and $r'(\mathcal{V})$ differ from their relativistic counterparts because of Eqs. (6.9.5) and (6.9.7). We are trying to compute the energy of a star, first according to general relativity, then according to Newtonian theory, and then calling the difference ΔE_{GTR}. How can we be sure, because of the ambiguities in the meaning of coordinates in general relativity, of what exactly is the same in the two cases? Equivalently, if given two identical stars and asked to compute one's energy as E and the other's as E_{Newt}, how do we know that the stars are identical?

The answer is that identical stars have the same number of baryons in a given proper volume (coordinate independent statement!). In other words, $\rho_0(\mathcal{V})$ is the same function in relativity and Newtonian theory.

Subtracting Eq. (6.9.10) from Eq. (6.9.9) gives

$$\Delta E_{\text{GTR}} = \int_0^R \rho_0 \, d\mathcal{V} \left[-u \frac{m}{r} - \frac{1}{2} \left(\frac{m}{r} \right)^2 + \frac{m'}{r'} - \frac{m}{r} \right]. \tag{6.9.13}$$

Exercise 6.18   Show from Eq. (6.9.7) that, to first order,

$$\mathcal{V} = \frac{4\pi r^3}{3} \left(1 + \frac{3}{r^3} \int_0^r mr \, dr \right). \tag{6.9.14}$$

Equations (6.9.12) and (6.9.14) give

$$r' - r = \frac{1}{r^2} \int_0^r mr \, dr. \tag{6.9.15}$$

Working again to first order, we find

$$m'(\mathcal{V}) - m(\mathcal{V}) = \int_0^{\mathcal{V}} d\mathcal{V} \left[\rho_0 - \rho \left(1 - \frac{2m}{r} \right)^{1/2} \right]$$

$$= -\int_0^{\mathcal{V}} \rho_0 \, d\mathcal{V} \left(u - \frac{m}{r} \right). \tag{6.9.16}$$

Now in Eq. (6.9.13) we can write

$$\frac{m'}{r'} - \frac{m}{r} = \frac{m' - m}{r'} - \frac{m(r' - r)}{rr'} \tag{6.9.17}$$

and substitute the results (6.9.15) and (6.9.16). Since we have kept all the second-order terms consistently, if we now evaluate the integrals with the *Newtonian* relations for ρ_0, r, $\mathcal{V}$, and so on, we will be making an error only at third order. Thus

$$\Delta E_{\text{GTR}} = I_1 + I_2 + I_3 + I_4 + I_5, \tag{6.9.18}$$

where

$$I_1 = -\int_0^M u \frac{m}{r} \, dm, \tag{6.9.19}$$

$$I_2 = -\frac{1}{2} \int_0^M \left(\frac{m}{r}\right)^2 dm, \tag{6.9.20}$$

$$I_3 = -\int_0^M \frac{dm}{r} \int_0^m u \, dm, \tag{6.9.21}$$

$$I_4 = \int_0^M \frac{dm}{r} \int_0^m \frac{m}{r} \, dm, \tag{6.9.22}$$

$$I_5 = -\int_0^M \frac{m \, dm}{r^4} \int_0^r mr \, dr. \tag{6.9.23}$$

These expressions can be simplified if we assume the mass distribution is that of a polytrope of index n; that is,

$$u = n \frac{P}{\rho_0} \tag{6.9.24}$$

and

$$\frac{1}{\rho_0} \frac{dP}{dr} = -\frac{m}{r^2}. \tag{6.9.25}$$

Exercise 6.19

(a) Show that when Eqs. (6.9.24) and (6.9.25) hold,

$$I_5 = \frac{1}{n} I_1. \tag{6.9.26}$$

Hint: $-m \, dm/r^4 = 4\pi \, dP$; integrate by parts.

(b) Similarly show that

$$I_4 = 2I_2 - \frac{2}{n} I_1 - \frac{3}{n} I_3. \tag{6.9.27}$$

Hint: $m \, dm/r = -4\pi r^3 \, dP$; integrate by parts to get two terms; integrate the first term by parts again and eliminate P using Eq. (6.9.24).

(c) Also show that

$$I_3 = I_1 - \frac{2n}{n+1}(I_2 + I_4). \tag{6.9.28}$$

Hint: Integrate $\int u \, dm$ by parts. Then show that $du = n \, dP/[\rho_0(n+1)] = nm \, d(1/r)/(n+1)$ and integrate by parts once more.

Combining the results of Eqs. (6.9.26)–(6.9.28), we get

$$\Delta E_{\text{GTR}} = \frac{5 + 2n - n^2}{n(5-n)} 2I_1 + \frac{n-1}{5-n} 3I_2. \tag{6.9.29}$$

Now reduce the integrals I_1 and I_2 to dimensionless form by making the appropriate substitutions for polytropes as in Section 3.3. The result is

$$\Delta E_{\text{GTR}} = -kM^{7/3}\rho_c^{2/3}, \tag{6.9.30}$$

where

$$k = \frac{(4\pi)^{2/3}}{(5-n)\left[\xi_1^2|\theta'(\xi_1)|\right]^{7/3}} \left[-\frac{5+2n-n^2}{(n+1)} 2\int_0^{\xi_1} \xi^3 \theta' \theta^{n+1} \, d\xi \right.$$

$$\left. + \frac{3}{2}(n-1)\int_0^{\xi_1} \xi^4 \theta'^2 \theta^n \, d\xi \right]. \tag{6.9.31}$$

Exercise 6.20 Verify Eqs. (6.9.30) and (6.9.31).

The integrals in Eq. (6.9.31) can be evaluated numerically, the polytropic functions being found either by simultaneously integrating the Lane–Emden equation or from the results of Service (1977). For the case $n = 3$, we obtain

$$\Delta E_{\text{GTR}} = -0.918294 \, M^{7/3}\rho_c^{2/3}. \tag{6.9.32}$$

6.10 White Dwarf Stability in General Relativity

To analyze the stability of a white dwarf taking into account the effects of general relativity, we write the total energy in the form

$$E = E_{\text{int}} + E_{\text{grav}} + \Delta E_{\text{int}} + \Delta E_{\text{GTR}}. \tag{6.10.1}$$

In a first approximation, only the first two terms are present. They can be evaluated for a polytropic density distribution:

$$E_{\text{int}} = \int u\, dm = \int \frac{nP}{\rho}\, dm$$

$$= K\rho_c^{1/n} M \frac{n}{|\xi_1^2 \theta'|} \int_0^{\xi_1} \xi^2 \theta^{n+1}\, d\xi. \tag{6.10.2}$$

$$E_{\text{grav}} = -G \int \frac{m}{r}\, dm$$

$$= (4\pi\rho_c)^{1/3} \frac{GM^{5/3}}{|\xi_1^2 \theta'|^{5/3}} \int_0^{\xi_1} \xi^3 \theta' \theta^n\, d\xi. \tag{6.10.3}$$

Now

$$\int_0^{\xi_1} \xi^3 \theta' \theta^n\, d\xi = \frac{1}{n+1} \int_0^{\xi_1} \xi^3 \frac{d}{d\xi} \theta^{n+1}\, d\xi$$

$$= -\frac{3}{n+1} \int_0^{\xi_1} \xi^2 \theta^{n+1}\, d\xi. \tag{6.10.4}$$

The last integral in Eq. (6.10.4) can be evaluated by suitable integrations by parts, but we have already essentially done so in deriving Eq. (3.2.11), where we found for a polytrope

$$E_{\text{grav}} = -\frac{3}{5-n} \frac{GM^2}{R}. \tag{6.10.5}$$

From the result of Exercise 3.4, we have

$$\frac{M}{R^3} = \frac{4\pi\rho_c |\theta'|}{\xi_1}, \tag{6.10.6}$$

and so

$$E_{\text{grav}} = -\frac{3}{5-n} GM^{5/3} \rho_c^{1/3} \left| \frac{4\pi\theta'}{\xi_1} \right|^{1/3}. \tag{6.10.7}$$

Comparing with Eqs. (6.10.3) and (6.10.4), we find

$$\int_0^{\xi_1} \xi^2 \theta^{n+1}\, d\xi = \frac{n+1}{5-n} \xi_1^3 |\theta'|^2. \tag{6.10.8}$$

Thus

$$E_{int} = k_1 K \rho_c^{1/n} M,$$

(6.10.9)

$$E_{grav} = -k_2 G \rho_c^{1/3} M^{5/3},$$

(6.10.10)

where

$$k_1 = \frac{n(n+1)}{5-n} \frac{|\xi_1^2 \theta'|}{\xi_1} = 1.75579,$$

(6.10.11)

$$k_2 = \frac{3}{5-n} \frac{|4\pi\xi_1^2\theta'|^{1/3}}{\xi_1} = 0.639001.$$

(6.10.12)

The numerical values of k_1, k_2 are for $n = 3$.

The term ΔE_{int} represents the departure of the equation of state from that of an $n = 3$ polytrope because the electrons are not completely relativistic. The internal energy per unit mass is

$$u = \frac{\varepsilon_e - m_e c^2 n_e}{\rho},$$

(6.10.13)

where

$$\rho = \rho_0 = \mu_e m_u n_e.$$

(6.10.14)

Using Eq. (2.3.20), we find

$$u = \frac{3}{4} \frac{m_e c^2}{\mu_e m_u} \left(x - \frac{4}{3} + \frac{1}{x} + \cdots \right),$$

(6.10.15)

where $x \gg 1$ is the relativity parameter.

The term in Eq. (6.10.15) proportional to x is simply $3P/\rho$, and was used above to calculate E_{int}. The next term is a constant and can be dropped from the variational principle, and so ΔE_{int} is given by

$$\Delta E_{int} = \frac{3}{4} \frac{m_e c^2}{\mu_e m_u} \int \frac{1}{x} dm.$$

(6.10.16)

By Eqs. (6.10.14) and (2.3.4), we have

$$x = \left(\frac{3\pi^2 \rho \lambda_e^3}{\mu_e m_u} \right)^{1/3}.$$

(6.10.17)

We can evaluate ΔE_{int} by integrating over an $n = 3$ polytropic density distribution; the error will be of higher order. Thus

$$\Delta E_{\text{int}} = k_3 \frac{m_e^2 c^3}{\hbar(\mu_e m_u)^{2/3}} M\rho_c^{-1/3}, \tag{6.10.18}$$

where[11]

$$k_3 = \frac{3}{4} \frac{1}{(3\pi^2)^{1/3}} \frac{1}{|\xi_1^2\theta'|} \int_0^{\xi_1} \xi^2\theta^2 \, d\xi \tag{6.10.19}$$

$$= 0.519723. \tag{6.10.20}$$

The correction for general relativity is given by Eq. (6.9.32):

$$\Delta E_{\text{GTR}} = -k_4 \frac{G^2}{c^2} M^{7/3}\rho_c^{2/3}, \tag{6.10.21}$$

$$k_4 = 0.918294. \tag{6.10.22}$$

Thus the total energy, Eq. (6.10.1), can be written

$$E = (AM - BM^{5/3})\rho_c^{1/3} + CM\rho_c^{-1/3} - DM^{7/3}\rho_c^{2/3}, \tag{6.10.23}$$

where

$$A = k_1 K, \quad B = k_2 G, \quad C = k_3 \frac{m_e^2 c^3}{\hbar(\mu_e m_u)^{2/3}}, \quad D = k_4 \frac{G^2}{c^2}. \tag{6.10.24}$$

Equilibrium is achieved when $\partial E/\partial\rho_c = 0$. This gives

$$(AM - BM^{5/3})\tfrac{1}{3}\rho_c^{-2/3} - \tfrac{1}{3}CM\rho_c^{-4/3} - \tfrac{2}{3}DM^{7/3}\rho_c^{-1/3} = 0. \tag{6.10.25}$$

To leading order, we can ignore the terms proportional to C and D. We then recover the Chandrasekhar expression for the mass,

$$M = \left(\frac{A}{B}\right)^{3/2} = 1.457 \left(\frac{\mu_e}{2}\right)^{-2} M_\odot, \tag{6.10.26}$$

[11]The integral in Eq. (6.10.19) has the value 4.32670 for $n = 3$.

where we have used Eq. (2.3.23) for K. Keeping the terms C and D gives small corrections to the value of M, which depend on ρ_c.

The onset of instability occurs when $\partial^2 E/\partial\rho_c^2 = 0$. This gives

$$-\tfrac{1}{3}\tfrac{2}{3}(AM - BM^{5/3})\rho_c^{-5/3} + \tfrac{1}{3}\tfrac{4}{3}CM\rho_c^{-7/3} + \tfrac{2}{3}\tfrac{1}{3}DM^{7/3}\rho_c^{-4/3} = 0. \quad (6.10.27)$$

Solve Eq. (6.10.25) for $AM - BM^{5/3}$ and substitute in Eq. (6.10.27). Since all quantities are then of the same order of smallness, one can replace M by $(A/B)^{3/2}$. Thus find

$$\rho_c = \frac{CB^2}{DA^2}$$

$$= \frac{16k_3k_2^2}{(3\pi^2)^{2/3}k_4k_1^2}\frac{m_u^2\mu_e^2}{\lambda_e^3 m_e}$$

$$= 2.646 \times 10^{10}\left(\frac{\mu_e}{2}\right)^2 \text{ g cm}^{-3}. \quad (6.10.28)$$

This is the *critical density for the onset of instability due to general relativity in a white dwarf*. Note that for ^{56}Fe ($\mu_e = 2.154$), $\rho_c = 3.07 \times 10^{10}$ g cm^{-3}. This is higher than the inverse β-decay threshold of 1.14×10^9 g cm^{-3} (Table 3.1), and so general relativity is irrelevant for iron white dwarfs. For ^{4}He or ^{12}C, however, ρ_c due to general relativity is 2.65×10^{10} g cm^{-3}, lower than the neutronization thresholds of 1.37×10^{11} g cm^{-3} and 3.90×10^{10} g cm^{-3}, respectively. In these cases, it is general relativity that limits the central density.[12]

Exercise 6.21 The energy of a star with Γ close to $\tfrac{4}{3}$ can also be written as [cf. Eq. (6.8.9)]

$$E = \alpha M\rho_c^{\Gamma-1} - k_2GM^{5/3}\rho_c^{1/3} - k_4\frac{G^2}{c^2}M^{7/3}\rho_c^{2/3}. \quad (6.10.29)$$

Here α is some constant. Show that the critical value of Γ for the onset of instability is modified because of the general relativity term to

$$\Gamma - \frac{4}{3} = 1.125\left(\frac{2GM}{Rc^2}\right). \quad (6.10.30)$$

Hint: Eliminate α by using $\partial E/\partial\rho_c = 0$.

[12] Pycnonuclear reactions limit carbon white dwarfs to $\rho_c \lesssim 1 \times 10^{10}$ g cm^{-3}, but the exact value of this limit is rather uncertain. See Section 3.7.

Exercise 6.22 What is the maximum gravitational redshift predicted for a spherical white dwarf made of ^{4}He, ^{12}C, or ^{56}Fe? Compare your results with Shapiro and Teukolsky (1976).

Exercise 6.23 Compute $\Delta M/M_{Ch}$, the fractional difference between the mass of the white dwarf and M_{Ch}, at the critical density (6.10.28).

The results of Eqs. (6.7.18) and (6.10.30) can be combined by writing the approximate formula

$$\omega^2 \simeq \frac{|W|}{I}\left[(3\bar{\Gamma}_1 - 4) - \beta\frac{|W|}{Mc^2}\right], \qquad (6.10.31)$$

where β is a numerical factor. For low-mass white dwarfs, the first term dominates. The fundamental oscillation frequency increases with increasing mass ($\omega^2 \sim G\rho$). The period decreases from about 20 s at $0.4M_\odot$ to 6 s at $1M_\odot$.

Exercise 6.24 Show that for an extremely relativistic, degenerate electron gas,

$$\Gamma_1 - \frac{4}{3} \simeq \frac{2}{3x^2}. \qquad (6.10.32)$$

From Eq. (6.10.32), we see that the first term in brackets in Eq. (6.10.31) goes as $\rho^{-2/3}$—that is, as R^2, as $M \to M_{Ch}$. But $|W|/I \sim R^{-3}$, so ω^2 continues to increase as $M \to M_{Ch}$; Newtonian white dwarfs are stable. However, the general relativity term in Eq. (6.10.31) goes as $1/R$ and causes ω^2 to go through a maximum and then to pass through zero—the star becomes unstable. The corresponding minimum period is about 2 s.[13] This number is important in ruling out pulsating white dwarfs as models for pulsars—pulsar periods as short as 1.56 ms are known.

---

[13] See, for example, Cohen, Lapidus, and Cameron (1969).

7

Rotation and Magnetic Fields

7.1 The Equations of Magnetohydrodynamics

We summarize here the equations of magnetohydrodynamics.[1] When a fluid is acted on by electromagnetic forces, Euler's equation (6.1.2) becomes

$$\rho \frac{d\mathbf{v}}{dt} = -\nabla P - \rho \nabla \Phi + \frac{1}{c} \mathbf{J} \times \mathbf{B}. \tag{7.1.1}$$

Here $\mathbf{J}$ is the current density and $\mathbf{B}$ the magnetic field strength. We are neglecting the term $\rho_e \mathbf{E}$, which would be present if the fluid had a charge density ρ_e, since astrophysical fluids or plasmas seldom have a net charge. If the fluid velocity $\mathbf{v}$ is much smaller than c (as is often the case in astrophysical applications), then in Maxwell's equation

$$\nabla \times \mathbf{B} - \frac{1}{c} \frac{\partial \mathbf{E}}{\partial t} = \frac{4\pi}{c} \mathbf{J} \tag{7.1.2}$$

we can neglect the displacement current term. Thus

$$\frac{1}{c} \mathbf{J} \times \mathbf{B} = \frac{1}{4\pi} (\nabla \times \mathbf{B}) \times \mathbf{B}. \tag{7.1.3}$$

With the vector identity

$$\tfrac{1}{2} \nabla (\mathbf{B} \cdot \mathbf{B}) = (\mathbf{B} \cdot \nabla)\mathbf{B} - (\nabla \times \mathbf{B}) \times \mathbf{B}, \tag{7.1.4}$$

[1] See, for example, Jackson (1975).

162

we can rewrite Eq. (7.1.1) as

$$\rho \frac{d\mathbf{v}}{dt} = -\nabla P - \rho \nabla \Phi - \frac{1}{8\pi} \nabla B^2 + \frac{1}{4\pi} (\mathbf{B} \cdot \nabla)\mathbf{B}. \qquad (7.1.5)$$

The **E** field is usually related to **J** and **B** by Ohm's law

$$\mathbf{J} = \sigma \left(\mathbf{E} + \frac{\mathbf{v}}{c} \times \mathbf{B} \right). \qquad (7.1.6)$$

Here σ is the conductivity, taken to be constant, and Eq. (7.1.6) is the generalization to first order in v/c of the rest-frame relation $\mathbf{J} = \sigma \mathbf{E}$.

The time-evolution of the **B** field is given by Faraday's law

$$\frac{1}{c} \frac{\partial \mathbf{B}}{\partial t} = -\nabla \times \mathbf{E}, \qquad (7.1.7)$$

which becomes on using Eqs. (7.1.6), (7.1.2), and $\nabla \cdot \mathbf{B} = 0$,

$$\frac{\partial \mathbf{B}}{\partial t} = \nabla \times (\mathbf{v} \times \mathbf{B}) + \frac{c^2}{4\pi\sigma} \nabla^2 \mathbf{B}. \qquad (7.1.8)$$

Often one can take the conductivity to be essentially infinite because ohmic dissipation timescales are long compared with other timescales of interest. Such fluids are called "perfect conductors," for which Eq. (7.1.6) gives

$$\mathbf{E} + \frac{\mathbf{v}}{c} \times \mathbf{B} = 0, \qquad (7.1.9)$$

and Eq. (7.1.8) becomes

$$\frac{\partial \mathbf{B}}{\partial t} = \nabla \times (\mathbf{v} \times \mathbf{B}). \qquad (7.1.10)$$

Equation (7.1.10) has the interpretation that the magnetic flux through any loop moving with the local fluid velocity of a perfectly conducting fluid is constant in time: the lines of force are "frozen in" the fluid.[2]

The scalar virial theorem will prove useful for discussing the effects of magnetic fields on white dwarfs. Dot the position vector **x** into Eq. (7.1.5) and

---

[2] More precisely, Eq. (7.1.10) implies that the magnetic flux $\phi_M \equiv \int_S \mathbf{B} \cdot d\mathbf{S}$ through any closed surface S moving with the fluid is constant. Furthermore, a fluid element threaded initially by a magnetic line of force continues to be threaded by a line of force.

integrate over the volume V of the star. Since $\mathbf{v} = d\mathbf{x}/dt$, we have

$$\mathbf{x} \cdot \frac{d\mathbf{v}}{dt} = \frac{d}{dt}(\mathbf{x} \cdot \mathbf{v}) - v^2$$

$$= \frac{1}{2}\frac{d^2}{dt^2}(x^2) - v^2. \qquad (7.1.11)$$

Thus the left-hand side of Eq. (7.1.5) becomes

$$\frac{1}{2}\frac{d^2 I}{dt^2} - 2T, \qquad (7.1.12)$$

where

$$I = \int \rho x^2 \, d^3x$$

is the generalized moment of inertia, and

$$T = \frac{1}{2}\int \rho v^2 \, d^3x \qquad (7.1.13)$$

is the kinetic energy. We have used the fact that

$$\frac{d}{dt}\int_V Q\rho \, d^3x = \int_V \rho \frac{dQ}{dt} d^3x \qquad (7.1.14)$$

for any attribute Q of the fluid.

The pressure term becomes

$$-\int \mathbf{x} \cdot \nabla P \, d^3x = -\int \nabla \cdot (\mathbf{x}P) \, d^3x + \int P\nabla \cdot \mathbf{x} \, d^3x$$

$$= 0 + 3\Pi, \qquad (7.1.15)$$

where

$$\Pi = \int P \, d^3x, \qquad (7.1.16)$$

and where the vanishing of P on the boundary of V has been used.

The gravitational term can be written as

$$-\int \rho \mathbf{x} \cdot \nabla \Phi \, d^3x = G \iint d^3x \, d^3x' \, \rho(\mathbf{x}) \mathbf{x} \cdot \nabla \frac{\rho(\mathbf{x}')}{|\mathbf{x} - \mathbf{x}'|}$$

$$= -G \iint d^3x \, d^3x' \, \rho(\mathbf{x}) \rho(\mathbf{x}') \frac{\mathbf{x} \cdot (\mathbf{x} - \mathbf{x}')}{|\mathbf{x} - \mathbf{x}'|^3}$$

$$= -\frac{1}{2} G \iint d^3x \, d^3x' \, \rho(\mathbf{x}) \rho(\mathbf{x}') \frac{(\mathbf{x} - \mathbf{x}') \cdot (\mathbf{x} - \mathbf{x}')}{|\mathbf{x} - \mathbf{x}'|^3}$$

$$= \frac{1}{2} \int d^3x \, \rho(\mathbf{x}) \Phi(\mathbf{x})$$

$$= W, \tag{7.1.17}$$

where W is the gravitational potential energy.

Analogously to Eq. (7.1.15), the term proportional to ∇B^2 contributes $3\mathfrak{M}$, where

$$\mathfrak{M} = \frac{1}{8\pi} \int B^2 \, d^3x \tag{7.1.18}$$

is the magnetic energy and we have taken the boundary of V to infinity to justify dropping the surface term.

Finally, we have

$$\mathbf{x} \cdot (\mathbf{B} \cdot \nabla)\mathbf{B} = (\mathbf{B} \cdot \nabla)(\mathbf{x} \cdot \mathbf{B}) - \mathbf{B} \cdot (\mathbf{B} \cdot \nabla)\mathbf{x}$$

$$= \nabla \cdot [\mathbf{B}(\mathbf{x} \cdot \mathbf{B})] - B^2, \tag{7.1.19}$$

since $\nabla \cdot \mathbf{B} = 0$. The divergence term in Eq. (7.1.19) vanishes on integration over all space, and so we get a contribution of $-2\mathfrak{M}$ from this term. Putting all the terms together, we get

$$\frac{1}{2} \frac{d^2I}{dt^2} = 2T + W + 3\Pi + \mathfrak{M}. \tag{7.1.20}$$

Note that Π is $\frac{2}{3}$ of the thermal energy of the nonrelativistic particles in the fluid plus $\frac{1}{3}$ of the thermal energy of the relativistic particles.

Assuming equilibrium, we have the scalar virial equation,

$$2T + W + 3\Pi + \mathfrak{M} = 0. \tag{7.1.21}$$

7.2 Magnetic White Dwarfs

Let us first restrict ourselves to nonrotating white dwarfs. Then the rotational kinetic energy $T = 0$ and the virial theorem (7.1.21) becomes

$$W + 3\Pi + \mathfrak{M} = 0, \tag{7.2.1}$$

or

$$W + 3M\left\langle \frac{P}{\rho} \right\rangle + \left\langle \frac{B^2}{8\pi} \right\rangle \frac{4}{3}\pi R^3 = 0, \tag{7.2.2}$$

where the brackets denote an average. Now in the limit of high conductivity, the magnetic flux

$$\phi_M \sim \langle B \rangle R^2 \tag{7.2.3}$$

is conserved when the radius changes. For nonrelativistic degeneracy (NRD), $P \sim \rho^{5/3}$, while for extreme relativistic degeneracy (ERD), $P \sim \rho^{4/3}$. Thus loosely scaling the bracketed quantities in Eq. (7.2.2) gives

$$0 = -\alpha_{3/2} \frac{GM^2}{R} + \beta_{3/2} \frac{M^{5/3}}{R^2} + \gamma_{3/2} \frac{\Phi_M^2}{R} \quad \text{(NRD)}, \tag{7.2.4}$$

$$0 = -\alpha_3 \frac{GM^2}{R} + \beta_3 \frac{M^{4/3}}{R} + \gamma_3 \frac{\Phi_M^2}{R} \quad \text{(ERD)}, \tag{7.2.5}$$

where the subscripts refer to the $n = \frac{3}{2}$ and $n = 3$ polytropic equations of state describing the two regimes. Here the quantities α, β, and γ are positive constants.

For both types of degeneracy the effect of the magnetic field is to *expand* the star slightly. Crudely, adding magnetic flux is equivalent to "reducing" the gravitational constant G to

$$G' = G - \frac{\gamma \Phi_M^2}{\alpha M^2} = G\left(1 - \frac{\mathfrak{M}}{|W|}\right). \tag{7.2.6}$$

For NRD white dwarfs, we can solve Eq. (7.2.4) for the radius of an equilibrium configuration:

$$R = \frac{\beta_{3/2}}{\alpha_{3/2} G' M^{1/3}} = \frac{R_0}{1 - \mathfrak{M}/|W|}, \tag{7.2.7}$$

where R_0 is the radius when $\mathbf{B} = 0$. For small values of

$$\delta \equiv \frac{\mathfrak{M}}{|W|},\tag{7.2.8}$$

the radius increases by a small amount.

For ERD dwarfs, we note first that the mass limit can increase by only a small amount for $\delta \ll 1$. Solving Eq. (7.2.5) for M, we find

$$M^{2/3} = \frac{\beta_3}{\alpha_3 G}\left(1 + \frac{\gamma_3 \Phi_M^2}{\beta_3 M^{4/3}}\right) \simeq \frac{\beta_3}{\alpha_3 G}\left(1 + \frac{\gamma_3 \Phi_M^2}{\alpha_3 G M^2}\right)\tag{7.2.9}$$

if $\delta \ll 1$, and hence,

$$M_{\max} = M_0\left(1 + \tfrac{3}{2}\delta\right)\tag{7.2.10}$$

where $M_0 = (\beta_3/\alpha_3 G)^{3/2}$ is the Chandrasekhar mass limit when $B = 0$ [cf. Eq. (3.3.17)].

However, the *radius* of an ERD configuration near $M_{\max}$ can increase considerably, even for small values of δ. The reason is simply that for $\delta \ll 1$, the star is very nearly an $n = 3$ polytrope, so that [cf. Eq. (3.2.10)] the energy satisfies

$$E = -\frac{3 - n}{3}|W| \ll |W|.\tag{7.2.11}$$

Thus a small change in E can result in a large change in R:

$$\frac{\Delta E}{E} = -\frac{\Delta R}{R}\tag{7.2.12}$$

implies, with $\Delta E = \Delta \mathfrak{M}$,

$$\frac{\Delta R}{R} = \frac{3}{3 - n}\frac{\Delta \mathfrak{M}}{|W|} = \frac{3}{3 - n}\Delta\delta.\tag{7.2.13}$$

Integrating, assuming n constant as δ increases, we find

$$R = R_0 \exp\left(\frac{3}{3 - n}\delta\right).\tag{7.2.14}$$

The radius increases significantly even for small δ.

Uniformly rotating magnetic white dwarfs have been constructed numerically by Ostriker and Hartwick (1968). They find that for a nonrotating $1.05 M_\odot$ star (representing Sirius B), the radius can increase by as much as a factor of $\exp(3.5\delta)$; compare Figure 7.1. For extreme cases in which $\delta \sim 0.1$ with a 40% change in R, the central fields approached $10^{12.3}G$, although the surface fields remained several orders of magnitude less.

Are such field strengths reasonable? Since flux is conserved during the contraction of a star to the white dwarf stage, the progenitors of white dwarfs must have had large fields to produce the required white dwarf field strengths:

$$R_i^2 B_i = R_{\text{wd}}^2 B_{\text{wd}} \qquad (7.2.15)$$

implies progenitor central field strengths of magnitude

$$B_i \sim 10^{12}\left(\frac{10^9}{10^{11}}\right)^2 \sim 10^8 G, \qquad (7.2.16)$$

assuming $R_i \sim R_\odot$. Hence

$$\frac{\mathfrak{M}_i}{|W_i|} \sim \frac{\mathfrak{M}_{\text{wd}}}{|W_{\text{wd}}|} \lesssim \text{few percent.} \qquad (7.2.17)$$

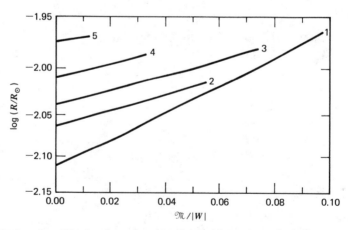

Figure 7.1 Radius of a white dwarf as a function of magnetic energy and angular momentum. For a $1.05 M_\odot$ star (representing Sirius B), we plot the radius R versus the ratio of magnetic to gravitational energies $|\mathfrak{M}/W|$; R is the larger of the equatorial and polar radii. The curves labeled 1, 2, 3, 4, and 5 represent uniformly rotating sequences having angular momenta equal to $(0, 1, 2, 3, 4) \times 1.92 \times 10^{49}$ g cm^2 s^{-1}, respectively. [Reprinted courtesy of Ostriker and Hartwick and *The Astrophysical Journal*, published by the University of Chicago Press; © 1968 The American Astronomical Society.]

Although there is no evidence for such field strengths,[3] they are not ruled out by observations. Moreover, the contraction of a "typical" interstellar cloud with radius ~ 0.1 pc, mass $\sim 1 M_\odot$, and frozen-in interstellar B field of magnitude $3 \times 10^{-6} G$ would result in a field strength of $10^8 G$ in the star formed from this material.[4]

The results of Ostriker and Hartwick show that with increasing field, the radius increases while ρ_c decreases. Moderate magnetic field strengths can in principle play a significant role in influencing the mass-radius relation, especially for more massive white dwarfs.

So far, however, there does not appear to be any compelling observational evidence in favor of such high interior field strengths. The radius of Sirius B falls right on the zero-field Hamada–Salpeter curve (cf. Section 3.6). About 5% of observed white dwarfs have surface magnetic fields in the range 10^6–$10^8 G$.[5] The implied internal field strengths are model-dependent, but presumably are in the range 10^8–$10^{10} G$. These are too small to be dynamically significant if $M \lesssim 1 M_\odot$.

7.3   Rotating Configurations: The Maclaurin Spheroids

Most of our intuition about rotating, self-gravitating configurations comes from studying uniform density ellipsoids, which can be analyzed relatively simply.[6] The simplest such homogeneous ellipsoids are the *Maclaurin spheroids*, which rotate with uniform angular velocity Ω.

The gravitational potential at any point (x, y, z) inside a uniform ellipsoid is a quadratic function of the coordinates:

$$\Phi = -\pi G \rho \left[A - A_1 x^2 - A_2 y^2 - A_3 z^2 \right], \qquad (7.3.1)$$

where the A's depend only on the shape of the ellipsoid and $A_1 + A_2 + A_3 = 2$. This follows from Poisson's equation

$$\nabla^2 \Phi = 4 \pi G \rho \qquad (7.3.2)$$

with ρ constant. The A's can be found by elegant geometrical arguments given, for example, in Chandrasekhar (1969). For the case of a spheroid, one can get Φ alternatively from the Green's function solution of Eq. (7.3.2),

$$\Phi = -G \rho \int \frac{d^3 x'}{|\mathbf{x} - \mathbf{x}'|}. \qquad (7.3.3)$$

---

[3] Magnetic A-type stars have *surface* fields $\lesssim 10^4 G$.
[4] Spitzer (1978), Chapter 13.3e.
[5] See, for example, Landstreet (1979); Angel (1978).
[6] The standard reference is Chandrasekhar (1969).

In spherical polar coordinates,

$$\frac{1}{|\mathbf{x} - \mathbf{x}'|} = \sum_{l=0}^{\infty} \frac{r_<^l}{r_>^{l+1}} P_l(\cos\theta) P_l(\cos\theta') + [\phi\text{-dependent terms}], \quad (7.3.4)$$

where P_l denotes a Legendre polynomial and $r_<$ ($r_>$) is the lesser (greater) of r and r'. By azimuthal symmetry, the ϕ-dependent terms do not contribute to Eq. (7.3.3). The surface of the spheroid has the polar equation $R = R(\theta)$, where

$$\frac{\sin^2\theta}{a^2} + \frac{\cos^2\theta}{c^2} = \frac{1}{R^2}, \quad (7.3.5)$$

and where a and c are the semimajor and semiminor axes, respectively. Thus

$$\Phi = -2\pi G\rho \sum_{l=0}^{\infty} P_l(\cos\theta) \int_0^{\pi} \sin\theta' \, d\theta' P_l(\cos\theta')$$

$$\times \left(\int_0^r \frac{(r')^{l+2} \, dr'}{r^{l+1}} + \int_r^R \frac{r^l \, dr'}{(r')^{l-1}} \right). \quad (7.3.6)$$

To evaluate Eq. (7.3.6) note first that all terms with odd l vanish immediately by parity, since in this case P_l is an odd function of $\cos\theta'$ while R is an even function. Examination of the integrals shows further that only the $l = 0$ and $l = 2$ terms contribute in Eq. (7.3.6). For the other terms with even $l \geq 4$, the exponents of the $\cos\theta'$ factors arising from R never exceed $l - 2$; so these terms vanish by orthogonality upon integration. We thus get

$$\Phi = -2\pi G\rho \left\{ -\frac{1}{3} r^2 + \int_0^1 \frac{dx}{1/a^2 + (1/c^2 - 1/a^2)x^2} \right.$$

$$\left. - (3\cos^2\theta - 1)\frac{r^2}{4} \int_0^1 dx\,(3x^2 - 1)\log\left[\frac{1}{a^2} + \left(\frac{1}{c^2} - \frac{1}{a^2}\right)x^2\right] \right\}, \quad (7.3.7)$$

where $x = \cos\theta'$. The integrals are elementary, and after some simplifications we obtain Eq. (7.3.1) with

$$A_1 = A_2 = \frac{(1 - e^2)^{1/2}}{e^3} \sin^{-1}e - \frac{1 - e^2}{e^2},$$

$$A_3 = \frac{2}{e^2} - \frac{2(1 - e^2)^{1/2}}{e^3} \sin^{-1}e,$$

$$A = \frac{2a^2(1 - e^2)^{1/2}}{e} \sin^{-1}e, \quad (7.3.8)$$

where the eccentricity is defined by

$$e^2 \equiv 1 - \frac{c^2}{a^2}. \tag{7.3.9}$$

A uniformly rotating spheroid in hydrostatic equilibrium satisfies

$$\frac{d\mathbf{v}}{dt} = -\frac{1}{\rho}\nabla P - \nabla\Phi, \tag{7.3.10}$$

with

$$\mathbf{v} = \mathbf{\Omega} \times \mathbf{r}. \tag{7.3.11}$$

Now

$$\frac{dv_i}{dt} = \left(v_j \partial_j\right)v_i$$

$$= \left(\varepsilon_{jkl}\Omega_k x_l \partial_j\right)\varepsilon_{imn}\Omega_m x_n$$

$$= \varepsilon_{jkl}\Omega_k x_l \varepsilon_{imj}\Omega_m$$

$$= \left(\delta_{ki}\delta_{lm} - \delta_{km}\delta_{li}\right)\Omega_k x_l \Omega_m$$

$$= \Omega_i\left(x_m \Omega_m\right) - \Omega^2 x_i. \tag{7.3.12}$$

If we choose $\mathbf{\Omega}$ along the z axis, Eq. (7.3.12) becomes

$$\frac{d\mathbf{v}}{dt} = -\Omega^2\left(x\mathbf{e}_x + y\mathbf{e}_y\right) = \mathbf{\Omega} \times \left(\mathbf{\Omega} \times \mathbf{r}\right), \tag{7.3.13}$$

which we recognize as simply the centripetal acceleration of the fluid. Thus the z component of Eq. (7.3.10) gives

$$0 = -\frac{1}{\rho}\frac{\partial P}{\partial z} - \frac{\partial \Phi}{\partial z}, \tag{7.3.14}$$

while the x component gives

$$-\Omega^2 x = -\frac{1}{\rho}\frac{\partial P}{\partial x} - \frac{\partial \Phi}{\partial x}. \tag{7.3.15}$$

Since Φ is a quadratic function of the coordinates, P must be a quadratic function, too. Since P vanishes on the surface of the spheroid, we have

$$P = P_c \left(1 - \frac{x^2 + y^2}{a^2} - \frac{z^2}{c^2} \right),$$ (7.3.16)

where P_c is the central pressure. Using Eq. (7.3.1), we find from Eq. (7.3.14) that

$$P_c = \pi G \rho^2 c^2 A_3,$$ (7.3.17)

and then from Eq. (7.3.15) that

$$\Omega^2 = 2\pi G \rho \left(A_1 - \frac{A_3 c^2}{a^2} \right)$$

$$= 2\pi G \rho \left[\frac{(1 - e^2)^{1/2}}{e^3} (3 - 2e^2) \sin^{-1} e - \frac{3(1 - e^2)}{e^2} \right].$$ (7.3.18)

The moment of inertia of the spheroid about the rotation axis is

$$I = \tfrac{2}{5} M a^2,$$ (7.3.19)

where

$$M = \tfrac{4}{3} \pi a^3 (1 - e^2)^{1/2} \rho.$$ (7.3.20)

The angular momentum is

$$J = I \Omega,$$ (7.3.21)

while the kinetic energy is

$$T = \tfrac{1}{2} I \Omega^2.$$ (7.3.22)

Exercise   7.1   Show that the gravitational potential energy of a Maclaurin spheroid is

$$W = \frac{1}{2} \rho \int \Phi \, d^3 x$$

$$= -\frac{3}{5} \left(\frac{4\pi}{3} \right)^2 G \rho^2 a^5 \frac{\sin^{-1} e}{e} (1 - e^2).$$ (7.3.23)

A useful ratio for parametrizing Maclaurin spheroids is

$$\frac{T}{|W|} = \frac{3}{2e^2} \left(1 - \frac{e(1 - e^2)^{1/2}}{\sin^{-1} e} \right) - 1.$$ (7.3.24)

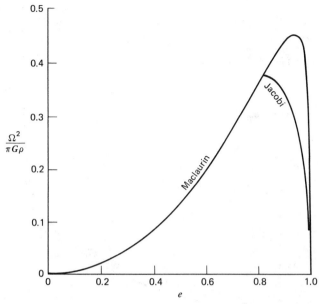

Figure 7.2   The square of the angular velocity (in the unit $\pi G\rho$) along the Maclaurin and the Jacobian sequences. The abcissa, in both cases, is the eccentricity defined by Eq. (7.3.9). [Reproduced, with permission, from *Ellipsoidal Figures of Equilibrium* by S. Chandrasekhar, published by Yale University Press. © 1969 by Yale University.]

The stability of Maclaurin spheroids can be investigated by a normal-mode analysis, or by the tensor virial method.[7] There are two instabilities of interest, both setting in via nonradial "toroidal" modes with azimuthal dependence $\exp(\pm 2i\phi)$. This pair of modes goes *dynamically* unstable ($\omega^2 < 0$) for $e >$ 0.952887, corresponding to $T/|W| > 0.2738$. *Secular* instability sets in earlier along the Maclaurin sequence, at $e = 0.812670$, $T/|W| = 0.1375$. This is a point where one of the modes has $\omega = 0$, but ω is positive on either side of this point. When $\omega = 0$, there is another nearby equilibrium configuration. In fact, $T/|W| = 0.1375$ is a *point of bifurcation*, where a whole new sequence of equilibrium configurations branches off the Maclaurin sequence. This new sequence consists of the *Jacobi ellipsoids*: uniformly rotating, homogeneous configurations with ellipsoidal surfaces ("rotating footballs"). The angular velocity and angular momentum of Maclaurin spheroids and Jacobi ellipsoids are plotted in Figures 7.2 and 7.3.

[7]Chandrasekhar (1969).

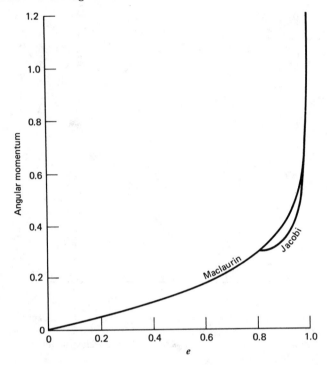

Figure 7.3   The angular momentum [in the unit $(GM^3\bar{a})^{1/2}$] along the Maclaurin and the Jacobian sequences. Here $\bar{a}$ is related to the three semimajor axes by $\bar{a} \equiv (abc)^{1/3}$; for a Maclaurin spheroid $a = b$. The abscissa in both cases is the eccentricity defined by Eq. (7.3.9). [Reproduced, with permission, from *Ellipsoidal Figures of Equilibrium* by S. Chandrasekhar, published by Yale University Press. © 1969 by Yale University.]

The Jacobi ellipsoid of a given angular momentum, mass, and volume has a lower energy than the corresponding Maclaurin spheroid (cf. Fig. 7.4). This suggests that Maclaurin spheroids beyond the point of bifurcation should be unstable to becoming Jacobi ellipsoids. However, the dynamical equations conserve energy unless dissipative terms are added. An instability that requires the presence of dissipation is called a secular instability, to distinguish it from a dynamical instability. The addition of dissipative terms to the dynamical perturbation equations causes ω to become complex beyond the bifurcation point, with the imaginary part of ω being proportional to the strength of the dissipation (e.g., the viscosity coefficient). The growth time of a dynamical instability, by contrast, is always the dynamical timescale $\sim (G\rho)^{-1/2}$.

Since in fact $\omega = 0$ is a double root at $T/|W| = 0.1375$, another sequence of equilibrium configurations also bifurcates from the Maclaurin sequence at the same point as the Jacobi ellipsoids: the *Dedekind ellipsoids*. These have the same

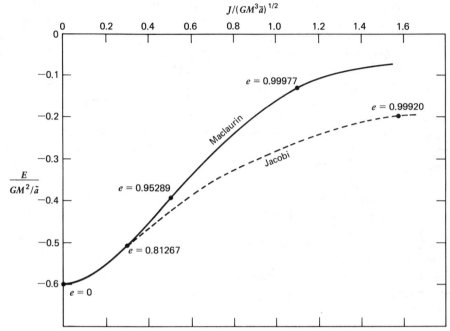

Figure 7.4   Energy $E = T + W$ as a function of angular momentum for Maclaurin spheroids and Jacobi ellipsoids. Here $\bar{a}$ is related to the three semimajor axes by $\bar{a} = (abc)^{1/3}$; for a Maclaurin spheroid $a = b$. Shown are the eccentricities [defined by Eq. (7.3.9)] of selected configurations, including the point of bifurcation ($e = 0.81267$) and point of onset of dynamical instability ($e = 0.95289$).

shape as the Jacobi ellipsoids, but their shape is stationary; the ellipsoidal surface is supported by the circulation of fluid inside ("stationary footballs").

Maclaurin spheroids beyond the point of bifurcation are also secularly unstable to becoming Dedekind ellipsoids. However, viscosity damps out differential rotation and so Dedekind ellipsoids are not equilibrium configurations in the presence of viscosity. Instead, the dissipation necessary is provided by gravitational radiation reaction.[8]

The situation becomes more complicated if both viscosity and gravitational radiation are operative.[9] The Jacobi mode, which deforms a Maclaurin spheroid into a Jacobi ellipsoid and which is unstable in the presence of viscosity, tends to be stabilized by gravitational radiation. The Dedekind mode, which is unstable

[8] Chandrasekhar (1970).
[9] Detweiler and Lindblom (1977).

due to gravitational radiation, tends to be stabilized by viscosity. In fact, by a suitable choice of the strength of viscosity relative to gravitational radiation, it is possible to stabilize the Maclaurin sequence all the way up to the point of dynamical instability. Probably the viscosity in astrophysical applications is not large enough for this to be important, but this is not certain.

As we shall see in the next section, compressible configurations with *uniform* rotation are not particularly interesting. As one considers objects that are more and more centrally condensed, one finds the allowed range of rotation, as measured, for example, by $T/|W|$, is severely limited by the condition of no mass shedding at the equator. Centrally condensed objects in uniform rotation lose mass before they are rotating fast enough to encounter "interesting" instabilities.

When one considers *differentially* rotating objects, however, $T/|W|$ in equilibrium can vary essentially over the full range allowed by the virial theorem, $0 \leqslant T/|W| \leqslant \frac{1}{2}$ [cf. Eq. (7.1.21)]. So again the question of nonaxisymmetric instabilities must be faced. For rough estimates, we adopt the criteria $T/|W| \gtrsim 0.14$ for secular instability and $T/|W| \gtrsim 0.26$ for dynamical instability. These values are close to the exact Maclaurin spheroid values, and seem to hold for a wide range of angular momentum distributions and equations of state. For further discussion, see Section 7.5.

7.4 Rotating White Dwarfs

Let us now consider white dwarf configurations that possess no magnetic fields but that are rotating. The rotational kinetic energy is

$$T \sim M\Omega^2 R^2 \sim \frac{J^2}{MR^2}, \tag{7.4.1}$$

where J is the (conserved) angular momentum. The virial equation (7.1.21) for rotating stars now becomes [cf. Eqs. (7.2.4) and (7.2.5)]

$$0 = -\alpha_{3/2}\frac{GM^2}{R} + \kappa_{3/2}\frac{J^2}{MR^2} + \beta_{3/2}\frac{M^{5/3}}{R^2} \quad \text{(NRD)}, \tag{7.4.2}$$

$$0 = -\alpha_3\frac{GM^2}{R} + \kappa_3\frac{J^2}{MR^2} + \beta_3\frac{M^{4/3}}{R} \quad \text{(ERD)}. \tag{7.4.3}$$

The fact that the rotational energy has a steeper dependence on the radius than the internal energy in the relativistic limit is quite significant. Equilibrium can *always* be achieved for any mass by *decreasing* the radius of the configuration.

Thus an equilibrium model of *any* mass can be constructed provided the angular momentum is nonzero. Of course, such models might not be physical, if, say, the radius were so small that the density increased to 10^{15} g cm^{-3}, thereby invalidating the assumed equation of state. Moreover, they may not be stable to inverse β-decay, fission, rapid evolution through viscous momentum transport, and other calamities.

James (1964) has numerically constructed *uniformly rotating* white dwarfs obeying the Chandrasekhar equation of state. For such configurations, the overall structure of the models is not very different from the nonrotating cases. In particular, the mass limit can be increased by only 3.5% for a given μ_e. The reason for this is clear, however, since, as we will show below, uniform rotation in centrally condensed bodies requires the ratio $T/|W|$ to be less than 0.007, which will not seriously alter the structure of a centrally condensed object. In fact, Eqs. (7.4.3) and (7.1.21) give

$$ 0 = -\alpha_3 \frac{GM^2}{R}\left(1 - \frac{2T}{|W|}\right) + \beta_3 \frac{M^{4/3}}{R}. \tag{7.4.4} $$

Solving for M, we get

$$ M = \left[\frac{\beta_3}{\alpha_3 G(1 - 2T/|W|)}\right]^{3/2} $$

$$ \simeq M_0\left(1 + \frac{3T}{|W|}\right) \simeq 1.02 M_0. \tag{7.4.5} $$

where M_0 is the Chandrasekhar mass limit in the absence of rotation. This estimate is in rough agreement with the findings of James.

We now derive the restriction $0 \leqslant T/|W| \leqslant 0.00744$ for a uniformly rotating, $n = 3$ polytrope, following the discussion of Zel'dovich and Novikov (1971). Note that in an $n = 3$ polytrope, the central density is about 54 times the mean density (Exercise 3.4), and so the object is certainly centrally condensed.

Consider first a *spherical* star rotating with break-up velocity at the equator:

$$ v^2 = \Omega^2 R^2 = \frac{GM}{R}. \tag{7.4.6} $$

Now for an $n = 3$ polytrope [cf. Eq. (3.2.11)]

$$ |W| = \frac{3}{5 - n}\frac{GM^2}{R} = \frac{3}{2}\frac{GM^2}{R}. \tag{7.4.7} $$

Also,

$$T = \tfrac{1}{2}I\Omega^2,$$ (7.4.8)

where

$$I = \frac{2}{3}M\langle r^2\rangle = \frac{2}{3}\int_0^M r^2\,dm$$

$$= \frac{2}{3}\frac{MR^2}{\xi_1^4|\theta'(\xi_1)|}\int_0^{\xi_1}\theta^n\xi^4\,d\xi.$$ (7.4.9)

The integral in Eq. (7.4.9) has the value 10.851, and so

$$\langle r^2\rangle = 0.11303R^2.$$ (7.4.10)

Thus, using Eq. (7.4.6) for Ω^2, we find at break-up

$$\frac{T}{|W|} = \frac{\tfrac{1}{2}\tfrac{2}{3}M\langle r^2\rangle\Omega^2}{\tfrac{3}{2}GM^2/R} = 0.025.$$ (7.4.11)

By contrast, for an incompressible fluid ($n \to 0$),

$$\langle r^2\rangle = \frac{3}{5}R^2, \qquad |W| = \frac{3}{5}\frac{GM^2}{R},$$ (7.4.12)

and so at break-up

$$\frac{T}{|W|} = \frac{1}{3}.$$ (7.4.13)

This illustrates the role of central mass concentration in reducing the rotational limit for polytropic configurations.

The above analysis has ignored the fact that in reality the rotating configuration is *spheroidal*, with radius $R_1 > R$ at the equator. We can analyze the shape approximately by adopting the Roche model. This model assumes that the distribution of the *bulk* of the mass is unchanged by the rotation, which is a good approximation because of the central condensation. In the outer layers, therefore, the gravitational potential remains $\Phi = -GM/r$. Now for constant angular velocity about the z axis, we can introduce the centrifugal potential

$$\Phi_c = -\tfrac{1}{2}\Omega^2(x^2 + y^2) = -\tfrac{1}{2}\Omega^2 r^2\sin^2\theta.$$ (7.4.14)

Then

$$\frac{d\mathbf{v}}{dt} = \nabla \Phi_c \tag{7.4.15}$$

[cf. Eq. (7.3.13)], and so the hydrostatic equilibrium equation (7.3.10) becomes

$$0 = \frac{1}{\rho} \nabla P + \nabla (\Phi + \Phi_c), \tag{7.4.16}$$

or

$$h + \Phi + \Phi_c = K, \tag{7.4.17}$$

where K is a constant and

$$h = \int \frac{dP}{\rho} = \frac{\Gamma}{\Gamma - 1} \frac{P}{\rho} \tag{7.4.18}$$

is the enthalpy per unit mass. We assume that K in Eq. (7.4.17) has the same value as in the nonrotating case. Since $h(R) = 0$, we have

$$K = -\frac{GM}{R}. \tag{7.4.19}$$

The effective potential $\Phi_{\text{eff}} = \Phi + \Phi_c$ is plotted in Figure 7.5. Along an equatorial radius, Φ_{eff} has a maximum at $r_c = (GM/\Omega^2)^{1/3}$, where $\Phi_{\text{max}} = -3GM/(2r_c)$. Now Eq. (7.4.17) has a meaningful solution only when h goes to zero at some radius $r = R_1$, the surface of the configuration. Since $h(r)$ is the distance from $\Phi_{\text{eff}}(r)$ to the line $\Phi_{\text{eff}} = K$ in Figure 7.5, we see that for a meaningful solution, K must be less than Φ_{max}. When $K = \Phi_{\text{max}}$, R_1 takes on its maximum value—namely, $r_c = 3R/2$. Thus the maximum expansion of a uniformly rotating star along the equator is a factor of $\frac{3}{2}$. The corresponding maximum value of Ω is

$$\Omega = \left(\frac{GM}{r_c^3} \right)^{1/2} = \left(\frac{2}{3} \right)^{3/2} \left(\frac{GM}{R^3} \right)^{1/2}, \tag{7.4.20}$$

which is a factor of $(\frac{2}{3})^{3/2} = 0.544$ smaller than the maximum value given by Eq. (7.4.6) for the spherical case. Hence

$$\left. \frac{T}{|W|} \right|_{\text{max}} = \left(\frac{2}{3} \right)^3 \times \text{spherical case} = 0.00744. \tag{7.4.21}$$

Contrast this situation with Maclaurin spheroids, which exist (although unstably) all the way to $T/|W| = 0.5$.

James (1964) shows that for polytropic indices $n > 0.808$, mass shedding occurs before a point of bifurcation is reached along a uniformly rotating sequence. However, this mass-shedding limit is probably not of much physical significance. Realistic stars are likely to be *differentially* rotating, at least in their outer layers, and the above analysis simply shows that uniformly rotating configurations are not adequate models of rapidly rotating stars.

Since stellar contraction without angular momentum loss requires that $T/|W|$ increase as $1/R$, rotation is likely to be more important in a compact object than in its main sequence progenitor. Detailed models of differentially rotating white dwarfs have been constructed by Ostriker and Bodenheimer (1968) in the mass range $0.5-4.1M_\odot$. (Models with $T/|W| \lesssim 0.14$ have $M \lesssim 2.4M_\odot$.) Each model is determined by specifying its mean molecular weight $\mu_e(=2)$, total mass M, total angular momentum J, and the distribution $j(m)$ of angular momentum per unit mass.

For the models considered, differential rotation is never very extreme since $\Omega(\text{equator})/\Omega(\text{center}) \gtrsim 0.2$ for all cases. The radii of the configurations are typically within a factor of 2 of 10^9 cm and do not distinguish the models significantly from nonrotating models in the mass range $0.4-0.9M_\odot$. Since central densities are well below 10^9 g cm^{-3}, the models are stable to inverse β-decay, and the adopted Chandrasekhar equation of state remains valid. Surface velocities of the massive models above $1.4M_\odot$ are in the range 3000–7000 km/s. Since the H lines observed in most white dwarf spectra exhibit narrow cores, such highly rotating objects must be exceptional.

Such stars (if they exist at all) are likely to be quite luminous. Durisen (1973) has shown that viscous dissipation of energy (which occurs on a timescale $> 10^9$

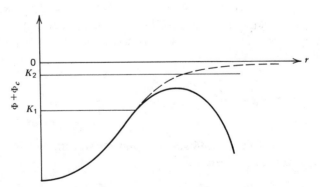

Figure 7.5 Shape of the potential $\Phi + \Phi_c$ along a radial direction located in the equatorial plane (*solid* line) and in the polar (z) axis (*dashed* line). Horizontal lines K_1 and K_2 correspond to different values of the constant K in Eq. (7.4.17). [After Zel'dovich and Novikov (1971).]

yr) in massive models results in high internal temperatures and luminosities above $10^{-1}L_\odot$. Although DC white dwarfs show no spectral lines (cf. Appendix A) and could therefore be rapidly rotating, their luminosity is well below this value.

Rapidly rotating white dwarfs will be stable against implosion via inverse β-decay provided $\rho_c \lesssim 3 \times 10^9$ g cm^{-3}, depending on the composition (cf. Table 3.1). We will assume they are secularly stable if $T/|W| \leq 0.14$ and dynamically stable if $T/|W| \leq 0.26$ (see Section 7.5). We now determine approximately the mass limit for such stars.

We shall again use an energy variational principle, a generalization of Eq. (6.10.23) to include rotation (cf. Zel'dovich and Novikov, 1971). We make two simplifying assumptions: (i) that the density is constant on similar spheroids; and (ii) that a spherical surface of constant density in the nonrotating case transforms into a spheroidal surface enclosing the same volume. Assumption (ii) is only strictly valid for an incompressible fluid. However, in the central regions of a compressible star, which contain most of the mass, it is not a bad approximation. If the equation of state is of the form $P = P(\rho)$, assumption (i) is also only true for an incompressible fluid.[10] Once again, the approximation yields reasonable numerical estimates when most of the mass is in a core region, as we shall see.

Our assumptions imply that the density as a function of mass inside a given layer in the spheroidal star is the same function as in a spherical star with the same central density. Thus, as in Eq. (6.10.9),

$$E_{\text{int}} = \int u \, dm$$

$$= k_1 K \rho_c^{1/3} M. \tag{7.4.22}$$

Now the gravitational potential energy of a sphere of constant density ($n = 0$ polytrope) is

$$W = -\frac{3}{5}\frac{GM^2}{R} = -\frac{3}{5}\left(\frac{4\pi}{3}\right)^{1/3} GM^{5/3}\rho^{1/3}. \tag{7.4.23}$$

The corresponding result for a spheroid of constant density is [cf. Eq. (7.3.23)]

$$W = -\frac{3}{5}\frac{GM^2}{a}\frac{\sin^{-1}e}{e} = -\frac{3}{5}\left(\frac{4\pi}{3}\right)^{1/3} GM^{5/3}\rho^{1/3}\frac{\sin^{-1}e}{e}(1 - e^2)^{1/6}. \tag{7.4.24}$$

For a spherical $n = 3$ polytrope, Eq. (6.10.10) gives

$$W = -k_2 GM^{5/3}\rho_c^{1/3}. \tag{7.4.25}$$

----

[10] Tassoul (1978), Section 4.4.

Under our assumptions, we now show that W for an $n = 3$ polytrope is modified by rotation in the same way as it is modified for the homogeneous star—namely,

$$W = -k_2 GM^{5/3}\rho_c^{1/3}\frac{\sin^{-1}e}{e}(1 - e^2)^{1/6}. \tag{7.4.26}$$

This result follows from Newton's theorem that the potential inside an ellipsoidal shell of constant density is constant. This can be shown by subtracting two expressions of the form (7.3.1).[11]

Imagine constructing a spheroidal star by starting with the outermost spheroidal layer of constant density. Inside this layer, place the next layer of slightly higher constant density, and continue in this way. Each layer is placed in a cavity where the potential is constant. The total potential energy will be smaller than for a spherical star by the same factor as the potential inside the spheroidal cavity is smaller than for the spherical cavity of the same volume and with the same external mass—that is, by the factor

$$\frac{\sin^{-1}e}{e}(1 - e^2)^{1/6}. \tag{7.4.27}$$

This proves the result (7.4.26). It will be convenient to introduce the oblateness parameter

$$\lambda \equiv \left(\frac{c^2}{a^2}\right)^{1/3} = (1 - e^2)^{1/3}. \tag{7.4.28}$$

Then the expression (7.4.27) becomes

$$g(\lambda) \equiv \lambda^{1/2}(1 - \lambda^3)^{-1/2}\cos^{-1}(\lambda^{3/2}), \tag{7.4.29}$$

and

$$W = -k_2 GM^{5/3}\rho_c^{1/3}g(\lambda). \tag{7.4.30}$$

Finally, we will need to evaluate the rotational energy

$$T = \frac{J^2}{2I}. \tag{7.4.31}$$

Now $I \propto Ma^2$ for a spheroid. Hence, compared with a sphere of the same volume,

$$\frac{I}{I_{\text{sphere}}} = \frac{a^2}{(a^2 c)^{2/3}} = \frac{1}{\lambda}. \tag{7.4.32}$$

[11]A simple geometrical proof is given in Chandrasekhar (1969), Section 17.

From Eqs. (7.4.9) and (7.4.31), we find

$$T = k_5 \lambda J^2 M^{-5/3} \rho_c^{2/3}, \qquad (7.4.33)$$

where

$$k_5 = \frac{3(4\pi)^{2/3} \xi_1^2 |\theta'(\xi_1)|^{5/3}}{4 \int_0^{\xi_1} \theta^n \xi^4 \, d\xi} = 1.2042. \qquad (7.4.34)$$

Thus

$$E = k_1 K M \rho_c^{1/3} - k_2 G M^{5/3} \rho_c^{1/3} g(\lambda) + k_5 \lambda J^2 M^{-5/3} \rho_c^{2/3}. \qquad (7.4.35)$$

The terms ΔE_{int} and ΔE_{GTR} can be ignored since they are small and only relevant for testing *radial* stability.

Equilibrium is determined by setting $\partial E / \partial \rho_c = 0 = \partial E / \partial \lambda$, keeping M and J fixed. The condition $\partial E / \partial \lambda = 0$ yields

$$g'(\lambda) = \frac{k_5 J^2 \rho_c^{1/3}}{k_2 G M^{10/3}} = \frac{T}{|W|} \frac{g(\lambda)}{\lambda}. \qquad (7.4.36)$$

Using Eq. (7.4.29), we get

$$\frac{T}{|W|} = \frac{1}{2} \left[1 + \frac{3\lambda^3}{1 - \lambda^3} - \frac{3\lambda^{3/2}}{(1 - \lambda^3)^{1/2} \cos^{-1} \lambda^{3/2}} \right], \qquad (7.4.37)$$

which is equivalent to the relation (7.3.24) between e and $T/|W|$ for a Maclaurin spheroid.

The condition $\partial E / \partial \rho_c = 0$ yields

$$\tfrac{1}{3} k_1 K M \rho_c^{-2/3} - \tfrac{1}{3} k_2 G M^{5/3} g(\lambda) \rho_c^{-2/3} + \tfrac{2}{3} k_5 \lambda J^2 M^{-5/3} \rho_c^{-1/3} = 0, \qquad (7.4.38)$$

or

$$\tfrac{1}{3} k_1 K M \rho_c^{-2/3} - \tfrac{1}{3} k_2 G M^{5/3} g(\lambda) \rho_c^{-2/3} \left(1 - \frac{2T}{|W|} \right) = 0. \qquad (7.4.39)$$

Thus

$$M = \left(\frac{k_1 K}{k_2 Gg(\lambda)(1 - 2T/|W|)} \right)^{3/2}$$

$$= \frac{M_0}{[g(\lambda)(1 - 2T/|W|)]^{3/2}}, \tag{7.4.40}$$

where $M_0 = M_{Ch} = 1.457(2/\mu_e)^2 M_\odot$.

The maximum equilibrium mass for secularly stable rotating configurations follows on setting $T/|W| \simeq 0.14$ in Eqs. (7.4.37) and (7.4.40). We find

$$\lambda = 0.693, \qquad g(\lambda) = 0.974,$$

$$M = 1.70 M_0 = 2.5 \left(\frac{2}{\mu_e} \right)^2 M_\odot. \tag{7.4.41}$$

This is exactly the value obtained by Durisen (1975) and Durisen and Imamura (1981) by detailed numerical calculations for the secularly stable mass limit. This good agreement presumably arises because the degree of differential rotation in Durisen's models is not very extreme.

The dynamically stable mass limit is found by setting $T/|W| \simeq 0.26$ in Eqs. (7.4.37) and (7.4.40). This gives

$$\lambda = 0.475, \qquad g(\lambda) = 0.902,$$

$$M = 3.51 M_0 = 5.1 \left(\frac{2}{\mu_e} \right)^2 M_\odot. \tag{7.4.42}$$

This is in reasonable agreement with Durisen's value of $4.6 M_\odot$.

The relevant dissipative mechanism for $T/|W| \gtrsim 0.14$ is gravitational radiation, which acts on a timescale $\sim 10^3 - 10^7$ yr (Friedman and Schutz, 1975; cf. Exercise 16.7), compared with $\sim 10^9$ yr for viscosity.[12]

Recently, Greenstein et al. (1977) concluded from observational studies that (DA) white dwarfs do *not* rotate appreciably ($v_{rot} \leqslant 40$ km s^{-1} and probably below 10–20 km s^{-1}), and that very rapidly rotating, massive white dwarfs have not been observed. Apparently, a large fraction of the angular momentum in the precursor red giant star is transported outward prior to final white dwarf formation.

---

[12] Chandrasekhar (1970) and Durisen (1975).

Exercise 7.2 Repeat Exercise 6.21, adding the rotational terms as in Eq. (7.4.35). Show that the critical value of $\Gamma - \frac{4}{3}$ is *lowered* by an amount

$$2\left(\tfrac{5}{3} - \Gamma\right)\frac{T}{|W|}.$$

This shows that rotation tends to *stabilize* the radial modes.

Exercise 7.3

(a) The density distribution of the sun is approximately that of an $n = 3$ polytrope. What is its moment of inertia? (Detailed solar models give 5.7×10^{53} g cm^2.)

(b) The surface angular velocity of the sun is 2.9×10^{-6} s^{-1}. Assuming the sun rotates as a rigid body, what is its angular momentum?

(c) What is $T/|W|$ for the sun?

(d) Assume the sun eventually collapses to a white dwarf with the same values of J and M. What will the ratio $T/|W|$ be then?

(e) According to Allen (1973), a typical B5 main sequence star has $M \sim 6M_\odot$, $R \sim 3.8R_\odot$, surface angular velocity $\sim 9 \times 10^{-5}$ s^{-1}. Again assuming an $n = 3$ polytropic density distribution, repeat calculations (a)–(d). Conclude that the star must lose both mass *and* angular momentum to form a white dwarf under these assumptions.

7.5 Stability Criteria for Rotating Stars

We will not be able to give in this book a detailed discussion of this somewhat technical subject. Instead, we summarize the results below, focusing on secular stability, which is the most relevant question for compact objects.

An important initial step towards developing a secular stability criterion for compressible rotating stars was the variational principle of Lynden-Bell and Ostriker (1967). They described a small perturbation of the star by a Lagrangian displacement ξ (cf. Section 6.2). The star was then claimed to be secularly stable if and only if a certain operator C (analogous to our V_2 of Section 6.7) was positive definite. It was thought that this was equivalent to having the total energy of the perturbation be positive definite for all initial data.

In the period 1968–1973, the tensor virial method was used[13] to investigate the stability of differentially rotating stars. In this method, one takes moments of the perturbation equations. The second moment is equivalent to evaluating the operator C for a trial displacement ξ that is linear in the coordinates. In the case of Maclaurin spheroids, the tensor virial method yields *exact* stability information because the unstable eigenfunction is in fact linear in the coordinates. For compressible stars, according to the tensor virial method, secular instability also

---

[13] For example, Tassoul and Ostriker (1968); Ostriker and Tassoul (1969); Ostriker and Bodenheimer (1973).

sets in when $T/|W| \simeq 0.14$ for a wide range of angular momentum distributions and equations of state. The insensitivity of the critical $T/|W|$ is a most remarkable result.

Since the tensor virial method is equivalent to choosing a certain trial function in the Lynden-Bell–Ostriker variational principle, it is at best a sufficient condition for instability. Unfortunately, it is not even that! The reason is that the Lynden-Bell–Ostriker criterion is not quite correct as originally formulated.[14] There exist "trivial" Lagrangian displacements ξ that do not change the physical configuration of the star. They correspond to a relabeling of particles, with the physical Eulerian perturbations $\delta\rho$, δs, and $\delta\mathbf{v}$ remaining zero. One can choose "trivial" ξ's that make the operator C negative—clearly this has nothing to do with stability.

A revised stability criterion has been developed by Bardeen, Friedman, Schutz, and Sorkin.[15] One restricts the allowed displacements ξ to those "orthogonal" to the "trivials," in a sense that can be made mathematically precise, and uses these ξ's in the operator C. In general, the value of C turns out to be not the total energy of the perturbation, but the "canonical" energy E_c (i.e., the value of the perturbation Hamiltonian). The canonical energy is not the total energy unless "trivial" displacements are excluded.

A further problem with the Lynden-Bell–Ostriker criterion was that no dissipative mechanism was explicitly identified. In the case of Maclaurin spheroids, secular instability to both viscosity and gravitational radiation occurs at the same point. This appears to be accidental—it is not true for analogous compressible models. In general, one must specify the nature of the dissipation as part of the stability criterion.[16] The condition that E_c be positive definite for all nontrivial displacements guarantees stability against gravitational radiation, but not viscosity. The tensor virial trial function is not orthogonal to the "trivials" and so gives no information in this case.

There is probably no rigorous way to formulate a criterion for viscosity in a differentially rotating star, since such a star would not be in equilibrium. The growth time for any purported instability would be the same as the timescale for the unperturbed star to adjust itself under the action of the viscosity. (This is probably not true for accretion disks, which have small radial velocities; but this question has not been looked at rigorously.)

The criterion for stability of *uniformly* rotating stars against viscosity turns out to be that the canonical energy in the co-rotating frame, $E_{c,r}$, be positive definite. ("Trivial" displacements are harmless in this case.) The tensor virial method is equivalent to a statement about E_c and so is again invalid in the general case.

---

[14] Bardeen, Friedman, Schutz, and Sorkin (1977). See also Friedman and Schutz (1978a, b) and Hunter (1977).

[15] Bardeen, Friedman, Schutz, and Sorkin (1977). See also Friedman and Schutz (1978a, b).

[16] See also the papers in footnote 14 for references to papers of Jeans and Lyttleton.

Friedman and Schutz (1978b) made the remarkable discovery that *all* rotating stars are secularly unstable to gravitational radiation. However, for slowly rotating stars the instability sets in via a very high mode number m, where the azimuthal dependence of ξ is $\exp(im\phi)$. The growth time of the instability is then much longer than the age of the Universe. The physically relevant instability usually sets in via the $m = 2$ mode.

Durisen and Imamura (1981) have looked for the onset of instability via an $m = 2$ mode in rotating white dwarfs and rotating polytropes using a trial function orthogonal to the "trivials." They find that instability to gravitational radiation still sets in for $T/|W|$ near 0.14. In fact, their values are only 1–7% higher than the tensor virial estimates. They are able to construct stable rapidly rotating white dwarfs up to $\sim 2.5 M_\odot$.

We conclude that for rough estimates, the value $T/|W| \simeq 0.14$ can be used for the onset of secular instability for a wide variety of angular momentum distributions and equations of state.

8
Cold Equation
of State Above
Neutron Drip

8.1 Introduction

In this chapter we shall resume our analysis, begun in Chapter 2, of the equation of state of cold, dense matter. Here we shall be concerned with the properties of condensed matter at densities above neutron drip, where $\rho_{\text{drip}} \approx 4.3 \times 10^{11}$ g cm^{-3}.[1] It is this high-density domain $\rho \gtrsim \rho_{\text{drip}}$ that exists in the interiors of neutron stars.

Our treatment of the equation of state will be of necessity introductory in nature, since a complete discussion requires the full apparatus of many-body quantum theory. Nevertheless, we will be able to discuss most of the important physical principles, and illuminate them in a way similar to our treatment in Chapter 2 of the equation of state in the low-density regime $\rho \lesssim \rho_{\text{drip}}$.

We will divide the determination of the equation of state above ρ_{drip} into two parts. The first part (Section 8.2) covers the intermediate density regime from ρ_{drip} to nuclear density, $\rho_{\text{nuc}} = 2.8 \times 10^{14}$ g cm^{-3}, the density at which nuclei begin to dissolve and merge together. The properties of dense matter and the associated equation of state are reasonably well understood in this density regime. The second part of our analysis focuses on the high-density range above ρ_{nuc}, where the physical properties of matter are still uncertain. The bulk of this chapter (Sections 8.3–8.14) will be devoted to a discussion of this high-density regime.

We must first state that in spite of considerable progress during recent years, our understanding of condensed matter (summarized in Box 8.1) is far from complete, particularly above ρ_{nuc}. Not only is the correct form of the nuclear

---

[1]See Section 2.7.

potential still uncertain, but a totally satisfactory many-body computational method for solving the Schrödinger equation, given the potential, remains to be developed. Moreover, laboratory data are almost entirely lacking for condensed matter above ρ_{nuc}. Indeed, because various properties of neutron stars prove to be sensitive to the adopted equation of state,[2] it may turn out that refined observations of neutron stars may provide the best measurements, albeit indirect, of the properties of matter at such high densities.

8.2 The Baym–Bethe–Pethick Equation of State

In the domain from ρ_{drip} to ρ_{nuc}, matter is composed of nuclei, electrons, and free neutrons. The nuclei disappear at the upper end of this density range because their binding energy decreases with increasing density. We can understand this in part since the strong attractive "tensor" force between two unlike nucleons in the $^3S_1$ state, which is crucial in binding the deuteron (cf. Section 8.3), does not act between neutrons because of the Pauli principle. In fact, a system of pure neutrons is unbound at any density. So, as the density increases and the nuclei become more neutron rich, their stability decreases until a critical value of neutron number is reached, at which point the nuclei dissolve, essentially by merging together.

We shall focus on the equation of state of Baym, Bethe, and Pethick (1971a, hereinafter BBP) in this regime, and mention some alternatives at the end of this section. Their treatment is a considerable improvement over those based only on a semiempirical mass formula, such as HW.[3] BBP used a mass formula, but incorporated in it results obtained from detailed many-body computations.

First, since the nuclei present are very neutron-rich, the matter inside nuclei is very similar to the free neutron gas outside. In the earlier calculations using a semiempirical mass formula, however, the energy of the interior nucleons was calculated from formulae fitted to nuclear matter—that is, "ordinary" nuclei with $Z/A \simeq 0.5$, while the energy of the neutron gas was found from calculations of neutron-neutron interactions. That the two approaches did not give similar results for the energy in the limit of dilute solutions of protons in neutrons is not surprising.

Second, the nuclear surface energy had previously been assumed to be that of a nucleus with no matter outside. However, the external neutron gas reduces the surface energy appreciably. This is to be expected: when the matter inside nuclei becomes identical to that outside, the surface energy must vanish.

Third, BBP included the effect of the nuclear lattice Coulomb energy more carefully.

[2] We will discuss these properties in Chapters 9–11.
[3] Compare Section 2.6.

The BBP treatment is based on a "compressible liquid drop" model of the nuclei. They write the total energy density as

$$\varepsilon = \varepsilon(A, Z, n_N, n_n, V_N)$$

$$= n_N(W_N + W_L) + \varepsilon_n(n_n)(1 - V_N n_N) + \varepsilon_e(n_e). \qquad (8.2.1)$$

Here n_N is the number density of nuclei, n_n the number density of neutrons outside of nuclei ("neutron gas"), and the new feature is the dependence on V_N, the volume of a nucleus. The quantity V_N decreases in response to the outside pressure of the neutron gas and so must be treated as a variable. The quantity W_N is the energy of a nucleus, including the rest mass of the nucleus, and depends on A, Z, n_n, and V_N. The lattice energy is denoted by W_L, while ε_n and ε_e are the energy densities of the neutron gas and electron gas respectively. Note that $V_N n_N$ is the fraction of volume occupied by nuclei, and $1 - V_N n_N$ the fraction occupied by the neutron gas. In terms of these quantities, we have

$$n_e = Z n_N \quad \text{(charge neutrality)}, \qquad (8.2.2)$$

while the baryon density is

$$n = A n_N + (1 - V_N n_N) n_n. \qquad (8.2.3)$$

Note that n_n is defined in terms of the number N_n of free neutrons in a volume V_n outside of nuclei:

$$n_n = \frac{N_n}{V_n} = \frac{N_n}{V(1 - V_N n_N)}, \qquad (8.2.4)$$

where V is a volume containing N_n neutrons and $n_N V$ nuclei.

Equilibrium is determined by minimizing ε at fixed n. Since ε depends on five variables, this leads to four independent conditions.

The first condition comes from considering a unit volume with a fixed number $n_N Z$ of protons, a fixed number $n_N(A - Z)$ of neutrons in nuclei, a fixed fraction $n_N V_N$ of the volume occupied by nuclei, and a fixed number $n_n(1 - V_N n_N)$ of neutrons outside of nuclei. What is the optimal A of the nuclei? This is determined by minimizing ε with respect to A at fixed $n_N Z$, $n_N A$, $n_N V_N$, and n_n. This implies that ε_n is fixed and, since n_e is fixed, so is ε_e. Define

$$x \equiv \frac{Z}{A}. \qquad (8.2.5)$$

Then, since $n_N = \text{constant}/A$ in this variation, Eq. (8.2.1) gives

$$\frac{\partial}{\partial A}\left(\frac{W_N + W_L}{A}\right)_{x,\, n_N A,\, n_N V_N,\, n_n} = 0. \tag{8.2.6}$$

Physically, this says that the energy per nucleon inside nuclei must be a minimum.

The second condition is that the nuclei must be stable to β-decay—that is, changes in Z. Thus ε is a minimum with respect to Z at fixed A, n_N, V_N, and n_n. Note that

$$\frac{\partial}{\partial Z}\varepsilon_e(n_e) = \frac{d\varepsilon_e}{dn_e}\frac{\partial n_e}{\partial Z} = \mu_e n_N, \tag{8.2.7}$$

by Eqs. (2.6.8) and (8.2.2), where μ_e is the electron chemical potential. We therefore get from Eq. (8.2.1)

$$\mu_e = -\frac{\partial}{\partial Z}(W_N + W_L)_{A,\, n_N,\, V_N,\, n_n}. \tag{8.2.8}$$

Equation (8.2.8) can be rewritten in terms of the nuclear chemical potentials as follows. The chemical potential of the neutrons in the nuclei, $\mu_n^{(N)}$, is the minimum energy to add a neutron to the nucleus—that is,

$$\mu_n^{(N)} = \frac{\partial}{\partial A}(W_N + W_L)_{Z,\, n_N,\, V_N,\, n_n}. \tag{8.2.9}$$

Similarly the proton chemical potential involves a fixed neutron number, $A - Z$:

$$\mu_p^{(N)} = \frac{\partial}{\partial Z}(W_N + W_L)_{A-Z,\, n_N,\, V_N,\, n_n}$$

$$= \frac{\partial}{\partial Z}(W_N + W_L)_{A,\, n_N,\, V_N,\, n_n} + \frac{\partial}{\partial A}(W_N + W_L)_{Z,\, n_N,\, V_N,\, n_n}, \tag{8.2.10}$$

since $\partial A/\partial Z = 1$.[4] Thus the β-stability condition (8.2.8) can be written in the familiar form

$$\mu_e = \mu_n^{(N)} - \mu_p^{(N)}. \tag{8.2.11}$$

The third condition is that the free neutron gas be in equilibrium with the neutrons in the nuclei; that is, it must cost no energy to transfer a neutron from the gas to the nucleus. This involves minimizing ε with respect to A at fixed Z, n_N,

---

[4]Note that all our chemical potentials include rest masses. BBP subtract the nucleon rest masses in their analysis.

V_N, and n. Differentiating Eq. (8.2.3) with respect to A under these conditions gives

$$\frac{\partial n_n}{\partial A} = -\frac{n_N}{1 - V_N n_N}.$$ (8.2.12)

In differentiating $W_N + W_L$ in Eq. (8.2.1), we can use

$$\left.\frac{\partial}{\partial A}\right|_{Z, n_N, V_N, n} = \left.\frac{\partial}{\partial A}\right|_{Z, n_N, V_N, n_n} + \frac{\partial n_n}{\partial A}\left.\frac{\partial}{\partial n_n}\right|_{Z, n_N, V_N, A}.$$ (8.2.13)

Since W_L is independent of n_n, we find, recalling Eq. (8.2.9),

$$n_N \mu_n^{(N)} - \frac{n_N}{1 - V_N n_N}\left[n_N \left.\frac{\partial W_N}{\partial n_n}\right|_{Z, A, n_N, V_N} + (1 - V_N n_N)\frac{d\varepsilon_n}{dn_n}\right] = 0,$$ (8.2.14)

or

$$\mu_n^{(N)} = \mu_n^{(G)},$$ (8.2.15)

where the free neutron chemical potential is defined by

$$\mu_n^{(G)} \equiv \frac{n_N}{1 - V_N n_N}\left.\frac{\partial W_N}{\partial n_n}\right|_{Z, A, n_N, V_N} + \frac{d\varepsilon_n}{dn_n}.$$ (8.2.16)

The term $d\varepsilon_n / dn_n$ is the usual term corresponding to a change in the bulk energy of the neutron gas. The other term in Eq. (8.2.16) corresponds to a change in the nuclear surface energy when a neutron is added to the gas. In fact, the energy of the nuclei per unit of volume occupied by the outside neutron gas is [cf. Eq. (8.2.4)]

$$\frac{(n_N V) W_N}{V_n} = \frac{n_N}{1 - V_N n_N} W_N,$$ (8.2.17)

which is exactly the term differentiated in Eq. (8.2.16).

The fourth equilibrium condition is that there be pressure balance between the neutron gas and the nuclei:

$$P^{(G)} = P^{(N)}.$$ (8.2.18)

This follows from minimizing ε with respect to V_N at fixed Z, A, n_N, and $N_n/V = n_n(1 - V_N n_N)$. Since $n_n = \text{constant}/(1 - V_N n_N)$, we have

$$\frac{\partial n_n}{\partial V_N} = \frac{n_n n_N}{1 - V_N n_N}.$$ (8.2.19)

Also

$$\frac{\partial}{\partial V_N}\bigg|_{Z,A,n_N,n_n(1-V_Nn_N)} = \frac{\partial}{\partial V_N}\bigg|_{Z,A,n_N,n_n} + \frac{\partial n_n}{\partial V_N}\frac{\partial}{\partial n_n}\bigg|_{Z,A,n_N,V_N}. \tag{8.2.20}$$

Thus differentiating Eq. (8.2.1) gives, on writing $1 - V_Nn_N = n_n(1 - V_Nn_N)/n_n$,

$$0 = n_N\frac{\partial}{\partial V_N}(W_N + W_L)_{Z,A,n_N,n_n} + n_N\frac{\partial n_n}{\partial V_N}\frac{\partial W_N}{\partial n_n}\bigg|_{Z,A,n_N,V_N}$$

$$+ n_n(1 - V_Nn_N)\frac{\partial n_n}{\partial V_N}\frac{\partial}{\partial n_n}\left(\frac{\varepsilon_n}{n_n}\right)_{Z,A,n_N,V_N}. \tag{8.2.21}$$

Now

$$P^{(N)} = -\frac{\partial}{\partial V_N}(W_N + W_L)_{Z,A,n_N,n_n}, \tag{8.2.22}$$

and so using Eqs. (8.2.16) and (8.2.19), Eq. (8.2.21) gives Eq. (8.2.18), where

$$P^{(G)} = n_n\mu_n^{(G)} - \varepsilon_n. \tag{8.2.23}$$

Note that the definition of pressure as in Eq. (8.2.23) follows from Eq. (2.1.21) (cf. also Exercise 2.4).

Exercise 8.1   Show that the total pressure is

$$P \equiv n^2\frac{\partial}{\partial n}\left(\frac{\varepsilon}{n}\right) = P^{(G)} + P_e + P_L, \tag{8.2.24}$$

where

$$P_e = n_e^2\frac{\partial}{\partial n_e}\left(\frac{\varepsilon_e}{n_e}\right), \qquad P_L = n_N^2\frac{\partial W_L}{\partial n_N}\bigg|_{Z,A,V_N,n_n}. \tag{8.2.25}$$

To calculate the equation of state, one must now specify the functional forms of W_N, W_L, ε_n, and ε_e. BBP used a compressible liquid drop model of nuclear matter, and wrote

$$W_N = A\big[(1 - x)m_nc^2 + xm_pc^2 + W(k, x)\big] + W_C + W_S. \tag{8.2.26}$$

Here W_C is the Coulomb energy, W_S the surface energy, and $W(k, x)$ the energy per nucleon of bulk nuclear matter of nucleon number density

$$n \equiv \frac{2k^3}{3\pi^2}. \tag{8.2.27}$$

The bulk nuclear energy $W(k, x)$ includes the effects of nucleon-nucleon interactions but excludes surface effects and Coulomb interactions. Inside the nuclei, $n = A/V_N$. For a consistent description, the same function $W(k, x)$ is used for the neutron gas, with $x = 0$; that is,

$$\varepsilon_n = n_n \left[W(k_n, 0) + m_n c^2 \right], \tag{8.2.28}$$

where[5]

$$n_n \equiv \frac{2k_n^3}{3\pi^2}. \tag{8.2.29}$$

The quantity $W(k, x)$ is found by smoothly interpolating the results of many-body calculations done in various limits of k and x. Parameters in $W(k, x)$ are determined by fitting to nuclear data analogously to a semiempirical mass formula.[6] Unlike such a formula, however, $W(k, x)$ depends on the density through k. A semiempirical formula implicitly assumes that the density is that of normal laboratory nuclei. The determination of W from a nuclear potential will be the subject of Sections 8.5–8.11.

The surface energy W_S used by BBP is constructed to vanish explicitly when the density of the neutron gas equals the density of the nucleus.

The leading order term in W_C is $3Z^2e^2/5r_N$, the Coulomb energy of a uniformly charged sphere of radius r_N, where $V_N \equiv 4\pi r_N^3/3$. BBP add several small corrections to this. They combine this leading order term with the lattice energy W_L, to get the result

$$W_{C+L} = \frac{3}{5} \frac{Z^2 e^2}{r_N} \left(1 - \frac{r_N}{r_c} \right)^2 \left(1 + \frac{r_N}{2r_c} \right), \tag{8.2.30}$$

where

$$\frac{4\pi}{3} n_N r_c^3 \equiv 1. \tag{8.2.31}$$

Exercise 8.2 Derive the result (8.2.30) and discuss the limit $r_c/r_N \to 1$. Consider a lattice of nuclei in the Wigner–Seitz cell approximation. Assume each nucleus is a uniformly charged sphere of radius r_N, and that electrons penetrate the nucleus uniformly.

The electrons are to excellent approximation an ideal degenerate Fermi gas. The leading correction to this is already incorporated in W_L, and so one takes for

---

[5] Note that the BBP definition of k_n differs by $2^{1/3}$ from the usual definition in terms of a Fermi wave number k_F.

[6] The semiempirical mass formula for nuclear matter was discussed in Section 2.6.

ε_e the usual extreme relativistic result

$$\varepsilon_e = \frac{\hbar c}{4\pi^2}\left(3\pi^2 n_e\right)^{4/3}. \qquad (8.2.32)$$

Given explicit functions for W_S and $W(k, x)$, the equation of state can now be constructed using the equilibrium conditions (8.2.6), (8.2.11), (8.2.15), and (8.2.18), and determining the pressure from Eq. (8.2.24). Note that the condition for neutron drip to occur (i.e., for a neutron gas to exist) is $\mu_n^{(G)} \geq m_n c^2$. The resulting BBP equation of state is shown in Figures 2.2, 8.1, and 8.5a and the adiabatic index Γ in Figures 2.3 and 8.2.

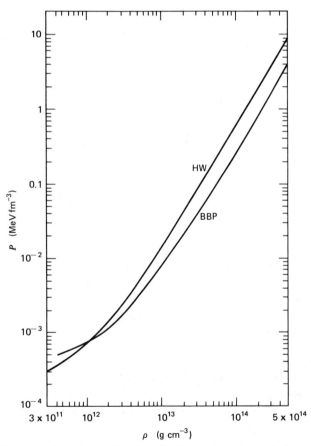

Figure 8.1   The BBP equation of state. The HW (1958) equation of state is shown for comparison. [After Baym et al. (1971a), by permission of North-Holland Publishing Company, Amsterdam.]

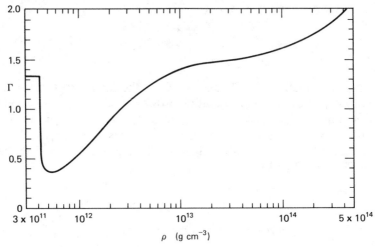

Figure 8.2 The adiabatic index $\Gamma = d\ln P/d\ln \rho$ as a function of ρ for the BBP equation of state. [After Baym et al. (1971a), by permission of North-Holland Publishing Company, Amsterdam.]

The key features of the BBP results are as follows: First, the free neutrons supply an increasingly larger fraction of the total pressure as the density increases. At neutron drip, the pressure is almost entirely due to electrons, but at $\rho = 1.5 \times 10^{12}$ g cm^{-3}, $P_n/P = 0.20$, while at $\rho = 1.5 \times 10^{13}$ g cm^{-3}, $P_n/P = 0.80$.

Second, as neutron drip is approached, $\Gamma \simeq \frac{4}{3}$ (ERD electron gas). Slightly above neutron drip, Γ drops sharply, as can be seen in Figure 8.2. BBP show that this drop is given by

$$\Gamma = \frac{4}{3}\left[1 - a(\rho - \rho_{\text{drip}})^{1/2}\right], \qquad (8.2.33)$$

where a is a positive constant. The reason for the drop is that the low-density neutron gas contributes appreciably to ρ but not much to P. The adiabatic index does not rise above $4/3$ again until $\rho \gtrsim 7 \times 10^{12}$ g cm^{-3}. As we shall see in Chapter 9, this result has the important consequence that *no stable stars can have central densities in the range explored in this section.*[7]

Third, BBP find that the nuclei survive in the matter up to $\rho \sim 2.4 \times 10^{14}$ g cm^{-3}. At this point, the nuclei are essentially touching and at higher densities the lattice disappears and one has a nuclear liquid.

Fourth, above 2.4×10^{14} g cm^{-3}, the electron chemical potential satisfies $\mu_e \gtrsim 104$ MeV $\approx m_\mu$, where m_μ is the muon rest mass. One then has to include

---

[7]There is no problem with having surface layers with densities in this range: it is the *mean* value of Γ that enters the stability criterion; see Eq. (6.7.11)

the muon contribution to the equation of state. BBP terminate their calculations at $\rho \sim 5 \times 10^{14}$ g cm^{-3}, beyond which density standard nuclear-matter theory is no longer applicable.

The BBP equation of state has been criticized in some aspects.[8] In particular, they predict a monotonic increase of Z with A, while other authors favor Z remaining roughly constant at ~ 40 with a better treatment of the surface energy. The resulting $P(\rho)$ relation is not changed much, while Γ is changed somewhat more.

8.3   The Nucleon–Nucleon Interaction

In a review paper, Bethe once estimated that in the preceding quarter century, more man hours of work had been devoted to understanding the problem of the nucleon–nucleon interaction than to any other scientific question in mankind's history—and this remark was made in 1953![9]

We begin our analysis of this problem with a general discussion in this section of the spin and isospin dependence of the potential. This is followed in Section 8.4 by a review of some of the general properties that the potential must have in order to reproduce experimental data. We then discuss in Sections 8.5 and 8.6 the dependence of the potential on the radial separation of the interacting nucleons. We shall focus on one particular model for the potential—the Yukawa potential—which may be regarded as a prototype for more detailed analyses. We employ the Yukawa potential both classically in Section 8.6 and in a quantum mechanical many-body wave equation in Sections 8.7 and 8.8 to illustrate how one obtains an equation of state from a nuclear potential. In the course of this analysis we demonstrate how one calculates the bulk energy $W(k, x)$ from a potential. Finally, we summarize more detailed work along these lines in Sections 8.9 and 8.10 and examine some of the "unresolved issues" in Sections 8.11–8.14. In the forthcoming chapters it will be seen how the microphysical properties of condensed matter influence the structure of equilibrium neutron stars.

Exercise 8.3   Use the results of Section 2.5 to estimate what fraction of nuclear matter at $\rho_{nuc} = 2.8 \times 10^{14}$ g cm^{-3} is composed of neutrons.

As in electromagnetism, in the nonrelativistic limit we may assume that the nuclear force is conservative and independent of the velocity of the nucleus. The force may therefore be derived from a static potential. Nuclear forces, unlike electrostatic forces, do not obey the superposition principle, however. The total interaction in a many-body nuclear system does not reduce to the sum of pairwise

---

[8]See, for example, Canuto (1974) for a discussion and review.
[9]Bethe (1953).

interactions. However, at densities below and near ρ_{nuc}, three-body and higher-order interactions seem to be less important than two-body interactions and so we will ignore them initially.[10]

The potential energy of the interaction between two nucleons depends not only on their separation r but also on their spins. The spin dependence follows from simple considerations of symmetry and spin operator mechanics.[11] The static potential V can depend on only three vectors: $\mathbf{n}$, a unit vector along the radial direction between the particles, and $\mathbf{s}_1$ and $\mathbf{s}_2$, the spin vectors of the two nucleons. We assume that nuclear forces are invariant with respect to rotations, inversions, and time reversal. (A small amount of parity violation has been observed, but it is understood in terms of the "weak" nuclear force, responsible for β-decay, which we are not interested in here.) Thus the potential must be a scalar under rotations, not a pseudoscalar. It also cannot involve the gradient vector, since that is equivalent to a velocity dependent force.

Now any function of a spin $\frac{1}{2}$ operator reduces to a linear function of the spin.[12] There are only two independent scalars that are linear in $\mathbf{s}_1$ and $\mathbf{s}_2$ and formed out of $\mathbf{n}$, $\mathbf{s}_1$, and $\mathbf{s}_2$: $\mathbf{s}_1 \cdot \mathbf{s}_2$ and $(\mathbf{n} \cdot \mathbf{s}_1)(\mathbf{n} \cdot \mathbf{s}_2)$. (Note that $\mathbf{n} \cdot \mathbf{s}_1$ by itself is a pseudoscalar.) If we were to allow velocity dependent forces, we could have terms like $\mathbf{L} \cdot \mathbf{S}$ and $(\mathbf{L} \cdot \mathbf{S})^2$, where $\mathbf{L}$ is the total orbital angular momentum and $\mathbf{S} = \mathbf{s}_1 + \mathbf{s}_2$ is the total spin.

Thus the most general spin-dependent potential is of the form

$$V_{\mathrm{ord}} = V_1(r) + V_2(r)(\boldsymbol{\sigma}_1 \cdot \boldsymbol{\sigma}_2) + V_3(r)\left[3(\boldsymbol{\sigma}_1 \cdot \mathbf{n})(\boldsymbol{\sigma}_2 \cdot \mathbf{n}) - \boldsymbol{\sigma}_1 \cdot \boldsymbol{\sigma}_2\right]. \quad (8.3.1)$$

Here we have written $\mathbf{s}_i = \boldsymbol{\sigma}_i/2$, where $\boldsymbol{\sigma}_i$ are the Pauli spin operators. The third term has been arranged so that it goes to zero on averaging over the directions of $\mathbf{n}$; the forces given by this term are called "tensor forces" and the dependence on $\mathbf{n}$ means that these are noncentral forces.

The suffix "ord" in Eq. (8.3.1) means that these are "ordinary" forces, which do not change the charge state of the nucleon. It is well established experimentally that, ignoring small electromagnetic effects and antisymmetry requirements, the nuclear interaction is the same for two protons as it is for two neutrons or for a neutron and a proton. This charge symmetry is called *isotopic invariance*. Formally, one can regard the proton and the neutron as two different charge states of one particle, the nucleon. This symmetry under interchange of protons and neutrons can be described mathematically with a formalism completely

---

[10] But see below for a discussion of the crucial role of three-nucleon interactions in achieving saturation.

[11] See, for example, Landau and Lifshitz (1977), Sections 116 and 117.

[12] See, for example, Landau and Lifshitz (1977), Section 55.

analogous to the rotation group. The nucleon is represented by a two-component spinor in an abstract group space. The property analogous to ordinary spin is called *isotopic spin*, or *isospin*, denoted by t. The proton has isospin $+\frac{1}{2}$, the neutron $-\frac{1}{2}$. The operator $\tau = 2t$ is simply a Pauli spin matrix acting on spinors in isospin space. The total isospin of a system of nucleons is $\mathbf{T} = \mathbf{t}_1 + \mathbf{t}_2 + \cdots$, with z-component $T_3 = (t_1)_3 + (t_2)_3 + \cdots$. Since the eigenvalue of the operator t_3 is $\frac{1}{2}$ for a proton and $-\frac{1}{2}$ for a neutron, we have $T_3 = Z - A/2$ for a system of Z protons and $A - Z$ neutrons.

Now the *total* wave function of a two-fermion system—that is, the product $\psi(\mathbf{r}_1, \mathbf{s}_1; \mathbf{r}_2, \mathbf{s}_2)\omega(\mathbf{t}_1, \mathbf{t}_2)$, where ω is the isospin part of the wave function—must be *antisymmetrical* with respect to a simultaneous interchange of the r's, s's, and t's. The absolute magnitude of the total isospin $\mathbf{T} = \mathbf{t}_1 + \mathbf{t}_2$ determines the symmetry of the isospin part ω just as $\mathbf{S} = \mathbf{s}_1 + \mathbf{s}_2$ determines the symmetry of the spin part. For two nucleons, we can have $T = 0$ or 1. The triplet state $T = 1$ is symmetrical, with $T_3 = 1$, 0, or -1. It can therefore describe a *pp*, *pn*, or *nn* system. For the singlet state $T = 0$, ω is antisymmetrical, $T_3 = 0$, and one has an *np* system only.

Since the value of T determines the symmetry of ω and, hence, because of the antisymmetry of the total wave function, the symmetry of ψ, conservation of T is equivalent to the existence of a definite symmetry for ψ. This appears to be an exact symmetry of the strong interaction (i.e., when electromagnetic forces are ignored). Note that conservation of T_3 is equivalent to conservation of charge for a fixed number of baryons, and so is an exact symmetry even in the presence of Coulomb forces.

We can use the isospin operators to construct the exchange operator P^τ (sometimes called the "Heisenberg operator") that interchanges $\mathbf{r}_1, \boldsymbol{\sigma}_1$ and $\mathbf{r}_2, \boldsymbol{\sigma}_2$ for two particles. Since $(P^\tau)^2 = 1$, the eigenvalues of P^τ are ± 1, depending on whether the operator acts on a symmetrical or antisymmetrical wave function $\psi(\mathbf{r}_1, \boldsymbol{\sigma}_1; \mathbf{r}_2, \boldsymbol{\sigma}_2)$. Now ψ_{sym} corresponds to an antisymmetric ω—that is, $T = 0$. Similarly ψ_{ant} corresponds to ω with $T = 1$. Thus we can write P^τ in a form that acts on the *isospin* variables of the wave function if it has the properties

$$P^\tau \omega_0 = +\omega_0, \qquad P^\tau \omega_1 = -\omega_1, \tag{8.3.2}$$

where the subscript denotes the value of T. Since $\mathbf{T}^2$ has eigenvalues $T(T + 1)$, we see that

$$P^\tau = 1 - \mathbf{T}^2 = 1 - (\mathbf{t}_1 + \mathbf{t}_2)^2 = -\tfrac{1}{2} - 2\mathbf{t}_1 \cdot \mathbf{t}_2, \tag{8.3.3}$$

where we have used the fact that t_1^2 and t_2^2 have the same definite values $t(t + 1) = \frac{3}{4}$. Thus finally

$$P^\tau = -\tfrac{1}{2}(1 + \boldsymbol{\tau}_1 \cdot \boldsymbol{\tau}_2). \tag{8.3.4}$$

The operator that exchanges the spins of two particles, leaving their coordinates unchanged ("Bartlett operator"), also has eigenvalues ± 1:

$$P^B \psi_{S=0} = -\psi_{S=0}, \qquad P^B \psi_{S=1} = +\psi_{S=1}. \qquad (8.3.5)$$

Comparing with Eq. (8.3.2), we see that

$$P^B = S^2 - 1 = \tfrac{1}{2}(1 + \sigma_1 \cdot \sigma_2). \qquad (8.3.6)$$

The operator that exchanges the particle positions only, leaving the spins unchanged ("Majorana operator"), is

$$P^M = P^B P^\tau = -\tfrac{1}{4}(1 + \sigma_1 \cdot \sigma_2)(1 + \tau_1 \cdot \tau_2). \qquad (8.3.7)$$

Now note that V_{ord} in Eq. (8.3.1) does contain some exchange-type interaction since $\sigma_1 \cdot \sigma_2$ can be rewritten in terms of P^B. As we shall discuss in more detail in Section 8.4, exchange interactions appear necessary to achieve "saturation" of nuclear forces. The most general velocity-independent exchange potential is

$$V(r) = V_{\text{ord}}(r) + V_{\text{exch}}(r), \qquad (8.3.8)$$

where

$$V_{\text{exch}}(r) = \{V_4(r) + V_5(r)(\sigma_1 \cdot \sigma_2) + V_6(r)[3(\sigma_1 \cdot n)(\sigma_2 \cdot n) - \sigma_1 \cdot \sigma_2]\} P^\tau.$$

Exercise 8.4 By rewriting the spin parts of Eq. (8.3.8) in terms of the total spin **S**, show that S^2 but not **S** commutes with $V(r)$. Hence only the magnitude, not the direction, of **S** is conserved by the interaction.

From the result of Exercise 8.4, we conclude that although $J = L + S$ is conserved,[13] **L** need not be separately conserved. It is the tensor force that is responsible for this distinction.

We can classify the allowed states of the two-nucleon system according to the values of T and S. For example, if $T = 1$ and $S = 1$, the coordinate part of the wave function must be antisymmetric under particle interchange (odd parity), and so L must be odd. If $L = 1$, we have the possibilities $J = 0$, 1, or 2. Thus the states[14] are $^3P_0$, $^3P_1$, or $^3P_2$. If $L = 3$, we have $J = 2$, 3, or 4, giving states $^3F_2$, $^3F_3$, $^3F_4$, and so on. Because only J, and not L separately, is in general conserved, the states $^3P_2$ and $^3F_2$ will be mixed. The states $^3P_0$ and $^3P_1$ have no states with which

---

[13] This is why we choose the V's to be functions of r only.

[14] Here we employ conventional spectroscopic notation to specify the states: $^{2S+1}L_J$, where $L = S$, P, D, F, G, and so on corresponds to $L = 0$, 1, 2, 3, 4, and so forth, respectively. The parity of a system of two particles is $(-1)^L$.

to mix. In these cases, conservation of parity allows L to be separately conserved. Table 8.1 shows the lowest few states of the two-nucleon system.

An application of these results is provided by the deuteron, the only bound two-nucleon system (binding energy 2.225 MeV). The ground state has $J = 1$, $T = 0$, and $S = 1$. According to Table 8.1, the lowest L assignment would lead to a mixture of $^3S_1$ and $^3D_1$ for the ground state. Since the magnetic moment of the deuteron is close to the sum of the magnetic moments of the neutron and proton, the ground state must be largely the S state. However, the deuteron has an electric quadrupole moment, which would be absent for a spherically symmetric ground state. This is nicely explained by the small admixture of the D state, and is direct evidence for the existence of the tensor force. (A vector interaction would have selection rules $\Delta L = 0, \pm 1$ and would be unable to mix $L = 0$ and $L = 2$ states.)

Exercise 8.5 By considering the action of $V_1(r) + \sigma_1 \cdot \sigma_2 V_2(r)$ on $S = 0$ and $S = 1$ states, show that one can have binding in the triplet state and not in the singlet state if V_2 is sufficiently attractive. How might this relate to the fact that nn has no bound state, while np does?

8.4   Saturation of Nuclear Forces

Experimental data show that, apart from Coulomb and surface effects, the energy and volume of nuclei increase in direct proportion to A. This property of nuclear forces, already built in to the semiempirical mass formula [Eq. (2.6.4)], is called *saturation* for reasons that will become clear in this section.

Saturation imposes severe constraints on nuclear forces. For example, a rather simple attractive potential of the form

$$V(r) = V_1(r) + V_3(r)\big[3(\sigma_1 \cdot \mathbf{n})(\sigma_2 \cdot \mathbf{n}) - \sigma_1 \cdot \sigma_2\big] \qquad (8.4.1)$$

can be fitted reasonably well to all data for NN (nucleon-nucleon) systems in S

Table 8.1

States of the Two-Nucleon System

T	S	Parity	Possible States	Possible NN
1	1	Odd	$^3P_0, {}^3P_1, ({}^3P_2 + {}^3F_2), {}^3F_3, \ldots$	$nn, pp,$
1	0	Even	$^1S_0, {}^1D_2, {}^1G_4, \ldots$	or np
0	1	Even	$(^3S_1 + {}^3D_1), {}^3D_2, ({}^3D_3 + {}^3G_3), \ldots$	np
0	0	Odd	$^1P_1, {}^1F_3, \ldots$	

states. However, such simple attractive potentials between nucleons *cannot* be the basic law for the nuclear force, as we now show.

The total energy of a nucleus is

$$E = T + W, \tag{8.4.2}$$

where T is the kinetic energy and W the potential energy. The potential energy would be a sum over all pairs of nucleons of $V(|\mathbf{r}_i - \mathbf{r}_j|)$—that is, $A(A - 1)/2$ times a negative function. The quantity T is, because of the exclusion principle, given essentially by Fermi statistics:

$$T \sim AE'_\mathrm{F} \sim An^{2/3} \sim A\left(\frac{A}{R^3}\right)^{2/3} = A^{5/3}R^{-2}, \tag{8.4.3}$$

where R is the radius of the nucleus.

The ground-state value of R can be found by minimizing E. For large A, we have $W \sim A^2 \gg T$, and so it is essentially the minimum of W that determines the ground state. The minimum will therefore occur at R of order the range of nuclear forces, independent of A. (This argument is given in more detail below.) Moreover, the binding energy $E_b \equiv -E_{\min}$ will be proportional to A^2. This is in contradiction with experiment: $R \sim A^{1/3}$ (i.e., constant *density* of nuclei, independent of A), and $E_b \sim A$.

Nuclear forces must have a property that brings about "saturation"; that is, $E_b \sim A$. This occurs in nuclei as well as in assemblies of nucleons at sufficient densities. The force must be attractive for a small number of nucleons, but become repulsive for a larger number. Chemical forces show saturation: two hydrogen atoms combine to form a hydrogen molecule, but a third atom cannot be bound. In the hydrogen molecule the two electrons can occupy the same orbital with opposite spins, but a third electron cannot because of the Pauli principle. For nuclear forces, saturation occurs as a result of several effects. These include the Pauli principle, so-called *exchange forces* arising from terms proportional to P^B and P^τ, and the *repulsive core*, a strongly repulsive potential $V(r)$ at small distances.

A simple model calculation illustrates quantitatively how a purely attractive nuclear potential leads to nuclear collapse as A increases, while the presence of a repulsive core helps bring about saturation.

The average kinetic energy of a system of A nucleons, treated as an ideal nonrelativistic gas, is[15]

$$T = \frac{3}{5}AE'_\mathrm{F} = \frac{3}{10}\left(\frac{3\pi^2}{2}\right)^{2/3}\frac{\hbar^2}{m}An^{2/3}. \tag{8.4.4}$$

---

[15]See Exercise 2.6.

Here we are treating the nucleons as identical particles, so there are $(2t + 1)(2s + 1) = 4$ spin states for each momentum state.

We approximate the interaction between two nucleons to be an attractive square well of range b and depth $-V_0$ (V_0 positive). If p is the probability for any two nucleons to be less than a distance b apart, then the total potential energy is

$$W = -\frac{A(A-1)}{2} p V_0. \tag{8.4.5}$$

If the nucleons are in a spherical nucleus of radius R with uniform and uncorrelated spatial distribution, then

$$P(b, R) = \frac{1}{\Omega^2} \int \int H(b - |\mathbf{r}_1 - \mathbf{r}_2|) \, d\mathbf{r}_1 \, d\mathbf{r}_2, \tag{8.4.6}$$

where H is the step function,

$$H(x) = \begin{cases} 1, & x > 0 \\ 0, & x < 0, \end{cases} \tag{8.4.7}$$

and

$$\Omega = \frac{4}{3}\pi R^3, \qquad n = \frac{A}{\Omega}. \tag{8.4.8}$$

The integral (8.4.6) is carried out in Appendix C. Clearly if $b > 2R$, the step function is always unity and $P = 1$. The general result is

$$p(b, R) = \left(\frac{b}{R}\right)^3 \left[1 - \frac{9}{16}\frac{b}{R} + \frac{1}{32}\left(\frac{b}{R}\right)^3\right], \qquad R > \frac{b}{2},$$

$$= 1, \qquad R < \frac{b}{2}. \tag{8.4.9}$$

Exercise   8.6   Deduce the limit $R \gg b$ of Eq. (8.4.9) by a simple argument.

Exercise   8.7   Estimate the magnitude of $V_0 b^2$ by considering the ground state of the deuteron. Suppose the deuteron potential is a three-dimensional, spherical square well of depth $-V_0$ for $r < b$, and zero for $r > b$. Find the ground state of such a potential and, making the approximation that the ground state energy satisfies $|E| \ll V_0$, show that

$$V_0 b^2 \simeq \frac{\pi^2 \hbar^2}{8\mu} \simeq 100 \text{ MeV fm}^2, \tag{8.4.10}$$

where μ is the reduced mass. (Since $b \sim 1.4$ fm, the pion Compton wavelength, and since, for the deuteron, $E = 2.225$ MeV, the approximation is justified.)

The total energy $E = T + W$ is plotted in Figure 8.3 as a function of R. Clearly there is no minimum for $R < b/2$. For $R > b/2$, we can write

$$E = \frac{\alpha A^{5/3}}{R^2} - \frac{A(A-1)}{R^2} V_0 b^2 \frac{b}{2R}\left[1 - \frac{9}{16}\left(\frac{b}{R}\right) + \frac{1}{32}\left(\frac{b}{R}\right)^3\right], \quad (8.4.11)$$

where

$$\alpha = \frac{3}{10}\left(\frac{9\pi}{8}\right)^{2/3}\frac{\hbar^2}{m} \simeq 30 \text{ MeV fm}^2. \quad (8.4.12)$$

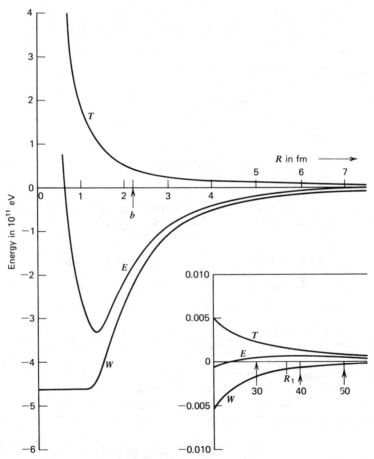

Figure 8.3 A plot of the potential energy W, kinetic energy T, and total energy $E = T + W$ as a function of nuclear radius R for a system of identical neutrons interacting via a purely attractive nuclear potential. The point R_1 is the radius at which E attains its maximum value. [After Blatt and Weisskopf (1952).]

Since the coefficient of the square bracket in Eq. (8.4.11) is much bigger than T, even for moderate values of A, the minimum of E will occur for R only slightly greater than $b/2$. In fact, setting $dE/dR = 0$ gives, with $A - 1 \simeq A$,

$$\frac{3V_0 b^2 A^{1/3}}{4\alpha} \frac{b}{R} \left(1 - \frac{3}{4} \frac{b}{R} + \frac{1}{16} \frac{b^3}{R^3} \right) = 1, \tag{8.4.13}$$

which may be solved numerically for R. In addition to the minimum near $b/2$ [where the bracketed quantity in Eq. (8.4.13) is small], there is also a maximum at $R = R_1 \gg b$, where the factor b/R is small.

The conclusion is clear: for V given by Eq. (8.4.5), nuclei compressed to densities above $n = 3A/4\pi R_1^3$ collapse to a stable state of radius $R \sim b/2$, where b is the range of the nuclear interaction. This state has $E_b = -E \propto A(A - 1) \sim A^2$, in violent disagreement with experiment.

Now suppose there is a repulsive core at short ranges. This results in a "forbidden zone" around each nucleon where the relative nuclear wave function must vanish. Since for a given density the volume available for occupation is thus diminished, the momentum and kinetic energy of the nucleus are increased. By increasing the size of T relative to W, we can shift the location of the minimum of E to more realistic values.

Quantitatively, we write

$$R = r_0 A^{1/3}, \tag{8.4.14}$$

so that $2r_0$ is the average spacing between nucleons. If r_c is the radius of the repulsive core, we can modify the previous expression for the kinetic energy by the replacement

$$T = \frac{\alpha A^{5/3}}{R^2} = \frac{\alpha A}{r_0^2} \rightarrow \frac{\alpha A}{(r_0 - r_c)^2}. \tag{8.4.15}$$

The expectation value of W is not appreciably modified because $r_0 > r_c$. Minimizing the modified version of Eq. (8.4.11) with respect to r_0, we find

$$\frac{3V_0 b^2}{4\alpha} \frac{b}{r_0} \left(1 - \frac{3}{4} \frac{b}{r_0 A^{1/3}} + \frac{1}{16} \frac{b^3}{r_0^3 A} \right) \left(1 - \frac{r_c}{r_0} \right)^3 = 1. \tag{8.4.16}$$

When Eq. (8.4.16) is solved numerically for r_0, one finds for $b = 1.8$ fm and $r_c = 0.4$ fm (a value consistent with high-energy scattering data)

$$0.9 \leqslant r_0 \, (\text{fm}) \leqslant 1.5 \qquad \text{for } 4 \leqslant A \leqslant 216. \tag{8.4.17}$$

This value of r_0 should be compared with the experimental value of ~ 1.2 fm.[16] More importantly, we see that r_0 varies little with A—the repulsive core produces saturation.[17]

To summarize, whatever form is chosen for the nuclear interaction in calculating an equation of state, a key experimental constraint is that the chosen potential reproduce the observed properties of nuclear matter at saturation. In particular, there are four important parameters to be reproduced:

1.   The density at which saturation occurs,

$$n_0 \approx 0.16 \text{ nucleons fm}^{-3}. \tag{8.4.18}$$

2.   and 3.   The energy and *compressibility* of symmetric nuclear matter. These are defined in terms of the quantity $W(k, x)$ introduced in Section 8.2 for the bulk energy of nuclear matter per nucleon. Symmetric nuclear matter ($Z = A/2$) has $x = \frac{1}{2}$. In the vicinity of the saturation density, we can write

$$W\left(k, \tfrac{1}{2}\right) = -W_V + \frac{1}{2}K\left(1 - \frac{k}{k_0}\right)^2, \tag{8.4.19}$$

where experimentally

$$W_V \approx 16 \text{ MeV}, \qquad K \equiv k^2 \left. \frac{\partial^2 W(k, \frac{1}{2})}{\partial k^2} \right|_{k=k_0} \approx 240 \text{ MeV}. \tag{8.4.20}$$

Here $k_0 = 1.33$ fm^{-1} follows from Eqs. (8.2.27) and (8.4.18). Note that the value of K, called the *compression modulus*, has only recently been determined experimentally.[18] Prior to this, a value of 300 MeV was often used as an input parameter for various equation of state calculations.

4.   Finally, the *volume symmetry coefficient* S_V, which measures the curvature of $W(k, x)$ due to changes[19] in x:

$$S_V \equiv \frac{1}{8} \left. \frac{\partial^2 W(k_0, x)}{\partial x^2} \right|_{x=1/2} \approx 30 \text{ MeV}. \tag{8.4.21}$$

---

[16] See, for example, de Shalit and Feshbach (1974) and references therein.

[17] For a quantitative discussion of the role of exchange forces in assisting in the generation of saturation in this simple model, the reader is referred to Blatt and Weiskopf (1952), Chapter 3.

[18] Blaizot et al. (1976); Youngblood et al. (1977), (1978); Lui et al. (1980).

[19] A *surface symmetry coefficient* S_s is sometimes defined in connection with the liquid drop surface energy term $W_s \equiv \omega_s A^{2/3}$, where $\omega_s(x) = \omega_s(\frac{1}{2}) - 4S_s(x - \frac{1}{2})^2$ near $x = \frac{1}{2}$. Mass formula fits to experimental data give $\omega_s(\frac{1}{2}) \approx 20$ MeV rather consistently, but differ appreciably in their estimates of S_s.

Accordingly, we can write for k near k_0 and x near $\frac{1}{2}$,

$$W(k, x) = -W_V + \frac{1}{2}K\left(1 - \frac{k}{k_0}\right)^2 + 4S_V\left(x - \frac{1}{2}\right)^2. \quad (8.4.22)$$

Exercise 8.8 Using the fact that a pure neutron gas is unbound at all densities, discuss the validity of Eq. (8.4.22) as $x \to 0$. Sketch a plausible curve for $W(k,0)$ as a function of k.

Hint: To what must W reduce as k, and hence n, $\to 0$?

Present-day two-nucleon nuclear potentials generally saturate at a density about twice the observed value. The repulsive core is determined to be at $r_c \simeq 0.4$ fm from fitting phase shifts to scattering data above 300 MeV, and cannot be shifted to change the predicted saturation density. In addition to all the effects mentioned previously, it is possible that the Δ resonance of the $N\pi$ system at 1236 MeV may improve the situation, but this has not been reliably calculated yet.

Recent variational calculations[20] of Pandharipande and his co-workers show that while it is not possible to obtain the ground-state energy W_V, equilibrium density n_0 and compressibility K of nuclear matter using *two*-nucleon potentials fit to laboratory scattering data alone, good agreement between theory and experiment *can* be achieved by the introduction of additional *three*-nucleon interactions (TNI). The required TNI contributions are added phenomenologically to the nuclear matter energy in a physically plausible manner.

8.5 Dependence of the NN Potential on the Nucleon Separation

Next we turn to the important (and unresolved) issue of the correct dependence of the dominant nucleon–nucleon (NN) potential on the separation of nucleons. In the absence of a fundamental theory, one chooses a reasonable form for the potential and fits it to phase shift data from low-energy NN-scattering experiments (~ 0–350 MeV), together with other experimentally known properties of nuclear matter (saturation energy, saturation density, symmetry energy, and compressibility of symmetric nuclear matter; deuteron properties, etc.).

Among the most successful potentials for fitting phase shift data is the Reid (1968) potential. This is a superposition of Yukawa-type terms (defined below), which is purely phenomenological in character: it *cannot* be written in the general form of Eq. (8.3.8), but consists of unrelated forms for each partial wave. The

[20]Legaris and Pandharipande (1981); Friedman and Panharipande (1981). Their variational calculations with realistic *two*-nucleon interactions alone yield the equilibrium value $k_0 = 1.7$ fm^{-1} and $-W_V = 17.5$ MeV, which are higher then the empirical values. The TNI contributions can be adjusted to remove the discrepancy.

Reid potential leads to a relatively "soft" equation of state since the average system energy is *attractive* at nuclear densities.

In recent years, more detailed analyses of the NN interaction by Bethe and Johnson (1974), Walecka (1974), Pandharipande and Smith (1975a,b) and others have suggested that the equation of state of condensed matter in the range $10^{14}-10^{15}$ g cm^{-3} is significantly stiffer than that derived from the Reid potential. Such "hard" equations of state result from potentials for which the average system interaction energy is dominated by the attractive part of the potential at nuclear densities, but the repulsive part at higher densities. The equation of state derived by Pandharipande and his colleagues incorporating three- as well as two-nucleon interactions, while less stiff than the earlier two-nucleon models of Pandharipande and Smith, are still stiff compared to the "soft" Reid equation of state.

The harder equations of state give rise to important changes in the structure and masses of heavy neutron stars. In particular, as the interaction energy becomes repulsive above nuclear densities, the corresponding pressure forces are better able to support stellar matter against gravitational collapse. The result is that the maximum masses of stars based on stiff equations of state are greater than those based on soft equations of state. Also, stellar models based on stiffer equations have a lower central density, a larger radius and a thicker crust. Such differences are important in determining mass limits for neutron stars, their surface potentials, moments of inertia, precession frequencies, and so on, which can be indirectly related to observational data.

Before addressing these issues in Chapters 9 and 10, we shall first describe in a naive, but quantitative, fashion how an NN potential can be chosen and then incorporated in a many-body calculation to yield an equation of state. Both the *choice* of the potential and the *technique* for incorporating it in a valid many-body calculation are still unresolved issues. Our discussion will be at best pedagogical in nature. We will choose a simple form for $V(r) = V_1(r)$, ignoring spin- and isospin-dependent interactions. In lieu of a rigorous calculation of $W(k, x)$, we will calculate the interaction energy density $\varepsilon \equiv (W + mc^2)n$ for a system of identical nucleons of mass m and density n, ignoring the dependence of W on x.

8.6 The Yukawa Potential

In 1935 Yukawa made the bold suggestion that just as the electromagnetic force arises from the exchange of a virtual photon, the nuclear force might arise from the exchange of virtual particles, called *mesons*. To explain the finite range of nuclear forces, the meson would have to have a nonzero mass, in contrast with the long-range electromagnetic force mediated by the massless photon.

In Appendix D we discuss the classical version of massive scalar and vector field theories. The scalar field (one field component) describes quanta of spin 0, while the vector theory (three independent field components) describes quanta of spin 1. In the limit of slowly moving particles of "charge" g interacting via scalar or vector fields, we show that the interaction energy is

$$V_{12} = \pm g^2 \frac{e^{-\mu r}}{r}, \tag{8.6.1}$$

where μ is the inverse of the Compton wavelength of the field quanta. This energy corresponds to the term $V_1(r)$ in Eq. (8.3.8) with $V_i(r) = 0$ for $2 \leqslant i \leqslant 6$. Here the plus sign (repulsive force) applies to the vector field, while the minus sign (attraction) is for the scalar field. Note that to have a range $1/\mu \sim 1.4$ fm requires the field quanta to have a mass ~ 140 MeV (assuming the range is comparable to the Compton wavelength of the meson). This value is just the *pion* mass. Pions have spin 0, and so pion exchange presumably is a major component of the attractive nuclear force. On fitting an expression like (8.6.1) to low-energy experimental data, one finds[21] that $g^2/\hbar c \sim 10$. This is the origin of the name "strong nuclear force" (cf. electromagnetism, where $e^2/\hbar c \sim \frac{1}{137}$).

We will now show how the Yukawa potential

$$\phi = \pm g \frac{e^{-\mu r}}{r} \tag{8.6.2}$$

can be used to give the equation of state of nuclear matter in various approximations. Of course, Eq. (8.6.2) is an extremely oversimplified starting point. We use it to show how, given a choice of potential, various approximations are still needed to solve the many-body problem.

We start with a simple classical analysis, based on Zel'dovich (1962). We can calculate the total energy of a system of particles classically by summing over the interactions of all pairs of particles. To facilitate the calculation, we assume that the macroscopic assembly is *uniformly* distributed, thereby neglecting the influence of the interaction on the mean interparticle separation. In other words, we ignore any "correlations" between particle positions due to their mutual interaction. Finally, we assume that the number of particles is sufficiently large that we can replace sums by integrals, and that the characteristic size of the assembly R satisfies $R \gg 1/\mu$.

The interaction energy in a volume $\mathcal{V}$ is then

$$E_{\mathcal{V}} = \frac{1}{2} \sum_{i \neq j} V_{ij} = \pm \frac{1}{2} n^2 g^2 \int \int \frac{e^{-\mu r_{ij}}}{r_{ij}} d\mathcal{V}_i \, d\mathcal{V}_j. \tag{8.6.3}$$

We evaluate Eq. (8.6.3) by first choosing a particle at r_j as the origin, and then

---

[21]For example, the data on low-energy (~ 100 MeV) pp and πN elastic scattering can be fit to a one pion exchange potential (OPEP) model with $g^2/\hbar c \approx 15$ (see, e.g., Perkins, 1972).

integrating over spherical shells of radius $r = r_{ij}$. Since $R \gg 1/\mu$, we make a very small error by ignoring surface effects and taking the integral to infinity:

$$\int_0^\infty \frac{e^{-\mu r}}{r} 4\pi r^2 \, dr = \frac{4\pi}{\mu^2},$$

(8.6.4)

and so, integrating over r_j,

$$E_{\mathrm{cV}} = \pm \frac{1}{2} n^2 g^2 \frac{4\pi}{\mu^2} \mathcal{V}.$$

(8.6.5)

Thus the total energy density is

$$\varepsilon = \varepsilon_{\mathrm{kin}} \pm \frac{2\pi n^2 g^2}{\mu^2}.$$

(8.6.6)

The kinetic energy density can be approximated by the ideal Fermi gas relations in the nonrelativistic and extreme relativistic degeneracy limits (cf. Section 2.3).

$$\varepsilon_{\mathrm{kin}} = nmc^2 + \tfrac{3}{10}(3\pi^2)^{2/3} \frac{\hbar^2}{m} n^{5/3} \quad (\text{NRD}),$$

$$= \frac{(9\pi)^{2/3}}{4} \hbar c n^{4/3} \quad (\text{ERD}).$$

(8.6.7)

In the notation of Sections 8.2 and 8.4, the above model provides a crude calculation of the bulk energy of nuclear matter $W = \varepsilon/n - mc^2$. The equation of state is calculated from

$$P = n^2 \frac{d}{dn}\left(\frac{\varepsilon}{n}\right),$$

(8.6.8)

giving

$$P = P_{\mathrm{kin}} \pm \frac{2\pi n^2 g^2}{\mu^2},$$

(8.6.9)

where

$$P_{\mathrm{kin}} = K n^\Gamma,$$

(8.6.10)

and

$$\Gamma = \tfrac{5}{3} \quad (\text{NRD})$$

$$= \tfrac{4}{3} \quad (\text{ERD})$$

(8.6.11)

[cf. Eq. (2.3.26)].

We thus see that for low densities ($\rho \lesssim \rho_{\text{nuc}}$), where the nuclear force is expected to be attractive, the pressure is *softened* somewhat by the inclusion of the interaction. For very high densities, however, the equation of state is *hardened* due to the dominance of the "repulsive core" Yukawa potentials. As $\rho \equiv \varepsilon/c^2 \to \infty$ (i.e., as $n \to \infty$), the equation of state satisfies

$$P \to \rho c^2. \tag{8.6.12}$$

The sound speed approaches

$$c_s = \left(\frac{dP}{d\rho}\right)^{1/2} \to c, \tag{8.6.13}$$

in contrast to an ideal relativistic gas, for which

$$P \to \frac{1}{3}\rho c^2, \qquad c_s \to \frac{1}{\sqrt{3}}c. \tag{8.6.14}$$

The relevance of this will be discussed further in Chapter 9.

8.7 Hartree Analysis

The zero-order quantum mechanical generalization of the classical calculation described in the previous section is achieved in the nonrelativistic limit via the *Hartree* equations.[22] In this approach, we describe the many-body nucleon system by a product wave function.

$$\Psi = u_1(\mathbf{r}_1)u_2(\mathbf{r}_2) \cdots u_N(\mathbf{r}_N), \tag{8.7.1}$$

where each nucleon i is completely described by its own normalized wave function $u_i(\mathbf{r}_i)$ ($i = 1, 2, \ldots, N$). The assumed form of Ψ therefore does not include the effects of spin, nor of particle correlations since the wave function of particle i is u_i regardless of the position of the other particles.

The omission of spin effects can be overcome by the Hartree–Fock approach, described in the next section. The correlation effects can be included by introducing "correlation functions" into the Hartree–Fock wave function Ψ, and solving the resulting wave equation by a suitable approximation scheme (see Section 8.9).

[22]A detailed description of the Hartree analysis can be found, for example, in Bethe and Jackiw (1968), Chapter 4.

In the Hartree approximation, the ground-state energy of the system is

$$\langle H \rangle = \langle \Psi | H | \Psi \rangle$$

$$= \sum_i \int d\mathcal{V} \, u_i^*(\mathbf{r}) \left(-\frac{\hbar^2}{2m} \nabla^2 \right) u_i(\mathbf{r})$$

$$+ \sum_{i<j} \int \int d\mathcal{V}_1 \, d\mathcal{V}_2 \, V_{12} |u_i(\mathbf{r}_1)|^2 |u_j(\mathbf{r}_2)|^2, \qquad (8.7.2)$$

where H is the total Hamiltonian and, in our case, V_{12} is given by Eq. (8.6.1). Note that in the Hartree analysis,

$$\int |u_i(\mathbf{r})|^2 \, d\mathcal{V} = 1, \qquad (8.7.3)$$

but different u_i are not necessarily orthogonal.

One can derive the Hartree equation by using Eq. (8.7.2) as a variational principle, allowing arbitrary variations in the functions u_i and u_i^*, subject to Eq. (8.7.3). One finds

$$-\frac{\hbar^2}{2m} \nabla^2 u_i + V_i u_i = \varepsilon_i u_i, \qquad i = 1, 2, \ldots, N, \qquad (8.7.4)$$

where the effective potential on particle i is

$$V_i(\mathbf{r}_1) = \sum_{j \neq i} \int d\mathcal{V}_2 \, V_{12}(\mathbf{r}_{12}) |u_j(\mathbf{r}_2)|^2. \qquad (8.7.5)$$

Rather than solve Eq. (8.7.4) for the u_i's self-consistently, which is difficult, we shall find the ground-state energy of the system by a perturbation calculation. We assume that to lowest order the u_i's may be described by *free-particle, plane-wave* states of momentum $\mathbf{p} = \hbar \mathbf{k}$:

$$u_i = \frac{1}{\mathcal{V}^{1/2}} e^{i\mathbf{k} \cdot \mathbf{r}_i}. \qquad (8.7.6)$$

Here we are using box normalization, $\mathcal{V} = L^3$, where $L \gg \mu^{-1}$ is the linear dimension of the box.

Consistent with Fermi statistics, we assume that the lowest energy levels are doubly occupied up to $p = p_F$ or $k = k_F$. Thus sums over i become integrals over $\mathbf{k}$, up to k_F. The product wave function Ψ of Eq. (8.7.1) now represents a uniform density degenerate system. We can therefore think of the u_i's as a rather restricted set of trial functions in Eq. (8.7.2), only appropriate for interaction potentials that are rather weak. This is the lowest-order quantum mechanical analog of the Zel'dovich calculation.

Equation (8.7.2) becomes, using Eq. (8.7.6),

$$\langle H \rangle = \sum_k \frac{p^2}{2m} \pm \frac{1}{2\mathcal{V}^2} \sum_{k,k'} \int\int d\mathcal{V}_1 \, d\mathcal{V}_2 \, g^2 \frac{\exp(-\mu r_{12})}{r_{12}}. \qquad (8.7.7)$$

The double integral in Eq. (8.7.7) can be evaluated exactly as in Eq. (8.6.3), giving

$$\langle H \rangle = \sum_k \frac{p^2}{2m} \pm \sum_{k,k'} \frac{2\pi g^2}{\mathcal{V}\mu^2}. \qquad (8.7.8)$$

Now as usual

$$\frac{1}{\mathcal{V}}\sum_k \rightarrow \frac{2}{h^3}\int d^3p = 2\int \frac{d^3k}{(2\pi)^3}. \qquad (8.7.9)$$

Thus

$$\sum_k 1 = \frac{2}{h^3}\mathcal{V}\int_0^{p_F} 4\pi p^2 \, dp = n\mathcal{V},$$

$$\sum_k \frac{p^2}{2m} = \mathcal{V}\frac{2}{h^3}\int_0^{p_F} \frac{p^2}{2m} 4\pi p^2 \, dp = \mathcal{V}\frac{3}{10}(3\pi^2)^{2/3}\frac{\hbar^2}{m}n^{5/3}. \qquad (8.7.10)$$

Hence we obtain for the total energy density

$$\varepsilon \equiv \frac{\langle H \rangle}{\mathcal{V}} + nmc^2$$

$$= \varepsilon_{\text{kin}} \pm \frac{2\pi n^2 g^2}{\mu^2}, \qquad (8.7.11)$$

exactly the same as the classical result in the nonrelativistic limit [cf. Eqs. (8.6.6) and (8.6.7)].

8.8   Hartree–Fock Analysis

The wave function of an N-fermion system should be antisymmetric under interchange of any pair of particles. This can be accomplished by writing the wave function in the form of a *Slater determinant*:

$$\Psi = \frac{1}{(N!)^{1/2}} \begin{vmatrix} u_1(1) & u_1(2) & \cdots & u_1(N) \\ u_2(1) & & & \\ \vdots & & & \\ u_N(1) & & & u_N(N) \end{vmatrix}. \qquad (8.8.1)$$

Each term in the sum for Ψ is a product of single-particle wave functions of the form

$$u_i(j) = u_i(\mathbf{r}_j)\chi_i(\sigma_j), \qquad (8.8.2)$$

where the spinor $\chi(\sigma)$ is either χ_1 (spin "up") or χ_2 (spin "down"):

$$\chi_1 \equiv \begin{pmatrix} 1 \\ 0 \end{pmatrix}, \qquad \chi_2 \equiv \begin{pmatrix} 0 \\ 1 \end{pmatrix}. \qquad (8.8.3)$$

Here σ is the argument of χ in spin space and takes on the values 1 or 2. If $\chi = \chi_1$, then $\chi(1) = 1$, $\chi(2) = 0$; and likewise if $\chi = \chi_2$, then $\chi(1) = 0$, $\chi(2) = 1$. Unlike the Hartree case, we require orthonormality:

$$\sum_{\sigma_1} \int d\mathcal{V}_1 u_i^*(1) u_j(1) = \delta_{ij}. \qquad (8.8.4)$$

The variational condition

$$\delta\langle\Psi|H|\Psi\rangle = 0 \qquad (8.8.5)$$

gives the usual *Hartree–Fock* equations, which we will not bother to write out.[23] Instead, we will use plane-wave states as trial functions in the ground-state energy. To calculate this energy, note that Ψ is normalized so that any operator of the form

$$F = \sum_i f_i, \qquad (8.8.6)$$

where f_i is a one-fermion operator, will have expectation value

$$\langle\Psi|F|\Psi\rangle = \sum_i \langle i|f|i\rangle, \qquad (8.8.7)$$

---

[23]See, for example, Bethe and Jackiw (1968), Chapter 4, whose notation we adopt in this section.

where $|i\rangle = u_i$. For an operator of the form

$$F = \sum_{i<j} g_{ij}, \qquad (8.8.8.)$$

where g_{ij} is a symmetric two-fermion operator,

$$\langle\Psi|F|\Psi\rangle = \sum_{i<j}[\langle ij|g|ij\rangle - \langle ij|g|ji\rangle]$$

$$= \frac{1}{2}\sum_{i,j}[\langle ij|g|ij\rangle - \langle ij|g|ji\rangle]. \qquad (8.8.9)$$

For the Hamiltonian, we have

$$f_i = -\frac{\hbar^2}{2m}\nabla_i^2, \qquad g_{ij} = V_{ij}. \qquad (8.8.10)$$

The f term and the first g term in Eq. (8.8.9) (called the "direct" term) combine to give

$$\langle H\rangle_{\text{Hartree}},$$

which we have already calculated in Section 8.7. In addition there is the "exchange" term

$$I = -\frac{1}{2}\sum_{i,j}\langle ij|g|ji\rangle$$

$$= -\frac{1}{2}\sum_{i,j}\sum_{\sigma_1,\sigma_2}\int d\mathcal{V}_1\,d\mathcal{V}_2\,u_i^*(\mathbf{r}_1)u_j^*(\mathbf{r}_2)V_{12}u_i(\mathbf{r}_2)u_j(\mathbf{r}_1)$$

$$\times \chi_i^*(\sigma_1)\chi_j^*(\sigma_2)\chi_i(\sigma_2)\chi_j(\sigma_1). \qquad (8.8.11)$$

Since

$$\sum_\sigma \chi_i^*(\sigma)\chi_j(\sigma) = \delta(m_{si}, m_{sj}), \qquad (8.8.12)$$

where $m_s = \pm\frac{1}{2}$ is the z component of spin, we get

$$I = -\frac{1}{2}\sum_{i,j}\delta(m_{si}, m_{sj})\int d\mathcal{V}_1\,d\mathcal{V}_2\,u_i^*(\mathbf{r}_1)u_j^*(\mathbf{r}_2)V_{12}u_i(\mathbf{r}_2)u_j(\mathbf{r}_1)$$

$$= -\frac{1}{2}2\int d\mathcal{V}_1\,d\mathcal{V}_2\,V_{12}|\rho(\mathbf{r}_1,\mathbf{r}_2)|^2, \qquad (8.8.13)$$

where we have defined

$$\rho(\mathbf{r}_1,\mathbf{r}_2) \equiv \sum_{j=1}^{N} u_j^*(\mathbf{r}_2)u_j(\mathbf{r}_1), \tag{8.8.14}$$

and the factor 2 comes from the two possible values of $m_{si} = m_{sj}$.

For plane-wave states [Eq. (8.7.6)], we have

$$\rho(\mathbf{r}_1,\mathbf{r}_2) = \frac{1}{\mathcal{V}} \sum_{\mathbf{k}} e^{i\mathbf{k}\cdot(\mathbf{r}_1 - \mathbf{r}_2)}$$

$$\to \frac{1}{(2\pi)^3} \int e^{i\mathbf{k}\cdot\mathbf{r}_{12}} d^3k. \tag{8.8.15}$$

Note that there is no factor 2 for spin in Eq. (8.8.15); the factor 2 has already been accounted for in Eq. (8.8.13). Writing $\mathbf{k}\cdot\mathbf{r}_{12} = kr_{12}\cos\theta$ and $d^3k = 2\pi\, d(\cos\theta)\, k^2\, dk$ in Eq. (8.8.15), we find

$$\rho(\mathbf{r}_1,\mathbf{r}_2) = \frac{1}{2\pi^2} \frac{1}{r_{12}^3}(\sin k_F r_{12} - k_F r_{12}\cos k_F r_{12}). \tag{8.8.16}$$

Exercise 8.9 Show that $\rho(\mathbf{r}_1,\mathbf{r}_1) = \tfrac{1}{2}n$, where $n \equiv k_F^3/3\pi^2$, and explain why this result is reasonable.

On inserting Eq. (8.6.1) for V_{12} in Eq. (8.8.13), and defining

$$\mathbf{R} = \tfrac{1}{2}(\mathbf{r}_1 + \mathbf{r}_2),$$

$$\mathbf{r} = \mathbf{r}_1 - \mathbf{r}_2 = \mathbf{r}_{12}, \tag{8.8.17}$$

we find

$$I = \mp g^2 \int d^3R\, d^3r\, \rho^2(r)\frac{e^{-\mu r}}{r}. \tag{8.8.18}$$

The integral over $\mathbf{R}$ gives a factor $\mathcal{V}$. The integral over $\mathbf{r}$ can be written in dimensionless form by defining

$$x = k_F r, \qquad \alpha = \frac{\mu}{k_F}. \tag{8.8.19}$$

Then

$$I = \mp g^2 \mathcal{V}\frac{k_F^4}{\pi^3}I(\alpha), \tag{8.8.20}$$

where

$$I(\alpha) = \int_0^\infty \frac{dx}{x^5}(\sin x - x\cos x)^2 e^{-\alpha x}. \tag{8.8.21}$$

The integral can be evaluated by repeated integrations by parts on the term $1/x^n$, until tabulated integrals are obtained. The result is

$$I(\alpha) = \frac{1}{4} - \frac{\alpha^2}{24} - \frac{\alpha}{3}\tan^{-1}\left(\frac{2}{\alpha}\right) + \left(\frac{\alpha^2}{8} + \frac{\alpha^4}{96}\right)\ln\left(1 + \frac{4}{\alpha^2}\right). \quad (8.8.22)$$

Limiting forms are

$$I(\alpha) \to \frac{1}{4}, \qquad \alpha \to 0, \quad (8.8.23)$$

$$\to \frac{1}{9\alpha^2}, \qquad \alpha \to \infty. \quad (8.8.24)$$

Exercise 8.10 Derive the result (8.8.24) by replacing $\sin x - x\cos x$ by its small-x approximation.

Note that $\alpha \sim$ interparticle separation $\div$ interaction range. Our perturbation treatment is therefore strictly applicable only in the limit $\alpha \gg 1$. We thus find

$$I = \mp g^2 \frac{1}{9\pi^3}\frac{k_F^6}{\mu^2}\mathcal{V}$$

$$= \mp g^2 \frac{1}{9\pi^3}\frac{(3\pi^2 n)^2}{\mu^2}\mathcal{V}$$

$$= \mp \frac{g^2 \pi n^2 \mathcal{V}}{\mu^2}. \quad (8.8.25)$$

This is opposite in sign and precisely $\frac{1}{2}$ the magnitude of the direct term in Eq. (8.7.11). The reason is that in this limit only particles with opposite spins can get close enough to interact by the Pauli principle, so the total interaction energy is $\frac{1}{2}$ what it would be without spin effects.

The physical effect of the exchange term is to lessen the role of particle interactions and thus to lower the energy for a repulsive force and increase the energy for an attractive force. Hence the exchange term contributes an effective attraction ($\varepsilon_{exch} < 0$) for repulsive forces and an effective repulsion ($\varepsilon_{exch} > 0$) for attractive forces. Recalling that $\rho \equiv \varepsilon/c^2$, we have in the limit $\alpha \gg 1$

$$\rho = nm + \frac{3}{10}(3\pi^2)^{2/3}\frac{\hbar^2}{mc^2}n^{5/3} \pm \frac{\pi n^2 g^2}{\mu^2 c^2}, \quad (8.8.26)$$

$$P = Kn^{5/3} \pm \frac{\pi n^2 g^2}{\mu^2}. \quad (8.8.27)$$

Equations (8.8.26) and (8.8.27) thus constitute the equation of state obtained from a Yukawa-type interaction in the Hartree–Fock approximation.

Exercise 8.11   The limit $\alpha \ll 1$ is the limit in which the Yukawa potential becomes the Coulomb potential. Use the results of this section to find the exchange correction to the equation of state of a cold, nonrelativistic electron gas (cf. Section 2.4 and Salpeter, 1961).

8.9   Correlation Effects

The Hartree–Fock analysis described in the previous section cannot properly account for *correlations* between nucleons. One way of writing a many-fermion wave function that incorporates two-body correlations is

$$\Psi = F\Phi, \tag{8.9.1}$$

where Φ is a Slater determinant of plane-wave states as in Eq. (8.8.1) and where F is a symmetrized product of two-body correlation functions,

$$F = \prod_{i<j} f_{ij}. \tag{8.9.2}$$

More specifically, Eq. (8.9.1) is

$$\Psi(\mathbf{r}_1,\ldots,\mathbf{r}_N) = A \prod_{i<j} f_{ij}\big(|\mathbf{r}_i - \mathbf{r}_j|\big) \prod_m \phi_m(\mathbf{r}_m), \tag{8.9.3}$$

where A is the antisymmetrization operator acting on spins, isospins, and coordinates and the ϕ_m are plane-wave states, including spin and isospin:

$$\phi_m(\mathbf{r}) = \frac{1}{\mathcal{V}^{1/2}} e^{i\mathbf{k}_m \cdot \mathbf{r}} \chi(\sigma_m) \omega(\tau_m). \tag{8.9.4}$$

A wave function of the form (8.9.3) is called a *Jastrow trial function* and can be used as the basis of a variational calculation to determine the ground-state energy of the system. The Jastrow factors f_{ij} serve to prevent two particles from approaching too close in the presence of a repulsive, hard-core interaction. Accordingly, f_{ij} decreases from unity at large interparticle separations to nearly zero as $|\mathbf{r}_i - \mathbf{r}_j|$ decreases to the core radius, r_c. Other components in the interparticle force (e.g., tensor terms) can induce additional correlations.

A general variational computation consists of minimizing the expectation value of the Hamiltonian by varying the f's:

$$\delta \langle H \rangle = 0, \quad \text{where } \langle H \rangle = \frac{\langle \Psi | H | \Psi \rangle}{\langle \Psi | \Psi \rangle}. \tag{8.9.5}$$

Usually, such variational calculations are done by decomposing $\langle H \rangle$ into a cluster expansion:

$$\langle H \rangle = \sum_{n=1}^{N} E_n. \tag{8.9.6}$$

The n-body cluster contribution E_n involves a $3n$-fold spatial integral from the matrix element in Eq. (8.9.5). If the density of the system is not too high and correlations are of short range, $\langle H \rangle$ can be evaluated in terms of the low-order cluster contributions.

There are many schemes for ordering, truncating, and summing the cluster expansion.[24] Typically, the detailed form of the ground-state energy involves complicated multiple integrals in terms of the ϕ_m and f_{ij}. Accordingly, we will not present them here, but will, in the next section, summarize the results of one such variational calculation (Bethe–Johnson).

An alternative to the variational approach for treating strongly interacting fermion systems is based on the nuclear matter theory of Brueckner, Bethe, and Goldstone (BBG).[25] In lowest order, BBG theory sums the contributions from *two-body* scattering processes. At this order, the theory yields the Hartree–Fock expression for the ground-state energy, but with the "base" potential $V(r)$ replaced by a "dressed" potential $\tilde{V}(r)$, which corrects for multiple virtual particle exchange. The whole approach may be viewed as a perturbation expansion in the parameter nr_c^3, where n is the nucleon density and r_c is the radius of the "hard core" (cf. Section 8.4). For small values of this parameter the independent-pair approximation[26] on which the theory is based is valid; for large values the approximation breaks down.

Traditionally, BBG theory has been accepted as reliable for densities $\lesssim 2\rho_{\text{nuc}}$. Above this density the many-body clusters become more important and a variational treatment is essential. We should mention that at the time of writing, considerable progress is being made with variational treatments that incorporate three- as well as two-nucleon interactions.[27] These calculations can now repro-

---

[24] See reviews by Clark (1979b) and Day (1978) and recent work by Pandharipande; for example, Friedman and Pandharipande (1981), Legaris and Pandharipande (1981), and references therein.

[25] See, for example, de Shalit and Feshbach (1974) for a general discussion of BBG theory.

[26] The independent-pair approximation focuses on the motion of two interacting fermions in the presence of other "spectator" fermions that affect the particle pair only via the Pauli principle.

[27] Legaris and Pandharipande (1981); Friedman and Pandharipande (1981).

duce the results obtained from lowest-order BBG theory in a regime of supposedly joint validity. In addition, as mentioned below, saturation can now be achieved.

8.10   The Bethe–Johnson Equation of State

As an example of a variational calculation, we describe the results of Bethe and Johnson (1974; hereafter BJ), who used the lowest-order "constrained variational technique" of Pandharipande.[28] They calculated the equation of state both for pure neutron matter and for matter with hyperons with masses below the Δ resonance (1236 MeV). They adopted a form for the nuclear interaction similar to the potential used earlier by Reid[29]—that is, a sum of Yukawa functions of different ranges and strengths. The coefficients in the potential were separately adjusted in each partial wave to fit experimental nucleon–nucleon scattering data.

As we saw earlier, the exchange of vector mesons induces repulsive NN forces, while the exchange of scalar mesons induces attractive forces. The three lowest mass vector mesons are the ρ (769 MeV), ω (783 MeV), and the ϕ (1019 MeV). Of these, the ω couples most strongly to nucleons ($g_\omega^2/\hbar c \sim 10 \pm 2$ from high-energy experimental data). Accordingly, BJ included only the ω, with a range $\sim \mu_\omega^{-1} = \hbar/m_\omega c = 0.25$ fm. Since the ω is an isoscalar, the repulsive core should be independent of total isospin T in the NN system. One of the major differences between the Reid and BJ potentials is the incorporation of the repulsive core via ω-exchange.

The BJ potential is thus taken to be of the form

$$V_{\text{BJ}}(r) = \sum_j C_j \frac{e^{-jx}}{x} + V_T(r), \qquad (8.10.1)$$

where

$$x \equiv \mu r, \qquad \mu \equiv \frac{m_\pi c}{\hbar} = 0.7 \text{ fm}^{-1}. \qquad (8.10.2)$$

The coefficients C_j for $j \neq 1$ are chosen to fit experimental data, while C_1 and the tensor interaction V_T (cf. Section 8.3) are taken from a theoretical one-pion-exchange potential model (OPEP).[30] In Eq. (8.10.1), the exchange of (pseudo) scalar π-mesons is employed to get the long-range attractive part of the potential (range j^{-1} times $\mu_\pi^{-1} \simeq 5.5\mu_\omega^{-1}$, where the main attraction is due to 2-π exchange,

---

[28] Pandharipande (1971).
[29] Reid (1968).
[30] de Shalit and Feshbach (1974), Chapter 1, Eq. (3.5).

i.e., $j = 2$), which results in Yukawa terms with *negative* coefficients C_j. The dominant repulsive term is given by

$$V_\omega \equiv g_\omega^2 \frac{e^{-\mu_\omega r}}{r}, \tag{8.10.3}$$

where $g_\omega^2/\hbar c = 29.6$. This disturbingly large value of g_ω^2, about three times the value obtained from high-energy scattering data, results from the BJ fit to low-energy data when the range of repulsion is set exactly equal to μ_ω^{-1}. Note that, unlike the Reid potential, not all of the j's can be integers. V_ω corresponds to $j = 5.5$ in Eq. (8.10.1).

The BJ potentials reproduce experimental phase shifts, nucleon matter binding energy, and the deuteron quadrupole moment as accurately as the Reid potential. Their results for the simplest pure neutron calculation ("Model I") are typical of their general findings, although the range of repulsion in this model is the same as Reid, $j = 7$ in Eq. (8.10.1). The equation of state can be summarized as follows (n in fm^{-3}):

$$\frac{\varepsilon}{n} \equiv W(k,0) + m_n c^2,$$

$$W(k,0) = 236 n^a \text{ MeV/particle}, \tag{8.10.4}$$

$$P = n^2 \frac{d(\varepsilon/n)}{dn} = 364 n^{a+1} \text{ MeV fm}^{-3}$$

$$= 5.83 \times 10^{35} n^{a+1} \text{ dyne cm}^{-2}, \tag{8.10.5}$$

$$c_s^2 = \frac{dP}{d\rho} = \frac{n^a}{1.01 + 0.648 n^a} c^2, \tag{8.10.6}$$

where

$$a = 1.54, \quad 0.1 \leq n \leq 3 \text{ fm}^{-3} \quad \text{or} \quad 1.7 \times 10^{14} \leq \rho \leq 1.1 \times 10^{16} \text{ g cm}^{-3}.$$

Exercise 8.12 Use Eq. (8.10.3) in the plane-wave Hartree–Fock analysis of Section 8.8 to obtain a pure neutron "liquid" equation of state. Compare your results with those of BJ in the domain $0.1 \leq n \leq 3$ fm^{-3}.

Exercise 8.13 Use the expression (8.10.6) for c_s to establish a density above which the BJ equation of state is certain to be invalid.

The key point about the BJ equation of state is its comparative stiffness, corresponding to an adiabatic index $\Gamma = 2.54$. This is considerably stiffer than equations of state obtained from the Reid potential because of the more realistic repulsive core. The greater stiffness results in larger maximum neutron star masses.[31]

Bethe and Johnson also employed their many-body technique to study a hyperonic liquid composed of n, p, Λ, Σ, and Δ particles. They found that indeed the light hyperons with masses < 1250 MeV appear at typical neutron star densities, $n \lesssim 2$ fm^{-3}, depending on the model. However, as in previous studies, the resulting equation of state is not very different from the pure neutron equation of state. There is a slight softening of the equation of state due to the availability of new cells in phase space and the corresponding lowering of the Fermi sea. On the other hand, such calculations are highly uncertain. The potential between hyperons is largely unknown, and is generally taken to be the same as between nucleons.

To show how the appearance of new particles is incorporated in the equation of state, let us first study the appearance of muons in an ideal n-p-e gas. Normally, muons decay to electrons via

$$\mu^- \rightarrow e^- + \nu_\mu + \bar{\nu}_e. \tag{8.10.7}$$

When the Fermi energy of the electrons is high enough, it is energetically favorable for electrons to turn into muons, so that muons and electrons are in equilibrium:

$$\mu^- \leftrightarrow e^-. \tag{8.10.8}$$

In writing Eq. (8.10.8), we have as usual assumed that the neutrinos leave the system. Note that, while a detailed calculation of rates is necessary to verify that the process (8.10.8) is in equilibrium on the timescale of interest, once we know it is in equilibrium, thermodynamics does not require us to know any details of the process. We simply write down the equation for chemical equilibrium,

$$\mu_\mu = \mu_e, \tag{8.10.9}$$

making sure that the appropriate quantities are conserved (in this case, charge). Equilibrium between n, p, and e requires

$$\mu_n = \mu_p + \mu_e, \tag{8.10.10}$$

while charge neutrality requires

$$n_p = n_e + n_\mu. \tag{8.10.11}$$

Given expressions for the chemical potentials and number densities in terms of the density, then Eqs. (8.10.9)–(8.10.11) and the equation for the density are a

---

[31]See Chapter 9.

sufficient set of equations to solve for all the properties of the gas as a function of density. For example, for ideal gases we have

$$m_\mu c^2 \left(1 + x_\mu^2\right)^{1/2} = m_e c^2 \left(1 + x_e^2\right)^{1/2}, \tag{8.10.12}$$

$$m_n c^2 \left(1 + x_n^2\right)^{1/2} = m_p c^2 \left(1 + x_p^2\right)^{1/2} + m_e c^2 \left(1 + x_e^2\right)^{1/2}, \tag{8.10.13}$$

$$\left(m_p x_p\right)^3 = \left(m_e x_e\right)^3 + \left(m_\mu x_\mu\right)^3, \tag{8.10.14}$$

$$\rho = \frac{m_n}{\lambda_n^3} \chi(x_n) + \frac{m_p}{\lambda_p^3} \chi(x_p) + \frac{m_e}{\lambda_e^3} \chi(x_e) + \frac{m_\mu}{\lambda_\mu^3} \chi(x_\mu), \tag{8.10.15}$$

(cf. Section 2.3). Here the x's are the usual dimensionless Fermi momenta. The threshold condition for muons to appear is $n_\mu = 0$; that is, $x_\mu = 0$. Since the electrons are highly relativistic, we can take $x_e \gg 1$. Equations (8.10.12)–(8.10.14) become

$$m_\mu = m_e x_e, \tag{8.10.16}$$

$$m_n \left(1 + x_n^2\right)^{1/2} = m_p \left(1 + x_p^2\right)^{1/2} + m_e x_e, \tag{8.10.17}$$

$$m_p x_p = m_e x_e. \tag{8.10.18}$$

Thus

$$x_n = \left\{ \left[\frac{\left(m_p^2 + m_\mu^2\right)^{1/2} + m_\mu}{m_n} \right]^2 - 1 \right\}^{1/2} = 0.4986,$$

and so $x_p = 0.1126$, $x_e = 206.8$ and $\rho = 8.21 \times 10^{14}$ g cm^{-3}.

Exercise 8.14   Show that there is a range of density below 8.21×10^{14} g cm^{-3} where a muon inserted into an ideal *n-p-e* gas would be stable against the reaction (8.10.7), even though it is not in equilibrium.

Hint: Consider energy and momentum conservation in the muon decay.

Exercise 8.15   Write down the relations between chemical potentials satisfied by an ideal gas containing n, p, e, μ^-, Λ^0, Σ^-, Σ^0, and Σ^+ in equilibrium. Explain why, even though Σ^- has the highest mass of the four hyperons, it appears at the lowest density. (Masses are given in Table E.1.) What is this density?

Computer Exercise 8.16   Construct the equation of state of Exercise 8.15, and plot the number densities of the various species as a function of n in the range $0 \leqslant n \leqslant 10$ fm^{-3}. Thus verify the results of Ambartsumyan and Saakyan (1960), plotted in Figure 8.4. Compare the equation of state with the ideal *n-p-e* equation of state.

The result of Exercise 8.16, that neutrons remain the dominant species in neutron star interiors, explains why the BJ hyperonic equation of state is similar to their pure neutron equation of state.

Like all many-body calculations of the equation of state, the BJ calculation is by no means the final word on the issue. Its limitations include:

1. Failure of the many-body computational technique they used to reproduce the "exact" results of a Monte Carlo calculation by Cochran and Chester (1973) for a hypothetical repulsive potential of the Reid form.
2. Failure of all phenomenological NN potentials to achieve saturation, even with improved calculational techniques.
3. The fitted ω-meson coupling constant $g_\omega^2/\hbar c \geq 20$, which contradicts experimental values of 10 ± 2.

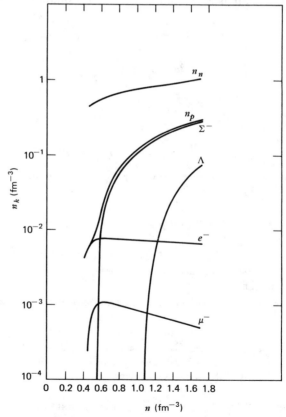

Figure 8.4 The concentrations n_k in a free hyperonic gas as a function of total baryon density n. [After Canuto (1975). Reproduced, with permission, from the *Annual Review of Astronomy and Astrophysics* Volume 13. © 1975 by Annual Reviews Inc.]

4. Naive and inadequate treatment of hyperonic forces, and omission of hyperon "mass shifts" due to the ambient dense medium. Such mass shifts may, for example, postpone the appearance of the Δ until densities exceed 10^{16} g cm^{-3}.

5. Causality violation, $P > \rho c^2$, at high densities (cf. Exercise 8.13).

6. Naive treatment of the Δ resonance as an independent "undressed" stable elementary particle, and the neglect of pion condensation (see Sections 8.11 and 8.12).

In spite of the nontrivial problems, the BJ equation of state remains one of the best to date. The repulsive nuclear core leads to a class of stiff equations of state, which seem to be somewhat more consistent with current observed neutron star data (cf. Chapter 9).

It is interesting that the recent TNI equation of state of Pandharipande and his collaborators, which incorporates three- as well as two-nucleon interactions and leads to good agreement between theory and experiment, is quite similar to the BJ equation of state. Both the TNI and BJ equations of state are stiffer than those based on Reid potentials, but not as stiff as the so-called TI (tensor interaction) model of Pandharipande and Smith (1975a, b) or the RMF (relativistic mean field) model of Walecka (1974). The various models are compared in Box 8.1, Table 8.2, and Figures 8.5a and 8.5b.

8.11   Unresolved Issues: The Δ Resonance

There are a number of nagging problems, requiring difficult calculations, that remain to be resolved before the equation of state can be understood even at densities around $2\rho_{nuc}$. Some of these (e.g., many-body computational schemes) have been mentioned already. While it would be inappropriate in this book to go into detail on all these topics, we do at least wish to summarize some of the problems, and indicate the expected sign of the various effects on the equation of state.

One of the unresolved issues involves the Δ resonance, an excited state of the nucleon with mass 1236 MeV, and quantum numbers $t = \frac{3}{2}, J = \frac{3}{2}$. Pion exchange between two nucleons can produce virtual intermediate states which are NN, NΔ, or ΔΔ. As pointed out by Green and Haapakoski (1974), such attractive pion exchange processes will be suppressed in a dense nuclear medium due to modifications of the intermediate state energy as well as the Pauli exclusion principle (i.e., many of the states are occupied). Accordingly, some of the attractive components of the usual phenomenological two-body potentials, which were fitted to *free-space* NN phase-shift data, should be suppressed. The equation of state should thus be *stiffer* than that given by the free-space potentials.

Sawyer (1972) has given an alternative explanation of this effect. The ambient dense medium tends to alter the self-energy of the Δ state and leads to an *increase* of ≥ 200 MeV in its energy. One can write the chemical potential as

$$\mu_\Delta = \left(p_F^2 c^2 + m_\Delta^2 c^4 \right)^{1/2} + U(p_F) \equiv \left(p_F^2 c^2 + m_\Delta^{*2} c^4 \right)^{1/2}, \quad (8.11.1)$$

where $U(p_F)$ includes the self-energy corrections, and $m_\Delta^* > m_\Delta$ is the effective mass. Increasing the effective mass of the Δ decreases its concentration and so stiffens the equation of state at high densities.

The effect is more pronounced in pure neutron matter as opposed to symmetric nuclear matter. The reason is that nn systems have $T = 1$, while np systems have $T = 1$ or 0 (Table 8.1). However, in the dominant $N\Delta$ process T can take on the values 1 or 2 (Δ has $t = \frac{3}{2}$), and so the process occurs only via the $T = 1$ channel.

The Δ resonance helps lower the saturation density for symmetric nuclear matter, since now the $T = 1$ states become repulsive for $\rho \gtrsim \rho_{nuc}$. The implication of a stiffer equation of state for neutron stars is to lower the density and increase the radius for a given mass.

8.12 Unresolved Issues: Pion Condensation

If one ignores the effects of the strong interactions between pions and nucleons, the criterion for the formation of negatively charged pions in dense nuclear matter via the reaction

$$n \rightarrow p + \pi^- \qquad (8.12.1)$$

is that $\mu_n - \mu_p = \mu_e$ exceed the π^- rest mass, $m_\pi = 139.6$ MeV. As we have seen, $\mu_e \sim 100$ MeV at $\rho \sim \rho_{nuc}$, so one might expect π^- to appear at somewhat higher densities. This would have at least two important consequences: the equation of state would be *softened*, and the rate of neutron star cooling via neutrinos would be enhanced.[32]

Because pions are hadrons and do interact strongly, their properties are greatly influenced by the background matter. Although the lowest order s-wave interactions with nucleons in nuclear matter increase their effective mass, impeding π^- formation, higher-order p-wave interactions have the opposite sign. Present calculations[33] imply that π^- does indeed appear at $\rho \sim 2\rho_{nuc}$, but these results must be regarded as highly tentative.

An interesting and important consequence of the possible appearance of pions, which have spin 0, is that they can form a *Bose–Einstein condensate* at sufficiently low temperatures. An ideal condensate consists of a large number of bosons in a state of zero kinetic energy. To find the critical temperature T_c, recall that the

[32] Neutron star cooling is discussed in Chapter 11.
[33] Migdal (1978).

Box 8.1

Cold Equation of State Above Neutron Drip: Summary

1. The cold equation of state above "neutron drip" ($\rho_{\text{drip}} \approx 4 \times 10^{11}$ g cm^{-3}) may be conveniently divided into two regimes. The range from $\rho \sim \rho_{\text{drip}}$ to $\rho \sim \rho_{\text{nuc}}$, where $\rho_{\text{nuc}} \approx 2.8 \times 10^{14}$ g cm^{-3}, is reasonably well understood (e.g., BBP). Matter in nuclear equilibrium consists of neutron-rich nuclei in a Coulomb lattice, electrons, and free neutrons. As the density increases, free neutrons supply an increasingly larger fraction of the total pressure. At $\rho \sim \rho_{\text{nuc}}$, the nuclei begin to dissolve and merge together.

The high-density range above ρ_{nuc} is not so well understood. The pressure in this regime is dominated by nucleons (mainly neutrons) interacting via strong forces. In addition to neutrons and a small percentage of protons and electrons, other elementary particles and resonance states are likely to be present.

2. The calculation of the equation of state in the nonrelativistic domain from $\rho \sim \rho_{\text{nuc}}$ to $\rho \sim 10^{15}$ g cm^{-3} is subject to two distinct complications: (1) determining the nuclear potential for the nucleon-nucleon interaction, and (2) finding an appropriate technique for solving the many-body problem. The potential is constrained somewhat by nucleon-nucleon scattering data and nuclear matter results.

3. At very high densities above 10^{15} g cm^{-3}, the composition is expected to include an appreciable number of hyperons and the nucleon interactions must be treated relativistically. Unfortunately, relativistic many-body techniques for strongly interacting matter are not fully developed.

4. Presently available nuclear equations of state are subject to many uncertainties, including the possibility of neutron and proton superfluidity, of pion condensation, of neutron solidification, of phase transitions to "quark matter," and the consequences of the Δ resonance.

5. The properties of neutron stars are sensitive to the equilibrium equation of state in the density regime above "neutron drip" (cf. Chapter 9).

6. Representative models for the equation of state above neutron drip are depicted in Figures 8.5a and 8.5b and described in Table 8.2.

Table 8.2

Representative Equations of State Above Neutron Drip

Equation of State	Density Range (g cm^{-3})	Composition	Interactions	Many-Body Theory
Ideal neutron gas (Oppenheimer and Volkoff, 1939; OV)	$0 \leqslant \rho \leqslant \infty$	n	None	Noninteracting neutrons
Baym–Bethe–Pethick (1971a; BBP)	$4.3 \times 10^{11} < \rho \leqslant 5 \times 10^{14}$	e^-, n, and equilibrium nuclide	Reid soft core	Mass formula for nuclei constructed from compressible liquid-drop model
Reid (Pandharipande, 1971; R)	$\rho > 7 \times 10^{14}$	n	Reid soft core adapted to nuclear matter	Variational principle applied to correlation function
Bethe–Johnson (1974; BJ)	$1.7 \times 10^{14} \leqslant \rho \leqslant 3.2 \times 10^{16}$	$n, \rho, \Lambda, \Sigma^{\pm,0}$ $\Delta^{\pm,0}$, and Δ^{++}	Modified Reid soft core	Constrained variational method

Tensor-Interaction (Pandharipande and Smith, 1975a; TI)	n	$\rho > 8.4 \times 10^{13}$	Nuclear attraction due to pion exchange tensor interactions	Constrained variational method
Three-nucleon interaction (Friedman and Pandharipande 1981; TNI)	n	$\rho > 1.7 \times 10^{14}$	Two- and three-nucleon interactions	Constrained variational method
Mean field (Pandharipande and Smith, 1975b; MF)	n	$\rho > 4.4 \times 10^{11}$	Nuclear attraction due to scalar exchange	Mean field approximation for scalar; variational method
Relativistic mean field (Walecka, 1974; RMF)	n	$\rho > 1.7 \times 10^{14}$	Relativistic mean field scalar plus vector exchange fitted to nuclear matter	Relativistic mean field approximation

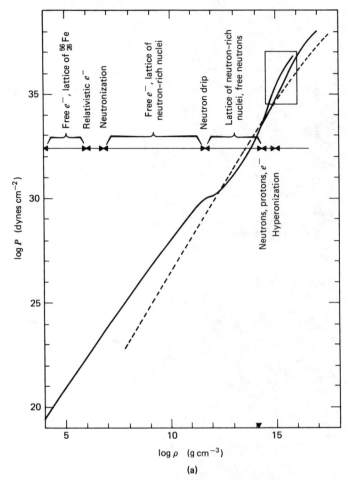

Figure 8.5a The equilibrium equation of state of cold degenerate matter. The *solid* line shows the BPS (1971) equation of state in the region $\rho \leqslant \rho_{\text{drip}} \approx 4.3 \times 10^{11}$ g cm^{-3}, matched smoothly onto the BBP (1971) equation of state in the region $\rho_{\text{drip}} \leqslant \rho \leqslant \rho_{\text{nuc}} \approx 2.8 \times 10^{14}$ g cm^{-3}. The *dashed* line shows for comparison the OV (1939) equation of state for a free neutron gas. Representative equations of state for the region above ρ_{nuc} reside in the rectangular *box* (upper right corner) and are shown enlarged in Figure 8.5b.

maximum value of the chemical potential for bosons of mass m is $\mu = mc^2$. (A larger value would imply negative occupation numbers for some momentum states.) For a given number density n, T_c is determined by setting $\mu = mc^2$. This gives (cf. Section 2.2)

$$n = \frac{g}{h^3} \int \frac{1}{e^{(E-mc^2)/kT_c} - 1} d^3p. \tag{8.12.2}$$

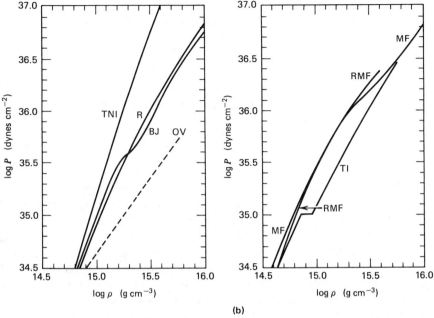

(b)

Figure 8.5b Representative equations of state for cold degenerate matter above $\rho_{nuc} = 2.8 \times 10^{14}$ g cm^{-3}. See Table 8.2 for definitions of letters labeling different curves. [After Arnett and Bowers (1977). Reprinted courtesy of the authors and *The Astrophysical Journal*, published by the University of Chicago Press; © 1977 The American Astronomical Society.]

At low temperatures we can make the nonrelativistic approximation,

$$E - mc^2 = \frac{p^2}{2m}. \tag{8.12.3}$$

Defining a dimensionless variable

$$z = \frac{p^2}{2mkT_c}, \tag{8.12.4}$$

we find

$$n = \frac{g}{\hbar^3 2^{1/2} \pi^2} (mkT_c)^{3/2} \int_0^\infty \frac{z^{1/2} \, dz}{e^z - 1}. \tag{8.12.5}$$

The integral in Eq. (8.12.5) has the value $\pi^{1/2} \zeta(\frac{3}{2})/2$, where ζ is the Riemann zeta function, and so

$$T_c = \frac{3.31}{mk} \left(\frac{n}{g}\right)^{2/3} \hbar^2. \tag{8.12.6}$$

For $T < T_c$, particles with positive kinetic energy are distributed according to Eq. (8.12.2) with T_c replaced by T. By Eq. (8.12.5), $n \sim T^{3/2}$, so

$$n(z > 0) = n \left(\frac{T}{T_c} \right)^{3/2}. \tag{8.12.7}$$

The remaining particles are all in the lowest state with $z = 0$:

$$n(z = 0) = n \left[1 - \left(\frac{T}{T_c} \right)^{3/2} \right]. \tag{8.12.8}$$

Particles in the $z = 0$ state have no momentum (the condensation is in *momentum* space, not physical space), and hence do not contribute to the pressure. As $T \to 0$, virtually all of the bosons reside in this state. Thus it is clear why pion condensation leads to a *softening* of the equation of state. One detailed but very preliminary calculation by Au (1976) gives a 75% reduction in the overall pressure at $\rho \sim 3\rho_{\text{nuc}}$. The detailed manner in which pion condensation softens the equation of state depends quite sensitively on the overall nucleon-nucleon interaction. A fully reliable calculation of the equation of state incorporating both the stiffening effects of isobars (Section 8.11) *and* the softening effect of pion condensation has yet to be performed.

For a simple treatment of the effects of pion condensation, consider the ideal n-p-e gas at $T = 0$ once more, but allow π^- to be produced above threshold. Equilibrium requires that

$$\mu_n - \mu_p = \mu_e = \mu_\pi, \tag{8.12.9}$$

which gives

$$m_n \left(1 + x_n^2 \right)^{1/2} - m_p \left(1 + x_p^2 \right)^{1/2} = m_e \left(1 + x_e^2 \right)^{1/2}, \tag{8.12.10}$$

$$m_e \left(1 + x_e^2 \right)^{1/2} = m_\pi. \tag{8.12.11}$$

In Eq. (8.12.11) we have used the fact that at $T = 0$ the condensed pions all have zero kinetic energy. Charge neutrality requires

$$n_e + n_\pi = n_p, \tag{8.12.12}$$

so

$$\frac{1}{3\pi^2 \lambda_e^3} x_e^3 + n_\pi = \frac{1}{3\pi^2 \lambda_p^3} x_p^3. \tag{8.12.13}$$

The baryon density, mass density, and pressure may be obtained from [cf. Eqs. (2.5.8)–(2.5.10)]

$$n = \frac{1}{3\pi^2\lambda_p^3}x_p^3 + \frac{1}{3\pi^2\lambda_n^3}x_n^3, \tag{8.12.14}$$

$$\rho = \frac{m_e}{\lambda_e^3}\chi(x_e) + \frac{m_p}{\lambda_p^3}\chi(x_p) + \frac{m_n}{\lambda_n^3}\chi(x_n) + m_\pi n_\pi, \tag{8.12.15}$$

$$P = \frac{m_e c^2}{\lambda_e^3}\phi(x_e) + \frac{m_p c^2}{\lambda_p^3}\phi(x_p) + \frac{m_n c^2}{\lambda_n^3}\phi(x_n). \tag{8.12.16}$$

Given, for example, ρ, then Eqs. (8.12.10), (8.12.11), (8.12.13), and (8.12.15) provide four equations for x_n, x_p, x_e, and n_π, and so all the quantities can be determined.

From Eq. (8.12.11) with $x_e \gg 1$, we see that the threshold for π^- production occurs when

$$x_e = \frac{m_\pi}{m_e} = 273.2. \tag{8.12.17}$$

At threshold, $n_\pi = 0$, so Eq. (8.12.13) gives

$$x_p = \frac{m_\pi}{m_p} = 0.1488. \tag{8.12.18}$$

Then Eq. (8.12.10) gives

$$x_n = 0.5843, \tag{8.12.19}$$

and so from Eq. (8.12.15)

$$\rho \equiv \rho_\pi = 1.36 \times 10^{15} \text{ g cm}^{-3}. \tag{8.12.20}$$

For $\rho < \rho_\pi$, the equation of state is exactly as obtained in Section 2.5. For $\rho > \rho_\pi$, x_e remains constant, so n_e and P_e remain constant as ρ increases. An increasing fraction of the negative charge consists of pions that contribute rest-mass density but no pressure.

Computer Exercise 8.17 Compare the cold ideal *n-p-e-π* equilibrium equation of state with the *n-p-e* equation of state without pion condensation. Plot the two curves on a graph of log P (dynes/cm^2) vs. log ρ (g/cm^3). In addition, plot the adiabatic index

$\Gamma \equiv d\ln P/d\ln n$ vs. $\log \rho$ (g/cm^3) for the two cases (cf. Figs. 2.2 and 2.3). Comment on the relative stiffness of the two equations of state above threshold.

Note that in equilibrium we have

$$\pi^+ + \pi^- \leftrightarrow 2\gamma \leftrightarrow \pi^0, \qquad (8.12.21)$$

so that

$$\mu_{\pi^0} = 0, \qquad (8.12.22)$$

$$\mu_{\pi^+} = -\mu_{\pi^-} = -\mu_e < 0. \qquad (8.12.23)$$

Thus for both π^0 and π^+, the distribution function satisfies

$$f = \frac{1}{e^{(E-\mu)/kT} - 1} \to 0 \quad \text{for all } p \geqslant 0 \text{ as } T \to 0.$$

(Recall that E includes the particle rest-mass energy.) Thus π^0 and π^+ are not present in an ideal gas at $T = 0$. Similarly, the existence of π^- precludes the existence of K^- and of *all* other mesons, positive or negative. (Recall that the π^- is the lowest mass negatively charged boson.) Similarly, positrons and antibaryons are excluded. If interactions are taken into account, however, these conclusions are no longer strictly valid. In particular, the possibility of π^0 condensation has been considered.[34]

Pion condensation may also enhance the possibility that neutron matter at sufficiently high density *solidifies*. It is not inconceivable that the short-range, repulsive core of the NN potential might be strong enough to lock neutrons into regularly arranged lattice sites.[35] The result would be neutron stars with solid inner cores as well as solid outer crusts. Anticipating our discussion of neutron stars below, we note that such structure would lead to several direct observational consequences:

1. "Starquakes" induced by the release of elastic energy in seismically active cores; this might result in occasional speedups in the rotation rate of pulsars, as observed, for example, in the Vela pulsar.[36]
2. Gravitational radiation from coherent solid core oscillations[37].

---

[34] See, for example, Baym and Pethick (1979) and references therein.
[35] Recall our discussion of white dwarf solidification via the Coulomb repulsive field between nuclei in Section 4.3.
[36] Ruderman (1969).
[37] Dyson (1969).

3. Neutron star wobble due to an oblate, rotating solid crust or core; such wobble might turn X-ray emission from accreting gas on and off at regular intervals, as is observed in the 35-day cycle of the X-ray binary, Hercules X-1.[38]

Now it is clear that an *infinite* "hard-core potential" by construction gives rise to a solid structure as soon as the interparticle spacing approaches the core radius, r_c. At these densities, the infinitely strong repulsion effectively "cages" each neutron in its own crystal lattice site. Of more relevance, however, is the question of whether or not a more realistic Yukawa-type repulsive soft core (e.g., the Bethe–Johnson potentials) can result in solidification. In this case it is not clear whether or not the potential is too soft, allowing particles to tunnel through it, never producing the required localization for crystallization except possibly at extremely high densities.[39]

Work on solidification encounters the same difficulties discussed previously in determining the nuclear equation of state: the form of potential and the many-body computational technique. Results of several investigations,[40] though very uncertain, suggest that in the absence of pion condensation, solidification does *not* take place. However, Pandharipande and Smith (1975a) have demonstrated that π^0 condensation may provide a possible mechanism for solidification. The π^0 field serves to enhance the effective tensor force in dense matter leading to spatial ordering. The solid or liquid nature of dense nuclear matter, however, is not resolved as of this writing.

8.13 Unresolved Issues: Ultrahigh Densities

At densities significantly greater than ρ_{nuc} (e.g., $\gtrsim 10\rho_{nuc}$), it is no longer possible to describe nuclear matter in terms of a nonrelativistic many-body Schrödinger equation or in terms of a potential. The "meson clouds" surrounding the nucleons begin to overlap and the picture of localized individual particles interacting via two-body forces becomes invalid. Even before this breakdown of the *concept* of a potential, different potentials, all of which reliably reproduce low-energy phase shift data, result in different equations of state. The reason is that for $\rho \sim 10^{15}$ g cm^{-3}, the potentials are sensitive to the repulsive core region unaffected by the phase-shift data.

A typical approach to the equation of state in this regime (Walecka, 1974) is to construct a *relativistic* Lagrangian that allows "bare" nucleons to interact attractively via scalar meson exchange and repulsively via the exchange of a more

---

[38]Lamb et al. (1975).
[39]See, for example, Canuto (1975) for a review and references.
[40]See, for example, Baym and Pethick (1975) for a review and references.

massive vector meson ω. In the nonrelativistic limit, both the classical and quantum theories yield Yukawa-type potentials.[41] Making a suitable "mean-field approximation," Walecka finds that at the highest densities, the vector meson exchange dominates and one still has the Zel'dovich result

$$P \to \rho c^2, \qquad c_s \to c. \qquad (8.13.1)$$

Though very promising, these and similar calculations are at best preliminary forays into the ultrahigh density domain. It is not clear whether *any* theory based on interacting nucleons and mesons will ultimately prove satisfactory, or whether a theory based directly on quark interactions will be required.

An alternative approach to the equation of state at ultrahigh densities is based on the assumption that a whole host of baryonic resonant states arise at high density. Statistical hadron models have been developed by several investigators.[42] The pioneering model of Hagedorn is typical: a baryon resonance mass spectrum is assumed to be given by

$$N(m)\,dm \sim m^a e^{m/m_0}\,dm, \qquad (8.13.2)$$

where $N(m)\,dm$ is the number of resonances between masses m and $m + dm$. Existing data on baryonic resonances can be fitted to Eq. (8.13.2) with $m_0 \simeq 160$ MeV and $-\frac{7}{2} \lesssim a \lesssim -\frac{5}{2}$.

In equilibrium, the threshold condition for a mass m to appear is $mc^2 = \mu_n$, so the neutron chemical potential sets the limit on the maximum resonance mass at every density. For asymptotically large densities, Eq. (8.13.2) gives

$$n = \int_0^{\mu_n} N(m)\,dm \sim m_0 \mu_n^a e^{\mu_n/m_0}. \qquad (8.13.3)$$

Since new resonances are appearing exponentially fast, the highest mass states present are nonrelativistic. Thus the density is asymptotically

$$\rho \sim \int_0^{\mu_n} mN(m)\,dm \sim m_0 \mu_n^{a+1} e^{\mu_n/m_0} \qquad (8.13.4)$$

$$\sim n\mu_n. \qquad (8.13.5)$$

The pressure is

$$P = n^2 \frac{d}{dn}\left(\frac{\rho c^2}{n}\right) \sim n^2 c^2 \frac{d\mu_n}{dn} \sim \frac{n^2 c^2}{dn/d\mu_n}. \qquad (8.13.6)$$

[41] Recall the results of Sections 8.6 and 8.7.
[42] Hagedorn (1970); Leung and Wang (1973); Frautschi et al. (1971).

Now Eq. (8.13.3) gives asymptotically

$$\frac{dn}{d\mu_n} \sim \mu_n^a e^{\mu_n/m_0} \sim \frac{n}{m_0},$$ (8.13.7)

while Eq. (8.13.4) gives

$$\ln \rho \sim \frac{\mu_n}{m_0}.$$ (8.13.8)

Thus Eq. (8.13.6) gives, with Eqs. (8.13.7) and (8.13.8),

$$P \sim \frac{\rho c^2}{\ln \rho}.$$ (8.13.9)

Hagedorn actually obtains

$$P = \frac{\rho c^2}{\ln(\rho/\rho_0)},$$ (8.13.10)

where $\rho_0 = 2.5 \times 10^{12}$ g cm^{-3} in one calculation.

As one might expect from an equation of state that creates new "particles" continuously with increasing density, rather than enlarging the Fermi sea of a single species, the Hagedorn equation of state is remarkably soft. Moreover, no repulsive interactions were included in the model. Thus the velocity of sound in such matter is given by

$$c_s^2 = \frac{dP}{d\rho} = \frac{c^2}{\ln(\rho/\rho_0)} \left[1 - \frac{1}{\ln(\rho/\rho_0)} \right].$$ (8.13.11)

Note that $c_s \to 0$ as $\rho/\rho_0 \to \infty$, in striking contrast to the mean-field treatment that gave $c_s \to c$.

There is no hard theoretical or experimental evidence yet to decide between these two extremes. However, (i) the use of data for "free" decaying baryons to determine a resonance mass formula as in Eq. (8.13.2) is almost certainly wrong. In principle, the mass shift due to the dense medium [cf. Eq. (8.11.1)] can become sufficiently large to *eliminate* virtually all of the higher baryon resonance states.[43] The net effect would be to substantially stiffen the Hagedorn equation of state. (ii) Preliminary estimates of neutron star masses from observations tend to rule out soft Hagedorn-type equations of state for densities below a few times ρ_{nuc}. These predict maximum neutron star masses $\lesssim 0.7 M_\odot$, well below "observed"

---

[43] Canuto (1975); Sawyer (1972).

values (see Chapter 9). Of course, such observations say nothing about the validity of the Hagedorn equation of state at very high densities $\rho \gg \rho_{nuc}$, where the existence of stable neutron stars is unlikely.

8.14 Unresolved Issues: Quark Matter

There is growing evidence that the fundamental building blocks of all strongly interacting particles (e.g., N, Δ, π, ρ,...) are *quarks*. If this is true, any fundamental description of nuclear matter at high density must involve quarks. Nuclei begin to "touch" at baryon density $\sim (4\pi r_n^3/3)^{-1} \sim$ few ρ_{nuc}; here $r_n \sim 1$ fm, the characteristic nucleon radius. Just above this density, one might speculate that matter will undergo a phase transition at which quarks would begin to "drip" out of the nucleons. The result would be quark matter, a degenerate Fermi liquid.

Since no quarks have been observed in the free state in any experiment, they are believed to be permanently confined to the interior of hadrons by a force that increases as one tries to separate the quarks. However, the current theory of quark-quark interactions ("quantum chromodynamics") suggests that quark interactions become arbitrarily weak as the quarks are squeezed closer together ("asymptotic freedom"). Collins and Perry (1975) have therefore suggested that at sufficiently high densities, quark matter may be treated to first approximation as an ideal, relativistic Fermi gas. What would the asymptotic form of the equation of state be if this were true? We address this question immediately below.

Quarks can be transformed into other kinds of quarks (new "flavors") by weak interactions. In neutron stars, one is likely to be above threshold only for the three lightest quarks, u, d, and s (cf. Appendix E). One has

$$d \rightarrow u + l + \bar{\nu},$$

$$s \rightarrow u + l + \bar{\nu}. \tag{8.14.1}$$

Here l denotes e^- or μ^-, and $\bar{\nu}$ is the associated antineutrino. Assuming β-equilibrium and neglecting the neutrinos as usual, we have

$$\mu_d = \mu_u + \mu_l, \tag{8.14.2}$$

$$\mu_s = \mu_u + \mu_l. \tag{8.14.3}$$

Now for an extremely relativistic degenerate Fermi gas,

$$n_i \propto g_i \mu_i^3, \tag{8.14.4}$$

where

$$g_i = 2, \quad i = l,$$

$$g_i = 6, \quad i = u, d, \text{ or } s. \tag{8.14.5}$$

The 6 for quarks is the product of two spin states and three "color" states for each kind of quark.[44] Thus Eqs. (8.14.2)–(8.14.4) give

$$\mu_d = \mu_s \quad \text{and} \quad n_d = n_s. \tag{8.14.6}$$

Equilibrium between μ^- and e^- gives[45]

$$\mu_\mu = \mu_e \equiv \mu_l,$$

$$n_\mu = n_e \equiv n_l. \tag{8.14.7}$$

Charge neutrality implies

$$\tfrac{2}{3} n_u - \tfrac{1}{3} n_d - \tfrac{1}{3} n_s - n_\mu - n_e = 0, \tag{8.14.8}$$

which reduces to

$$n_u - n_s - 3n_l = 0. \tag{8.14.9}$$

Now, using Eqs. (8.14.3) and (8.14.4),

$$\left(\frac{n_s}{6}\right)^{1/3} = \left(\frac{n_u}{6}\right)^{1/3} + \left(\frac{n_l}{2}\right)^{1/3}. \tag{8.14.10}$$

Defining

$$x = \frac{n_e}{n_s}, \qquad y = \frac{n_u}{n_s}, \tag{8.14.11}$$

we find from Eqs. (8.14.9) and (8.14.10)

$$y - 1 = 3x, \tag{8.14.12}$$

$$1 = y^{1/3} + (3x)^{1/3}. \tag{8.14.13}$$

The only real solution to Eqs. (8.14.12) and (8.14.13) is $y = 1$, $x = 0$. Thus asymptotically we have

$$n_u = n_s = n_d, \qquad n_e = n_\mu = 0. \tag{8.14.14}$$

---

[44] See Appendix E for a brief discussion of quarks and elementary particles.
[45] When $\mu_e < m_\mu c^2$, no muons are present and Eq. (8.14.7) does not hold a priori. However, the equilibrium composition, Eq. (8.14.14), is unchanged.

The key feature of this model is that at high density the extreme relativistic free particle result applies; that is,

$$P \rightarrow \tfrac{1}{3}\rho c^2, \qquad \rho \rightarrow \infty. \qquad (8.14.15)$$

This is a relatively soft equation of state.

For finite densities, quark interactions must be taken into account. At moderately high densities, one could use a perturbation expansion in the strong interaction coupling constant α_s since the quarks are asymptotically free. At lower densities, however, confinement occurs. A popular phenomenological model is the MIT "bag" model of Chodos et al. (1974). Here quarks in the nucleon are confined to a finite region of space or "bag" whose volume is held finite by a confining pressure $B > 0$ called the "bag constant" (i.e., B is the energy density needed to "inflate" the bag). Reasonable fits to observed masses of hadrons are obtained in the model for $m_u = m_d \simeq 0$, $B \simeq 55$ MeV fm^{-3}, and $\alpha_s \equiv g_s^2/16\pi\hbar c$ $\simeq 0.55$, where g_s is the quark-gluon coupling constant. The energy density of quark matter is then given by the noninteracting Fermi contribution ($\propto n^{4/3}$) plus B.

Calculations of the pure neutron to quark phase transition in the bag model have been performed by a number of groups.[46] For all the neutron equations of state considered, the phase transition occurs *above* the maximum density for a stable neutron star.[47] However, the bag model is only phenomenological, and our understanding of strong interactions is at present only tentative. The existence of stable "quark stars" is still an unresolved issue.

---

[46] Baym and Chin (1976b); Chapline and Nauenberg (1976); Keister and Kisslinger (1976).
[47] But see, for example, Fechner and Joss (1978) for an alternative model.

9

Neutron Star Models: Masses and Radii

In this chapter we shall discuss equilibrium models for neutron stars. Cold nonrotating models are constructed by solving the OV equation of hydrostatic equilibrium, Eq. (5.7.6), for a specified equation of state $P = P(\rho)$. One can construct a sequence of models parametrized by ρ_c; models with $dM/d\rho_c > 0$ are stable in this density range, while those with $dM/d\rho_c < 0$ are unstable (cf. Sections 6.8 and 6.9).

The key uncertainty in neutron stars models is the equation of state of nuclear matter, particularly above nuclear densities (recall $\rho_{\mathrm{nuc}} = 2.8 \times 10^{14}$ g cm^{-3}). As we have discussed in Chapter 8, in spite of considerable progress in recent years, there are still some important theoretical issues that remain to be resolved. It is nevertheless interesting that our present understanding of condensed matter is already adequate to place rather stringent limits on the masses of stable neutron stars, as we shall see below.

For white dwarfs, observations of masses and radii are used as confirmations of the astrophysical models.[1] For neutron stars, because of the uncertainties in the equation of state, observations of masses and radii are instead used to test theories of nuclear physics!

9.1 Neutron Stars: History of the Idea and Discovery

In 1934 Baade and Zwicky proposed the idea of neutron stars, pointing out that they would be at very high density and small radius, and would be much more

----

[1]Section 3.6.

gravitationally bound than ordinary stars. They also made the remarkably prescient suggestion that neutron stars would be formed in supernova explosions.[2]

The first calculation of neutron star models was performed by Oppenheimer and Volkoff (1939), who assumed matter to be composed of an ideal gas of free neutrons at high density. Work on neutron stars at this time focused mainly on the idea that neutron cores in massive normal stars might be a source of stellar energy. When this motivation faded as the details of thermonuclear fusion become understood, neutron stars were generally ignored by the astronomical community for the next 30 years. However, this was by no means universally so. For example, the papers of Harrison, Wakano, and Wheeler (1958), Cameron (1959a), Ambartsumyan and Saakyan (1960), and Hamada and Salpeter (1961) contain detailed discussions of the equation of state and neutron star models, and the book by Harrison, Thorne, Wakano, and Wheeler (1965) contains an extensive discussion. A reason often given for the neglect of the neutron star idea is that because of their small area, their residual thermal radiation would be too faint to observe at astronomical distances with optical telescopes.

However, the discovery of cosmic, nonsolar X-ray sources by Giacconi et al.[3] in 1962 did generate a great flurry of interest in neutron stars. A sizeable number of theorists independently speculated that the X-ray telescope was observing a young, warm neutron star, and they feverishly began to calculate the cooling of neutron stars.[4] The identification of the first "quasi-stellar object" (QSO, or quasar) by Schmidt at Mt. Palomar in 1963 triggered further interest in neutron stars. This interest stemmed from the possibility that the large redshifts of spectral lines observed for quasars[5] might be attributed to the gravitational redshift at the surface of a compact object. Arguments showing[6] that the largest quasar redshift already exceeded the maximum gravitational redshift from a stable neutron star soon dispelled any connection between quasars and (isolated) neutron stars.

In any case, with the discovery of X-ray sources and quasars, dozens of theoreticians focused their attention on the equilibrium properties of compact stars and on star collapse. But in spite of this mounting theoretical effort, most

---

[2] Baade and Zwicky (1934): "With all reserve we advance the view that supernovae represent the transitions from ordinary stars into *neutron stars*, which in their final stages consist of extremely closely packed neutrons."

According to Rosenfeld (1974), on the day that word came to Copenhagen from Cambridge telling of Chadwick's discovery of the neutron in 1932, he, Bohr, and Landau spent the evening discussing possible implications of the discovery. It was then that Landau suggested the possibility of cold, dense stars composed principally of neutrons. Landau's only publication on the subject was concerned with neutron cores (Landau, 1938).

[3] Giacconi, Gursky, Paolini, and Rossi (1962).

[4] Chapter 11 is devoted to this subject.

[5] The first QSO identified by Schmidt, 3C273, had a redshift $\delta\lambda/\lambda = 0.158$, which was unprecedented for a normal "star."

[6] Salpeter (1965); in addition to this argument there was strong evidence that quasar redshifts were cosmological in origin.

physicists and astronomers did not take the possibility of neutron stars (let alone black holes!) very seriously. Probably the vast extrapolation from familiarly known physics was the most important reason for their attitude!

All this changed when pulsars were discovered in late 1967.[7] Gold (1968) proposed that they were rotating neutron stars, and this is generally accepted today (see Chapter 10).

Since 1968, there has been much theoretical work on properties of neutron stars. This was further stimulated by the discovery of pulsating, compact X-ray sources ("X-ray pulsars") by the UHURU satellite in 1971. These are believed to be neutron stars in close binary systems, accreting gas from their normal companion stars. Although the idea of accreting binary systems for X-ray sources had been proposed earlier, the first conclusive evidence for periodicity was found in the sources Cen X-3 and Her X-1.[8]

The near simultaneous discoveries[9] of the Crab and Vela pulsars in the late Fall of 1968, both of which are situated in supernova remnants, provided evidence for the formation of neutron stars in supernova explosions. The Crab nebula, for example, is the remnant of the supernova explosion observed by Chinese astronomers in 1054 A.D.[10]

Optical and X-ray observations of binary X-ray sources allow one to determine the neutron star masses in some of these systems. The discovery of the first binary pulsar by Hulse and Taylor (1975) also provides an opportunity to measure the mass of a neutron star and, as we shall see later, to test for the existence of gravitational radiation.

At this time of writing, about 350 pulsars are known, three of which are in binary systems. Over 300 compact X-ray sources are known, about 19 of which show periodicity, and so are probably in binary systems.

9.2   Ideal Gas Equation of State in the Nuclear Domain

A first approximation to the structure of a neutron star is provided by assuming that the degenerate gas consists of noninteracting particles. Indeed, the dimensional argument of Landau presented in Section 3.4 assumed an ideal degenerate gas of neutrons and gave $M_{\max} \sim 1.5 M_\odot$ at $R \sim 3$ km for the maximum neutron star mass and corresponding radius. In this section, we consider the properties of a neutron star based on successively more sophisticated *ideal* gas equations of

---

[7] Hewish, Bell, Pilkington, Scott, and Collins (1968).

[8] Schreier et al. (1972); Tananbaum et al. (1972a).

[9] Vela: Large, Vaughan, and Mills (1968); Crab: Staelin and Reifenstein (1968) and Richards and Comella (1969) (radio pulses); Cocke, Disney, and Taylor (1969) (optical pulses).

[10] Duyvendak (1942) and Mayall and Oort (1942). Baade (1942) and Minkowski (1942) identified the "south preceding star," near the center of the Crab nebula, as the likely (collapsed) remnant of the star that exploded in 1054.

state. In the next section we will consider models based on more realistic equations of state.

First, one can treat the case of a pure, ideal neutron gas. The equation of state was derived in Chapter 2. The first numerical treatment of neutron star structure, by Oppenheimer and Volkoff, adopted this equation of state together with the general relativistic equation of hydrostatic equilibrium [the OV equation; Eq. (5.7.6)]. They obtained the equilibrium configurations shown in Figure 9.1. Note that

$$M_{max} = 0.7 M_\odot, \qquad R = 9.6 \text{ km}, \qquad \rho_c = 5 \times 10^{15} \text{ g cm}^{-3} \qquad (9.2.1)$$

are the parameters of the maximum mass configuration. As discussed in Section 6.8, configurations with $\rho_c > 5 \times 10^{15}$ g cm^{-3} are unstable to gravitational collapse. The Chandrasekhar mass limit for a neutron star (i.e., a Newtonian $n = 3$ polytrope of infinite density) is $5.73 M_\odot$. Relativity lowers this value for two main reasons: (i) the maximum mass occurs at a *finite* value of ρ_c where the neutrons are becoming relativistic, but are not extremely relativistic; (ii) $5.73 M_\odot$ is the *rest mass* of the neutrons—the total mass is smaller than this because of the gravitational binding energy of the star.

Exercise   9.1   Use Eqs. (5.3.3) and (5.6.8) to show that the maximum redshift z of a spectral line emitted from the surface of a stable neutron star is

$$z \equiv \frac{\Delta\lambda}{\lambda} = 0.13, \qquad (9.2.2)$$

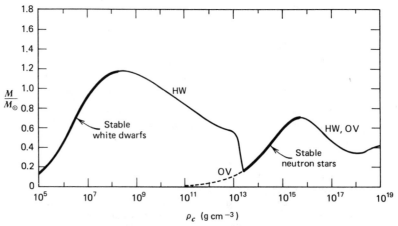

Figure 9.1   Gravitational mass versus central density for the HW (1958) and OV (1939) equations of state. The stable white dwarf and neutron star branches of the HW curve are designated by a *heavy solid* line.

assuming an ideal neutron equation of state. Use the Oppenheimer–Volkoff results for M and R.

Low-density neutron stars with the ideal neutron gas equation of state can be approximated by $n = \frac{3}{2}$ Newtonian polytropes [Eq. (2.3.27)]. Using the formulas for polytropes in Section 3.3, we find

$$R = 14.64 \left(\frac{\rho_c}{10^{15} \text{ g cm}^{-3}} \right)^{-1/6} \text{km}, \tag{9.2.3}$$

$$M = 1.102 \left(\frac{\rho_c}{10^{15} \text{ g cm}^{-3}} \right)^{1/2} M_\odot, \tag{9.2.4}$$

$$= \left(\frac{15.12 \text{ km}}{R} \right)^3 M_\odot. \tag{9.2.5}$$

Thus in the Oppenheimer–Volkoff calculation, there is no minimum neutron star mass: $M \rightarrow 0$, $R \rightarrow \infty$ as $\rho_c \rightarrow 0$. In reality, of course, neutrons become unstable to β-decay at a sufficiently low density.

We can give an approximate derivation of the OV results using the energy variational principle of Chapter 6. Since ρ_c at $M_{\max}$ is at the transition from nonrelativistic to relativistic neutrons, it is not clear what kind of polytrope best approximates a neutron star near $M_{\max}$. Exercise 9.2 shows that using $n = 3$ in our variational principle is inconsistent; we accordingly adopt $n = \frac{3}{2}$. The results accurately describe low-density neutron stars, and give an estimate of $M_{\max}$.

From Eqs. (6.10.9) and (6.10.11), we have

$$E_{\text{int}} = k_1 K \rho_c^{2/3} M, \qquad k_1 = 0.795873. \tag{9.2.6}$$

The value of K is given in Eq. (2.3.27). Equations (6.10.10) and (6.10.12) give

$$E_{\text{grav}} = -k_2 G \rho_c^{1/3} M^{5/3}, \qquad k_2 = 0.760777. \tag{9.2.7}$$

To compute ΔE_{int}, we first compute

$$u = \frac{\varepsilon_n - m_n c^2 n_n}{\rho_0}, \tag{9.2.8}$$

where

$$\rho_0 = m_n n_n = \frac{m_n x^3}{3\pi^2 \lambda_n^3}, \tag{9.2.9}$$

and $x \ll 1$ is the relativity parameter. Using Eq. (2.3.19), we find

$$u = c^2 \left(\frac{3}{10} x^2 - \frac{3}{56} x^4 \right). \tag{9.2.10}$$

The first term in Eq. (9.2.10) gives E_{int}, and it is the second term that gives ΔE_{int}:

$$\Delta E_{\text{int}} = -\frac{3}{56} c^2 \int x^4 \, dm. \tag{9.2.11}$$

Substituting for x in terms of ρ_0 from Eq. (9.2.9), and then making the usual substitutions for a polytrope in the integral, we get

$$\Delta E_{\text{int}} = -k_3 \frac{\hbar^4}{m_n^{16/3} c^2} M \rho_c^{4/3}, \tag{9.2.12}$$

where

$$k_3 = \frac{3}{56} (3\pi^2)^{4/3} \frac{1}{|\xi_1^2 \theta'(\xi_1)|} \int_0^{\xi_1} \theta^{3.5} \xi^2 \, d\xi$$

$$= 1.1651. \tag{9.2.13}$$

Finally, Eqs. (6.9.30) and (6.9.31) give

$$\Delta E_{\text{GTR}} = -k_4 \frac{G^2}{c^2} M^{7/3} \rho_c^{2/3}, \tag{9.2.14}$$

where

$$k_4 = 0.6807. \tag{9.2.15}$$

The total energy of the star is then

$$E = A M \rho_c^{2/3} - B M^{5/3} \rho_c^{1/3} - C M \rho_c^{4/3} - D M^{7/3} \rho_c^{2/3}, \tag{9.2.16}$$

where

$$A = k_1 K, \qquad B = k_2 G, \qquad C = \frac{k_3 \hbar^4}{m_n^{16/3} c^2}, \qquad D = k_4 \frac{G^2}{c^2}. \tag{9.2.17}$$

Equilibrium is achieved when $\partial E / \partial \rho_c = 0$. This relation simplifies to

$$2 A \rho_c^{-1/3} - B M^{2/3} \rho_c^{-2/3} - 4 C \rho_c^{1/3} - 2 D M^{4/3} \rho_c^{-1/3} = 0. \tag{9.2.18}$$

Keeping only the first two terms, one recovers the $n = \frac{3}{2}$ polytrope result (9.2.4). Keeping all the terms gives a better approximation to the M vs. ρ_c relation of Oppenheimer and Volkoff.

The onset of instability occurs when $\partial^2 E/\partial \rho_c^2 = 0$. This relation becomes, on simplifying,

$$-2A\rho_c^{-1/3} + 2BM^{2/3}\rho_c^{-2/3} - 4C\rho_c^{1/3} + 2DM^{4/3}\rho_c^{-1/3} = 0. \quad (9.2.19)$$

Add Eqs. (9.2.18) and (9.2.19) to get

$$\rho_c = \frac{BM^{2/3}}{8C}. \quad (9.2.20)$$

Substitute this back in Eqs. (9.2.18), and write

$$y = M^{4/9}. \quad (9.2.21)$$

The resulting equation is a cubic equation in y:

$$2A - 3B^{2/3}C^{1/3}y - 2Dy^3 = 0. \quad (9.2.22)$$

The positive root of this occurs at $y = 6.605 \times 10^{14}$ cgs, so that

$$M = 1.11M_\odot, \qquad \rho_c = 7.43 \times 10^{15} \text{ g cm}^{-3}. \quad (9.2.23)$$

Note that M determined in Eq. (9.2.23) is really the rest mass of the neutrons (the polytropic formulas use $dm = 4\pi r^2 \rho_0\, dr$). If we substitute the values of M and ρ_c from Eq. (9.2.23) back into Eq. (9.2.16), we find

$$\frac{E}{c^2} = -0.08M_\odot, \quad (9.2.24)$$

and so the maximum *total* mass of the neutron star is predicted to be $1.03M_\odot$, about 40% too high.

Exercise 9.2

(a) Repeat the calculation of $M_{\max}$ using the $n = 3$ polytropic relations as in Section 6.10. You should find $M = 0.741M_\odot$ at $\rho_c = 3.46 \times 10^{17}$ g cm^{-3}.

(b) Recall that in this case you dropped a constant term in the calculation of ΔE_{int} [cf. Eq. (6.10.15)]. The omitted term is exactly $-m_n c^2 N$, the negative of the rest-mass energy of the star. Hence E/c^2 is now the total mass of the star. Compute this quantity using the results of part (a), and show that it is *bigger* than $0.741M_\odot$. This means that the star is unbound, and the approximation scheme has failed. To get some idea of why this might be so, compute the ratio $\Delta E_{\text{GTR}}/E_{\text{grav}}$ at the values of M and ρ_c found in part (b).

Exercise 9.3 Consider a neutron star supported by an ideal, degenerate neutron gas assumed to be *nonrelativistic at all densities*; that is, $P \equiv K\rho_0^{5/3}$ for all ρ_0.

(a) Use the post-Newtonian variational method to derive the M vs. ρ_c relation for equilibrium configurations and show that it agrees with Eq. (9.2.4) in the low-density, Newtonian domain.

(b) Show that in the post-Newtonian approximation *all* equilibrium configurations are radially stable and that $M \rightarrow 3.38 M_\odot$ as $\rho_c \rightarrow \infty$.

(c) A numerical integration of the OV integration for this equation of state actually gives $M_{max} = 0.84 M_\odot$ for the maximum total mass at a central density $\rho_{c,max} = 5.4 \times 10^{15}$ g cm^{-3}, with $dM/d\rho_c < 0$ for all $\rho_c > \rho_{c,max}$. Discuss the origin of the discrepancy.

Hint: Consider E/c^2 and M/R for large ρ_c in the post-Newtonian approximation.

A more realistic ideal gas equation of state considers an equilibrium mixture of noninteracting neutrons, protons, and electrons. This equation of state was analyzed in Section 2.5. Recall that neutrons are present only at densities exceeding 1.2×10^7 g cm^{-3}. The neutron/proton ratio reaches a maximum at about 7.8×10^{11} g cm^{-3}, and then decreases to 8 as $\rho \rightarrow \infty$. As always for an ideal gas, $P \rightarrow \rho/3$ as $\rho \rightarrow \infty$.

Stellar models constructed with this equation of state do not differ much from the OV results, since $P(\rho)$ is very similar to the pure neutron case. For example,

$$M_{max} = 0.72 M_\odot, \qquad R = 8.8 \text{ km}, \qquad \rho_c = 5.8 \times 10^{15} \text{ g cm}^{-3}. \quad (9.2.25)$$

Stars with $\rho_c \lesssim 7.8 \times 10^{11}$ g cm^{-3} really belong to the high-density branch of white dwarfs, and are therefore unstable. We thus expect there to be a local minimum in the M vs. ρ_c curve near this value of ρ_c. The corresponding minimum neutron star mass is, from Eq. (9.2.4),

$$M_{min} \simeq 0.03 M_\odot, \qquad R \simeq 48 \text{ km}. \quad (9.2.26)$$

More realistically, recall that the Harrison–Wheeler equation of state (Section 2.6) smoothly joins the white dwarf equation of state (including the neutron drip regime) to the ideal n-p-e gas at $\rho = 4.5 \times 10^{12}$ g cm^{-3}. The resulting M vs ρ_c curve is shown in Figure 9.1. The minimum neutron star mass according to HW is

$$M_{min} \simeq 0.18 M_\odot, \qquad R \simeq 300 \text{ km}, \qquad \rho_c \simeq 2.6 \times 10^{13} \text{ g cm}^{-3}.$$

Analogous results would apply for the BPS equation of state (Section 2.7) matched to an ideal n-p-e gas. In Sections 9.3 and 9.4 we shall give better estimates of the minimum and maximum neutron star masses based on more realistic equations of state.

Exercise 9.4 Suppose that at densities above nuclear ($> 10^{14}$ g cm^{-3}), neutrons aggregate into droplets containing N nucleons per drop. Assume that all droplets obey ideal gas Fermi–Dirac statistics, and use the Landau argument of Section 3.4 to do the following:

(a)   Estimate the maximum mass of a neutron star supported by cold nucleon droplets and compare this mass with the corresponding value for a star composed of free neutrons.

(b)   Estimate the radius of the maximum mass configuration composed of nucleon droplets and compare this radius with the corresponding value for a star composed of free neutrons.

(c)   Determine the ratio of the surface potentials for the two neutron star models.

(d)   Suppose now that instead of coagulating into droplets, the neutrons divide into N equal mass, "fundamental" particles at high densities. How do your answers to (a), (b), and (c) change? Can the maximum mass of a neutron star be increased indefinitely by an infinite succession of subdivisions?

9.3   Realistic Theoretical Models

Having sketched in Chapter 8 some of the ingredients of a realistic equation of state for cold, dense matter, we can now compare the actual neutron star models

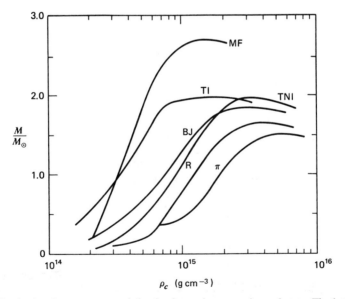

Figure 9.2   Gravitational mass vs. central density for various equations of state. The letters labeling the different curves are identified in Table 8.2 with the exception of π, which denotes a Reid equation of state modified by charged-pion condensation. The ascending portions of the curves represent stable neutron stars. [After Baym and Pethick (1979). Reproduced with permission, from the *Annual Review of Astronomy and Astrophysics*, Vol. 17. © 1979 by Annual Reviews Inc.]

that result when different equations of state are employed in the OV equation. Figure 9.2 shows the M vs. ρ_c curves for six representative equations of state; Figure 9.3 shows the corresponding M vs. R curves. Table 9.1 lists the maximum masses calculated for four of these models. Several general features emerge:

1. Stars calculated with a stiff equation of state (e.g., BJ, TNI, TI, and MF in Figure 9.2) have greater maximum masses than stars derived from a soft equation of state (e.g., R and π).

2. Stars derived from a stiff equation of state have a lower central density, a larger radius, and a much thicker crust (cf. Figure 9.4 and discussion below) than do stars of the same mass computed from a soft equation of state.

3. Pion condensation, if it occurs, tends to contract neutron stars of a given mass as well as decrease M_{max}.

   In Figure 9.4 we show cross-sectional slices of two $1.4M_\odot$ stars computed from the soft Reid and stiff TI equations of state. The "layering" of the configurations is simply a consequence of the onset of different regimes in the equation of state

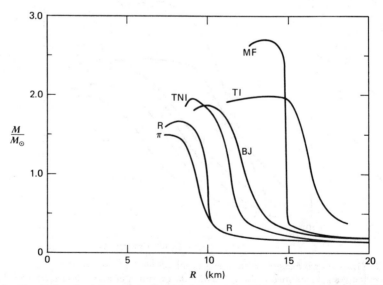

Figure 9.3 Gravitational mass versus radius for the same equations of state depicted in Figure 9.2. [After Baym and Pethick (1979). Reproduced with permission, from the *Annual Review of Astronomy and Astrophysics*, Vol. 17. © 1979 by Annual Reviews Inc.]

Table 9.1
The Maximum Mass of a Neutron Star
For Various Equations of State

Equation of State[a]	Maximum Mass $(M_\odot)$
π	1.5
R	1.6
BJ	1.9
TNI	2.0
TI	2.0
MF	2.7

[a]See Table 8.2 and the caption to Figure 9.2 for identification of the equation of state abbreviations.

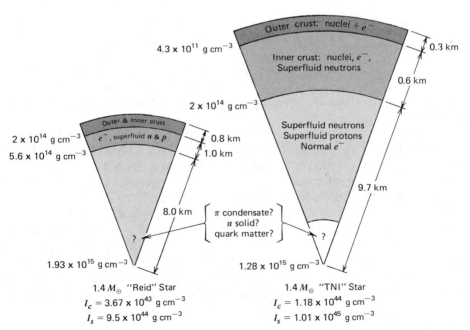

Figure 9.4 Cross sections of $1.4 M_\odot$ Reid and TNI stars illustrating the various regions discussed in the text. The moments of inertia of the entire crust, I_c, and of the superfluid interior, I_s, are also shown for comparison. [After Pines (1980a).]

as one proceeds to higher densities. The layers may be identified as follows[11]:

1. The *surface* ($\rho \lesssim 10^6$ g cm^{-3}), a region in which the temperatures and magnetic fields expected for most neutron stars can significantly affect the equation of state.

2. The *outer crust* (10^6 g cm$^{-3} \lesssim \rho \lesssim 4.3 \times 10^{11}$ g cm^{-3}), a solid region in which a Coulomb lattice of heavy nuclei coexists in β-equilibrium with a relativistic degenerate electron gas (cf. white dwarf equations of state).

3. The *inner crust* (4.3×10^{11} g cm$^{-3} \lesssim \rho \lesssim (2$–$2.4) \times 10^{14}$ g cm^{-3}), which consists of a lattice of neutron-rich nuclei together with a superfluid neutron gas and an electron gas.[12]

4. The *neutron liquid* ($(2$–$2.4) \times 10^{14}$ g cm$^{-3} \lesssim \rho \lesssim \rho_{core}$), which contains chiefly superfluid neutrons with a smaller concentration of superfluid protons and normal electrons.

5. A *core* region ($\rho > \rho_{core}$), which may or may not exist in some stars and depends on whether or not pion condensation occurs, or whether there is a transition to a neutron solid or quark matter or some other phase physically distinct from a neutron liquid at densities above some critical value ρ_{core}.

If the equation of state is stiff, the central density of a relatively massive $1.4M_\odot$ neutron star is $\lesssim 10^{15}$ g cm^{-3}; in fact, even the most massive, stable neutron stars have $\rho_c \lesssim$ few $\times 10^{15}$ g cm^{-3} (Figure 9.2). Thus, as pointed out previously, the possibility of a transition to quark matter or some other exotic form of matter seems unlikely. However, the existence of a third stable branch of "quark stars" on the M vs. ρ_c diagram beyond white dwarfs and neutron stars remains a possibility.

Neutron stars with masses near the Chandrasekhar limit of $1.4M_\odot$ may be favored in nature.[13] Such stars with moderately stiff (e.g., TNI) equations of state are not likely to form pion condensates, which seem to require $\rho \gtrsim 2\rho_{nuc}$ if they form at all. However, given the uncertainties in the equation of state described in the previous chapter, these conclusions are at best tentative.

The *minimum* mass of a stable neutron star is determined by setting the *mean* value of the adiabatic index Γ equal to the critical value $\sim \frac{4}{3}$ for radial stability against collapse. As we discussed in Section 8.2, $\Gamma(\rho)$ for the BBP equation of state drops rapidly below $\frac{4}{3}$ at $\rho_{drip} \approx 4.3 \times 10^{11}$ g cm^{-3} and does not rise above $\frac{4}{3}$ again until ρ exceeds 7×10^{12} g cm^{-3}. The result is that the minimum neutron star mass calculated using the BBP equation of state matched to the BPS equation

---

[11]Pandharipande, Pines, and Smith (1976).
[12]For a brief discussion of the superfluid properties of neutron star matter, see Section 10.9.
[13]See, for example, Section 9.4 and Chapter 18.

of state below neutron drip is

$$M_{\min} = 0.0925 M_\odot, \qquad \rho_c = 1.55 \times 10^{14} \text{ g cm}^{-3}, \qquad R = 164 \text{ km.} \qquad (9.3.1)$$

The reason that ρ_c for the minimum mass configuration is very much larger than 7×10^{12} g cm^{-3} is first that it is the *mean* value of Γ that is relevant for stability, and second that this mean value must be *greater* than $\frac{4}{3}$ for stability in general relativity [cf. Eq. (6.9.2)]. Since the equation of state is reasonably well understood for all $\rho \lesssim \rho_c$, we can regard the above parameters for the minimum mass configuration as rather well established.

The same positive statement cannot yet be made for the *maximum* mass equilibrium configuration, due to uncertainties in the equation of state above $\rho_{\text{nuc}} \approx 2.4 \times 10^{14}$ g cm^{-3}.[14] Note, however, that all present microscopic calculations of the equation of state lead to neutron stars with $M_{\max} \lesssim 3M_\odot$. This result will have important implications for the identification of black holes (Section 13.5). We shall return to the crucial issue of the maximum neutron star mass in Section 9.5.

9.4   Observations of Neutron Star Masses

As we have seen, global neutron star parameters such as masses, radii, moments of inertia, and so on are sensitive to the adopted microscopic model of the nucleon-nucleon interaction. Thus, astronomical observations of these parameters can shed light on hadronic physics. Of key importance is the determination of neutron star masses from direct observations. Recent observations have determined the masses of a few neutron stars by the following means:

(a)   X-Ray Binaries

The most reliable means of determining astronomical masses is via Kepler's Third Law, and neutron stars are no exception. Consider two spherical masses M_1 and M_2 in circular orbit about their center of mass.[15] Figure 9.5 shows such a system as seen in the orbital plane. The separation of the two masses is a, and their distances from the center of mass are a_1 and a_2:

$$a = a_1 + a_2, \qquad (9.4.1)$$

$$M_1 a_1 - M_2 a_2 = 0, \qquad (9.4.2)$$

---

[14] Recall Figure 9.2 and Table 9.1.
[15] The case of elliptical orbits is treated in Section 16.5.

The angle i is called the inclination of the orbital plane to the line of sight. From the diagram, it is clear that any spectral feature emitted from M_1 will be Doppler shifted. The amplitude of the variation is v_1, the projection of the orbital velocity of M_1 along the line of sight:

$$v_1 = \frac{2\pi}{P} a_1 \sin i, \tag{9.4.3}$$

where P is the orbital period. Thus if the spectrum of M_1 shows periodic variations, then P and v_1 can be measured and hence one gets $a_1 \sin i$. Alternatively, for X-ray pulses one can measure periodic variations in the time of arrival of pulses. The amplitude of these variations is simply the light travel time across the projected orbit—that is, $(a_1 \sin i)/c$.

Now Kepler's Law says that

$$\frac{G(M_1 + M_2)}{a^3} = \left(\frac{2\pi}{P}\right)^2. \tag{9.4.4}$$

From Eqs. (9.4.1) and (9.4.2) we have

$$a = \frac{M_1 + M_2}{M_2} a_1, \tag{9.4.5}$$

and so

$$f(M_1, M_2, i) \equiv \frac{(M_2 \sin i)^3}{(M_1 + M_2)^2} = \frac{Pv_1^3}{2\pi G}. \tag{9.4.6}$$

The quantity f is called the "mass function" and depends only on the observable quantities P and v_1 (or $a_1 \sin i$).

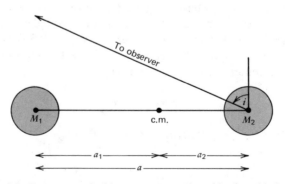

Figure 9.5 Parameters of a binary system as viewed in the orbital plane.

If only one mass function can be measured for a binary system, then one cannot proceed further than Eq. (9.4.6) without additional assumptions. For six X-ray binaries, it has been possible to measure both the mass function for the optical companion as well as the X-ray mass function. One has

$$f_X = \frac{(M_O \sin i)^3}{(M_X + M_O)^2}, \quad f_o = \frac{(M_X \sin i)^3}{(M_X + M_O)^2}, \quad (9.4.7)$$

where X refers to the X-ray source and O to the optical companion. The ratio of these two expressions gives the mass ratio,

$$q \equiv \frac{M_X}{M_O}, \quad (9.4.8)$$

and then we can write from Eq. (9.4.7)

$$M_X = \frac{f_X q(1 + q)^2}{\sin^3 i}. \quad (9.4.9)$$

A unique value for M_X still depends on knowing $\sin i$. In practice, observations of the X-ray eclipse duration and/or variations in the optical light curve are used[16] to set geometrical constraints on $\sin i$. In this way, the mass determinations shown in Figure 9.6 were made. Included in Figure 9.6 are estimates of the uncertainties of the mass determinations.

(b) Binary Pulsars

In the case of the Hulse–Taylor binary pulsar (PSR 1913 + 16), only one mass function is available from the radio observations. The companion is a compact object of some kind, but has not been directly observed. However, the high precision of pulsar observations combined with the relatively high orbital velocity has allowed measurements of the general relativistic periastron advance and the second-order Doppler shift. These have given two more relations between M_1, M_2, a_1, and $\sin i$, in addition to Eqs. (9.4.3) and (9.4.6), and allowed a complete solution for all the parameters.[17] The best values for the masses are[17,18]:

$$M_{\text{pulsar}} = 1.41 \pm 0.06 M_\odot,$$

$$M_{\text{companion}} = 1.41 \pm 0.06 M_\odot. \quad (9.4.10)$$

---

[16] See, for example, Bahcall (1978a).

[17] See Section 16.5 and Footnote 12 therein for a full discussion; the observation of the effect of gravitational radiation from the system is discussed there.

[18] Taylor and Weisberg (1982).

Two other binary pulsars have since been detected. However, the observations suggest that relativistic effects will not be important in these and so only the mass function is likely to be determined from the data.

Limits on the masses of two *nonpulsating* X-ray binaries have been estimated (Cyg X-1: $9 \lesssim M/M_\odot \lesssim 15$; 3U $1700-37$: $0.6 \lesssim M/M_\odot$). However, as will be discussed in Section 13.5, at least one of these objects, Cyg X-1, is likely to be a black hole.

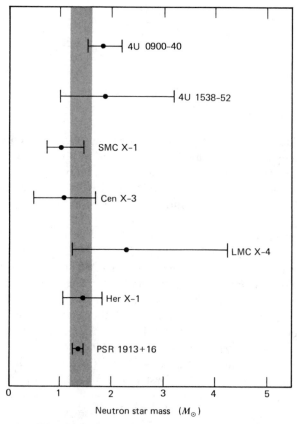

Neutron star mass $(M_\odot)$

Figure 9.6 Observational determinations of neutron star masses. The first six masses are derived from observations of binary X-ray pulsars. PSR 1913 + 16 is a binary radio pulsar. The most probable value of the mass of each neutron star is indicated by the *filled circle*; the bars indicate error estimates. The *shaded* region represents the smallest range of neutron star masses consistent with all of the data $(1.2-1.6 M_\odot)$. This range might be expected on the basis of current theoretical scenarios for neutron-star formation (see text). [After Rappaport and Joss (1983), in *Accretion Driven Stellar X-Ray Sources*, edited by W. H. G. Lewin and E. P. J. van den Heuvel, Cambridge University Press.]

The above mass determinations lead to several significant conclusions.[19] First, present observational data are consistent with standard theories of gravity and hadronic matter. Although the data are not yet sufficiently precise to accurately distinguish between competing models, the lower limit to the mass of PSR 1913 + 16, $M \gtrsim 1.35M_\odot$, is already significant for it rules out many soft equations of state (e.g., ideal degenerate neutron gas, Harrison–Wheeler, Hagedorn, etc.). A single reliable mass determination above $1.6M_\odot$ would rule out all but the stiffest equations of state depicted in Figure 9.2.

Second, present mass determinations for neutron stars are all consistent with the speculation based on advanced stellar evolution theory[20] that the degenerate cores of all evolved stars (and so by implication all neutron stars) should have masses $\sim M_{Ch} \sim 1.4M_\odot$. The results for PSR 1913 + 16 are particularly intriguing in this regard. From Figure 9.6, we see that requiring all neutron stars to have the same mass implies $1.2 \lesssim M/M_\odot \lesssim 1.8$, again ruling out very soft equations of state.

Direct radius determinations for neutron stars are nonexistent. Observational data combined with special theoretical assumptions do yield some information, however. For example, the assumptions that 10 well-observed X-ray bursters (1) radiate like blackbodies (2) have peak luminosities near the "Eddington limit" $L_E = 1.3 \times 10^{38}(M/M_\odot)$ erg s^{-1} for a $1.4M_\odot$ star,[21] and (3) are located symmetrically about the Galactic center at a distance of ~ 9 kpc, lead Van Paradijs (1978) to the conclusion that the emitting surface has a radius ~ 8.5 km. A number of effects ignored in this naive analysis (e.g., strong B fields, electron scattering, etc.), indicate that this may be an underestimate by about a factor of 2.

Exercise 9.5 Examine the consequences of the "result" that neutron stars may have masses of $1.4M_\odot$ and radii $\gtrsim 8.5$ km in light of Figure 9.3.

9.5 The Maximum Mass

The possibility of identifying some compact objects as black holes relies in part on being able to state categorically that the observed object has a mass larger than the maximum allowed mass of a stable neutron star (or white dwarf). We have seen in Section 9.3., however, that the maximum neutron star mass is a sensitive function of the as-yet-unknown equation of state for nuclear matter. Are there *any* upper limits that we can put on the mass of a neutron star that do not depend on details of the equation of state in the unknown high-density regime? Surprisingly, the answer is yes!

---

[19]Joss and Rappaport (1976); Bahcall (1978a); Pines (1980a).
[20]For example, Iben (1974); Arnett (1979).
[21]The Eddington limit is discussed in Section 13.7.

Our discussion of mass limits in this section will assume no rotation. We will show in the next section that this is probably not a severe restriction. Unlike white dwarfs, rotation cannot increase the maximum neutron star mass dramatically.

A general treatment of the maximum mass of a stable neutron star was given by Rhoades and Ruffini (1974), who made the following set of assumptions:

1. General relativity is the correct theory of gravity. In particular, this means that the OV equation determines the equilibrium structure.
2. The equation of state satisfies the "microscopic stability" condition,

$$\frac{dP}{d\rho} \geq 0. \tag{9.5.1}$$

   If this condition were violated, small elements of matter would spontaneously collapse.
3. The equation of state satisfies the causality condition

$$\frac{dP}{d\rho} \leq c^2; \tag{9.5.2}$$

   that is, the speed of sound is less than the speed of light.
4. The equation of state below some "matching density" ρ_0 is known.

Rhoades and Ruffini then performed a variational calculation to determine which equation of state above ρ_0, subject to the inequality constraints (9.5.1) and (9.5.2), maximizes the mass. They found the plausible result that it is the one for which equality holds in Eq. (9.5.2); that is,

$$P = P_0 + (\rho - \rho_0)c^2, \qquad \rho \geq \rho_0. \tag{9.5.3}$$

They chose $\rho_0 = 4.6 \times 10^{14}$ g cm^{-3} and adopted the HW equation of state for $\rho < \rho_0$. This is not crucial. The result for M_{max} can be scaled roughly as $\rho_0^{-1/2}$. Also, only a few percent of the mass is contributed from the region with $\rho < \rho_0$. There the HW equation of state is in reasonable agreement with more realistic equations of state; if anything, realistic equations of state are stiffer.

A numerical integration of the OV equation with the HW equation of state below ρ_0 and Eq. (9.5.3) above ρ_0 yields

$$M_{max} \simeq 3.2 \left(\frac{\rho_0}{4.6 \times 10^{14} \text{ g cm}^{-3}} \right)^{-1/2} M_\odot. \tag{9.5.4}$$

We can recover this result with an approximate analytic calculation by considering uniform density spheres in general relativity.[22] Assume that ρ and n are constant in the interior of the star. The equation of state is defined by $\rho(n)$, the pressure being given by the First Law:

$$\frac{d\rho}{dn} = \frac{\rho + P/c^2}{n}.$$ (9.5.5)

The mass of the star is

$$M = 4\pi \int_0^R \rho r^2 \, dr = \frac{4}{3}\pi R^3 \rho,$$ (9.5.6)

while the total baryon number is (with $c = G = 1$) [cf. Eq. (5.7.9.)]

$$A = 4\pi \int_0^R \frac{nr^2 \, dr}{[1 - 2m(r)/r]^{1/2}}$$

$$= 2\pi n \left(\frac{3}{8\pi\rho}\right)^{3/2} (\chi - \sin\chi \cos\chi).$$ (9.5.7)

Here we have defined χ by

$$\sin\chi = \left(\frac{8\pi\rho}{3}\right)^{1/2} R,$$ (9.5.8)

so that

$$\sin^2\chi = \frac{2M}{R}.$$ (9.5.9)

Now according to the energy variational principle (Section 6.9), an equilibrium configuration must extremize the energy at fixed baryon number:

$$\left.\left|\frac{\partial M}{\partial \chi}\right|\right|_A = 0.$$ (9.5.10)

Here we are taking χ as the independent variational parameter. Using Eqs. (9.5.5)–(9.5.8) in Eq. (9.5.10), we find

$$\frac{P}{\rho} = \zeta(\chi),$$ (9.5.11)

[22] Nauenberg and Chapline (1973).

where

$$\zeta(\chi) \equiv \frac{6 \cos \chi}{9 \cos \chi - 2 \sin^3 \chi / (\chi - \sin \chi \cos \chi)} - 1. \qquad (9.5.12)$$

Note that this solution differs from that given in Exercise 5.3. There one found the *exact* solution of the OV equation for a single equation of state, $\rho = $ constant. Here one is finding an *approximate* solution for any equation of state by using the restricted class of trial functions $\rho = $ constant, $n = $ constant (and hence $P = $ constant). For a given equation of state $\rho(n)$, Eqs. (9.5.5), (9.5.11), (9.5.8), (9.5.6), and (9.5.7) determine P, χ, R, M, and A as functions of n.

Exercise 9.6   Derive Eq. (9.5.11).

Exercise 9.7   Show that $0 \leqslant P/\rho \leqslant 1$ requires $0 \leqslant \chi \leqslant 72°94$, and that $P/\rho \to \infty$ as $\chi \to 80°03$.

Exercise 9.8   Discuss the Newtonian limit of the above equations. In particular, show that Eq. (9.5.11) reduces to

$$P \mathcal{V} = \frac{|W|}{3}, \qquad (9.5.13)$$

where $|W|$ is the gravitational potential energy of a uniform density sphere of volume $\mathcal{V}$ [cf. Eqs. (7.3.17), (7.3.8), and (7.3.23) in the spherical limit, $e = 0$].

Exercise 9.9   Show that in the Newtonian limit $M \propto P^{3/2}/\rho^2$, and hence M increases with n provided P increases faster than $\rho^{4/3}$. Show that in the high-density limit in which $P/\rho \sim$ constant, one obtains instead $M \propto \rho^{-1/2}$.

Exercise 9.9 shows the scaling behavior, $M \sim \rho^{-1/2}$, referred to earlier. Also, it shows that M achieves a maximum, typically, at some intermediate value of n. As usual, the existence of the maximum is related to the stability condition: stability requires

$$\left| \frac{\partial^2 M}{\partial \chi^2} \right|_A > 0. \qquad (9.5.14)$$

After a tedious calculation, this condition can be written as a condition on the adiabatic index,

$$\Gamma > \Gamma_c(\chi), \qquad (9.5.15)$$

where

$$\Gamma \equiv \frac{\partial \ln P}{\partial \ln n} = \left(1 + \frac{\rho}{P}\right) \frac{dP}{d\rho}, \qquad (9.5.16)$$

and the critical adiabatic index is

$$\Gamma_c(\chi) = (\zeta + 1)\left\{1 + \frac{(3\zeta + 1)}{2}\left[\frac{(\zeta + 1)}{6\zeta} \tan^2 \chi - 1\right]\right\}. \qquad (9.5.17)$$

Exercise   9.10   Show that in the Newtonian limit

$$\Gamma_c = \frac{4}{3} + \frac{89}{105}\frac{M}{R}, \qquad (9.5.18)$$

and compare this uniform density result with Eq. (6.10.30) for an $n = 3$ polytrope.

The dependence of Γ_c on $\zeta = P/\rho$ is shown in Figure 9.7. The adiabatic index Γ for any given equation of state can also be plotted, and the intersection with Γ_c determines the point of instability. Also shown in the figure is the causality constraint $dP/d\rho \leqslant c^2$. This requires $\zeta \leqslant 0.364$, $\chi \leqslant 64°19$, and so

$$\frac{M}{R} \leqslant 0.405 \quad \text{(causality + stability constraints)}. \qquad (9.5.19)$$

This should be contrasted with the weaker limits implied by the results of Exercise 9.7.

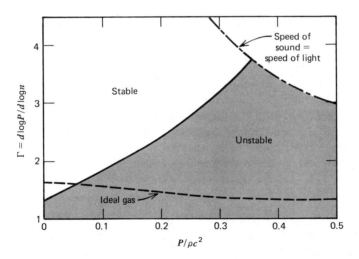

Figure 9.7 The stability domain in the uniform-density approximation. The curve separating the stable and unstable regions is the function Γ_c given in Eq. (9.5.17). For illustration, the adiabatic index Γ of a free neutron gas is shown by the *dashed* line. [Reprinted courtesy of Nauenberg and Chapline and *The Astrophysical Journal*, published by the University of Chicago Press; © 1973 The American Astronomical Society.]

Exercise 9.11 What is the maximum gravitational redshift due to emission from the surface of a stable neutron star, according to Eq. (9.5.19)?

We are now prepared to determine the maximum neutron star mass. Since we assume we know the equation of state for $\rho \leqslant \rho_0$, with $P_0 \equiv P(\rho_0)$, the causality constraint $dP/d\rho \leqslant c^2$ allows us to write

$$P \leqslant P_0 + v^2(\rho - \rho_0), \qquad \rho \geqslant \rho_0, \qquad (9.5.20)$$

where $v \leqslant 1$ is the maximum speed of sound above ρ_0. Now for a fixed value of ρ, M increases with increasing P, and hence with increasing v [Eqs. (9.5.6), (9.5.8), (9.5.11), and (9.5.20)]. The maximum occurs when P/ρ reaches a critical value, denoted ζ_c, at which point $\Gamma = \Gamma_c(\chi_c)$; that is,

$$\frac{\zeta_c + 1}{\zeta_c} v^2 = \Gamma_c \qquad (9.5.21)$$

from Eq. (9.5.16). Substituting the limiting value of P/ρ from Eq. (9.5.21) into Eq. (9.5.20) sets an upper limit on the mean density of the star:

$$\rho_c \geqslant \frac{\rho_0 - P_0/v^2}{1 - (\zeta_c + 1)/\Gamma_c}. \qquad (9.5.22)$$

This in turn determines an upper mass limit via Eqs. (9.5.6) and (9.5.8):

$$M \leqslant \frac{1}{2}\left(\frac{3}{8\pi}\right)^{1/2}\left(\frac{1 - (\zeta_c + 1)/\Gamma_c}{\rho_0 - P_0/v^2}\right)^{1/2} \sin^3 \chi_c. \qquad (9.5.23)$$

The resulting mass limit has been determined by Nauenberg and Chapline as a function of v. They use the BBP equation of state to set the largest reliable value of $\rho_0 - P_0/v^2$. Their value of $\rho_0 = 5 \times 10^{14}$ g cm^{-3} is close to the matching density of Rhoades and Ruffini and, not surprisingly, their predicted maximum mass for $v = 1$,

$$M_{\text{max}} \sim 3.6 M_\odot \quad \text{(semianalytic)}, \qquad (9.5.24)$$

is close to the Rhoades–Ruffini value (Table 9.2). Note that (i) the maximum mass increases with v and (ii) $M_{\text{max}} \sim \rho_0^{-1/2}$ since $\rho_0 - P_0/v^2 \sim \rho_0$ near $v = 1$.

Finally, we note that abandoning the causality constraint still leads to a severe mass limit, assuming general relativity to be valid. Letting $v \to \infty$ in the above analysis, we see from Eq. (9.5.21) that $\Gamma_c \to \infty$. From Eq. (9.5.17), we see that $\Gamma_c \sim \zeta^2$ in this limit, and $\zeta \to \infty$ as $\chi \to 80°.03$ (Exercise 9.7). Thus Eq. (9.5.23)

Table 9.2

Maximum Mass[a] of Stable Neutron Star ($M_\odot$)

$\dfrac{v}{c}$	$\rho_0 = 5 \times 10^{14}$ g cm^{-3},[b] $P_0 = 7 \times 10^{33}$ dyne cm^{-2}	$\rho_0 = 1 \times 10^{15}$ g cm^{-3},[c] $P_0 = 5 \times 10^{34}$ dyne cm^{-2}
1.00	3.6	2.6
0.75	3.0	2.2
0.50	2.0	1.6
0.25	0.68	Unstable

[a]From Nauenberg and Chapline (1973)
[b]Equation of state matching parameters from BBP (see Table 8.2).
[c]Equation of state matching parameters from R (see Table 8.2).

implies

$$M_{max} \sim \frac{1}{2}\left(\frac{3}{8\pi}\right)^{1/2}\left(\frac{1}{\rho_0}\right)^{1/2} \sin^3 \chi_c$$

$$= 6.05 \left(\frac{4.6 \times 10^{14} \text{ g cm}^{-3}}{\rho_0}\right)^{1/2} M_\odot. \qquad (9.5.25)$$

In fact, the exact limit in general relativity is attained for an incompressible star. From Eq. (5.7.13), we have[23]

$$\left(\frac{M}{R}\right)_{max} = \frac{4}{9}. \qquad (9.5.26)$$

Then Eq. (9.5.6) gives

$$M_{max} = \frac{8}{27}\left(\frac{3}{4\pi\rho_0}\right)^{1/2} \sim 5.3\left(\frac{4.6 \times 10^{14} \text{ g cm}^{-3}}{\rho_0}\right)^{1/2} M_\odot \quad \text{(no causality)}.$$

$$(9.5.27)$$

A more careful calculation,[24] which allows for the envelope where $\rho < \rho_0$, gives $5.2 M_\odot$.

[23]For a proof that this is in fact the maximum, see, for example, Weinberg (1972). Higher values can be achieved if one allows the density to increase outwards; see, for example, Bondi (1964).
[24]See Hartle and Sabbadini (1977) and references therein.

In summary, stiff equations of state currently predict M_{max} in the range ~ 1.5–$2.7M_\odot$. Depending on the assumptions one is willing to make about constraints on possible equations of state, general relativity predicts absolute upper limits in the range ~ 3–$5M_\odot$. The higher values are only possible if one abandons the causality constraint $dP/d\rho \leqslant c^2$.[25]

9.6   The Effects of Rotation

In setting limits on the maximum neutron star mass, we assumed that the configurations were nonrotating and spherical. This is not true in general; we know that pulsars rotate. Moreover, we know that in the case of white dwarfs, rotation can in principle substantially increase the maximum mass (Section 7.4). We shall show in this section that, in contrast to white dwarfs, existing calculations indicate that rotation *cannot* increase the maximum neutron star mass appreciably ($\lesssim 20\%$). Also, even this increase occurs only for rotation frequencies much higher than all observed pulsar frequencies.

Rapidly rotating configurations in general relativity are technically difficult to construct. In addition, no simple stability criteria are known. Existing calculations assume (i) slow rotation[26], or (ii) uniform rotation and homogeneous configurations[27], or (iii) post-Newtonian gravity and an ideal Fermi gas equation of state.[28] The numerical calculations of Butterworth and Ipser are the only excursion into the regime of strong gravity and rapid rotation. We shall follow the analytic treatment of Lightman and Shapiro for simplicity and reach essentially the same conclusions as the strong field treatment.

We start with the energy variational principle, Eq. (9.2.16) modified to include rotation as in Eq. (7.4.35):

$$E = AM\rho_c^{2/3} - Bg(\lambda)M^{5/3}\rho_c^{1/3} - CM\rho_c^{4/3} - DM^{7/3}\rho_c^{2/3} + k_5\lambda J^2 M^{-5/3}\rho_c^{2/3}.$$

$$(9.6.1)$$

Here E is the energy of a rotating neutron star, assumed to be close to an $n = \frac{3}{2}$ polytrope, in the post-Newtonian limit. The effect of rotation has been included via the approximation described in Section 7.4. The quantities A, B, C, and D are defined in Eq. (9.2.17), while the oblateness parameter λ and the function $g(\lambda)$

---

[25]Acoustic signals in a medium propagate with a speed $(dP/d\rho)^{1/2}$, which is in general a function of frequency. The equation of state for an equilibrium star gives essentially the zero-frequency limit of this, since one allows long enough times for equilibrium to be achieved. If one tries to have $dP/d\rho > c^2$ while maintaining causality, then the medium is spontaneously unstable. See, for example, Bludman and Ruderman (1970).

[26]Hartle (1967) and Hartle and Thorne (1968); Abramowicz and Wagoner (1976).

[27]Butterworth and Ipser (1975), (1976).

[28]Shapiro and Lightman (1976b).

are defined in Eqs. (7.4.28) and (7.4.29). The quantity k_5 [Eq. (7.4.34)] has the value 1.926 for $n = \frac{3}{2}$.

The equilibrium condition $\partial E/\partial \lambda = 0$ at fixed M and J gives the usual relation between $T/|W|$ and eccentricity [Eqs. (7.4.36) and (7.4.37)]. The second equilibrium condition is $\partial E/\partial \rho_c = 0$ (fixed M and J). Solving these two equations simultaneously gives the equilibrium relation between M and ρ_c for various values of J.[29]

The maximum mass is determined by setting $\partial^2 E/\partial \rho_c^2 = 0$ (condition for onset of radial instability).

Exercise   9.12   Show that the definition of $T/|W|$ and the two equations $\partial E/\partial \rho_c = \partial^2 E/\partial \rho_c^2 = 0$ can be rewritten

$$\frac{k_5 J^2 \rho_c^{1/3}}{B M^{10/3}} = \frac{T}{|W|} \frac{g(\lambda)}{\lambda}, \tag{9.6.2}$$

$$\rho_c = \frac{B g M^{2/3}}{8C}, \tag{9.6.3}$$

$$2A + (Bg)^{2/3} C^{1/3} \left(-3 + \frac{4T}{|W|} \right) M^{4/9} - 2D M^{4/3} = 0. \tag{9.6.4}$$

Equations (9.6.2)–(9.6.4) can be solved as follows: choose a value of λ, and hence find $g(\lambda)$ and $T/|W|$. Then solve Eq. (9.6.4), a cubic in $M^{4/9}$, for M. Equation (9.6.3) then gives ρ_c, and Eq. (9.6.2) gives J. The results are given in Table 9.3.

Computer Exercise   9.13   Verify the entries in Table 9.3 Note that you will have to add E/c^2 to M to get the total mass [see discussion after Eq. (9.2.23)].

Since no simple results are known for the onset of nonaxisymmetric instabilities in general relativity, we assume that the approximate Newtonian criteria, $T/|W| \lesssim 0.14$ for secular stability and $T/|W| \lesssim 0.26$ for dynamical stability, are still valid. Friedman and Schutz (1975) have shown that the typical gravitational radiation timescale for neutron stars with $T/|W| \sim 0.14$ is ≤ 1 yr.[30] Clearly such neutron stars will not live for a very long time.

From Table 9.3, we see that the maximum increase in M_{max} due to rotation is (using $T/|W| = 0.14$)

$$\frac{1.24 - 1.03}{1.03} \simeq 20\% \text{ increase.} \tag{9.6.5}$$

One should realize, however, that the above treatment of equilibrium models for

---

[29]See Shapiro and Lightman (1976b).

[30]Compare Exercise 16.7.

Table 9.3

Maximum Masses For Post-Newtonian Rotating Neutron Stars[a]

J $(10^{49}\ \text{g cm}^2\text{s}^{-1})$	$T/\lvert W\rvert$	$M_{\max}$ $(M_\odot)$	ρ_c $(10^{15}\ \text{g cm}^{-3})$
0.00	0.00	1.03	7.41
0.21	0.02	1.06	7.57
0.32	0.04	1.09	7.73
0.42	0.06	1.12	7.88
0.53	0.08	1.15	8.04
0.64	0.10	1.18	8.19
0.76	0.12	1.21	8.34
0.89	0.14	1.24	8.48
1.03	0.16	1.27	8.61
1.19	0.18	1.31	8.74
1.37	0.20	1.34	8.85
1.56	0.22	1.38	8.96
1.72	0.24	1.40	9.02
2.02	0.26	1.44	9.11

[a] The numbers quoted for $M_{\max}$ differ somewhat from those in Shapiro and Lightman (1976b), who quoted *rest* masses.

rotating neutron stars is probably reliable only for low mass configurations, with $M < M_\odot$. The $n = \frac{3}{2}$ approximation is not particularly good at the values of ρ_c in Table 9.3 [cf. Eq. (2.3.25)], and also $GM/Rc^2 > 0.2$ for all the entries [R is obtained from Eq. (3.3.9)], so the post-Newtonian approximation is also breaking down. Nevertheless, the qualitative result that the rotation-induced enhancement in $M_{\max}$ is $\sim$ 20% (compared with 70% for white dwarfs) is consistent with the more detailed numerical calculations of Butterworth and Ipser.

We can understand this difference between neutron stars and white dwarfs from Eq. (9.6.4), rewritten as

$$M = \left(\frac{2A - 2DM^{4/3}}{3B^{2/3}C^{1/3}}\right)^{9/4}\left[\frac{1}{g^{2/3}(1 - 4T/3\lvert W\rvert)}\right]^{9/4}. \qquad (9.6.6)$$

The term in square brackets represents the increase in M due to rotation, and is equally effective for white dwarfs and neutron stars. The term proportional to D represents the *decrease* in M due to relativistic gravity. The ratio of this term to the Newtonian term, $2A$, is $\sim GM/Rc^2$ and so is negligible for white dwarfs but not for neutron stars.

A more precise determination of $M_{\max}$ must await the detailed numerical analysis of rapidly rotating, fully relativistic neutron stars and their stability criteria.

10

Pulsars

10.1 History and Discovery

In 1967 a group of Cambridge astronomers headed by Anthony Hewish detected[1] astronomical objects emitting periodic pulses of radio waves. This discovery had a profound impact on subsequent astrophysical research. Its significance was highlighted by the award of the Nobel prize to Hewish in 1974.

The existence of stable equilibrium stars more dense than white dwarfs had been predicted by a number of theoreticians, including Baade and Zwicky (1934) and Oppenheimer and Volkoff (1939) (cf. Section 9.1). Baade and Zwicky (1934), Colgate and White (1966), and others suggested that such objects could be produced in supernova explosions. It was even surmised that initially they would be rapidly rotating,[2] with strong magnetic fields,[3] and that the energy source of the Crab nebula might be a rotating neutron star.[4] Preliminary details of a simple magnetic dipole model, capable of converting neutron star rotational energy into electromagnetic radiation and, subsequently, particle motions, were already in print[5] at the time of the discovery.

So the announcement by Hewish et al. (1968) of the discovery of a 1.377 s radio pulsar at 81.5 MHz did not occur in a theoretical vacuum. However, the identification of pulsars with neutron stars was not immediately obvious to most astrophysicists. The first argument that the observed pulsars were in fact rotating neutron stars, with surface magnetic fields of around 10^{12} G, was put forward by Gold (1968). He pointed out that such objects could account for many of the observed features of pulsars, such as the remarkable stability of the pulse period. Gold predicted a small increase in the period as the pulsar slowly lost rotational

---

[1] Hewish, Bell, Pilkington, Scott, and Collins (1968). For inspiration, the reader should read Jocelyn Bell Burnell's account of her role in the discovery (Burnell, 1977).

[2] Hoyle et al. (1964); Tsuruta and Cameron (1966).

[3] Woltjer (1964); Hoyle et al. (1964).

[4] Wheeler (1966).

[5] Pacini (1967).

energy. Shortly thereafter, the slowdown of the Crab pulsar was discovered. Gold (1969) showed that the implied energy loss was roughly the same as the energy required to power the Crab nebula. This success, together with the failure of alternative models, led to the general acceptance of the neutron star model.

A modernized version of the chain of reasoning leading to this historic identification is sketched in the next section.

10.2   Are Pulsars Really Rotating Neutron Stars?

The key observational facts that rule out all other possibilities are:

1. Pulsars have periods in the range 1.6 ms to 4.3 s.
2. Pulsar periods *increase* very slowly, and never decrease (except for occasional "glitches", discussed in Section 10.10).
3. Pulsars are remarkably good clocks. Some pulsar periods have been measured to 13 significant digits.

In a time 1.6 ms light travels roughly 500 km. This sets an upper limit to the size of the emitting region. No consistent models have been proposed in which the emitting region is part of a much bigger "source." In particular, it is hard to have such a good clock mechanism unless the emitting region is closely coupled to the whole "source." We may thus take 500 km as the maximum size of the source: we are dealing with a compact object. Is it a white dwarf, neutron star, or black hole?

Consider first white dwarf models. Three possibilities immediately suggest themselves for the underlying "clock" mechanism: rotation, pulsation, or a binary system. The shortest period for a rotating white dwarf occurs when it is rotating at break-up velocity:

$$\Omega^2 R \sim \frac{GM}{R^2}, \tag{10.2.1}$$

where Ω is the angular velocity of rotation. Equation (10.2.1) be rewritten in terms of the mean density ρ as

$$\Omega^2 \sim G\rho, \tag{10.2.2}$$

the universal relation between dynamical timescale and density for a self-gravitating system. Taking a maximum mean density of 10^8 g cm^{-3} gives a period

$$P = \frac{2\pi}{\Omega} \gtrsim 1 \text{ s}. \tag{10.2.3}$$

This result rules out rotating white dwarfs.

An estimate similar to Eq. (10.2.3) holds for pulsating white dwarfs. In fact, accurate calculations show that the shortest period for the fundamental mode is ~ 2 s (cf. Section 6.10). Higher harmonics would have shorter periods, but are ruled out for two reasons: first, it requires special conditions to excite a harmonic without exciting the fundamental mode. Any small nonlinearity will lead to mixing and destroy the sharpness of the period. Second, the loss of energy in a vibrating system usually leads to a *decrease* in the period.

Binary white dwarfs would satisfy

$$\Omega^2 r \sim \frac{GM}{r^2},$$

(10.2.4)

where r is the orbital radius. Since $r \gtrsim R$, we again recover the result (10.2.3).

A pulsating neutron star has a density $\sim 10^6$ times the density of a white dwarf. Thus the fundamental period is $\sim 10^{-3}$ s—typically much too short.

By a suitable choice of radius, one can arrange the orbital period of a binary neutron star system to lie in the observed range of 10^{-3}–4 s. However, as we shall see in Exercise 16.9, such a system would be a copious emitter of gravitational waves, with a lifetime

$$\tau \sim 10^{-3} \text{ yr} \left(\frac{P}{1 \text{ s}} \right)^{8/3}.$$

(10.2.5)

The timescale on which pulsar periods are observed to change is typically $P/\dot{P} \sim 10^7$ yr; moreover, gravitational wave emission causes the period to decrease, not increase.

As we shall see in Chapter 12, an isolated black hole has no structure on which to attach a periodic emitter. In particular, rotating black holes are axisymmetric. Any mechanism depending on accretion would not be periodic to such fantastic precision.

Only a rotating neutron star meets all the above objections. In addition, some further observational facts can be nicely explained by this interpretation, as we shall discover.

10.3 Observed Properties of Pulsars

Lack of space prevents us from giving a complete review of all the fascinating properties of pulsars. We confine ourselves to a brief summary of salient properties.[6]

[6]For a more complete treatment, see, for example, Groth (1975a); Manchester and Taylor (1977); Cordes (1979); Sieber and Wielebinski (1981); and Michel (1982).

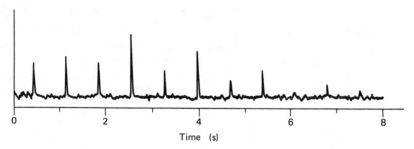

Time (s)

Figure 10.1 Chart record of individual pulses from one of the first pulsars discovered, PSR 0329 + 54. They were recorded at a frequency of 410 MHz and with an instrumental time constant of 20 ms. The pulses occur at regular intervals of about 0.714 s. [From *Pulsars* by Richard N. Manchester and Joseph H. Taylor, W. H. Freeman and Company. Copyright © 1977.]

(a) Pulse Shapes and Spectra

At present, about 350 pulsars are known. All exhibit broadband radio emission in the form of periodic pulses (Fig. 10.1). Pulse intensities vary over a wide range, and sometimes pulses are missing. The basic pulse record, however, is periodic. The duty cycle (fraction of period with measurable emission) is typically small (1–5%). On timescales ≤ 1 ms, the pulse shape is quite complex. Typically, a pulse consists of two or more subpulses. Subpulses exhibit complicated micro-structure that can be on timescales as short as 10 μs. However, the average of several hundred pulses is found to be remarkably stable (cf. Fig. 10.2). By measuring the arrival times of the same feature in successive averaged pulses, one can verify that the rotating neutron star is an excellent clock. Some pulsar periods are known to a part in 10^{13}.

The specific radio intensity[7] is a steep power law, $I_\nu \propto \nu^\alpha$, where a typical value is $\alpha \sim -1.5$ for $\nu < 1$ GHz and is even steeper at higher frequencies. A typical intensity (averaged over a period and integrated over solid angle) at 400 MHz is ~ 0.1 Jy (1 Jansky $\equiv 10^{-23}$ erg s^{-1} cm^{-2} Hz^{-1}).

Many pulsars show a high degree of linear polarization— up to 100% in some cases. The amount and position angle of the linear polarization frequently vary with time across the pulse. Circular polarization is not as frequent, and when present is not as strong as linear polarization.

(b) Periods

Observed pulsar periods lie between 1.558 ms (PSR 1937+214; Backer et al., 1982) and 4.308 s (PSR 1845 − 19). The second shortest period is that of the Crab

---

[7]The intensity I_ν of a luminous object is defined in Appendix I and is measured in erg s^{-1} cm^{-2} Hz^{-1} sr^{-1} [cf. Eqs. (I.1)–(I.3)].

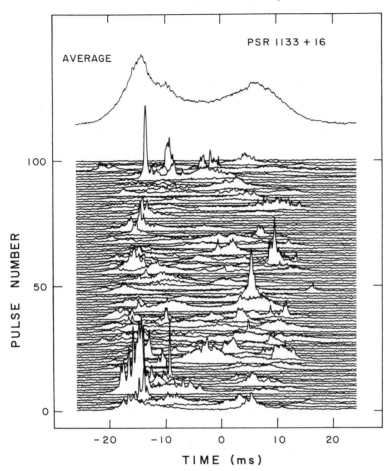

Figure 10.2   A sequence of 100 pulses from PSR 1133 + 16 recorded at 600 MHz. An average of 500 pulses is shown at the top. Consecutive pulses are plotted vertically to show variations in individual pulse shapes and arrival times; the *average* pulse behavior is quite stable and periodic, however. [From Cordes (1979). Copyright © 1979 by D. Reidel Publishing Company, Dordrecht, Holland.]

pulsar, PSR 0531 + 21 (0.0331 s). The third shortest period is that of the Hulse–Taylor binary pulsar, PSR 1913 + 16 (0.059 s), while the fourth shortest is for the Vela pulsar, PSR 0833 − 45 (0.089 s). The median period is 0.67 s.

In all cases where accurate observations have been made, pulsar periods are found to increase in a steady way (but see below for a discussion of "glitches"). Typically, $\dot{P} \sim 10^{-15}$ s s^{-1}. The implication from pulsar models is that pulsars are probably younger than a characteristic time $T \equiv P/\dot{P} \sim 10^7$ yr. Shorter period pulsars tend to have larger slowdown rates and shorter characteristic times.

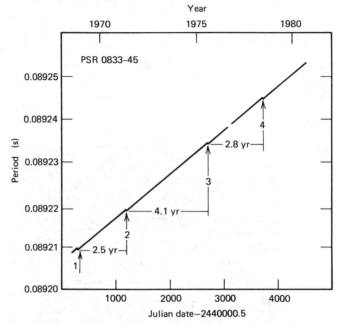

Figure 10.3 The pulse period of PSR 0833–45 (Vela) from late 1968 to mid-1980. The four large jumps are enumerated, and the time span in years between jumps is noted. [From Downs (1981). Reprinted courtesy of the author and *The Astrophysical Journal*, published by the University of Chicago Press; © 1981 The American Astronomical Society.]

For the Crab $T = 2486$ yr. However, the Hulse–Taylor binary pulsar has $T = 2.17 \times 10^8$ yr.

The most dramatic period irregularities in pulsars are the sudden spinups, or *glitches* observed in the Crab ($|\Delta P|/P \sim 10^{-8}$) and Vela ($|\Delta P|/P \sim 2 \times 10^{-6}$) pulsar periods. In each of the four "giant" Vela glitches observed so far (Fig. 10.3), the decrease in period was accompanied by an increase in period derivative ($\Delta \dot{P}/\dot{P} \sim 0.01$), which decayed away in about 50 days. The smaller Crab speed-ups (two of which have been unambiguously observed) varied in size by a factor of 4. They were also accompanied by a period derivative increase ($\Delta \dot{P}/\dot{P} \sim 2 \times 10^{-3}$), which decayed away in ~ 10 days. We will return to these data in Section 10.10, where we will discuss a theoretical model to explain the glitches.

In addition to the large period jumps or "macroglitches" observed for a few pulsars, smaller amplitude, irregular fluctuations have also been detected in the periods of numerous pulsars. This nonsecular jitter in the pulsar clock is known as *timing noise*.[8] Timing noise represents low-level fluctuations in the pulse repeti-

---

[8] Boynton et al. (1972). For a detailed discussion of pulsar timing noise observations, see Cordes and Helfand (1980) and references therein. For a critique of various theoretical models proposed to explain timing noise, see Cordes and Greenstein (1981) and references therein.

tion rate. It may be formed from small random events for which $|\Delta P|/P$ is at least two orders of magnitude below the macroglitch fluctuation amplitude. It has been shown for the Crab pulsar that the mathematical form of the noise spectrum is not comparable to the spectrum of the giant glitches, and so may represent a different phenomenon altogether.[9]

(c)   Distribution and Association with Supernova Remnants

We know that pulsars originate in our Galaxy because (i) they are concentrated in the Galactic plane (Figs. 10.4 and 10.5) and (ii) they show *dispersion* characteristic of signal propagation over Galactic distances. (This point will be elaborated further below.)

The angular distribution of pulsars in the Galactic plane is similar to that of supernova remnants and their presumed progenitors, OB stars, though the mean scale height above the plane is somewhat larger, ~ 300 pc. This is consistent with the hypothesis that pulsars arise in supernova explosions, possibly acquiring a relatively high velocity. More impressive is the association of several pulsars with known supernova remnants. PSR $0531+21$ and PSR $0833-45$ are associated with the Crab and Vela supernova remnants, respectively. The Crab nebula is the remnant of a supernova observed in the summer of 1054 A.D. by the Chinese. The nebula is about 2 kpc away and consists of outwardly expanding filaments emitting optical line radiation, and an amorphous region emitting synchrotron radiation. The discovery that the Crab pulsar was the heretofore unknown energy source of the nebula, and that the pulsar's theoretical age was in good agreement with the actual age of 926 yr (in 1980), has firmly established the rotating neutron star model for pulsars (see Section 10.5 for quantitative discussion).

The Vela supernova remnant has an estimated age of 30,000 to 50,000 yr, about the same order as the characteristic age of PSR $0833-45$, $T = 22000$ yr. This, combined with the positional and distance agreement ($d \sim 500$ pc), is the chief evidence for associating the pulsar with the supernova remnant.

In addition to the two definite pulsar supernova remnant associations, there is one probable and two possible associations. These associations are with young ($\lesssim 10^5$ yr) pulsars, so it is not surprising that more associations have not been found.[10] However, it is perhaps surprising that of the half dozen or so Galactic supernovae recorded in the past 2000 yr, only the Crab contains a pulsar. Perhaps the emission mechanism is quite directional, or perhaps not all supernovae make pulsars (cf. Chapters 1 and 11).

---

[9]Groth (1975b); Boynton (1981).

[10]Two pulsating X-ray sources have recently been discovered in supernova remnants. One is a fast X-ray pulsar ($P = 0.150$ s; Seward and Harnden, 1982) similar to the Crab, while the other is slower ($P = 3.5$ s; Fahlman and Gregory, 1981). Radio pulsations have been reported for the first of these sources (Manchester et al. 1982), and both are believed to be rotating neutron stars.

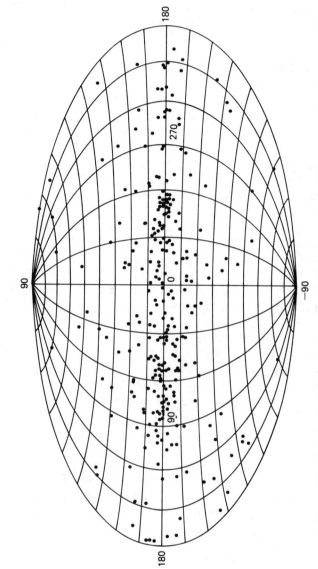

Figure 10.4 Galactic distribution of pulsars. In the adopted coordinates 0° latitude corresponds to the Galactic plane, while 0° longitude, 0° latitude corresponds to the direction of the Galactic center. [Courtesy of Y. Terzian.]

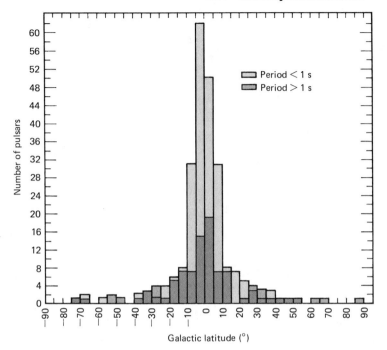

Figure 10.5 Distribution of pulsars in Galactic latitude. Latitude 0° corresponds to the Galactic plane. [Courtesy of Y. Terzian.]

10.4 The Dispersion Measure

Distances to pulsars are inferred from their *dispersion measures*, *DM*, defined by

$$DM \equiv \int_0^L n_e \, dl \equiv \langle n_e \rangle L, \qquad (10.4.1)$$

where L is the pulsar distance, n_e the electron number density, and l the path length along the line of sight. The dispersion measure is usually quoted in pc cm^{-3}. The name dispersion measure arises because electromagnetic waves are dispersed by the conducting interstellar medium. A broadband pulse arrives later at lower frequencies than at higher frequencies.

Quantitatively, we recall that the acceleration of an electron of charge $-e$ in a tenuous plasma by a propagating electromagnetic wave of frequency ω is given by

$$m\ddot{\mathbf{x}} = -e\mathbf{E}, \qquad (10.4.2)$$

where the electric field may be written as

$$\mathbf{E} = \mathbf{E}_0 e^{i\omega t}. \qquad (10.4.3)$$

Thus

$$\mathbf{x} = \frac{e}{m\omega^2}\mathbf{E}, \tag{10.4.4}$$

and so the polarization of the medium is

$$\mathbf{P} = n_e(-e)\mathbf{x} = -\frac{n_e e^2}{m\omega^2}\mathbf{E}. \tag{10.4.5}$$

But

$$\mathbf{P} = \frac{\varepsilon - 1}{4\pi}\mathbf{E}, \tag{10.4.6}$$

where ε is the dielectric constant. Thus

$$\varepsilon = 1 - \frac{\omega_p^2}{\omega^2}, \tag{10.4.7}$$

where

$$\omega_p^2 \equiv \frac{4\pi n_e e^2}{m} \tag{10.4.8}$$

is the plasma frequency.

Now for the propagation of an electromagnetic wave of wave number k, the phase velocity is

$$v_{\text{ph}} = \frac{\omega}{k} = \frac{c}{\varepsilon^{1/2}}. \tag{10.4.9}$$

Substituting Eq. (10.4.7) in (10.4.9), we get the dispersion relation

$$\omega^2 = \omega_p^2 + k^2 c^2. \tag{10.4.10}$$

Note that ω must be greater than ω_p for the wave to propagate.

The group velocity is

$$v_g = \frac{d\omega(k)}{dk} = c\left(1 - \frac{\omega_p^2}{\omega^2}\right)^{1/2}$$

$$\simeq c\left(1 - \frac{\omega_p^2}{2\omega^2}\right), \qquad \omega \gg \omega_p. \tag{10.4.11}$$

The arrival time for a pulse centered at frequency ω and travelling a distance L is

$$t_a(\omega) = \int_0^L \frac{dl}{v_g}$$

$$\simeq \frac{1}{c} \int_0^L \left(1 + \frac{\omega_p^2}{2\omega^2}\right) dl$$

$$= \frac{L}{c} + \frac{2\pi e^2}{mc\omega^2} DM, \tag{10.4.12}$$

where we have used the definition (10.4.1). The measured quantity is $\Delta t_a(\omega)$, the time delay of different frequency components of the pulse, and the relation

$$\frac{\Delta t_a}{\Delta \omega} = - \frac{4\pi e^2}{mc\omega^3} DM \tag{10.4.13}$$

gives DM.

If $\langle n_e \rangle$ is known (e.g., the dispersion measure for sources at *known* distances gives $\langle n_e \rangle \simeq 0.03$ cm^{-3} for the interstellar medium in the solar neighborhood[11]), then we can get the distance L for individual pulsars from the dispersion measure. Distances obtained in this way range from ~ 100 pc (PSR 0950 + 08) to ~ 18 kpc (PSR 1648 − 42). Uncertainties in how well the mean density $\langle n_e \rangle$ represents the actual density n_e at different places in the Galaxy imply that individual pulsar distance estimates may be in error by as much as a factor of 2, but the estimates are probably reasonably accurate on a statistical basis.

One pulsar, PSR 1929 + 10, is close enough for its distance to be determined reasonably well from its parallax during the course of a year. The value obtained in this way is about 50 pc.[12]

Exercise   10.1   Show that the assumption of a finite photon mass $m_\gamma \neq 0$ leads, in lowest order, to a dispersion law with the same dependence[13] on frequency as Eq. (10.4.13).

10.5   The Magnetic Dipole Model for Pulsars

We discuss in this section a very simply pulsar model that accounts for many of the observed properties of pulsars. The *magnetic dipole* model[14] explicitly demon-

---

[11]Spitzer (1978).

[12]Salter et al. (1979).

[13]By assuming that all the dispersion in the Crab pulsar results from massive photons, Feinberg (1969) has put a limit of $m_\gamma < 10^{-44}$ g on the photon mass!

[14]Pacini (1967), (1968); Gunn and Ostriker (1969).

strates how pulsar emission is derived from the kinetic energy of a rotating neutron star. In this *oblique rotator* version of the model, it is assumed that a neutron star rotates uniformly *in vacuo* at a frequency Ω and possesses a magnetic dipole moment **m** oriented at an angle α to the rotation axis. [This model will be contrasted with nonvacuum (e.g., *aligned rotator*) models below.] The rotation is assumed to be sufficiently slow that nonspherical distortions due to rotation can be ignored to lowest order.

Exercise   10.2 Estimate the ratio of centrifugal to gravitational acceleration at the equator of the Crab pulsar.

Independent of the internal field geometry, a pure magnetic dipole field at the magnetic pole of the star, B_p, is related to **m** by

$$|\mathbf{m}| = \frac{B_p R^3}{2},$$ (10.5.1)

where R is the radius of the star.[15] Such a configuration has a time-varying dipole moment as seen from infinity, and so radiates energy at a rate

$$\dot{E} = -\frac{2}{3c^3}|\ddot{\mathbf{m}}|^2.$$ (10.5.2)

Writing

$$\mathbf{m} = \tfrac{1}{2}B_p R^3\left(\mathbf{e}_{\parallel}\cos\alpha + \mathbf{e}_{\perp}\sin\alpha\cos\Omega t + \mathbf{e}_{\perp}'\sin\alpha\sin\Omega t\right),$$ (10.5.3)

where $\mathbf{e}_{\parallel}$ is a unit vector parallel to the rotation axis and $\mathbf{e}_{\perp}$ and $\mathbf{e}_{\perp}'$ are fixed mutually orthogonal unit vectors perpendicular to $\mathbf{e}_{\parallel}$, we find

$$\dot{E} = -\frac{B_p^2 R^6 \Omega^4 \sin^2\alpha}{6c^3}.$$ (10.5.4)

Note that the radiation is emitted at frequency Ω.

Equation (10.5.4) leads to several important consequences. First, the energy carried away by the radiation originates from the rotational kinetic energy of the neutron star,

$$E = \tfrac{1}{2}I\Omega^2,$$ (10.5.5)

[15]Compare Eq. (5.56) of Jackson (1975).

where I is the moment of inertia. Thus

$$\dot{E} = I\Omega\dot{\Omega}. \qquad (10.5.6)$$

Since $\dot{E} < 0$, $\dot{\Omega} < 0$; that is, the pulsar *slows down*. If we define a characteristic age T at the present time by

$$T \equiv -\left(\frac{\Omega}{\dot{\Omega}}\right)_0 = \frac{6Ic^3}{B_p^2 R^6 \sin^2\alpha\, \Omega_0^2}, \qquad (10.5.7)$$

then Eqs. (10.5.4) and (10.5.6) can be integrated to give

$$\Omega = \Omega_i \left(1 + \frac{2\Omega_i^2}{\Omega_0^2}\frac{t}{T}\right)^{-1/2}, \qquad (10.5.8)$$

where Ω_i is the initial angular velocity at $t = 0$. Setting $\Omega = \Omega_0$ in Eq. (10.5.8) gives the present age of the pulsar,

$$t = \frac{T}{2}\left(1 - \frac{\Omega_0^2}{\Omega_i^2}\right)$$

$$\simeq \frac{T}{2} \qquad \text{for } \Omega_0 \ll \Omega_i. \qquad (10.5.9)$$

Groth (1975b) and Gullahorn et al. (1977) report values for T for the Crab pulsar of 2486 yr in 1972, implying an age for the Crab of 1243 yr. This is in remarkably good agreement with the actual age of $1972 - 1054 = 918$ yr. Note that this estimate, first made in this way by Gunn and Ostriker (1969), does *not* depend on detailed properties of the underlying neutron star. It depends only on the general behavior of the function $\Omega(t)$ due to magnetic dipole radiation. The agreement can be improved by allowing for other mechanisms of energy loss, such as gravitational radiation as described below.

The magnetic dipole model can also be used to give a quantitative account of the energetics of the Crab pulsar. Following Gunn and Ostriker, assume the Crab is a spherical neutron star with $M = 1.4 M_\odot$, $R = 12$ km, and $I = 1.4 \times 10^{45}$ g cm^2. Then Eqs. (10.5.5) and (10.5.6) give

$$E = 2.5 \times 10^{49} \text{ erg}, \qquad \dot{E} = 6.4 \times 10^{38} \text{ erg s}^{-1}. \qquad (10.5.10)$$

(Only the value of I is really required to get these values.) Now Eqs. (10.5.5) and (10.5.6) do not depend on the detailed energy loss mechanism, but only on the assumption that the pulsar is a rotating neutron star with rotation as the energy source. It is therefore quite remarkable that $\dot{E}$ is comparable to the observed (kinetic plus radiation) energy requirements of the Crab nebula, which are approximately[16] 5×10^{38} erg s^{-1}. This agreement was first pointed out by Gold (1969), who also pointed out that a reasonably efficient mechanism for accelerating the emitting, relativistic electrons in the surrounding nebula would explain how they could maintain their high energy for over 900 years following the supernova explosion.

Note that $\dot{E}$ in Eq. (10.5.10) is much larger than the observed radiation in the radio *pulse*, which is $\sim 10^{31}$ erg s^{-1} for the Crab.

Using $\dot{E}$ for the Crab from Eq. (10.5.10), and adopting the magnetic dipole model, Eq. (10.5.4) gives

$$B_p = 5.2 \times 10^{12} \text{ G} \quad (\sin \alpha = 1). \tag{10.5.11}$$

Such a value arises naturally from the collapse of a main sequence star with a typical "frozen-in" surface magnetic field of 100 G. The decrease in radius by a factor of $\sim 10^5$ leads to an increase in B_p by a factor $\sim 10^{10}$. Most of the theoretically derived surface fields for other pulsars are of the same magnitude as Eq. (10.5.11).

It is encouraging that recent observations of binary X-ray pulsars (Chapter 13) yield comparable surface magnetic field strengths. Trümper et al. (1978) have observed a feature in the pulsed hard X-ray spectrum of Her X-1, while Wheaton et al. (1979) have observed a similar feature in 4U 0115 − 63. Interpreting the features as cyclotron lines implies

$$B \sim 4\text{–}6 \times 10^{12} \text{ G} \quad \text{in Her X-1,}$$

$$\sim 2 \times 10^{12} \text{ G} \quad \text{in 4U 0115} - 63. \tag{10.5.12}$$

The large magnetic field strengths characteristic of pulsars are presumably produced when the pulsars are formed. Can the field decay away?

The decay time is roughly

$$t_d \sim \frac{\sigma L^2}{c^2}, \tag{10.5.13}$$

where L is a characteristic length and σ is the conductivity [cf. Eq. (7.1.8)]. Dimensionally, an electrical conductivity can be formed from m_e, c, and e by

$$\sigma \sim \frac{m_e c^3}{e^2} \sim 10^{23} \text{ s}^{-1}. \tag{10.5.14}$$

---

[16] Manchester and Taylor (1977).

Using this value of σ and letting $L \sim R$ in Eq. (10.5.13), we estimate that for a "typical" (homogeneous) neutron star, $t_d \sim 10^6$ yr $\gg$ age of the Crab.

The actual value of σ depends on detailed electron-matter interactions and may differ wildly from Eq. (10.5.14). In any event, the decay of the pulsar magnetic field is small during the pulsar lifetime.[17]

As mentioned earlier, one way of making the theoretical and actual ages of the Crab pulsar agree is to invoke gravitational radiation. The lowest-order gravitational radiation is quadrupole (see Chapter 16), so the neutron star must have a time-varying quadrupole moment in order to radiate. Idealizing the neutron star as a slightly deformed, homogeneous ellipsoid with moment of inertia I and ellipticity ε, where

$$\varepsilon = \frac{\text{difference in equatorial radii}}{\text{mean equatorial radius}} = \frac{a - b}{(a + b)/2}, \qquad (10.5.15)$$

gives [cf. Eq. (16.6.9)]:

$$\dot{E}_{GW} = -\frac{32}{5}\frac{G}{c^5}I^2\varepsilon^2\Omega^6$$

$$= 1.4 \times 10^{38} \text{ erg s}^{-1}\left(\frac{I}{1.4 \times 10^{45} \text{ g cm}^2}\right)^2\left(\frac{P}{0.0331 \text{ s}}\right)^{-6}\left(\frac{\varepsilon}{3 \times 10^{-4}}\right)^2.$$

$$(10.5.16)$$

According to Eq. (10.5.16), small ellipticities can generate the emission required to account for the Crab pulsar deceleration, Eq. (10.5.10). Interior anisotropic magnetic fields of $\sim 10^{15}$ G, corresponding to internal main sequence fields of $\sim 10^5$ G prior to collapse, can generate such ellipticities.[18]

Note also that the deceleration law for ε constant is

$$\dot{E}_{GW} = I\Omega\dot{\Omega} \propto \Omega^6. \qquad (10.5.17)$$

Defining

$$T_{GW} \equiv -\left(\frac{\Omega}{\dot{\Omega}}\right)_0, \qquad (10.5.18)$$

---

[17]See Ewart et al. (1975), who emphasize that field decay, if it occurs at all, occurs only in the very outer regions of the star. Also, note that the persistence of strong magnetic fields in the pulsating X-ray sources (Chapters 13 and 15) suggests that magnetic field decay timescales in neutron stars probably exceed several million years even near the surface.

[18]The question of neutron star ellipticities is discussed further in Sections 10.11 and 16.6.

we can integrate Eq. (10.5.17) to obtain

$$\Omega = \Omega_i \left(1 + \frac{4\Omega_i^4}{\Omega_0^4} \frac{t}{T_{GW}} \right)^{-1/4}, \tag{10.5.19}$$

where Ω_i is the initial angular velocity at $t = 0$. Setting $\Omega = \Omega_0$ in Eq. (10.5.19) gives for the present age

$$t = \frac{T_{GW}}{4} \left(1 - \frac{\Omega_0^4}{\Omega_i^4} \right)$$

$$< \frac{2486}{4} = 621 \text{ yr.} \tag{10.5.20}$$

Thus gravitational radiation *cannot alone* be responsible for the Crab slowdown. However, a *combination* of gravitational and magnetic dipole radiation can be found which gives both the correct age and the observed pulsar deceleration rate.[19]

Exercise   10.3 In the combined model, write

$$I\Omega\dot{\Omega} = -\beta\Omega^4 - \gamma\Omega^6, \tag{10.5.21}$$

where β and γ are defined in Eqs. (10.5.4) and (10.5.16). Then

$$\frac{1}{T} \equiv -\frac{\dot{\Omega}}{\Omega}\bigg|_0 = \frac{\beta\Omega_0^2 + \gamma\Omega_0^4}{I}. \tag{10.5.22}$$

(a) Integrate Eq. (10.5.21) to find the age t:

$$(1 + \lambda)\left(1 - \mu + \lambda \log \frac{\lambda + \mu}{\lambda + 1} \right) = \frac{2t}{T}, \tag{10.5.23}$$

where

$$\lambda \equiv \frac{\gamma\Omega_0^2}{\beta}, \qquad \mu \equiv \frac{\Omega_0^2}{\Omega_i^2}. \tag{10.5.24}$$

(b) Using $t = 918$ yr, $T = 2486$ yr, show that the solution of Eq. (10.5.23) is $\lambda = 0.271$, insensitive to μ for $\mu \lesssim 0.01$.

(c) Hence show that

$$\varepsilon = 2.9 \times 10^{-4}, \qquad B_p \sin \alpha = 4.6 \times 10^{12} \text{ G.} \tag{10.5.25}$$

[19]Ostriker and Gunn (1969).

(d) Show that gravitational radiation dominates the energy loss for the first 130 years of the pulsar's life.

(e) The total energy radiated electromagnetically is

$$\Delta E_{\rm em} = \int \frac{\beta \Omega^4}{\dot{\Omega}} \, d\Omega. \tag{10.5.26}$$

Choose $\Omega_i = 10^4 \text{ s}^{-1}$ for definiteness, and show that

$$\Delta E_{\rm em} = 5.9 \times 10^{50} \text{ erg}, \tag{10.5.27}$$

while

$$\Delta E_{\rm GW} = \tfrac{1}{2} I \left(\Omega_i^2 - \Omega_0^2 \right) - \Delta E_{\rm em}$$

$$\simeq \tfrac{1}{2} I \Omega_i^2 = 7 \times 10^{52} \text{ erg}. \tag{10.5.28}$$

(f) Show that at the present time

$$\dot{E}_{\rm GW} = 1.4 \times 10^{38} \text{ erg s}^{-1}, \qquad \dot{E}_{\rm em} = 5.1 \times 10^{38} \text{ erg s}^{-1},$$

while with $\Omega_i = 10^4 \text{ s}^{-1}$,

$$\dot{E}_{\rm GW} = 2.9 \times 10^{48} \text{ erg s}^{-1}, \qquad \dot{E}_{\rm em} = 3.9 \times 10^{45} \text{ erg s}^{-1}$$

initially.

Exercise 10.4 What is the value of the polar eccentricity e and the ratio $T/|W|$ for the Crab pulsar initially if $\Omega_i = 10^4 \text{ s}^{-1}$? Is this configuration stable? (Assume Maclaurin spheroid relations; cf. Section 7.3).

10.6 Braking Index

For any power-law deceleration model, such as the magnetic dipole model, we can write

$$\dot{\Omega} = - \, (\text{constant}) \, \Omega^n, \tag{10.6.1}$$

where the parameter n is called the *braking index*. For the magnetic dipole model, $n = 3$. In general, one can define

$$n \equiv - \frac{\Omega \ddot{\Omega}}{\dot{\Omega}^2}, \tag{10.6.2}$$

which includes the case of Eq. (10.6.1). Thus the braking index can in principle be measured directly from the pulsar frequency and its derivatives. At present, only the Crab pulsar has a reliable determination; Groth (1975b) obtained

$$n = 2.515 \pm 0.005. \tag{10.6.3}$$

A number of factors[20] may be responsible for the deviation of n from the "canonical" value of 3. For example, Macy (1974) has argued that the alignment or counter-alignment of the magnetic axis of a pulsar with its symmetry axis (which is not necessarily its rotation axis) can occur if the magnetic axis wanders through the star (cf. "polar wandering" on the earth). In counteraligning models, n can be as low as 2 during counter-alignment, even though braking is due to magnetic dipole radiation. When counter-alignment is complete, $n = 3$.

Note that if gravitational radiation were the dominant cause of braking, then n would be 5. In the Gunn–Ostriker combined model for the Crab discussed in the previous section,

$$n = \frac{3 + 5\lambda}{1 + \lambda} = 3.43 \text{ now.} \tag{10.6.4}$$

10.7 Nonvacuum Pulsar Models: The Aligned Rotator

It is significant that Eq. (10.5.4),

$$\dot{E} \sim -\frac{B_p^2 R^6 \Omega^4}{c^3}, \tag{10.7.1}$$

may also hold in models other than the oblique, vacuum magnetic dipole model. In fact, as Goldreich and Julian (1969) first pointed out (and as we shall show below), strong electric fields parallel to the pulsar's surface will invariable rip off charged particles from the star. Consequently, a pulsar must possess a dense *magnetosphere*. Examining the case in which the magnetic dipole moment is aligned with the rotation axis of the star, Goldreich and Julian argued that those charged particles in the magnetosphere, threaded by magnetic field lines that close within the *light cylinder*, must corotate with the pulsar. Here the light cylinder is an imaginary cylinder with an axis along the pulsar rotation axis and with a radius extending out to the distance at which the corotation velocity attains the speed of light (see Fig. 10.6). This *corotation radius* is thus given by

$$R_c \equiv \frac{c}{\Omega} = 5 \times 10^9 P \text{ cm,} \tag{10.7.2}$$

[20] See Manchester and Taylor (1977) for a more complete discussion.

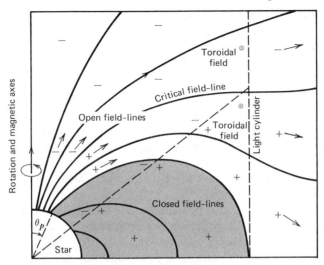

Figure 10.6   Sketch of the Goldreich–Julian (1969) model of the magnetosphere of a pulsar with parallel magnetic and rotation axes. Particles that are attached to closed magnetic-field lines corotate with the star and form the corotating magnetosphere. The magnetic-field lines that pass through the light cylinder (where the velocity of corotation equals the velocity of light) are open and are deflected back to form a toroidal field component. Charged particles stream out along these lines. The critical field line is at the same electric potential as the exterior interstellar medium. This line divides regions of positive and negative current flow from the star and the plus and minus signs indicate the charge of particular regions of space. The diagonal *dashed* line is the locus of $B_z = 0$, where the space charge changes sign. The angle subtended by the polar cap region containing open field lines is θ_p. [From *Pulsars* by Richard N. Manchester and Joseph H. Taylor. W. H. Freeman and Company. Copyright © 1977.]

where P is the pulsar period in seconds. As particles approach R_c in the equatorial plane, they become highly relativistic. Gold (1968) and others have proposed that the high-frequency radiation observed from pulsars may originate from the emission of relativistic particles near the light cylinder.

In the model of Goldreich and Julian and in virtually all other pulsar models, the magnetic field is largely dipolar in the near zone; that is, at distances less than the vacuum wavelength of the radiation emitted, $r < \lambda = c/\Omega = R_c$. Accordingly,

$$B \sim \frac{|\mathbf{m}|}{r^3} \sim B_p \left(\frac{R}{r}\right)^3, \qquad r < R_c, \qquad (10.7.3)$$

where B_p is the field strength at the magnetic pole and where R is the stellar radius. At greater distances the magnetic field is an outgoing-wave field associated with a perpendicular electric field of magnitude $E \sim B$ and an outward Poynting energy flux $S \sim cB^2/4\pi$. Thus the total rate of electromagnetic emission, obtained by joining the fields at $r \sim \lambda \sim R_c$, is given by

$$\dot{E} \sim - (4\pi r^2 S)_{r \sim R_c} \sim \frac{-c|\mathbf{m}|^2}{R_c^4} \sim \frac{-B_p^2 R^6 \Omega^4}{c^3},$$

which is just Eq. (10.7.1).

Hence, whether or not the dipole field is aligned with the rotation axis, plasma near the pulsar can in principle carry away sufficient angular momentum and energy to supply the necessary braking torque. Given the result Eq. (10.7.1), we again obtain $n = 3$ for the braking index. The details of this picture (e.g., ratio of particle to field energy flux across R_c; field geometry and flow pattern; etc.) are, however, a subject of much debate.[21]

Nonvacuum models *may* also provide a natural explanation for the magnetic fields in the surrounding Crab nebula, inferred from the observed synchrotron radiation. Although no reliable calculation exists for the flow of fields and particles beyond the light cylinder, it is clear that the energy density there is no longer strongly coupled to the rotating neutron star. If in this outer region the field energy dominates the particle energy—that is, $B^2 \gtrsim 8\pi \varepsilon_p$—then energy conservation implies

$$c\left(\frac{B^2}{8\pi} \right) 4\pi r^2 \sim \text{constant, independent of } r. \qquad (10.7.4)$$

Now for the Crab, $B \sim 10^6$ G at $r \sim R_c \sim 10^8$ cm. Thus

$$B \sim \frac{10^{14} \text{ G}}{r \text{ (cm)}}, \qquad r > R_c. \qquad (10.7.5)$$

This implies $B \sim 10^{-4}$ G at $r \sim 1$ pc, in agreement with estimated values for the nebular field.[22]

Exercise   10.5 Suppose, instead, $\varepsilon_p \gg B^2/8\pi$ for $r > R_c$. In this case the particle flow might "comb out" the magnetic field into a radial configuration far from the star. What is the resulting dependence of B vs. r, and what is the value of B at ~ 1 pc?

Answer: $\nabla \cdot \mathbf{B} = 0$ implies $B \sim 1/r^2$.

As mentioned above, the rotation of a neutron star possessing a magnetic field generates powerful electric fields in the space surrounding the star. Let us now consider the argument of Goldreich and Julian (1969) that, because of these electric fields, the surrounding region cannot be empty but must contain plasma.

---

[21]See, for example, articles in Sieber and Wielebinski (1981) and the excellent review by Michel (1982) for recent discussions and references.
[22]Ruderman (1972).

In their model, a spinning neutron star has an *aligned* dipole external magnetic field[23]:

$$\mathbf{B}^{(\text{out})} = B_p R^3 \left(\frac{\cos\theta}{r^3} \mathbf{e}_{\hat{r}} + \frac{\sin\theta}{2r^3} \mathbf{e}_{\hat{\theta}} \right). \tag{10.7.6}$$

The stellar material is assumed to be an excellent conductor, so that inside the star, an electric field will be present to satisfy Eq. (7.1.9):

$$\mathbf{E}^{(\text{in})} + \frac{\boldsymbol{\Omega} \times \mathbf{r}}{c} \times \mathbf{B}^{(\text{in})} = 0. \tag{10.7.7}$$

Assuming there are no surface currents, both the normal and tangential components of **B** are continuous across the stellar surface. Thus just inside the surface,

$$\mathbf{B}^{(\text{in})} = B_p \left(\cos\theta \, \mathbf{e}_{\hat{r}} + \frac{\sin\theta}{2} \mathbf{e}_{\hat{\theta}} \right). \tag{10.7.8}$$

Equation (10.7.7) then gives for **E** just inside the surface

$$\mathbf{E}^{(\text{in})} = \frac{R\Omega B_p \sin\theta}{c} \left(\frac{\sin\theta}{2} \mathbf{e}_{\hat{r}} - \cos\theta \, \mathbf{e}_{\hat{\theta}} \right). \tag{10.7.9}$$

The tangential component of **E** is continuous across the surface, so just outside the star Eq. (10.7.9) implies

$$E_{\theta}^{(\text{out})} = -\frac{\partial}{\partial\theta} \left(\frac{R\Omega B_p \sin^2\theta}{2c} \right) = \frac{\partial}{\partial\theta} \left[\frac{R\Omega B_p}{3c} P_2(\cos\theta) \right]. \tag{10.7.10}$$

Assume, for the moment, that the exterior region is a vacuum. Then

$$\mathbf{E}^{(\text{out})} = -\nabla\phi, \tag{10.7.11}$$

where

$$\nabla^2\phi = 0. \tag{10.7.12}$$

In order to satisfy the boundary condition (10.7.10) at $r = R$, the solution of Eq. (10.7.12) must be

$$\phi = -\frac{B_p\Omega}{3c} \frac{R^5}{r^3} P_2(\cos\theta), \tag{10.7.13}$$

that is, the external electric field is a quadrupole field.

[23]Since such an axisymmetric configuration will not pulse, *some* small departure from strict alignment must be imagined to explain the pulsed emission.

Exercise   10.6

(a)   Taking the dielectric constant to be unity everywhere, show that the discontinuity in the normal component of **E** at the stellar surface is associated with a surface charge density

$$\sigma = - \frac{B_p \Omega R}{4 \pi c} \cos^2 \theta. \tag{10.7.14}$$

(b)   Neglecting any macroscopic currents near the surface, so that $\nabla \times \mathbf{B}^{(\text{in})} = 0$ for a region near the surface, show that the internal **E** field is associated with an interior charge density

$$\rho_e = \frac{1}{4 \pi} \nabla \cdot \mathbf{E} = - \frac{1}{2 \pi c} \mathbf{\Omega} \cdot \mathbf{B}, \tag{10.7.15}$$

corresponding to

$$n_e = 7 \times 10^{-2} B_z P^{-1} \text{ cm}^{-3}, \tag{10.7.16}$$

where B_z is the z component of **B** in gauss and P is the pulsar period in seconds.

Now Eq. (10.7.7) implies that $\mathbf{E} \cdot \mathbf{B} = 0$ inside the star. However, outside the star Eqs. (10.7.6), (10.7.11), and (10.7.13) give

$$\mathbf{E} \cdot \mathbf{B} = - \frac{R \Omega}{c} \left(\frac{R}{r} \right)^7 B_p^2 \cos^3 \theta. \tag{10.7.17}$$

Thus the magnitude of the electric field parallel to **B** at the surface is approximately

$$E_{\parallel} \sim \frac{R \Omega}{c} B_p \sim 2 \times 10^8 P^{-1} B_{12} \text{ volt cm}^{-1}, \tag{10.7.18}$$

where B_{12} is the magnetic field strength in units of 10^{12} G. Such a strong field will impart a force to both electrons and ions at the surface that greatly exceeds the gravitational force. For protons,

$$\frac{\text{electric force}}{\text{gravitational force}} \sim \frac{e R \Omega B_p / c}{GMm/R^2} \sim 10^9 \gg 1. \tag{10.7.19}$$

Thus particles will be torn off the surface and create a region of plasma around the star—the *magnetosphere*. *A vacuum solution for the region surrounding a rotating neutron star is unstable.*

Inside the light cylinder, the plasma will corotate with the star because of the strong magnetic field. The magnetosphere acts as an extension of the perfectly conducting interior, so that $\mathbf{E} \cdot \mathbf{B} = 0$ and Eqs. (10.7.7), (10.7.15), and (10.7.16) remain valid. The region where the field lines close beyond the light cylinder ("open magnetosphere") receives particles which are permanently lost to the star and satisfies $\mathbf{E} \cdot \mathbf{B} \neq 0$ (see Fig. 10.6).

10.8 Pulsar Emission Mechanisms

The actual mechanism by which pulsars convert the rotational energy of the neutron star into the observed *pulses* is poorly understood. Many theoretical models have been proposed, but no single one is compelling.[24] This is so despite the seemingly universal characteristics of the radio emission from different pulsars; a single basic model probably applies to all pulsars.

On the other hand, the energy observed in pulses is only a small fraction[25] of the total rotational energy dissipated, so that ignorance of the actual pulsed emission process *may* be decoupled from the gross energetics of radiating neutron stars.

What are the fundamental observational requirements for the pulse emission mechanism? Some are:

1. The radiation must be emitted in a relatively narrow beam, fixed in orientation with respect to the neutron star. The beam must be $\leq 10°$ in longitude as seen by a distant observer and this width must be constant over many decades of frequency. Moreover, the beam shape and longitude must remain stable for many rotation periods.

2. The radiation mechanism must produce rather broad-band radiation at both radio and optical frequencies. The radio pulses have bandwidths $\gtrsim 100$ MHz.

3. The radiation process must generate the observed luminosities and brightness temperatures in the radio, optical, and X-ray bands.

4. At radio wavelengths, the emission should show strong linear polarization that is approximately independent of frequency and stable for long time intervals.

The brightness temperature T_b of an emitting region is defined by

$$I_\nu \equiv B_\nu(T_b), \qquad (10.8.1)$$

where I_ν is the specific intensity (erg s^{-1} cm^{-2} Hz^{-1} sr^{-1}) and B_ν is the Planck

---

[24]See Manchester and Taylor (1977), Ruderman (1980), Sieber and Wielebinski (1981), or Michel (1982) for a review and critique of some of them.

[25]The fraction is $\leq 10^{-9}$ for the Crab, and $\leq 10^{-2}$ for some old pulsars.

function (cf. Appendix I). For $h\nu \ll kT_b$, we have the familiar Rayleigh–Jeans law,

$$I_\nu = \frac{2\nu^2}{c^2} kT_b \quad (h\nu \ll kT_b). \tag{10.8.2}$$

Now observed pulsar radio luminosities, assuming conical beams and reasonable distances, range from 10^{25}–10^{28} erg s^{-1}. If we assume a source emitting area of magnitude $\sim (ct)^2 \sim 10^{15}$ cm^2, where $t \leq 10^{-3}$ s is a typical pulse duration, we get

$$I_\nu \sim 10^4 - 10^7 \text{ erg s}^{-1} \text{ cm}^{-2} \text{ Hz}^{-1} \text{ sr}^{-1},$$

$$T_b \sim 10^{23} - 10^{26} \text{ K},$$

$$kT_b \sim 10^{17} - 10^{22} \text{ eV}. \tag{10.8.3}$$

For incoherent emission, thermodynamics implies $kT_b \leq E_p$, where E_p is the particle energy. (This is simply a restatement of the fact that a blackbody is the most efficient radiator.) But the particle energies required by Eq. (10.8.3) are absurdly large. Even if such high-energy particles were available, they would radiate most of their power at very high frequencies and not in the radio band. We conclude that a *coherent* radiation mechanism is required, in which the total intensity is $\sim N^2$ times the intensity radiated by a single particle, where N is the number of particles emitting coherently.

Such coherence is not required for the observed X-ray or optical pulsed emission.

Exercise 10.7 Show that the brightness temperature for X-rays from the Crab pulsar (luminosity 10^{35} erg s^{-1}) is $\sim 10^{11}$ K, and argue that $E_p \sim 10$ MeV is a reasonable particle energy for the magnetosphere.

In most models, the emission is concentrated in a conical beam that corotates with the neutron star. In "polar cap" models, the radiation cone is aligned with the star's dipole magnetic field and lies within the corotating magnetosphere. In "light cylinder" models, the cone axis is tangential to the light cylinder and perpendicular to the rotation axis of the star; here the emission originates near R_c. Still other models have been proposed which place the emitting region beyond the light cylinder. Since the theoretical situation is yet to be resolved, we will not discuss these models further.

10.9 Superfluidity in Neutron Stars

Many-body fermion systems with an interaction that favors the formation of *pairs* of particles in two-body states may undergo a phase transition to a *superfluid* state. If the particles are charged, the state will be *superconducting*. Such is the case, for example, in the BCS theory for superconductivity; here, electrons in a metal with momentum $\mathbf{k}$ and spin $\mathbf{s}$ pair with electrons having momentum $-\mathbf{k}$ and spin $-\mathbf{s}$. The coupling is mediated by the electron-phonon interaction in the lattice.[26]

Since the basic nuclear interaction is attractive at large distances, the same long-range BCS pairing mechanism can arise in dense hadronic matter—for example, laboratory nuclei or neutron star matter. Although two neutrons cannot be bound in a vacuum, they can be bound when they are in the field of other nucleons. In highly degenerate Fermi systems, the pairing occurs mainly between states near the Fermi surface. In heavy nuclei or in neutron stars, where the ratio n_n/n_p is high, one need only consider n-n and p-p pairing; states of opposite momentum are at different Fermi surfaces for n and p.

Pairs of neutrons are bosons, so their behavior is presumably similar to that of ^{4}He atoms in liquid helium. ^{4}He is a superfluid below $T = 2.19$ K. In particular, it undergoes viscous-free flow: the kinetic energy of a pure superfluid is not dissipated by friction against the container walls or within the fluid itself.

In neutron stars, similar behavior may occur whenever the thermal energy kT is less than the latent heat Δ associated with the phase transition to a paired state ("superfluid energy gap"). The gap parameter Δ depends on the strength of the pairing interaction and, in turn, on the density. At nuclear densities ($\rho \lesssim 2.8 \times 10^{14}$ g cm^{-3}), we know from laboratory nuclei that both neutrons and protons have undergone a pairing transition in cold nuclear matter and that $\Delta \sim 1$–2 MeV. Hence at the relatively low temperatures ($\lesssim$ keV) expected for all but the youngest neutron stars (cf. Chapter 11), one expects to find neutron superfluidity in the crust and interior of neutron stars.

One also expects the remaining protons in the interior to be paired and hence superconducting. It is unlikely, however, that the electrons will be superconducting: electron-phonon coupling is too weak in the neutron star case.

Recent calculations[27] suggest that at least three distinct hadronic superfluids exist inside a neutron star:

1. In the inner crust (4.3×10^{11} g cm^{-3} $< \rho < 2 \times 10^{14}$ g cm^{-3}), the free neutrons may pair in a $^1S_0$ state to form a superfluid amid the neutron-rich nuclei.

2. In the quantum liquid regime ($\rho \gtrsim 2 \times 10^{14}$ g cm^{-3}), where the nuclei have dissolved into a degenerate fluid of neutrons and protons, the neutron fluid is likely to be paired in a $^3P_2$ state.

---

[26] See Weisskopf (1981) for an elementary discussion.
[27] See Pines (1980a) for a review.

3.   The protons in the quantum liquid are expected to be superconducting in a $^1S_0$ state.

There are a number of important consequences of hadron superfluidity and superconductivity, which may lead to several observational effects. It is important to note that superfluidity has little effect on the gross properties of neutron stars such as their masses and radii, and so on. The reason is that the pairing energy is $\lesssim 1\%$ of the total interaction energy in the neutron fluid region, and so it makes very little difference to the P vs. ρ relation discussed earlier for "normal" matter. Important physical consequences of superfluidity include the following:

(a) Thermal Effects

The heat capacity of a superfluid or superconductor is much lower than that of a normal degenerate gas at sufficiently low temperatures. In fact, the heat capacity varies as $\exp(-\Delta/kT)$, since to excite baryons one must first break up the BCS pair. This costs Δ in energy.[28]

This reduction in heat capacity shortens the cooling timescale for pulsars. However, the normal components of the fluid (e.g., the electrons) contribute fully to the heat capacity, so the effect of the superfluid is somewhat reduced.

Counterbalancing these effects, frictional interactions between normal and superfluid components can result in the thermal dissipation of rotational energy and a net increase in pulsar cooling times.[29] The issue is not resolved. Superfluidity also lowers the neutrino emission rates below their values for "normal" matter.

(b) Magnetic Effects

A laboratory superconductor generally exhibits a Meissner effect: a magnetic field is expelled from any region that becomes superconducting. Free electrons are, however, present in the case of a neutron star, and so we expect the magnetic field to thread the entire configuration. As a result, the charged components of the crust and core are tied together by the magnetic field and corotate.

The superfluid neutrons, however, are only weakly coupled to the crust and charged components. Now the charged particle component is constantly decelerated by the radiation reaction torque transmitted by the magnetic field [Eq. (10.5.4)]. The superfluid must then always be rotating *faster* than the pulsar, on average. Weak frictional forces between the normal outer crust and superfluid

---

[28] See Landau and Lifshitz (1969), Section 66 for a simple discussion.

[29] Greenstein (1979); see Chapter 11.

neutron interior couple the two components and convert some rotational energy into frictional heating.

Exercise   10.8   Establish an upper limit to the surface temperature T_s of the Crab pulsar by assuming that *all* of the rotational energy is dissipated as thermal blackbody radiation from the surface.
   Answer: $T_s \lesssim 3 \times 10^7$ K.

Relaxation of the charged and superfluid components to rigid-body rotation may take days or years after any sudden change in the pulsar period (i.e., period of the charged component). This dynamical behavior is the basis for some models for the Crab and Vela glitches (Section 10.10).

(c)   Hydrodynamic Effects

One unique consequence of neutron superfluidity is that because of the neutron star's rotation, the fluid will contain a discrete array of *vortices*. The vortices are parallel to the rotation axis and each have a quantized circulation

$$\oint \mathbf{v} \cdot \mathbf{d}l = \frac{h}{2m_n}, \tag{10.9.1}$$

where $\mathbf{v}$ is the fluid velocity and $2m_n$ the mass of a neutron pair.[30]
   Microscopically, the fluid moves irrotationally, $\nabla \times \mathbf{v} = 0$, everywhere except within the core of each vortex line. When averaged over many vortex lines, the average fluid velocity $\langle \mathbf{v} \rangle$ can satisfy the usual relation for uniform rotation:

$$\nabla \times \langle \mathbf{v} \rangle = 2\mathbf{\Omega}. \tag{10.9.2}$$

From eqs. (10.9.1) and (10.9.2) and Stokes' theorem, we find that the number of quantized vortex lines per unit area is

$$n_v = \frac{4\Omega m_n}{h} = 1.9 \times 10^5 \text{ cm}^{-2} \quad \text{for the Crab.} \tag{10.9.3}$$

The mean spacing between vortex lines is $n_v^{-1/2} \sim 10^{-2}$ cm, which is much smaller than the radius of the star. This justifies the macroscopic viewpoint: if one considers $\langle \mathbf{v} \rangle$, then the superfluid can be treated as rigidly rotating. The moment of inertia of the star has its classical value, too.

---

[30] See Feynman (1972), Chapter 11, for a discussion of why a rotating superfluid forms vortices, subject to Eq. (10.9.1).

It is still a subject of current research and debate whether or not the cores of the vortex lines are "pinned" to the crustal nuclei or instead thread the spaces between them. The answer to this is relevant to some models for pulsar glitches.

10.10   Pulsar Glitches and Hadron Superfluidity

It is quite remarkable that timing data for the Crab and Vela pulsars following their sudden spinups may provide evidence for superfluidity in neutron stars.

At the time of the first Vela glitch, which was characterized by a sudden change in $\dot{\Omega}$, $\Delta\dot{\Omega}/\dot{\Omega} \sim 10^{-2}$, together with a smaller increase in Ω, $\Delta\Omega/\Omega \simeq 2 \times 10^{-6}$, Baym et al. (1969) proposed a simple "two-component" neutron star model to explain the spinup phenomenon. We present this model, which is largely phenomenological, as a prototype to show how in principle the observational data for glitches can be incorporated in a theoretical picture. Later we shall compare some specific predictions of the model with the observations.

The model consists of a normal component, the crust and charged particles, of moment of inertia I_c, weakly coupled to the superfluid neutrons, of moment of inertia I_n. The charged component is assumed to rotate at the observed pulsar frequency $\Omega(t)$, since all charged particles are assumed to be strongly coupled to the magnetic field. The rotation of neutron superfluid is assumed to be quasi-"uniform" in the sense described in Section 10.9, with average angular frequency $\Omega_n(t)$. The coupling between the two components is described by a single parameter τ_c, the relaxation time for frictional dissipation.

In the model, it is imagined that the speed-up is triggered by a "starquake" occurring in the crust. Details of the starquake mechanism are unimportant here (see the next section for a model). All that counts are the assumptions that (1) the crust spinup is *rapidly* communicated to the *charged* particles in the interior by the strong magnetic field (timescale $\sim$ 100 s), while (2) the response of the neutron superfluid to the speedup of the crust-charged particle system is considerably slower ($\sim$ yr), due to a much weaker frictional coupling between the normal and superfluid components.

Exercise 10.9   The spinup of the crust is communicated to the charged particles in the interior by the magnetic field which threads them. Any distortion in the magnetic field due, for example, to differential rotation, generates magnetic "sound waves" or Alfvén waves, which travel with a speed

$$v_A \sim \left(\frac{P_B}{\rho} \right)^{1/2}, \qquad P_B \sim \frac{B^2}{8\pi}. \tag{10.10.1}$$

If differential rotation in the crust-charged component is smoothed out in the time τ_A for an Alfvén wave to traverse the neutron star, estimate τ_A for a typical pulsar.

   Answer: $\tau_A \sim 50$ s.

The glitch-free interaction between the two components after a starquake is governed by two linear equations:

$$I_c\dot{\Omega} = -\alpha - \frac{I_c(\Omega - \Omega_n)}{\tau_c},$$ (10.10.2)

$$I_n\dot{\Omega}_n = \frac{I_c(\Omega - \Omega_n)}{\tau_c}.$$ (10.10.3)

Here α is the external braking torque on the crust due, for example, to magnetic dipole radiation-reaction forces. Taking α and τ_c to be constant over the time-scales of interest, Eqs. (10.10.2) and (10.10.3) may be solved to give

$$\Omega = -\frac{\alpha}{I}t + \frac{I_n}{I}\Omega_1 e^{-t/\tau} + \Omega_2,$$ (10.10.4)

$$\Omega_n = \Omega - \Omega_1 e^{-t/\tau} + \frac{\alpha\tau}{I_c},$$ (10.10.5)

where

$$I \equiv I_c + I_n, \qquad \tau \equiv \tau_c I_n/I,$$ (10.10.6)

and Ω_1 and Ω_2 are arbitrary constants depending on initial conditions.

Note that the steady-state solution ($t/\tau \to \infty$) is

$$\Omega_n - \Omega = \frac{\alpha\tau}{I_c} = \frac{\alpha}{I}\frac{I_n}{I_c}\tau_c = \frac{I_n}{I_c}\frac{\tau_c}{T}\Omega,$$ (10.10.7)

where

$$\frac{1}{T} = -\frac{\dot{\Omega}}{\Omega} = \frac{\alpha}{I\Omega}$$ (10.10.8)

is the characteristic age defined in Eq. (10.5.7).

Post-glitch observations indicate $\tau \sim$ months for Vela and $\tau \sim$ weeks for the Crab (cf. discussion below). So, crudely assuming $I_n \sim I_c$, which is true for a massive 1.4 $M_\odot$ star built from a stiff TI or TNI equation of state (cf. Fig. 9.4), then Eqs. (10.10.6) and (10.10.7) predict $(\Omega_n - \Omega)/\Omega \sim 10^{-5}$ for Vela and Crab.

Equation (10.10.4) has been used to fit the post-glitch timing data for the Crab and Vela pulsars. Assume that a glitch occurs instantaneously in the observed (crust) angular velocity at $t = 0$: $\Omega(t) \to \Omega(t) + \Delta\Omega_0$. Replace Ω_1 and Ω_2 in Eq.

(10.10.4) by the constants $\Delta\Omega_0$ and the "healing parameter" Q, so that the equation takes the form

$$\Omega(t) = \Omega_0(t) + \Delta\Omega_0\left[Qe^{-t/\tau} + 1 - Q\right]. \qquad (10.10.9)$$

The healing parameter Q describes the degree to which the angular velocity relaxes back toward its extrapolated value: if $Q = 1$, $\Omega(t) \rightarrow \Omega_0(t)$ as $t \rightarrow \infty$. Here $\Omega_0(t) = \Omega_0 - \alpha t/I$ is the pulsar frequency in the absence of the glitch, where Ω_0 is a constant.

The behavior indicated by Eq. (10.10.9) and illustrated in Figure 10.7 gives a reasonable fit to all large glitches observed so far. Figure 10.3 shows the available data for the Vela pulsar, which clearly exhibits glitch behavior. Table 10.1 summarizes the results for the nine pulsar glitches for which a detailed analysis has been performed.

Clearly a crucial test for this phenomenological two-component model is whether or not the fitted post-glitch functions yield the *same* values of Q and τ for all the glitches of a single pulsar. Though there are obvious discrepancies, it is remarkable that to lowest order the values are reasonably constant. Analysis difficulties in the timing data may be as much responsible for the apparent discrepancies as the over-simplification of the two-component model. In general, however, the existence of at least two distinct components, with relaxation timescales comparable to theoretical predictions for crust-superfluid coupling, is supported by the data.

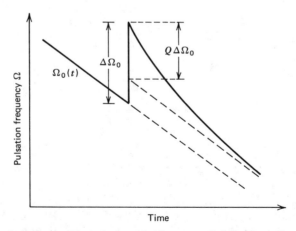

Figure 10.7 Time dependence of the pulsar angular frequency, Ω, following a discontinuous speedup. [See Eq. (10.10.9).] [From *Pulsars* by Richard N. Manchester and Joseph H. Taylor. W. H. Freeman and Company. Copyright © 1977.]

Table 10.1

Observations of Glitches and Post-Glitch Behavior for Three Pulsars

Date	Pulsar	Reference	$\Delta\Omega_0/\Omega_0$	$Q(\times 100)$	$\tau(d)$
3/69	Vela	a	2.34×10^{-6}	3.4 ± 1.0	75 ± 20
8/71	Vela	a	1.96×10^{-6}	3.5 ± 0.1	60 ± 10
12/71	Vela	a	1.17×10^{-8}	55^{+21}_{-12}	80 ± 20
10/75	Vela	a	2.01×10^{-6}	8.8 ± 0.8	40 ± 5
7/78	Vela	a	3.05×10^{-6}	2.4 ± 0.5	55 ± 5
10/81	Vela	b	1.14×10^{-6}	?	?
9/69	Crab	c	$\sim 10^{-8}$	~ 93	~ 4.1
10/71	Crab	d	2×10^{-9}	~ 96	15
2/75	Crab	d	3.7×10^{-8}	~ 96	15
$\sim 9/77$	$1641 - 45$	e	1.9×10^{-7}	?	31000

[a] Downs (1981)
[b] McCulloch et al. (1981)
[c] Boynton et al. (1972)
[d] Löhsen (1975)
[e] Manchester et al. (1978)

Can the data provide more physical insight into the nature of a neutron star, assuming the model to be correct? Decidedly yes, as a number of investigators have frequently argued.[31]

Consider the parameters Q and τ. Defining

$$\Delta\Omega(t) \equiv \Omega(t) - \Omega_0(t), \tag{10.10.10}$$

we find from eq. (10.10.9) that

$$Q = -\frac{\Delta\dot{\Omega}(t=0)}{\Delta\Omega_0}\tau \tag{10.10.11}$$

and

$$\tau = -\frac{\Delta\dot{\Omega}(t=0)}{\Delta\ddot{\Omega}(t=0)}. \tag{10.10.12}$$

Thus Q and τ can in principle be determined directly from the post-glitch data.

Now Q can also be related to the moments of inertia of the various components. For assume that the "starquake" responsible for the glitch results in

---

[31] See, for example, Alpar et al. (1981) and references therein to earlier work.

changes ΔI_c, $\Delta\Omega$, ΔI_n, $\Delta\Omega_n$, and so on, but leaves τ_c and α roughly constant. Then on the starquake timescale the angular momentum of *each* component is separately conserved, implying

$$\frac{\Delta I_c}{I_c} = -\frac{\Delta\Omega}{\Omega},$$

(10.10.13)

$$\frac{\Delta I_n}{I_n} = -\frac{\Delta\Omega_n}{\Omega_n}.$$

(10.10.14)

Differentiating Eq. (10.10.2) gives

$$\Delta\dot{\Omega} = \frac{\alpha\,\Delta I_c}{I_c^2} - \frac{\Delta\Omega - \Delta\Omega_n}{\tau_c}.$$

(10.10.15)

But Eqs. (10.10.13) and (10.10.8) give

$$\frac{\alpha\,\Delta I_c}{I_c^2} = -\frac{\alpha\,\Delta\Omega}{I_c\Omega} = -\frac{I}{I_c}\frac{\Delta\Omega}{T}.$$

(10.10.16)

Typically

$$\frac{I_c}{I}\frac{T}{\tau_c} \gg 1,$$

(10.10.17)

so the first term in Eq. (10.10.15) can be ignored in comparison with the second.

Exercise 10.10 Verify that for the Vela pulsar inequality (10.10.17) is satisfied for reasonable estimates of I_c/I (e.g., ≥ 0.01).

Thus Eq. (10.10.15) becomes, using Eq. (10.10.8),

$$\frac{\Delta\dot{\Omega}}{\dot{\Omega}} = \frac{\Delta\Omega}{\Omega}\frac{T}{\tau_c}\left(1 - \frac{\Delta\Omega_n}{\Delta\Omega}\right).$$

(10.10.18)

Exercise 10.11 Show that the two-component model predicts successfully that, following a glitch, fractional changes in $\dot{\Omega}$ are greater than fractional changes in Ω.

Now Eqs. (10.10.18) and (10.10.8) give for Eq. (10.10.11) (with $\Delta\Omega = \Delta\Omega_0$)

$$Q = \frac{\tau}{\tau_c}\left(1 - \frac{\Delta\Omega_n}{\Delta\Omega}\right).$$

(10.10.19)

Recalling Eqs. (10.10.6), (10.10.13), and (10.10.14), we get

$$Q = \frac{I_n}{I}\left(1 - \frac{\Delta I_n/I_n}{\Delta I_c/I_c}\frac{\Omega_n}{\Omega}\right). \tag{10.10.20}$$

Now $\Omega_n - \Omega \ll \Omega$, so typically we expect

$$Q \simeq \frac{I_n}{I}. \tag{10.10.21}$$

Table 10.1 thus provides a definite value for the moments of inertia of the normal and superfluid components of the neutron star.

Any comparison between the observed values of Q with theoretical values of I_n/I is complicated by the fact that even if the two-component model is correct, I_n/I depends on both the adopted equation of state and the mass of the neutron star. Values of I_n/I are tabulated in Table 10.2 for the soft Reid, moderately stiff BJ, and very stiff TI and MF equations of state examined previously in Section 9.3.[32] Typically, harder equations of state and smaller neutron star masses imply fractionally larger crusts and, consequently, smaller values of I_n/I. One possible interpretation of the post-glitch data, assuming the Reid interaction model to be correct, is that the Crab pulsar represents a rather massive neutron star ($M/M_\odot \gtrsim 1.3M_\odot$), while the Vela pulsar represents a light neutron star ($M/M_\odot \lesssim 0.15$). In any case, a superficial comparison of the data with the two-component model suggests that the Crab and Vela pulsars have very different masses. Alternatively, one might argue in favor of a moderately stiff, for example, TNI, equation of state to explain laboratory data for nuclear matter and the possibility that all neutron stars have approximately the same mass ($\sim 1.4M_\odot$, for which TNI gives $Q \sim I_n/I = 0.90$). Noting the discrepancy between this hypothesis and the observed values of Q for Vela, one would conclude[33] that variations of Q inferred from the post-glitch timing data might be due to different glitch origins and/or a more complex post-glitch dynamical behavior. The two-component model is thus probably too naive to reliably account for the coupling between the crust and superfluid neutron interior. Indeed, Downs (1981) has shown that the post-glitch behavior of the Vela pulsar is substantially different from that inferred from the simple two-component model of a weakly coupled crust and superfluid core. Obviously, further analysis is required,[34] but the model nevertheless illustrates how pulsar timing data might be employed as a powerful probe of neutron star structure and hadron physics.

[32] Recall that the TNI equation of state is comparable to the BJ equation of state.
[33] Pines (1980a).
[34] See, for example, Alpar et al. (1981).

Table 10.2

Properties of Neutron Stars Sensitive to the Hadron Equation
of State and Mass[a]

	Interaction Model	Mass $(M_\odot)$	I_n/I $(\approx Q)$	$t_q(\text{Crab})^b$ (yr)
hardness	Reid (R)	1.33	0.96	2,100
	Bethe–Johnson (BJ)	1.33	0.77	130
	Mean field (MF)	1.33	0.66	25
	Tensor-interaction (TI)	1.33	0.44	10
	Tensor-interaction (TI)	0.10	0.0	3.0×10^{-3}
		0.29	0.0	0.16
mass		0.73	0.06	0.92
		1.08	0.29	3.9
		1.33	0.44	10
		1.85	0.70	71
		1.93	0.66^c	600

[a] From Pandharipande, Pines, and Smith (1976), with $\rho_{crust} = 2 \times 10^{14}$ g cm^{-3}.
[b] $t_q = T(\omega_q^2/\Omega^2)|\Delta\varepsilon|$ [cf. Eq. (10.11.24)]. $T \equiv \Omega/\dot{\Omega} = 2260$ yr, $\Omega = 190$ s^{-1}, $\Delta\varepsilon = 0.9 \times 10^{-9}$, $\omega_q^2 = 2A^2/BI_0$ from Ref. a.
[c] This model contains a solid core.

10.11   The Origin of Pulsar Glitches: Starquakes

Numerous models have been proposed to explain the origin of sudden pulsar spinups since they were first observed in 1969.[35] These include starquakes (occurring in the crust and/or the core), vortex pinning, magnetospheric instabilities, and instabilities in the motion of the superfluid neutrons. As a prototype calculation of the interplay between microphysical properties of neutron stars and macrophysical, observable quantities, we shall examine in detail the starquake model.[36] Here a sudden cracking of the neutron star crust decreases the moment of inertia and hence increases Ω. We will see that this "crustquake" model serves as a reasonable explanation for the Crab spinups, but does not, without modification, apply to Vela.

Exercise   10.12   (based on Scargle and Pacini, 1971, and Roberts and Sturrock, 1972.) Spinup models based on magnetospheric instabilities are motivated by the apparent

[35] See Pines, Shaham, and Ruderman (1974) and Manchester and Taylor (1977) for a review.
[36] Ruderman (1969); Baym and Pines (1971).

correlation between the occurrence of some of the Crab glitches with new activity in the "wisps" in the Crab nebula. Suppose that the magnetic field in the closed field-line region about the pulsar can trap charged particles up to a maximum plasma moment of inertia

$$I_p \sim \frac{B_p^2 R^3}{6 \Omega^2}.$$ (10.11.1)

[Ruderman (1972); justify Eq. (10.11.1) on energetic grounds.] Calculate $\Delta \Omega / \Omega$ if all of this plasma were suddenly released without generating a torque on the star. Evaluate your answer for the Crab and Vela pulsars, assuming reasonable values for B_p, M, and R.

Answer: $\dfrac{\Delta \Omega}{\Omega} \sim \begin{cases} 10^{-6}, & \text{Vela} \\ 10^{-7}, & \text{Crab} \end{cases}$

The starquake mechanism is based on the idea that nuclei in the crust of the neutron star ($\rho < 2$–2.4×10^{14} g cm^{-3}) form a solid, Coulomb lattice (cf. Section 8.2). This crust is oblate in shape because of the star's rotation (cf. Section 9.6). As the star slows down, centrifugal forces on the crust decrease and a stress arises to drive the crust to a less oblate equilibrium shape. However, the rigidity of the solid crust resists this stress and the shape remains more oblate than the equilibrium value. Finally, when the crust stresses reach a critical value, the crust cracks. Some stress is relieved and the oblateness excess is reduced. As a result, the moment of inertia of the crust is suddenly decreased, and so by angular momentum conservation the frequency is suddenly increased. This gives the observed pulsar spinup.

Quantitatively, define a time-dependent oblateness parameter ε according to

$$I = I_0(1 + \varepsilon),$$ (10.11.2)

where I is the moment of inertia and I_0 is the nonrotating, spherical value. The parameter ε is related to the deformation parameter λ and the eccentricity e defined in Eqs. (7.4.28) and (7.4.32) by

$$1 + \varepsilon = \lambda^{-1} = (1 - e^2)^{-1/3}.$$ (10.11.3)

For small deviations from sphericity, we have

$$\varepsilon \simeq \frac{e^2}{3}.$$ (10.11.4)

Now consider the total energy of a rotating Newtonian neutron star,

$$E = E_{\text{int}} + W + T + E_{\text{strain}}.$$ (10.11.5)

Here, in addition to the usual terms [cf. Eq. (7.4.35)], we introduce a new term E_{strain} to account for the strain energy of the crust.

Expanding the expression (7.4.29) for λ close to 1, we find

$$g(\lambda) \simeq 1 - \frac{(1 - \lambda)^2}{5} \simeq 1 - \frac{\varepsilon^2}{5}. \tag{10.11.6}$$

Thus

$$E_{\text{int}} + W + T \simeq E_{\text{int}} + W_0 + A\varepsilon^2 + J^2\frac{(1 - \varepsilon)}{2I_0}, \tag{10.11.7}$$

where W_0 is the gravitational potential energy for a nonrotating neutron star, $A \equiv |W_0|/5$ and, as we assumed before, E_{int} is independent of ε.

The strain energy arises from the compression of a solid, bcc Coulomb lattice. From the BBP calculation (cf. Section 8.2), we know that the neutron star crust is composed of neutron-rich nuclei locked in such a lattice structure. The strain energy may be estimated crudely as follows: Imagine that the star has some initial deformation ε_0. At this reference oblateness, the star is strain-free. As Ω decreases, ε falls below ε_0, straining the crust. Consider the one-dimensional Coulomb lattice analyzed in Section 3.7. Compression of the lattice as the star readjusts its shape results in a strain or elastic energy

$$E_{\text{strain}} \sim N\left(\tfrac{1}{2}Kx^2\right), \tag{10.11.8}$$

where

$$N \sim n_A R^3 \tag{10.11.9}$$

is the total number of lattice sites (nuclei) in the crust, $n_A \sim 1/R_0^3$ is the mean ion density there, $R_0 \sim R/N^{1/3}$ is the ionic separation, K is the lattice "spring constant" [Eq. (3.7.19)], and

$$x \sim \delta R_0 \sim \frac{\delta R}{N^{1/3}} \sim \frac{R\,\delta\varepsilon}{N^{1/3}} \sim \frac{R}{N^{1/3}}|\varepsilon - \varepsilon_0| \tag{10.11.10}$$

is the rms displacement of an ion about its equilibrium position following compression. Equation (10.11.8) thus becomes

$$E_{\text{strain}} \sim \frac{Z^2 e^2}{R_0} n_A R^3 (\varepsilon - \varepsilon_0)^2. \tag{10.11.11}$$

A more accurate treatment by Baym and Pines (1971) gives for an unscreened, bcc Coulomb lattice

$$E_{\text{strain}} = B(\varepsilon - \varepsilon_0)^2, \tag{10.11.12}$$

where

$$B = 0.42 \left(\frac{4\pi}{3} R^3 \right) n_A \frac{Z^2 e^2}{a},$$

$$a \equiv \left(\frac{2}{n_A} \right)^{1/3}. \tag{10.11.13}$$

Exercise   10.13 Show that for typical pulsar parameters, $B \ll A$ in the crust. Assume for the crust $\rho \leq 2 \times 10^{14}$ g cm^{-3}, $Z \sim 10$, and $n_A \leq 5 \times 10^{34}$ cm^{-3}.

The mean stress σ in the crust is defined by

$$\sigma \equiv \left| \frac{1}{V_c} \frac{\partial E_{\text{strain}}}{\partial \varepsilon} \right| = \mu(\varepsilon_0 - \varepsilon), \tag{10.11.14}$$

where V_c is the volume of the crust and

$$\mu = \frac{2B}{V_c} \tag{10.11.15}$$

is the mean shear modulus of the crust.
  Using Eqs. (10.11.7) and (10.11.12), Eq. (10.11.5) becomes

$$E = E_0 + A\varepsilon^2 + J^2 \frac{(1 - \varepsilon)}{2I_0} + B(\varepsilon - \varepsilon_0)^2, \tag{10.11.16}$$

where E_0, the energy of the nonrotating star, is independent of ε. Minimizing Eq. (10.11.16) with respect to ε, at fixed M, J, and ρ_c, we obtain the equilibrium distortion

$$\varepsilon = \frac{I_0 \Omega^2}{4(A + B)} + \frac{B\varepsilon_0}{A + B}. \tag{10.11.17}$$

Since $\Omega^2 \sim G\rho\varepsilon$ and $A \gg B$, the first term in Eq. (10.11.17) dominates and the rigidity of the crust results in only a small deviation of ε from the value it would have if the star were a perfect liquid. Thus

$$\varepsilon \simeq \frac{I_0 \Omega^2}{4A}. \tag{10.11.18}$$

Exercise   10.14 Derive Eq. (10.11.18) for an incompressible spheroid directly from Eq. (7.3.18) in the limit of low J.

When the mean stress in the crust reaches some critical value, σ_c, the crust will crack and the net stress, strain and oblateness will suddenly be reduced. The crustquake thus leads to a sharp decrease $\Delta\varepsilon_0$ in the "reference" oblateness and a corresponding decrease $\Delta\varepsilon$ in the actual oblateness. Equation (10.11.17) gives

$$\Delta\varepsilon = \frac{B}{A+B}\Delta\varepsilon_0 \simeq \frac{B}{A}\Delta\varepsilon_0 \ll \Delta\varepsilon_0. \qquad (10.11.19)$$

This change is directly observable, since by Eq. (10.11.2)

$$\Delta\varepsilon = \frac{\Delta I}{I} = -\frac{\Delta\Omega}{\Omega} = -(1-Q)\left(\frac{\Delta\Omega}{\Omega}\right)_0. \qquad (10.11.20)$$

Here we have used the definition of the "healing parameter," Q: $\Delta\Omega$ in the second equality is the difference between the stable angular frequencies before the glitch and following complete relaxation after the glitch.

For Vela, Eq. (10.11.18) implies an equilibrium oblateness $\varepsilon \sim 10^{-4}$, while Eq. (10.11.20) gives for the spinup $\Delta\varepsilon \sim 10^{-6}$. Corresponding values for the Crab are $\varepsilon \sim 10^{-3}$, $\Delta\varepsilon \sim 10^{-8}$–$10^{-9}$. An observable effect arises from a shrinkage of the pulsar crust by only a fraction of a millimeter!

Following a quake, the pulsar continues to slow down in the usual way until the stress builds up again to the critical value. The time t_q between quakes is then given by

$$t_q \simeq \frac{|\Delta\sigma|}{\dot{\sigma}}. \qquad (10.11.21)$$

From Eqs. (10.11.14) and (10.11.19), the stress relieved in a quake is

$$\Delta\sigma = \mu(\Delta\varepsilon_0 - \Delta\varepsilon) \simeq \mu A \frac{\Delta\varepsilon}{B}. \qquad (10.11.22)$$

The subsequent build-up of stress occurs at the steady rate

$$\dot{\sigma} = -\mu\dot{\varepsilon} = -\frac{\mu}{2A}I_0\Omega\dot{\Omega} = \frac{\mu}{2A}\frac{I_0\Omega^2}{T}, \qquad (10.11.23)$$

where we have used Eq. (10.11.18) for $\dot{\varepsilon}$. Thus Eq. (10.11.21) becomes

$$t_q \simeq T\frac{\omega_q^2}{\Omega^2}|\Delta\varepsilon|, \qquad (10.11.24)$$

where

$$\omega_q^2 \equiv \frac{2A^2}{BI_0}. \qquad (10.11.25)$$

Now the parameter ω_q^2, and hence t_q, is a rapidly varying function of stellar mass as well as the hadronic equation of state (among other reasons, $B \propto V_c$, the crustal volume). This dependence is illustrated in Table 10.2, where t_q is listed for several different equations of state. According to the table, the assumption that the Crab pulsar is a $1.3 M_\odot$ TI star leads to the prediction of roughly a 10-year interval between quakes, in reasonable agreement with the observed spinups (cf. Table 10.1), given the uncertainties in the calculation of A and B. By contrast, a 10-year timescale between quakes requires a Crab pulsar mass $\lesssim 0.5 M_\odot$ with a BJ potential and a much smaller mass with a Reid potential.

The crustquake model is unable to account for the relatively short time interval between Vela spinups, however. Even the most favorable choice of parameters—for example, Vela as a light, $0.3 M_\odot$ TI neutron star—gives $t_q \sim 10^5$ yr, compared to the observed timescale of a few years. Whether or not *corequakes*, which might occur if the neutron star has a solid core, can explain the relatively large and frequent Vela spinups remains an open question at present.[37] Key difficulties with this model include (i) whether or not neutron star cores are solid, and (ii) nonobservation of X-ray emission due to the released strain energy.

Attempts have also been made to understand the Vela glitches from a detailed model of pinned vorticity.[38] In this model the giant glitches in Vela and other pulsars represent vorticity jumps following catastrophic "unpinning" of vortex lines in the pinned crustal neutron superfluid. Post-glitch behavior is attributed to milder, glitch-induced vortex "creep." Although forecasts based on this model did not correctly predict the time of the October 1981 Vela glitch, the pinned vorticity model might contain the basic ingredients of a more realistic theoretical explanation of the giant glitches.

Exercise 10.15

(a) Starting from Eq. (10.11.16), compute ΔE, the release of strain energy in a starquake. Use Eqs. (10.11.17) and (10.11.19) to write ΔE in the form

$$\Delta E = 2(A + B)(\varepsilon_0 - \varepsilon)\,\Delta\varepsilon. \qquad (10.11.26)$$

Do not assume $B \ll A$.

(b) The value of ΔE is sensitive to the value of $\varepsilon_0 - \varepsilon$ at the time of the quake. Estimate ΔE for a Vela *corequake*, assuming $\varepsilon_0 - \varepsilon \sim 2 \times 10^{-3}$. For a $1.93 M_\odot$ TI star with a solid core Pandharipande, Pines, and Smith give $A = 17.8 \times 10^{52}$ erg, $B_{core} = 14.4 \times 10^{52}$ erg.

Answer: $\Delta E \sim 1 \times 10^{45}$ erg

---

[37]Pines, Shaham, and Ruderman (1972); Pandharipande, Pines, and Smith (1976).
[38]Alpar et al. (1981).

11
Cooling of Neutron Stars

11.1 Introduction

The determination of surface temperatures of neutron stars by detecting thermal blackbody radiation can in principle yield significant information about the interior hadronic matter and neutron star structure. (We studied the analogous situation for white dwarfs in Chapter 4.) For example, the current upper limit of 2×10^6 K for the surface temperature of the Crab pulsar imposes a severe constraint on the thermal history of a neutron star for the first 930 years after its formation. Knowing the thermal evolution of a neutron star also yields information on such temperature-sensitive properties as transport coefficients, transition to superfluid states, crust solidification, internal pulsar heating mechanisms such as frictional dissipation at the crust-superfluid interfaces, and so on.

It is generally believed that neutron stars are formed at very high interior temperatures ($T \gtrsim 10^{11}$ K) in the core of a supernova explosion (cf. Chapters 1 and 18). The predominant cooling mechanism immediately after formation is neutrino emission, with an initial cooling timescale of seconds. After about a day, the internal temperature drops to 10^9–10^{10} K. Photon emission overtakes neutrinos only when the internal temperature falls to $\sim 10^8$ K, with a corresponding surface temperature roughly two orders of magnitude smaller. Neutrino cooling dominates for *at least* the first 10^3 years, and typically for much longer, in all standard cooling calculations performed recently. These theoretical calculations[1] provide curves of the neutron star surface temperature as a function of time, which in principle are subject to observational verification.

[1]See, for example, Glen and Sutherland (1980); Van Riper and Lamb (1981); Nomoto and Tsuruta (1981).

Thermal evolution calculations are sensitive to the adopted nuclear equation of state, the neutron star mass, the assumed magnetic field strength, the possible existence of superfluidity, pion condensation, quark matter, and so on. A wide range of thermal histories result when all the possibilities are considered.

Typically, one finds that surface temperatures fall to several times 10^6 K for objects approximately 300 years old, and remain in the vicinity of $(0.5–2) \times 10^6$ K for at least 10^4 yr. Such temperatures imply potentially detectable photon emission in the soft X-ray band, 0.2–3 keV. Indeed, prior to the discovery of pulsars, it was believed that young neutron stars might first be detected as discrete X-ray sources.[2] Detailed cooling rates[3] and cooling curves[4] were therefore constructed in anticipation of the observations.

Only after the discovery of pulsars were neutron stars also observed to be X-ray sources; these sources, however, were not isolated neutron stars cooling down after formation, but rather neutron stars in binary systems accreting gas from their companions (cf. Chapter 13). In such systems it is the infalling gas that is the source of the X-ray emission. For many years the only observational constraint on an isolated neutron star cooling by X-ray emission was the upper limit of $\sim 3 \times 10^6$ K for the Crab pulsar surface temperature, obtained from lunar occultation measurements. (The Crab emits X-ray pulses with the same period as the radio pulses, but these have nothing to do with the thermal properties of the pulsar.)

Recently, the thousandfold increase in flux sensitivity and the excellent angular resolution in the soft X-ray region afforded by the orbiting Einstein Observatory (HEAO-2) have made it possible to extend the observations considerably. With such a satellite, a search of nearby supernova remnants should either detect point sources or else provide upper limits to surface temperatures nearly an order of magnitude lower than previous measurements.[5] As a result, data from the Einstein Observatory and its successors may be able to determine whether or not, for example, pion condensates or free quarks actually exist in neutron star interiors. (As we shall see, these states dramatically accelerate the cooling.) Inspired by this prospect and by the data acquired so far, we discuss the physics of neutron star cooling in greater detail below.

11.2   Neutrino Reactions in Neutron Stars ($T \lesssim 10^9$ K)

We shall be interested in the thermal history of a neutron star *after* it has already cooled to an interior temperature below a few times 10^9 K. We shall discuss the relatively short epoch during which the temperature drops from $\sim 10^{11}$ K to

---

[2] See, for example, Chiu (1964); Chiu and Salpeter (1964).
[3] Bahcall and Wolf (1965).
[4] Tsuruta and Cameron (1965).
[5] Helfand (1981).

$\sim 10^9$ K in Chapter 18. This earlier epoch, initiated by stellar core collapse and a supernova explosion, is significantly shorter than the extended time spent cooling down from 10^9 K.

For internal temperatures below a few times 10^9 K, any neutrinos emitted during the cooling process escape freely from the neutron star, without interacting further with the neutron star matter. We shall verify this statement in Section 11.7. This fact, which distinguishes the low-temperature neutron star cooling epoch from the earlier high-temperature formation epoch, greatly simplifies the determination of the late thermal evolution.

At the very high temperatures ($T \gtrsim 10^9$ K) found in the cores of evolved, massive stars, the dominant mode of energy loss via neutrinos is from the so-called[6] URCA reactions:

$$n \to p + e^- + \bar{\nu}_e, \qquad e^- + p \to n + \nu_e. \tag{11.2.1}$$

These reactions also dominate during core collapse. In both cases the nucleons in the hot interior are nondegenerate. However, when the nucleons become degenerate as in a neutron star that has cooled below 10^9 K, these reactions are highly suppressed. We now demonstrate this important result.

Recall the discussion of Section 2.5: matter in the degenerate interior satisfies the β-equilibrium condition

$$\mu_n = \mu_p + \mu_e, \tag{11.2.2}$$

where to good approximation $[\mathbb{O}(kT/\mu_n)^2]$ the chemical potentials are just the Fermi energies. Thus

$$E_F(n) = E_F(p) + E_F(e), \tag{11.2.3}$$

where at nuclear densities

$$E_F(n) \simeq m_n c^2 + \frac{p_F^2(n)}{2m_n},$$

$$E_F(p) \simeq m_p c^2 + \frac{p_F^2(p)}{2m_p},$$

$$E_F(e) \simeq p_F(e)c. \tag{11.2.4}$$

Charge neutrality requires that [cf. Eq. (2.5.7)]

$$p_F(p) = p_F(e), \tag{11.2.5}$$

---

[6]Just as the URCA casino in Rio de Janeiro is a perfect sink for money, these reactions are a perfect sink for the star's energy.

so Eq. (11.2.3) becomes

$$\frac{p_F^2(n)}{2m_n} \simeq p_F(e)c\left(1 + \frac{p_F(p)}{2m_p c}\right) - Q, \tag{11.2.6}$$

where $Q = (m_n - m_p)c^2 = 1.293$ MeV is small in comparison to the other terms in Eq. (11.2.6) (cf. Exercise 11.1). From Eq. (11.2.6) we see that the neutron Fermi energy (minus rest mass) is very nearly equal to the electron Fermi energy:

$$E_F'(n) \equiv \frac{p_F^2(n)}{2m_n} \simeq p_F(e)c = E_F(e), \tag{11.2.7}$$

and so

$$p_F(e) = p_F(p) \ll p_F(n), \tag{11.2.8}$$

$$E_F'(p) \ll E_F'(n). \tag{11.2.9}$$

Now consider the possibility of a reaction such as neutron decay, Eq. (11.2.1). The only neutrons capable of decaying lie within $\sim kT$ of the Fermi surface, $E_F'(n)$. Hence, by energy conservation, the final proton and electron must also be within $\sim kT$ of their Fermi surfaces; the energy of the escaping neutrino must also be $\sim kT$. Now, according to inequality (11.2.8), the proton and electron must have small momenta compared to the neutron. But this is impossible: the decay cannot conserve momentum if it conserves energy.

In order for the process to work, a bystander particle must be present to absorb momentum. Chiu and Salpeter (1964) therefore proposed that "modified" URCA reactions

$$n + n \rightarrow n + p + e^- + \bar{\nu}_e, \tag{11.2.10}$$

$$n + p + e^- \rightarrow n + n + \nu_e, \tag{11.2.11}$$

would be important for neutron star cooling. Accompanying these reactions are the muon-neutrino emitting reactions

$$n + n \rightarrow n + p + \mu^- + \bar{\nu}_\mu, \tag{11.2.12}$$

$$n + p + \mu^- \rightarrow n + n + \nu_\mu, \tag{11.2.13}$$

which occur whenever $\mu_e > m_\mu c^2$ ($\rho \gtrsim 8 \times 10^{14}$ g cm^{-3}; cf. Section 8.10). Corresponding reactions involving τ-neutrinos do not occur at typical neutron star interior densities, since $m_\tau c^2 = 1784$ MeV $\gg \mu_e$.

In the succeeding sections we shall calculate the cooling rate due to these modified URCA reactions. The reader interested only in the results may skip directly to Eq. (11.5.22).

Exercise 11.1 Show that in the ideal Fermi gas approximation (cf. Section 2.5), the following relations hold approximately for $\rho \lesssim 2\rho_{\text{nuc}}$ (i.e., nonrelativistic nucleons):

$$n_n = 1.7 \times 10^{38}\left(\frac{\rho}{\rho_{\text{nuc}}}\right) \text{ cm}^{-3}$$

$$n_e = n_p = 9.6 \times 10^{35}\left(\frac{\rho}{\rho_{\text{nuc}}}\right)^2 \text{ cm}^{-3},$$

$$E_F'(n) = E_F(e) = 60\left(\frac{\rho}{\rho_{\text{nuc}}}\right)^{2/3} \text{ MeV},$$

$$E_F'(p) = 1.9\left(\frac{\rho}{\rho_{\text{nuc}}}\right)^{4/3} \text{ MeV},$$

$$p_F(n) = 340\left(\frac{\rho}{\rho_{\text{nuc}}}\right)^{1/3} \text{ MeV/c},$$

$$p_F(e) = p_F(p) = 60\left(\frac{\rho}{\rho_{\text{nuc}}}\right)^{2/3} \text{ MeV/c}.$$

Here $\rho_{\text{nuc}} = 2.8 \times 10^{14}$ g cm^{-3}.

11.3   Weak Interaction Theory

Our current understanding of weak interactions such as β-decay is provided by the *Weinberg-Salam-Glashow theory* (WSG).[7] In this theory, the weak force between fermions is mediated by the exchange of massive vector bosons, much as electromagnetic interactions are mediated by the exchange of photons (massless vector bosons; cf. Appendix D).

In the standard WSG model, there are two charged "intermediate" bosons W^+ and W^-, and one neutral intermediate boson, the Z^0. They are predicted to have masses given by

$$m_W c^2 = \frac{37.3 \text{ GeV}}{\sin\theta_w} = 78.1 \pm 1.7 \text{ GeV},$$

$$m_Z c^2 = \frac{m_W c^2}{\cos\theta_w} = 88.9 \pm 1.4 \text{ GeV},\qquad(11.3.1)$$

[7]A nice historical introduction to weak interaction theory is given by Coleman (1979).

where the *Weinberg angle* θ_w is experimentally determined to be

$$\sin^2\theta_w = 0.228 \pm 0.010. \tag{11.3.2}$$

At the low-interaction energies, $E^2/c^4 \ll m_W^2, m_Z^2$, which characterize all processes occuring in neutron star interiors, the WSG Hamiltonian encompasses the old "V − A" Hamiltonian proposed by Feynman and Gell-Mann in 1958. Here V stands for the vector part of the interaction and A the axial vector part. The effective Hamiltonian density has the form in both theories

$$\mathcal{H}_{WK} = \frac{G_F}{\sqrt{2}} J_\mu^\dagger J^\mu, \tag{11.3.3}$$

where J_μ is the 4-current density of the interacting fermions. The universal Fermi coupling constant G_F appearing in the effective low-energy Hamiltonian is related to WSG parameters by ($\hbar = c = 1$)

$$\begin{aligned}
G_F &= \frac{1}{4\sqrt{2}} \frac{\alpha}{m_W^2 \sin^2\theta_w} \\
&= 1.16632 \pm 0.00004 \times 10^{-5}\ \text{GeV}^{-2} \\
&= 1.43582 \pm 0.00005 \times 10^{-49}\ \text{erg cm}^3 \\
&\simeq \frac{1.0 \times 10^{-5}}{m_p^2}.
\end{aligned} \tag{11.3.4}$$

Here $\alpha = e^2/\hbar c \simeq \frac{1}{137}$ is the fine-structure constant.

An important difference between the old and new theories at low energy is that in the old theory only *charged current* reactions were allowed. These are reactions that, in the language of the new theory, are mediated by the charged vector bosons, W^+ or W^-. In the WSG model, additional *neutral current* reactions, mediated by the Z^0, can exist. In fact, the experimental confirmation of neutral current reactions in 1974 in part led to the award of a Nobel Prize to Weinberg, Salam, and Glashow in 1979.

Reactions (11.2.1) proceed by a charged current interaction (the neutron changes into a charged proton), while the scattering processes

$$\nu + n \to \nu + n, \qquad \nu + p \to \nu + p \tag{11.3.5}$$

proceed via a neutral current. Some processes, such as neutrino–lepton scattering

$$\nu_e + e^- \to \nu_e + e^-, \qquad \nu_\mu + \mu^- \to \nu_\mu + \mu^-, \tag{11.3.6}$$

occur via both charged and neutral currents (see Fig. 11.1).

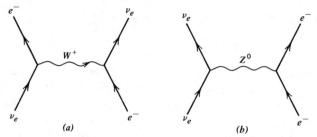

Figure 11.1 Feynman diagrams for electron–neutrino scattering $e^- + \nu_e \rightarrow e^- + \nu_e$: (*a*) charged current reaction and (*b*) neutral current reaction.

Cross sections in the old V − A theory typically go as $G_F^2 E^2$, where E is the center-of-mass energy in a two-body interaction. Thus as $E \rightarrow \infty$, the cross section diverges rapidly. In WSG theory cross sections converge faster than $\sigma \propto (\ln E)^n$, where $n \leqslant 2$. The WSG model, like QED, is renormalizable: divergent integrals arising in higher-order terms in perturbation expansions in the coupling constant can be removed in a well-defined manner.[8]

Finally, we note that the WSG model unifies, in a single Lagrangian, both the weak and electromagnetic interactions. More precisely, the field equations obtained from the Lagrangian relate electromagnetic to weak fields just as Maxwell's equations relate **E** to **B** fields. This triumph has led to new attempts at a "Grand Unification" of all the known forces into a single theory. A discussion of this would take us too far afield, however.

11.4 Free Neutron Decay

The modified URCA reactions involve both strong and weak interactions. For example, pions are exchanged between the colliding neutrons in reaction (11.2.10) before one of the neutrons decays into a proton. Before calculating the rate of these processes, we therefore consider a simpler reaction: pure neutron decay in vacuum, reaction (11.2.1), where strong interaction effects are very small. Much of the discussion in this section also applies to β-decay in nuclei.

The energy released in a typical β-decay is small ($Q \sim$ MeV) compared to the rest energy of the nucleons. Accordingly, we can employ the Golden Rule of time-dependent perturbation theory in the nonrelativistic limit to obtain the decay rate:

$$d\Gamma = \frac{2\pi}{\hbar} \left(\frac{1}{2} \sum_{\text{spins}} |H_{fi}|^2 \right) \rho_e \rho_{\bar{\nu}} \, dE_e \, dE_{\bar{\nu}} \, \delta(E_{\bar{\nu}} + E_e - Q). \qquad (11.4.1)$$

[8] t'Hooft (1971); Lee and Zinn-Justin (1972); t'Hooft and Veltman (1972a, b).

Here ρ_e and $\rho_{\bar{\nu}}$ are the densities of the final states of the e^- and $\bar{\nu}$, respectively, per unit energy interval, while E_e is the electron energy. The quantity H_{fi} is the weak interaction matrix element in the nonrelativistic limit,

$$H_{fi} = G_F \int \Psi_f^* \Psi_i \, d\mathcal{V} = G_F \int \psi_p^* \psi_e^* \psi_{\bar{\nu}}^* \psi_n \, d\mathcal{V}, \qquad (11.4.2)$$

where ψ_p, ψ_e, $\psi_{\bar{\nu}}$, and ψ_n are the proton, electron, anti-neutrino, and neutron wave functions, respectively. We imagine that the decaying neutron is in a box of unit volume and normalize all wave functions accordingly.

Note that the integral in Eq. (11.4.2) is the probability amplitude to find all four particles at the same point in space. Accordingly, the weak interaction in this low-energy domain is describable by a "contact potential":

$$H_{fi} = \int \Psi_f^*(\mathbf{r}) V_w(\mathbf{r}, \mathbf{r}') \Psi_i(\mathbf{r}') \, d\mathcal{V} d\mathcal{V}', \qquad (11.4.3)$$

where

$$V_w(\mathbf{r}, \mathbf{r}') = G_F \delta(\mathbf{r} - \mathbf{r}'). \qquad (11.4.4)$$

If we neglect the Coulomb distortion on the electron due to the proton, we can write ψ_e and $\psi_{\bar{\nu}}$ as plane-wave states. In addition, we note that the wavelengths of these states are much larger than the nuclear radius:

$$\lambda = \frac{1}{k} = \frac{\hbar}{p} \sim \frac{\hbar c}{E} \sim 10^{-11} \text{ cm}$$

for $E \sim$ MeV. We can thus expand the spatial part of the lepton wave functions as

$$\psi_e^* \psi_{\bar{\nu}}^* = \exp[-i(\mathbf{k}_e + \mathbf{k}_{\bar{\nu}}) \cdot \mathbf{r}] = 1 - i(\mathbf{k}_e + \mathbf{k}_{\bar{\nu}}) \cdot \mathbf{r} + \cdots \qquad (11.4.5)$$

and retain only the first term of the expansion (11.4.5) in the integral (11.4.2). Provided the remaining nucleon overlap integral does not vanish, it is said to describe an *allowed transition*; higher-order terms in the expansion (11.4.5) give rise to forbidden transitions. Clearly, no orbital angular momentum is carried off by an allowed transition.

The spin part of the combined lepton wave function can be either a singlet state (total spin zero) or a triplet state (total spin unity). In singlet transitions, the nucleon wave function does not change its spin or its total angular momentum J, so $\Delta J = 0$. Triplet transitions require a spin flip, and total angular momentum can be conserved if $\Delta J = \pm 1$ or 0. The singlet transition selection rule, $\Delta J = 0$, is called the *Fermi selection rule*, while triplet transitions obey the *Gamow–Teller selection rule.*[9] The spin flip in triplet transitions is effected by the pseudovector, or axial, part of the weak interaction, while singlet transitions occur via the pure vector part.

We can now write

$$\frac{1}{2} \sum_{\text{spins}} |H_{fi}|^2 = G_F^2 \left[C_V^2 |M_V|^2 + 3C_A^2 |M_A|^2 \right]. \qquad (11.4.6)$$

Here M_V and M_A are the vector and axial vector nuclear matrix elements, which are determined by the overlap integrals of the initial and final nuclear states; C_V and C_A are coupling constants that would both be unity if strong interactions had no effect on the interaction, and the factor of 3 arises from the statistical weight of the triplet transition. Experimentally, it is found that

$$|C_V| = 0.9737 \pm 0.0025$$

$$\left| \frac{C_A}{C_V} \right| \equiv a = 1.253 \pm 0.007. \qquad (11.4.7)$$

For neutron decay, it is a good approximation to take

$$M_V \simeq M_A \simeq 1, \qquad (11.4.8)$$

since the neutron and proton wave functions are very similar (isotopic spin symmetry). Thus Eq. (11.4.6) becomes

$$\frac{1}{2} \sum_{\text{spins}} |H_{fi}|^2 = G_F^2 C_V^2 (1 + 3a^2). \qquad (11.4.9)$$

On integrating over $dE_{\bar{\nu}}$ using the δ-function, Eq. (11.4.1) becomes

$$d\Gamma = \frac{2\pi}{\hbar} G_F^2 C_V^2 (1 + 3a^2) \rho_e \rho_{\bar{\nu}} \, dE_e. \qquad (11.4.10)$$

For an electron with a definite spin orientation, we have

$$\rho_e = \frac{4\pi p^2}{h^3} \frac{dp}{dE_e} = \frac{4\pi p E_e}{c^2 h^3}, \qquad (11.4.11)$$

---

[9]In nuclei undergoing beta decay, transitions in which both the initial and final nuclear states are spinless ($J = 0 \rightarrow J = 0$) can only proceed via Fermi transitions since $0 \rightarrow 0$ Gamow–Teller transitions are strictly forbidden.

where

$$p = \frac{\left(E_e^2 - m_e^2 c^4\right)^{1/2}}{c}. \tag{11.4.12}$$

Since the neutrino is massless,[10] we have by energy conservation

$$E_{\bar{\nu}} = Q - E_e = p_{\bar{\nu}} c, \tag{11.4.13}$$

and so

$$\rho_{\bar{\nu}} = \frac{(Q - E_e)^2}{2\pi^2 \hbar^3 c^3}. \tag{11.4.14}$$

Thus

$$d\Gamma = \frac{G_F^2 C_V^2 (1 + 3a^2)}{2\pi^3 \hbar^7 c^6} \left(E_e^2 - m_e^2 c^4\right)^{1/2} E_e (Q - E_e)^2 \, dE_e. \tag{11.4.15}$$

Defining dimensionless energies

$$\varepsilon \equiv \frac{E_e}{m_e c^2}, \qquad \varepsilon_0 \equiv \frac{Q}{m_e c^2} = 2.5312, \tag{11.4.16}$$

we get

$$d\Gamma = \frac{m_e^5 c^4 G_F^2 C_V^2 (1 + 3a^2)}{2\pi^3 \hbar^7} \, df, \tag{11.4.17}$$

where the energy spectrum of the decay electron arises purely from phase-space factors:

$$df = (\varepsilon^2 - 1)^{1/2} \varepsilon (\varepsilon_0 - \varepsilon)^2 \, d\varepsilon. \tag{11.4.18}$$

Integrating over the allowed range of ε from 1 to ε_0 gives

$$f = \frac{1}{60} (\varepsilon_0^2 - 1)^{1/2} (2\varepsilon_0^4 - 9\varepsilon_0^2 - 8) + \frac{1}{4} \varepsilon_0 \ln\left[\varepsilon_0 + (\varepsilon_0^2 - 1)^{1/2}\right]$$

$$= 1.6369, \tag{11.4.19}$$

[10] Even if the neutrino is found to have a small mass $m_\nu \lesssim$ few eV, the rates derived in this chapter are essentially unchanged.

and so the neutron decay rate is

$$\Gamma = \frac{m_e^5 c^4 G_F^2 C_V^2 (1 + 3a^2) f}{2\pi^3 \hbar^7} = \frac{1}{972} \, \text{s}. \qquad (11.4.20)$$

The observed rate is $1/(925 \pm 11)$ s; the discrepancy arises because we ignored Coulomb effects in the calculation of f (f increases to about 1.70), and there are also quantum electrodynamic effects ("radiative corrections") of another 2%.

11.5   The Modified URCA Rate

Having evaluated the weak interaction matrix element for free neutron decay, we are now in a position to calculate the rate of the modified URCA reaction (11.2.10). We shall follow the original approach of Bahcall and Wolf (1965).

Let the subscripts 1, 2, 1', p, e, and $\bar{\nu}$ denote the two initial neutrons, the final neutron, the proton, the electron, and the antineutrino, respectively. The reaction rate for given initial states 1 and 2 is

$$d\Gamma = \frac{2\pi}{\hbar} \delta(E_f - E_i) |H_{fi}|^2 \rho_p \rho_e \rho_{\bar{\nu}} \, dE_p \, dE_e \, dE_{\bar{\nu}}, \qquad (11.5.1)$$

where $|H_{fi}|^2$ must be summed over final spins and averaged over initial spins. We shall explicitly retain the normalization volume $\mathcal{V}$ in all phase-space factors and wave functions, so that in empty space for each species j we would have

$$\rho_j \, dE_j \equiv d^3 n_j = \mathcal{V} \frac{d^3 p_j}{h^3}. \qquad (11.5.2)$$

However, the reaction takes place in a dense gas and most of the low-energy cells in phase space are occupied. Hence each factor of $d^3 n_j$ must be multiplied by $1 - f_j$, where

$$f_j = \frac{1}{\exp\left[(E_j - \mu_j)/kT\right] + 1} \qquad (11.5.3)$$

is the fraction of phase space occupied at energy E_j (Fermi–Dirac distribution; cf. Chapter 2). Factors of $(1 - f_j)$ reduce the reaction rate and are called *blocking factors*.

Note that in the Golden Rule as used in Eq. (11.5.1), the number of phase-space factors is one less than the number of final particles because of momentum conservation. We can obtain a more symmetrical expression by adding a factor

$$\delta^3(\mathbf{p}_f - \mathbf{p}_i)\, d^3\mathbf{p}_{1'} = \delta^3(\mathbf{p}_f - \mathbf{p}_i)\, d^3 n_{1'} \frac{h^3}{\mathcal{V}}$$

$$= \delta^3(\mathbf{k}_f - \mathbf{k}_i)\, d^3 n_{1'} \frac{(2\pi)^3}{\mathcal{V}}.$$

On integrating over $\mathbf{p}_{1'}$, this factor gives unity.

The total antineutrino luminosity in a volume $\mathcal{V}$ is obtained by integrating the rate (11.5.1) over all initial states, multiplied by $E_{\bar{\nu}}$:

$$L_{\bar{\nu}} = \frac{(2\pi)^4}{\hbar \mathcal{V}} \int d^3 n_1\, d^3 n_2\, d^3 n_{1'}\, d^3 n_p\, d^3 n_e\, d^3 n_{\bar{\nu}}\, \delta(E_f - E_i)$$

$$\times \delta^3(\mathbf{k}_f - \mathbf{k}_i)\, S |H_{fi}|^2 E_{\bar{\nu}}. \tag{11.5.4}$$

Here

$$S = f_1 f_2 (1 - f_{1'})(1 - f_p)(1 - f_e), \tag{11.5.5}$$

the factors f_1 and f_2 accounting for the distribution of initial states.

Exercise   11.2   Why is there no blocking factor $(1 - f_{\bar{\nu}})$ in Eq. (11.5.5)? Under what circumstances should we include such a factor?

The all-important matrix element for the interaction can be written as

$$H_{fi} = \langle n, p, e, \bar{\nu} \, |V_w| \, n, n \rangle, \tag{11.5.6}$$

where V_w is the weak interaction "contact" Hamiltonian given previously for free-neutron decay [Eq. (11.4.4)]. Now it is again appropriate to represent the leptons as free-particle states in Eq. (11.5.6). However, it is not at all valid to do so for the nucleons. The reason is that the total nucleon Hamiltonian is

$$H_{\text{nuc}} = H_{\text{free}} + H_s + V_w \equiv H_0 + V_w, \tag{11.5.7}$$

where H_{free} is the free-particle contribution (kinetic energy of each particle) and H_s is the strong interaction Hamiltonian. Thus Eq. (11.5.6) gives the lowest-order transition rate only if the nucleon wave functions appearing there are already eigenfunctions of H_0. As discussed previously, the solution to the many-body Schrödinger equation including H_s is highly nontrivial and by no means resolved theoretically as yet. We shall simply follow Bahcall and Wolf and attempt to estimate the effect of H_s on the two-body nucleon wave function.

We therefore write

$$H_{fi} = \int d\mathcal{V} d\mathcal{V}' \psi_{np}^*(\mathbf{r}) \psi_e^*(\mathbf{r}) \psi_{\bar{\nu}}^*(\mathbf{r}) V_w(\mathbf{r}, \mathbf{r}') \psi_{nn}(\mathbf{r}')$$

$$= \frac{G_F}{\mathcal{V}} \int d\mathcal{V} \psi_{np}^*(\mathbf{r}) \psi_{nn}(\mathbf{r}). \tag{11.5.8}$$

Here we have used Eq. (11.4.4) and set

$$\psi_e(\mathbf{r}) = \frac{1}{\mathcal{V}^{1/2}} e^{i\mathbf{k}_e \cdot \mathbf{r}} \rightarrow \frac{1}{\mathcal{V}^{1/2}} \tag{11.5.9}$$

and similarly for $\psi_{\bar{\nu}}$ as in Eq. (11.4.5).

We assume that the interaction in the initial n-n state is dominated by s-wave scattering (i.e., $L = 0$).[11] Then the n-n state must have total spin $S = 0$ (cf. Table 8.1). The V part of the weak interaction couples this state to an $S = 0$ n-p state, while the A part couples it to an $S = 1$ n-p state. Thus [cf. Eq. (11.4.6)]

$$\sum_{\text{spins}} |H_{fi}|^2 = \frac{4G_F^2}{\mathcal{V}^2} \left(C_V^2 |M_V|^2 + 3C_A^2 |M_A|^2 \right), \tag{11.5.10}$$

where

$$M_V = \int d\mathcal{V} \left(\psi_{np}^0 \right)^* \psi_{nn}^0,$$

$$M_A = \int d\mathcal{V} \left(\psi_{np}^1 \right)^* \psi_{nn}^0. \tag{11.5.11}$$

The factor of 4 occurs in Eq. (11.5.10) because either neutron in the n-n pair can become a proton, giving a factor of 2 in amplitude and 4 in probability.

Since the range of the strong force is of order $\lambda_\pi = \hbar/m_\pi c$, we expect that the relative wave functions in Eq. (11.5.11) will overlap in a relatively small volume of order λ_π^3. Thus we expect $M \sim \lambda_\pi^3/\mathcal{V}$. Defining nondimensional matrix elements

$$\tilde{M}_V = \frac{\mathcal{V} M_V}{\lambda_\pi^3}, \qquad \tilde{M}_A = \frac{\mathcal{V} M_A}{\lambda_\pi^3}, \tag{11.5.12}$$

---

[11] This assumption is valid at low densities, $\rho \lesssim \rho_{\text{nuc}}$. In this low-density limit the hard core radius of the nuclear potential r_c is much less than $\hbar/p_f(n)$. Since the hard core dominates the distortion of the wave functions from plane-wave states, the angular momentum associated with a scattering event is $\sim p_f r_c \ll \hbar$, implying s-wave scattering.

we get for Eq. (11.5.4)

$$L_{\bar{\nu}} = 64\pi^4 \mathcal{V} G_F^2 \hbar^{-1} \lambda_\pi^{-9} \left(C_V^2 |\tilde{M}_V|^2 + 3C_A^2 |\tilde{M}_A|^2 \right) P, \qquad (11.5.13)$$

where the dimensionless phase-space factor P is

$$P = \mathcal{V}^{-6} \lambda_\pi^{15} \int \prod_{j=1}^6 d^3 n_j \, S E_{\bar{\nu}} \, \delta^3 (\mathbf{k}_f - \mathbf{k}_i) \delta (E_f - E_i). \qquad (11.5.14)$$

Note that since each factor $d^3 n_j$ is proportional to $\mathcal{V}$, P is independent of $\mathcal{V}$ and so $L_{\bar{\nu}}$ is proportional to $\mathcal{V}$. Numerically, we have for the antineutrino emissivity $\varepsilon_{\bar{\nu}}$

$$\varepsilon_{\bar{\nu}} \equiv \frac{L_{\bar{\nu}}}{\mathcal{V}} = \left(5.1 \times 10^{48} \, \text{erg cm}^{-3} \, \text{s}^{-1}\right) P \left(|\tilde{M}_V|^2 + 4.7 |\tilde{M}_A|^2\right). \qquad (11.5.15)$$

The phase-space factor is evaluated in Appendix F, where it is shown that

$$P \simeq 2.1 \times 10^{-30} \left(\frac{\rho}{\rho_{\text{nuc}}}\right)^2 T_9^8, \qquad (11.5.16)$$

where $\rho_{\text{nuc}} = 2.8 \times 10^{14}$ g cm^{-3} and T_9 is the temperature in units of 10^9 K. Note that all of the temperature dependence in $\varepsilon_{\bar{\nu}}$ is in the phase-space factor, a result that is generally true of neutrino cooling reactions. The 8 powers of T originate as follows: for each degenerate species, only a fraction $\sim kT/E_F$ can effectively contribute to the cooling rate. There are two such initial species and three such final species. The antineutrino phase space is proportional to $E_{\bar{\nu}}^2$, and the energy loss rate gives another factor $E_{\bar{\nu}}$. Since $E_{\bar{\nu}} \sim kT$, we have altogether 8 powers of T.

Evaluation of the nondimensional matrix elements is not so straightforward. We can define a fundamental length scale from the Fermi momentum of the dominant neutrons by

$$l = \frac{\hbar}{p_F(n)} \sim 0.4 \lambda_\pi \left(\frac{\rho}{\rho_{\text{nuc}}}\right)^{-1/3}. \qquad (11.5.17)$$

Bahcall and Wolf suggests that $\tilde{M}_V$ and $\tilde{M}_A$ might be expected to vary as $(l/\lambda_\pi)^3$ —that is, to be of order unity at $\rho \sim \rho_{\text{nuc}}$ and to decrease slowly as $1/\rho$. In any case they use some results of a nuclear matter calculation to formally calculate

$$|\tilde{M}_A|^2 = |\tilde{M}_V|^2 \simeq 1.0 \left(\frac{\rho_{\text{nuc}}}{\rho}\right)^{4/3}. \qquad (11.5.18)$$

Thus Eq. (11.5.15) becomes

$$\varepsilon_{\bar{\nu}} = \left(6.1 \times 10^{19} \text{ erg cm}^{-3} \text{ s}^{-1}\right)\left(\frac{\rho}{\rho_{\text{nuc}}}\right)^{2/3} T_9^8. \qquad (11.5.19)$$

To the expression (11.5.19) we must now add the rate of *neutrino* energy loss from the "inverse" reaction (11.2.11). By time-reversal invariance, the matrix elements M_A and M_V for reaction (11.2.11) are the complex conjugates of M_A and M_V for reaction (11.2.10). Since the phase-space factors are the same for both reactions (with the approximations we have made of neglecting all lepton momenta), the two reactions give the *same* energy loss rate.

The muon-neutrino reactions (11.2.12) and (11.2.13) must be considered when $\mu_e > m_\mu c^2$ (i.e., $\rho > 2.9\rho_{\text{nuc}}$ in the free-particle model; cf. Section 8.10). The ν_μ reactions differ from the ν_e reactions only in the phase-space factor P, where a factor $p_\mu^2 \, dp_\mu$ appears instead of $p_e^2 \, dp_e$. Thus the ratio of ν_μ to ν_e energy loss rate is

$$F = \frac{p_\mu^2 \, dp_\mu}{p_e^2 \, dp_e}. \qquad (11.5.20)$$

Each term in Eq. (11.5.20) is to be evaluated at the Fermi surface, since only that region of phase space contributes to the reaction rate.

Exercise 11.3 Use the equilibrium relation $E_F(\mu) = E_F(e)$ to show that $p_\mu \, dp_\mu = p_e \, dp_e$ at the respective Fermi surfaces, so that

$$F = \begin{cases} 0, & \rho \lesssim 2.9\rho_{\text{nuc}} \\[2ex] \left[1 - \left(\frac{m_\mu c^2}{E_F(e)}\right)^2\right]^{1/2}, & \rho \gtrsim 2.9\rho_{\text{nuc}} \end{cases}. \qquad (11.5.21)$$

Multiplying Eq. (11.5.19) by $2(1 + F)$ gives finally the total energy loss rate by modified URCA reactions:[12]

$$\varepsilon_\nu^{\text{URCA}} = \left(1.2 \times 10^{20} \text{ erg cm}^{-3} \text{ s}^{-1}\right)\left(\frac{\rho}{\rho_{\text{nuc}}}\right)^{2/3} T_9^8(1 + F). \qquad (11.5.22)$$

For a neutron star of mass M and uniform density ρ, this gives a luminosity

$$L_\nu^{\text{URCA}} = \left(8.5 \times 10^{38} \text{ erg s}^{-1}\right)\frac{M}{M_\odot}\left(\frac{\rho_{\text{nuc}}}{\rho}\right)^{1/3} T_9^8(1 + F). \qquad (11.5.23)$$

---

[12] Our numerical values differ slightly from those of Bahcall and Wolf (1965) because of different values for ρ_{nuc} and the β-decay coupling constants.

Recently, Friman and Maxwell (1979) have repeated the above calculations using a more realistic expression for the strong NN interaction. They obtain the same density dependence, but a numerical coefficient of 7.4×10^{20} in Eq. (11.5.22), nearly an order of magnitude larger. For a uniform density star with no muons, this gives

$$L_\nu^{\text{URCA}} = \left(5.3 \times 10^{39} \text{ erg s}^{-1}\right) \frac{M}{M_\odot} \left(\frac{\rho_{\text{nuc}}}{\rho}\right)^{1/3} T_9^8. \qquad (11.5.24)$$

11.6 Other Reaction Rates

Having sketched the derivation of Eq. (11.5.22), we now discuss briefly the rates of other possible cooling reactions.

(a) Nucleon Pair Bremsstrahlung

The most significant new cooling mechanism that becomes possible when neutral currents are considered is nucleon pair bremsstrahlung:

$$n + n \rightarrow n + n + \nu + \bar{\nu}, \qquad n + p \rightarrow n + p + \nu + \bar{\nu}. \qquad (11.6.1)$$

These reactions have been studied by Flowers et al. (1975) and later by Friman and Maxwell (1979), who found that while the rate also varies as T^8, it is smaller than the modified URCA rate by a factor of 30.

(b) Neutrino Pair Bremsstrahlung

If the neutrons are "locked" in a superfluid state (cf. Section 10.9), the rates for all the reactions discussed so far are cut down by a factor $\sim \exp(-\Delta/kT)$, where Δ is the superfluid energy gap. In this case, cooling from neutrino pair bremsstrahlung from nuclei in the crust can be important. The rate for the process

$$e^- + (Z, A) \rightarrow e^- + (Z, A) + \nu + \bar{\nu} \qquad (11.6.2)$$

is estimated to be

$$L_\nu^{\text{brem}} \sim \left(5 \times 10^{39} \text{ erg s}^{-1}\right)\left(M_{\text{cr}}/M_\odot\right) T_9^6, \qquad (11.6.3)$$

where M_{cr} is the mass of the crust.[13] Since this process goes as T_9^6, it decreases less rapidly than reaction (11.5.22) as the star cools.

[13] Maxwell (1979).

(c) Pionic Reactions

As Bahcall and Wolf originally pointed out, pion condensation can dramatically increase the cooling rate in neutron star interiors. If pion condensates exist (Section 8.12), then "quasi-particle" β-decay can occur via

$$N \to N' + e^- + \bar{\nu}_e \tag{11.6.4}$$

and its inverse. Here the quasi-particles N and N' are linear combinations of neutron and proton states in the pion sea. The pion condensate allows both energy and momentum to be conserved in the reaction, which is the analogue of the ordinary URCA process, Eq. (11.2.1).

Bahcall and Wolf considered a simplified version of reaction (11.6.4)—that is, cooling via the decay of free pions:

$$\pi^- + n \to n + e^- + \bar{\nu}_e, \tag{11.6.4a}$$

$$\pi^- + n \to n + \mu^- + \bar{\nu}_\mu, \tag{11.6.4b}$$

and the "inverse" processes

$$n + e^- \to n + \pi^- + \nu_e, \tag{11.6.4c}$$

$$n + \mu^- \to n + \pi^- + \nu_\mu. \tag{11.6.4d}$$

As in the modified URCA reactions, the total rate for all four processes is essentially four times the rate of reaction (11.6.4a) alone. (Recall that muons are already present when and if pions appear.)

We can make a rough estimate of the reaction rate as follows: Since there are two fewer fermions participating in the reactions than in the modified URCA reactions, the phase-space factor varies as T^6 rather than T^8. We therefore expect the total rate to be

$$L_\nu^\pi \sim L_\nu^{\text{URCA}} \times \left(\frac{E_f'(n)}{kT} \right)^2 \frac{n_\pi}{n_n}, \tag{11.6.5}$$

where n_π/n_n is the ratio of the number densities of pions and neutrons. Since $E_F'(n) \sim 60(\rho/\rho_{\text{nuc}})^{2/3}$ MeV, Eqs. (11.5.23) and (11.6.5) give

$$L_\nu^\pi \sim \left(8 \times 10^{44} \text{ erg s}^{-1} \right) \frac{M}{M_\odot} \frac{\rho}{\rho_{\text{nuc}}} T_9^6 \frac{n_\pi}{n_n}. \tag{11.6.6}$$

Bahcall and Wolf actually obtained a numerical factor 1×10^{46} and no ρ

dependence. The more recent calculation of Maxwell et al. (1977) gives

$$L_\nu^\pi = \left(1.5 \times 10^{46} \text{ erg s}^{-1}\right)\theta^2 \frac{M}{M_\odot} \frac{\rho_{\text{nuc}}}{\rho} T_9^6, \qquad (11.6.7)$$

where $\theta \sim 0.3$ is an angle measuring the degree of pion condensation (i.e., it replaces the factor n_π/n_n). Since L_ν^π is larger than L_ν^{URCA} by a hefty factor, it will dominate the cooling rate at all temperatures of interest, provided pion condensation occurs.

Exercise   11.4   Estimate an *upper limit* to the thermal neutrino cooling rate from a neutron star of radius 10 km (Bludman and Ruderman, 1975). Assume that the star radiates electron neutrinos and antineutrinos like a Fermi–Dirac "blackbody" at temperature T and chemical potential $\mu_\nu = \mu_{\bar\nu} = 0$. For sufficiently high T, all the luminosities calculated above exceed this rate. Explain this discrepancy.

   Hint: Kirchhoff's Law.

   Answer: $L_{\text{max}} \approx 6 \times 10^{44} T_9^4 \text{ erg s}^{-1}$.

(d)   Quark Beta Decay

If the core of a neutron star consists largely of quark matter, then there is the possibility of significant neutrino emission via the β-decay of degenerate, relativistic quarks.[14] Unlike ordinary neutron star matter in which the simple beta decay (i.e., URCA) processes described by Eq. (11.2.1) are suppressed, the corresponding processes can occur for quarks.

   Recall our earlier discussion in Section 8.14 of three-component (u, d, and s) quark matter in beta equilibrium. There we showed that if we ignored the masses of the quarks and their mutual interactions, the equilibrium composition was given by

$$n_u = n_d = n_s = n, \qquad n_e = n_\mu = 0, \qquad (11.6.8)$$

where $n = (n_u + n_d + n_s)/3$ is the baryon number density. According to Eq. (11.6.8), each quark species has the same Fermi momentum, given by[15]

$$p_f(q) = 235\left(\frac{n}{n_{\text{nuc}}}\right)^{1/3} \frac{\text{MeV}}{c}, \qquad (11.6.9)$$

where $n_{\text{nuc}} \equiv \rho_{\text{nuc}}/m_B = 0.17 \text{ fm}^{-3}$.

   The existence of finite quark masses and quark-quark interactions is likely to alter the above composition. For example, the s quark is thought to be rather

---

[14] Iwamoto (1980); Burrows (1980).
[15] Equation (11.6.9) shows that Fermi energies in neutron stars are well below the production thresholds for the heavy c, b, and t quarks.

heavy ($m_s \sim 100\text{–}300$ MeV) and, if present in a neutron star, it is not likely to be relativistic. On the other hand, the u and d quarks are probably very light ($m_u \sim m_d \sim 5\text{–}10$ MeV), so they will be highly relativistic and their masses can still be ignored. Resulting modifications to the equilibrium composition include the presence of leptons to carry some of the negative charge while preserving overall charge neutrality. In fact, since characteristic quark Fermi energies greatly exceed $m_e c^2$, any small difference between μ_d and μ_u (or between μ_s and μ_u) will invariably result in electrons with relativistic Fermi energy [cf. Eqs. (8.14.2) and (8.14.3)].

The simplest neutrino processes occurring in quark matter are the β-decay reactions involving the relativistic quarks,

$$d \rightarrow u + e^- + \bar{\nu}_e, \tag{11.6.10}$$

$$u + e^- \rightarrow d + \nu_e. \tag{11.6.11}$$

A detailed analysis of these reactions is given by Iwamoto (1980). He points out, for example, that if particle masses and quark-quark interactions are ignored, then the conservation of energy and momentum requires that the momentum of all the particles must be collinear for reaction (11.6.10) to occur at all. (Why?) However, in that case, the WSG matrix element for the reaction turns out to vanish identically.[16] Iwamoto then shows that if one takes into account quark-quark interactions, the particles need not be collinear to satisfy energy and momentum conservation and the matrix element is nonzero. Specifically, he notes that the participating quarks must reside close to their Fermi surfaces. To lowest order in the strong interaction (QCD) coupling constant α_s, the relation between quark chemical potential and Fermi momentum is[17]

$$\mu_{u,d} = \left(1 + \frac{8\alpha_s}{3\pi}\right) p_F(u,d) c. \tag{11.6.12}$$

It is easy to verify that this modification to the usual ideal gas relationship $\mu = p_F c$ for a relativistic fermion permits energy and momentum conservation for a *finite* angle between the momenta of the interacting particles in Eq. (11.6.10).

When interactions are included (but the small neutrino momentum ignored), the emissivities due to reactions (11.6.10) and (11.6.11) are equal and Iwamoto

---

[16] The same result does not apply to reaction (11.6.11), provided the small momentum $\sim kT/c$ of the emitted neutrino is accounted for. In this case the particles do not need to be precisely collinear and the matrix element is nonzero (Burrows, 1980). However, the reaction rate for (11.6.11), ignoring interactions, is suppressed by a factor $\sim kT/p_f(e)c$ from the rate calculated in (11.6.10) when interactions are included.

[17] Baym and Chin (1976a).

calculates their sum to be

$$\varepsilon_\nu^{\text{quark}} \approx \left(8.8 \times 10^{26} \text{ erg s}^{-1} \text{ cm}^{-3}\right) \alpha_s \frac{n}{n_{\text{nuc}}} Y_e^{1/3} T_9^6, \qquad (11.6.13)$$

where $Y_e = n_e/n$ is the number of electrons per baryon. The value of α_s is momentum dependent and not very well determined experimentally. As we discussed in Section 8.14, the M.I.T. bag model suggests the value $\alpha_s \approx 0.55$. However, the analysis of charmonium decay gives[18] $\alpha_s \approx 0.065$. Iwamato adopts the value $\alpha_s = 0.1$ in Eq. (11.6.13). He also sets $Y_e = 0.01$, a value typical of *normal* neutron star matter.[19] The resulting emissivity is

$$\varepsilon_\nu^{\text{quark}} \approx \left(1.9 \times 10^{25} \text{ erg s}^{-1} \text{ cm}^{-3}\right) \frac{n}{n_{\text{nuc}}} T_9^6. \qquad (11.6.14)$$

The corresponding luminosity for a uniform density star composed of quark matter is

$$L_\nu^{\text{quark}} \approx \left(1.3 \times 10^{44} \text{ erg s}^{-1}\right) \frac{M}{M_\odot} T_9^6. \qquad (11.6.15)$$

The presence of s quarks will lead to additional neutrino emission via β-decay reactions of the type

$$s \to u + e^- + \bar{\nu}_e, \qquad (11.6.16)$$

$$u + e^- \to s + \nu_e. \qquad (11.6.17)$$

Since the s quark is massive the momenta of the interacting particles can differ significantly from collinearity, thereby ensuring nonzero matrix elements. However, weak interactions coupling s and u quarks [e.g., reactions (11.6.16) and (11.6.17)] are "Cabibbo suppressed" in WSG theory relative to interactions coupling d and u quarks [e.g., reactions (11.6.10) and (11.6.11)]. More precisely, the former are proportional to the factor $\sin^2 \theta_C$ while the latter are proportional to $\cos^2 \theta_C$, where θ_C is called the *Cabbibo angle*. Experimentally θ_C is found to satisfy $\cos^2 \theta_C \approx 0.974$. Iwamoto thus concludes that s-quark neutrino processes will not alter the total emissivity (11.6.14) appreciably.

The question of whether quark matter exists in neutron star cores is by no means resolved. However, a comparison of Eqs. (11.5.24), (11.6.7), and (11.6.15)

---

[18]Charmonium $= c\bar{c}$ in a bound state; for an analysis of its decay, see Applequist and Politzer (1975).
[19]Compare Exercise 8.3; note that such a high value of Y_e is appropriate only if the s quark is heavy [recall Eqs. (11.6.8) and (11.6.9)]; see Duncan et al. (1983) for a full discussion.

shows that a star with quark matter would cool at a rate much faster than an ordinary neutron star and comparable to that of a star with a pion-condensed core.

11.7 Neutrino Transparency

Once the neutrino luminosities are known, we can calculate cooling timescales. Fundamental to the discussion that follows is the assumption that neutrinos, once created, escape from neutron stars without undergoing further interactions or energy loss. We now show that for $T \lesssim 10^9$ K, this assumption is valid.

Before the discovery of neutral currents in 1974, the argument went as follows:[20] Reactions of the type

$$\nu_e + n \rightarrow p + e^-, \qquad \bar{\nu}_e + p \rightarrow n + e^+,$$

$$\bar{\nu}_e + p + n \rightarrow n + n + e^+, \qquad (11.7.1)$$

and so on are all forbidden in neutron star interiors by conservation of energy and momentum. The most important interaction for neutrino energy loss would then be inelastic scattering off electrons (for ν_e and $\bar{\nu}_e$) and off muons (for ν_μ and $\bar{\nu}_\mu$). In a degenerate gas, the cross section for $\nu_e - e^-$ scattering, assuming only charged current interactions, is[21]

$$\sigma_e \simeq \chi\sigma_0 \left(\frac{E_\nu}{m_e c^2} \right)^2 \frac{E_\nu}{E_F(e)}, \qquad (11.7.2)$$

where

$$\sigma_0 \equiv \frac{4}{\pi} \left(\frac{\hbar}{m_e c} \right)^{-4} \left(\frac{G_F}{m_e c^2} \right)^2 = 1.76 \times 10^{-44} \text{ cm}^2, \qquad (11.7.3)$$

and where

$$\chi \approx 0.1 \quad \text{(V} - \text{A theory: \quad only charged-current interactions),}$$
$$\approx 0.06 \quad \text{(WSG theory: \quad charged- and neutral-current interactions).}$$

The cross sections for the other three scattering processes are comparable.

The mean free path of an electron neutrino is then

$$\lambda_e = (\sigma_e n_e)^{-1}. \qquad (11.7.4)$$

---

[20] Bahcall and Wolf (1965).
[21] Tubbs and Schramm (1975).

Using the results of Exercise 11.1, and the old V − A value for σ_e, we get

$$\lambda_e \sim (9 \times 10^7 \text{ km}) \left(\frac{\rho_{\text{nuc}}}{\rho}\right)^{4/3} \left(\frac{100 \text{ keV}}{E_\nu}\right)^3. \qquad (11.7.5)$$

Since $\lambda_e \gg 10$ km, the star is transparent to neutrinos.

With the possibility of neutral currents, the *effective* absorption mean free path is significantly reduced because of *n-ν elastic*[22] scattering reactions, such as

$$n + \nu_e \rightarrow n + \nu_e, \qquad n + \nu_\mu \rightarrow n + \nu_\mu, \qquad (11.7.6)$$

which only occur via neutral currents. Although elastic scattering does not degrade the neutrino energy, it does prevent the neutrino from escaping directly after emission. The neutrino scatters many times off neutrons in the interior, until it finally encounters an electron (or muon) and scatters *inelastically*, or else it random-walks its way to surface and escapes with its energy unchanged.

The cross section for elastic scattering off neutrons is[23]

$$\sigma_n = \frac{1}{4}\sigma_0 \left(\frac{E_\nu}{m_e c^2}\right)^2, \qquad (11.7.7)$$

and the associated mean free path is

$$\lambda_n = \frac{1}{\sigma_n n_n} \simeq 300 \text{ km} \frac{\rho_{\text{nuc}}}{\rho} \left(\frac{100 \text{ keV}}{E_\nu}\right)^2. \qquad (11.7.8)$$

Here we have used Exercise 11.1 for n_n. The *effective* mean free path for inelastic scattering off electrons, taking into account the increased probability of encountering an electron in the interior due to the zig-zag path of the neutrino, is[24]

$$\lambda_{\text{eff}} \sim (\lambda_n \lambda_e)^{1/2} \qquad (11.7.9)$$

$$\sim 2 \times 10^5 \text{ km} \left(\frac{\rho_{\text{nuc}}}{\rho}\right)^{7/6} \left(\frac{100 \text{ keV}}{E_\nu}\right)^{5/2}, \qquad (11.7.10)$$

where we have used the WSG value for σ_e. So although λ_{eff} is significantly less

---

[22] Here *elastic* means that, to excellent approximation, the neutrino energy $E_\nu \ll m_n c^2$ *before* scattering equals the neutrino energy *after* scattering in the star (i.e., $\approx$ neutron) frame.

[23] Tubbs and Schramm (1975).

[24] The random-walk relation (11.7.9) is essentially derived in Section 14.5 for the analogous case of photons diffusing through a hot plasma. In that example the interesting energy absorption process is inverse bremsstrahlung and the elastic scattering process is Thomson scattering off free electrons.

than λ_e, it is still much larger than the radius of a neutron star. Hence, the "old" conclusion regarding the transparency of a neutron star to low-energy neutrinos ($T \lesssim 10^9$ K, $kT \lesssim 100$ keV) is still valid.

11.8 Cooling Curves

The temperature of a neutron star can now be calculated as a function of time. The thermal energy of the star resides almost exclusively in degenerate fermions (neutrons or quarks). Neglecting interactions, the heat capacity of N such particles of mass m and relativity parameter $x = p_f/mc$ is[25]

$$C_v = Nc_v \equiv \frac{dU}{dT}\bigg|_{N,\mathcal{V}} = \frac{\pi^2(x^2 + 1)^{1/2}}{x^2} Nk\left(\frac{kT}{mc^2}\right), \qquad (11.8.1)$$

where c_v is the specific heat per particle.

Exercise 11.5
(a) Integrate Eq. (11.8.1) to find the total thermal energy U. (The quantity x remains constant to lowest order.)

(b) What is U for a normal neutron star of mass M, density ρ and temperature T?

Hint: Assume $x \ll 1$ and use the results of Exercise 11.1.

Answer:

$$U_n \simeq 6 \times 10^{47}\ \text{erg}\ (M/M_\odot)(\rho/\rho_{\text{nuc}})^{-2/3}T_9^2. \qquad (11.8.2)$$

(c) What is U for relativistic quark matter with $n_u = n_d = n_s = n$?

Hint: Use Eq. (11.6.9).

Answer:

$$U_q \simeq 9 \times 10^{47}\ \text{erg}\ (M/M_\odot)(n/n_{\text{nuc}})^{-1/3}T_9^2. \qquad (11.8.3)$$

The temperature T appearing in Eqs. (11.8.1)–(11.8.3) is the interior temperature. Neutron star interiors are to good approximation isothermal, because of the high thermal conductivity of the degenerate electron gas.[26] Just as for hot white dwarfs (cf. Chapter 4), it is only in the low-density, nondegenerate, outermost layer that an appreciable temperature gradient exists.[27]

[25] Chandrasekhar (1939), Eq. (X.218).

[26] Actually the quantity $T(-g_{00})^{1/2}$ is constant in thermal equilibrium; allowance must be made for the gravitational redshift of the thermal energy. See Lightman et al. (1975), Problem 14.2.

[27] But see Nomoto and Tsuruta (1981), who find that for neutron stars with a stiff (TI) equation of state, deviations from isothermality in the interior may persist for at least a few thousand years after formation.

The cooling equation is

$$\frac{dU}{dt} = C_v \frac{dT}{dt} = -(L_\nu + L_\gamma),$$ (11.8.4)

where L_ν is the total neutrino luminosity and L_γ is the photon luminosity. Assuming blackbody photon emission from the surface at an effective surface temperature T_e, we have

$$L_\gamma = 4\pi R^2 \sigma T_e^4 = 7 \times 10^{36} \text{ erg s}^{-1} \left(\frac{R}{10 \text{ km}}\right)^2 T_{e,7}^4,$$ (11.8.5)

where σ is the Stefan–Boltzmann constant and $T_{e,7}$ is the temperature in units of 10^7 K.

Inserting the appropriate luminosities into Eq. (11.8.4) and integrating gives the time for the star to cool from an initial interior temperature $T(i)$ to a final temperature $T(f)$.

Exercise   11.6

(a)   Assuming the modified URCA process dominates, calculate the cooling time. Use Eqs. (11.5.24) and (11.8.2).

Answer:

$$\Delta t(\text{URCA}) \simeq 1 \text{ yr} \left(\frac{\rho}{\rho_{\text{nuc}}}\right)^{-1/3} T_9^{-6}(f) \left\{1 - \left[\frac{T_9(f)}{T_9(i)}\right]^6\right\}.$$ (11.8.6)

(b)   Repeat the calculation for pion condensation [Eq. (11.6.7)], quarks [Eqs. (11.6.15) and (11.8.3)], and crust neutrino bremsstrahlung [Eq. (11.6.3)], in turn.

Answer:

$$\Delta t(\text{pions}) \simeq 20 \text{ s } \theta^{-2} \left(\frac{\rho}{\rho_{\text{nuc}}}\right)^{1/3} T_9^{-4}(f) \left\{1 - \left[\frac{T_9(f)}{T_9(i)}\right]^4\right\},$$ (11.8.7)

$$\Delta t(\text{quarks}) \simeq 1 \text{ hr} \left(\frac{n}{n_{\text{nuc}}}\right)^{-1/3} T_9^{-4}(f) \left\{1 - \left[\frac{T_9(f)}{T_9(i)}\right]^4\right\},$$ (11.8.8)

$$\Delta t(\text{brem}) \simeq 2 \text{ yr} \left(\frac{M}{M_{\text{cr}}}\right) \left(\frac{\rho}{\rho_{\text{nuc}}}\right)^{-2/3} T_9^{-4}(f) \left\{1 - \left[\frac{T_9(f)}{T_9(i)}\right]^4\right\}.$$ (11.8.9)

Detailed calculations by Tsuruta (1974, 1979) and Malone (1974) indicate that in general the surface and interior temperatures are related by

$$\frac{T_e}{T} \sim 10^{-2}\alpha, \qquad 0.1 \leq \alpha \leq 1.$$ (11.8.10)

We can understand this result in a simple way by adapting our discussion in Section 4.1 of the degenerate–nondegenerate transition region of a white dwarf to the corresponding region of a neutron star. Note that it is still the region where the *electrons* become nondegenerate that is relevant; everywhere interior to this the neutron star is roughly isothermal. We shall again assume that bound-free and free-free processes dominate the opacity, so that Eqs. (4.1.1)–(4.1.11) remain valid. (In reality, Thomson scattering dominates the opacity for $T \gtrsim 2 \times 10^8$ K.)[28] Taking the surface composition to be pure $^{56}_{26}$Fe, we have that $X = 0$, $Z = 1$, $\mu_e = \frac{56}{26}$, and $\mu = \frac{56}{27}$. Equating Eq. (4.1.11) for L in terms of T to Eq. (11.8.5) for L in terms of T_e gives

$$\frac{T_e}{T} \simeq 1 \times 10^{-2} T_9^{-1/8} \left(\frac{M}{M_\odot} \right)^{1/4} \left(\frac{R}{10 \text{ km}} \right)^{-1/2}. \qquad (11.8.11)$$

Equation (11.8.11) is consistent with the more detailed numerical treatments and shows the relative insensitivity of α in Eq. (11.8.10) to temperature.

Exercise 11.7   Redo the analysis leading to Eq. (11.8.11) for $T \gtrsim 2 \times 10^8$ K. Replace Eq. (4.1.3) by $\kappa = \kappa_T = 0.40/\mu_e$ cm^2 g^{-1}.

   Answer: $L \simeq (1.6 \times 10^{23} \text{ erg s}^{-1}) \mu \left(\dfrac{M}{M_\odot} \right) T^{3/2}$,

$$\frac{T_e}{T} \simeq 1 \times 10^{-2} T_9^{-5/8} \left(\frac{M}{M_\odot} \right)^{1/4} \left(\frac{R}{10 \text{ km}} \right)^{-1/2}.$$

   Now if photon emission is ever the dominant energy loss mechanism, then Eqs. (11.8.2), (11.8.5), and (11.8.10) give

$$\Delta t(\text{photons}) = 2 \times 10^3 \text{ yr } \alpha^2 \left(\frac{M}{M_\odot} \right)^{1/3} T_{e,7}^{-2}(f) \left\{ 1 - \left[\frac{T_{e,7}(f)}{T_{e,7}(i)} \right]^2 \right\}, \qquad (11.8.12)$$

where the relation $M = 4\pi \rho R^3 / 3$ has been used to eliminate R, and it is assumed that $\alpha \simeq$ constant for short time intervals.

Exercise 11.8   An approximate fit to Tsuruta's (1979) calculation is

$$T_e \simeq (10T)^{2/3}, \qquad (11.8.13)$$

where T_e and T are both measured in degrees K. Redo the analysis for $\Delta t(\text{photons})$ using this relation.

   Answer: $\Delta t(\text{photons}) = 6 \times 10^4 \text{ yr } \left(\dfrac{M}{M_\odot} \right)^{1/3} T_{e,7}^{-1}(f) \left\{ 1 - \left[\dfrac{T_{e,7}(f)}{T_{e,7}(i)} \right] \right\}$.

---

[28] Chiu and Salpeter (1964).

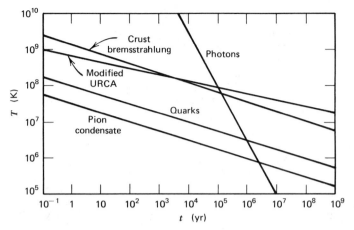

Figure 11.2 Schematic neutron star cooling curves of *interior* temperature versus time for various processes, were they to operate alone. [After Baym and Pethick (1979). Reproduced, with permission, from the *Annual Review of Astronomy and Astrophysics*, Vol. 17. © 1979 by Annual Reviews, Inc.]

The above results are summarized in Figure 11.2, where the relative importance of the various cooling processes is plotted as a function of time for a neutron star with $M = M_\odot$, $\rho = \rho_{\text{nuc}}$ and $\theta^2 = 0.1$. Each curve in the figure gives $T(t)$ for each process separately assuming the others to be absent. The most effective cooling process at any time will be that with the lowest $T(t)$; the interior temperature will be approximately T at that epoch.

These cooling timescales were determined assuming that the neutrons and protons comprised a normal fluid. Superfluidity would modify the results in two different ways. The specific heat capacity increases discontinuously as the temperature falls below the transition temperature, and then decreases exponentially at lower temperatures. So immediately above the transition temperature the cooling timescale increases, while at lower temperatures it decreases. Second, neutrino production processes are suppressed in a superfluid, thus increasing the cooling timescale.

We have also neglected the role of a magnetic field in reducing the photon opacity of a neutron star near the surface. For a given value of T, a decrease in opacity increases T_e, and hence the photon luminosity.

11.9   Comparison with Observations

Detailed neutron star cooling curves have been constructed by several authors.[29] In addition, X-ray satellite observations have recently been obtained for about 50

[29]See, for example, Glen and Sutherland (1980); Van Riper and Lamb (1981) and Nomoto and Tsuruta (1981) and references therein for recent calculations.

Table 11.1

Hot Neutron Stars in Supernova Remnants (SNR)

Name	Age (yr)	D^a (kpc)	R^b_{max} (arc min)	T_e (10^6 K)	Reference
Cas A	~ 300	2.8	0.4	< 1.5	Murray et al. (1979)
Kepler	375	8.0^c	0.2	< 2.1	Helfand et al. (1980)
Tycho	407	3.0^c	0.5	< 1.8	Helfand et al. (1980)
Crab	925	2.0	1.6	< 2.0	Harnden et al. (1979a)
SN 1006	973	1.0	3.3	< 0.8	Pye et al. (1981)
RCW 86	1794	2.5^c	2.5	< 1.5	Helfand et al. (1980)
W28	3400	2.3^d	5.0	< 1.8	Helfand (1981)
G350.0 − 18	~ 10^4	4.0^d	7.0	< 2.0	Helfand (1981)
G22.7 − 0.2	~ 10^4	4.8^d	7.0	< 2.2	Helfand (1981)
Vela	~ 10^4	0.4	1.4	< 1.5	Harnden et al. (1979b)
RCW 103	~ 10^3	3.3		< 2.2	Tuohy and Garmire (1980)

^{a}Distance of SNR from the sun.
^{b}Distance from SNR center within which a neutron star would be found, assuming it had a transverse velocity of < 1000 km s^{-1}.
^{c}Woltjer (1972).
^{d}Clark and Caswell (1976).

supernova remnants, including 7 historical ones.[30] The so-called "standard" model calculations without pion condensation or quark matter suggest a surface temperature T_e of *at least* $(1–2) \times 10^6$ K for an object around 300 years old, with minimum temperatures remaining in the range $(0.5–1.5) \times 10^6$ K for at least 10^4 yr [see Fig. 11.2 and Eq. (11.8.12)].

The upper limit of ~ 3×10^6 K for the Crab pulsar deduced from lunar occultations[31] is entirely consistent with these conventional calculations. Recently, the Einstein Observatory (the HEAO-2 X-ray satellite) has reduced this upper limit to ~ 2×10^6 K. This result and the observed upper limits to the surface temperatures from several other young supernova remnants are shown in Figure 11.3 and listed in Table 11.1. In Figure 11.3, the observational data are compared to theoretical predictions. Of the seven historical supernovae (the first seven entries in Table 11.1), only the Crab is known with certainty to contain a neutron star. The historical supernovae are important because their age is known accurately. The observed upper limits for T_e are close to the theoretical limits. Some of the observed upper limits (for Cas A, Tycho and especially SN 1006) even fall below the lowest theoretical limit, suggesting either that the neutron star cooled more rapidly than the standard calculations allow, or that no neutron star was left

[30] Helfand, Chanan, and Novick (1980); Helfand (1981); and references therein.
[31] Wolff et al. (1975); Toor and Seward (1977).

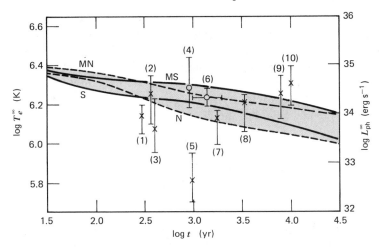

Figure 11.3 Neutron star cooling: comparison between observational data from the Einstein Observatory and typical theoretical cooling curves. The adopted neutron star model has a mass $M = 1.3 M_\odot$, a radius $R = 16$ km and is essentially governed by the TI equation of state. The observable quantities T_e^∞ and L_{ph}^∞ (measured at infinity) are plotted as functions of age for four cases: case S with superfluid neutrons and case N with nonsuperfluid (normal) nucleons, both for zero magnetic field; and cases MS (superfluid) and MN (nonsuperfluid), both for a magnetic field $B = 4.4 \times 10^{12}$ G. The numbers refer to the observational data for (1) Cas A, (2) Kepler, (3) Tycho, (4) Crab, (5) SN 1006, (6) RCW 103, (7) RGW 86, (8) W28, (9) G350.0 − 18, and (10) G22.7 − 0.2. *Crosses* indicate observed upper limits, *circles* refer to possible "detections" of point sources. Estimated error bars originate from uncertainties in interstellar absorption and source distance. [Reprinted courtesy of Nomoto and Tsuruta and *The Astrophysical Journal*, published by the University of Chicago Press; © 1981 The American Astronomical Society.]

behind by the supernova explosion. The first possibility could be accounted for by more exotic cooling mechanisms (pion condensates or quark matter), while the second might require a revision of our beliefs about supernovae and neutron stars.[32]

Out of the nearly 50 supernova remnants observed to date, only 4 show any evidence of a central point source of X-rays. However, in each instance, the identification of this emission with thermal radiation from the stellar surface is uncertain. Besides the Crab, Vela is the only other known pulsar associated with a supernova remnant. The Vela observations suggest a surface temperature of ~ 1.5×10^6 K. However, just as for the Crab, one cannot be sure without spectral information that the radiation is thermal, so this is again best interpreted as an upper limit. The other two point sources are in the remnant RCW 103 and

[32] The hypothesis that a Type II supernova leaves behind a neutron star while a Type I supernova disrupts the star completely and leaves behind no compact remnant may partially explain the X-ray data (Chevalier, 1981a; Nomoto and Tsuruta, 1981). See Chapter 18 for a discussion of supernova types.

in 3C58, the probable remnant of SN 1185. In the case of RCW 103, a temperature of 2.2×10^6 K is indicated while for 3C58 no reliable limit is available. The lack of detections in the remaining remnants, with varying upper limits, again suggests either rapid cooling or the absence of neutron stars in the majority of supernova remnants. At this stage we can only speculate on the ultimate outcome of this important investigation. Further observations and improved theoretical calculations are clearly required.

Exercise   11.9   The imaging proportional counter on the Einstein Observatory has a sensitivity of 2.5×10^{-5} photons cm^{-2} s^{-1} between 0.2–3.0 keV. Assuming a neutron star of radius 10 km at a distance of 100 pc, what upper limit can the satellite put on T_e?
   Answer: $T_e \lesssim 0.2 \times 10^6$ K

Exercise   11.10   Use the gravitational redshift of energy and time to show that a sphere of mass M and radius R radiating like a blackbody at temperature T and luminosity L is measured by a distant observer to be radiating at temperature T_∞ and L_∞, where

$$L_\infty = L(-g_{00}),$$

$$T_\infty = T(-g_{00})^{1/2},$$

and

$$-g_{00} = 1 - \frac{2M}{R}.$$

(You may use the fact that the distant observer also sees a blackbody photon distribution.)

12
Black Holes

12.1 Introduction

We have seen in previous chapters that both white dwarfs and neutron stars have a maximum possible mass. What happens to a neutron star that accretes matter and exceeds the mass limit? What is the fate of the collapsing core of a massive star, if the core mass is too large to form a neutron star? The answer, according to general relativity, is that nothing can halt the collapse. As the collapse proceeds, the gravitational field near the object becomes stronger and stronger. Eventually, nothing can escape from the object to the outside world, not even light. A black hole has been born.

A black hole is defined simply as a region of spacetime that cannot communicate with the external universe. The boundary of this region is called the surface of the black hole, or the *event horizon*.[1]

The ultimate fate of collapsing matter, once it has crossed the black hole surface, is not known. Densities for a $1 M_\odot$ object are $\sim 10^{17}$ g cm^{-3} as the black hole is formed, and are smaller for larger masses. How can we be sure that some hitherto unknown source of pressure does not become important above such extreme densities and halt the collapse? The answer is that by the time a black hole forms it is already too late to hold back the collapse: matter must move on worldlines *inside* the local light cone, and the spacetime geometry is so distorted that even an "outward" light ray does not escape. In fact, since all forms of energy gravitate in relativity, increasing the pressure energy only *accelerates* the late stages of the collapse.

If we extrapolate Einstein's equations all the way inside a black hole, they ultimately break down: a *singularity* develops. There is as yet no quantum theory of gravitation, and some people believe that the singularity would not occur in

---

[1]Strictly speaking, the event horizon is a three-dimensional hypersurface in spacetime (a 2-surface existing for some time interval). We shall, however, speak loosely of the event horizon or black hole surface as a 2-surface at some instant of time.

such a theory. It would be replaced by finite, though unbelievably extreme, conditions.

Exercise 12.1 Construct a density by dimensional analysis out of c, G, and $\hbar$. Evaluate numerically this "Planck density" at which quantum gravitational effects would become important.

Answer: $\rho \sim 10^{94}$ g cm^{-3}.

As long as the singularity is hidden inside the event horizon, it cannot influence the outside world. The singularity is said to be "causally disconnected" from the exterior world. We can continue to use general relativity to describe the observable universe, even though the theory breaks down inside the black hole.

One might expect that the solutions of Einstein's equations describing equilibrium black holes would be extremely complicated. After all, black holes can be formed from stars with varying mass distributions, shapes (multipole moments), magnetic field distributions, angular momentum distributions, and so on. Remarkably, the most general stationary black hole solution is known analytically. It depends on only three parameters: the mass M, angular momentum J, and charge Q of the black hole. All other information about the initial state is radiated away in the form of electromagnetic and gravitational waves during the collapse. The remaining three parameters are the only independent observable quantities that characterize a stationary black hole.[2] This situation is summarized by Wheeler's aphorism, "A black hole has no hair."

The mass of a black hole is observable, for example, by applying Kepler's Third Law for satellites in the Newtonian gravitational field far from the black hole. The charge is observable by the Coulomb force on a test charge far away. The angular momentum is observable by non-Newtonian gravitational effects. For example, a torque-free gyroscope will precess relative to an inertial frame at infinity (Lense–Thirring effect).

12.2 History of the Black Hole Idea

As early as 1795 Laplace (1795) noted that a consequence of Newtonian gravity and Newton's corpuscular theory of light was that light could not escape from an object of sufficiently large mass and small radius. In spite of this early foreshadowing of the possibility of black holes, the idea found few adherents, even after the formulation of general relativity.

In December of 1915 and within a month of the publication of Einstein's series of four papers outlining the theory of general relativity, Karl Schwarzschild (1916) derived his general relativistic solution for the gravitational field surround-

[2] See Carter (1979) for a complete discussion.

ing a spherical mass. Schwarzschild sent his paper to Einstein to transmit to the Berlin Academy. In replying to Schwarzschild, Einstein wrote, "I had not expected that the exact solution to the problem could be formulated. Your analytical treatment of the problem appears to me splendid." Although the significance of the result was apparent to both men, neither they nor anyone else knew at that time that Schwarzschild's solution contained a complete description of the external field of a spherical, electrically neutral, nonrotating black hole. Today we refer to such black holes as *Schwarzschild black holes*, in honor of Schwarzschild's great contribution.

As we described in Chapter 3, Chandrasekhar (1931b) discovered in 1930 the existence of an upper limit to the mass of a completely degenerate configuration. Remarkably, Eddington (1935) realized almost immediately that if Chandrasekhar's analysis was to be accepted, it implied that the formation of black holes would be the inevitable fate of the evolution of massive stars. He thus wrote in January 1935: "The star apparently has to go on radiating and radiating and contracting and contracting until, I suppose, it gets down to a few kilometers radius when gravity becomes strong enough to hold the radiation and the star can at last find peace." But he then went on to declare, "I felt driven to the conclusion that this was almost a reductio ad absurdum of the relativistic degeneracy formula. Various accidents may intervene to save the star, but I want more protection than that. I think that there should be a law of Nature to prevent the star from behaving in this absurd way."

As is clear from his concluding remarks, Eddington never accepted Chandrasekhar's result of the existence of an upper limit to the mass of a cold, degenerate star. This in spite of Eddington's being one of the first to understand and appreciate Einstein's theory of general relativity! (His book *The Mathematical Theory of Relativity* (1922) was the first textbook on general relativity to appear in English.) In fact, Eddington subsequently proceeded to modify the equation of state of a degenerate relativistic gas so that finite equilibrium states would exist for stars of arbitrary mass.[3]

But Eddington was not alone in his misgivings about the inevitability of collapse as the end product of the evolution of a massive star. Landau (1932), in the same paper giving his simple derivation of the mass limit (cf. Section 3.4), acknowledged that for stars exceeding the limit, "there exists in the whole quantum theory no cause preventing the system from collapsing to a point." But rather then follow the sober advice put forth at the beginning of his paper ("It seems reasonable to try to attack the problem of stellar structure by methods of theoretical physics"), Landau, in the end, retreats and declares, "As in reality such masses exist quietly as [normal] stars and do not show any such tendencies,

---

[3] Chandrasekhar (1980) has recently lamented Eddington's shortsightedness regarding black holes, declaring, "Eddington's supreme authority in those years effectively delayed the development of fruitful ideas along these lines for some thirty years."

we must conclude that all stars heavier than $1.5M_\odot$ certainly possess regions in which the laws of quantum mechanics (and therefore quantum statistics) are violated."

In 1939 Oppenheimer and Snyder (1939) revived the discussion by calculating the collapse of a homogeneous sphere of pressureless gas in general relativity. They found that the sphere eventually becomes cut off from all communication with the rest of the Universe. This was the first rigorous calculation demonstrating the formation of a black hole.

Black holes and the problem of gravitational collapse were generally ignored until the 1960s, even more so than neutron stars. However, in the late 1950s, J. A. Wheeler and his collaborators began a serious investigation of the problem of collapse.[4] Wheeler (1968) coined the name "black hole" in 1968.

In 1963 R. Kerr (1963) discovered an exact family of charge-free solutions to Einstein's vacuum field equations. The charged generalization was subsequently found as a solution to the Einstein–Maxwell field equations by Newman et al. (1965). Only later was the connection of these results to black holes appreciated. We know today that the *Kerr–Newman geometry* described by these solutions provides a unique and complete description of the external gravitational and electromagnetic fields of a stationary black hole.

A number of important properties of black holes were discovered and several powerful theorems concerning black holes were proved during this period. The discovery of quasars in 1963, pulsars in 1968, and compact X-ray sources in 1962 helped motivate this intensive theoretical study of black holes. Observations of the binary X-ray source Cygnus X-1 in the early 1970s (cf. Section 13.5) provided the first plausible evidence that black holes might actually exist in space.

We turn now from history to a discussion of the physics of black holes. We shall begin our treatment with a discussion of the simplest black hole, one with $J = Q = 0$.

12.3   Schwarzschild Black Holes

We repeat here the Schwarzschild solution from Eq. (5.6.8):

$$ds^2 = -\left(1 - \frac{2M}{r}\right) dt^2 + \left(1 - \frac{2M}{r}\right)^{-1} dr^2 + r^2 d\theta^2 + r^2 \sin^2\theta \, d\phi^2.$$

$$(12.3.1)$$

We are using the geometrized units ($c = G = 1$) of Section 5.5.

---

[4]See Harrison et al. (1965) for an account of these investigations.

A static observer in this gravitational field is one who is at fixed r, θ, ϕ. The lapse of proper time for such an observer is given by Eq. (12.3.1) as

$$d\tau^2 = -ds^2 = \left(1 - \frac{2M}{r}\right) dt^2, \tag{12.3.2}$$

or

$$d\tau = \left(1 - \frac{2M}{r}\right)^{1/2} dt. \tag{12.3.3}$$

This simply shows the familiar gravitational time dilation (redshift) for a clock in the gravitational field compared with a clock at infinity (i.e., $d\tau < dt$). Note that Eq. (12.3.3) breaks down at $r = 2M$, which is the *event horizon* ($\equiv$ *surface of the black hole* $\equiv$ *Schwarzschild radius*). Another name for this is the *static limit*, because static observers cannot exist inside $r = 2M$; they are inexorably drawn into the central singularity, as we shall see later.

A static observer makes measurements with his or her local orthonormal tetrad (Section 5.1). Using carets to denote quantities in the local orthonormal frame, we have from Eq. (12.3.1)

$$\vec{e}_{\hat{t}} = \left(1 - \frac{2M}{r}\right)^{-1/2} \vec{e}_t,$$

$$\vec{e}_{\hat{r}} = \left(1 - \frac{2M}{r}\right)^{1/2} \vec{e}_r,$$

$$\vec{e}_{\hat{\theta}} = \frac{1}{r}\vec{e}_\theta,$$

$$\vec{e}_{\hat{\phi}} = \frac{1}{r\sin\theta}\vec{e}_\phi. \tag{12.3.4}$$

This is clearly an orthonormal frame, since[5]

$$\vec{e}_{\hat{t}} \cdot \vec{e}_{\hat{t}} = \left(1 - \frac{2M}{r}\right)^{-1} \vec{e}_t \cdot \vec{e}_t = \left(1 - \frac{2M}{r}\right)^{-1} g_{tt} = -1, \text{ etc.} \tag{12.3.5}$$

12.4 Test Particle Motion

To explore the Schwarzschild geometry further, let us consider the motion of freely moving test particles. Recall from Eq. (5.2.21) that such particles move

---

[5] The reader may wish to review the last part of Section 5.2, which discusses the relationship between an orthonormal frame and a general coordinate system.

along geodesics of spacetime, the geodesic equations being derivable from the Lagrangian

$$2L = -\left(1 - \frac{2M}{r}\right)\dot{t}^2 + \left(1 - \frac{2M}{r}\right)^{-1}\dot{r}^2 + r^2\dot{\theta}^2 + r^2\sin^2\theta\,\dot{\phi}^2, \quad (12.4.1)$$

where $\dot{t} \equiv dt/d\lambda = p^t$ is the t-component of 4-momentum, and so on. Here we have chosen the parameter λ to satisfy $\lambda = \tau/m$ for a particle of mass m.

The Euler–Lagrange equations are

$$\frac{d}{d\lambda}\left(\frac{\partial L}{\partial \dot{x}^\alpha}\right) - \frac{\partial L}{\partial x^\alpha} = 0, \quad x^\alpha = (t, r, \theta, \phi). \quad (12.4.2)$$

For θ, ϕ, and t these are, respectively,

$$\frac{d}{d\lambda}(r^2\dot{\theta}) = r^2\sin\theta\cos\theta\,\dot{\phi}^2, \quad (12.4.3)$$

$$\frac{d}{d\lambda}(r^2\sin^2\theta\,\dot{\phi}) = 0, \quad (12.4.4)$$

$$\frac{d}{d\lambda}\left[\left(1 - \frac{2M}{r}\right)\dot{t}\right] = 0. \quad (12.4.5)$$

Instead of using the r-equation directly, it is simpler to use the fact that

$$g_{\alpha\beta}p^\alpha p^\beta = -m^2. \quad (12.4.6)$$

In other words, in Eq. (12.4.1) L has the value $-m^2/2$.

Now Eq. (12.4.3) shows that if we orient the coordinate system so that initially the particle is moving in the equatorial plane (i.e., $\theta = \pi/2$, $\dot{\theta} = 0$), then the particle remains in the equatorial plane. This result follows from the uniqueness theorem for solutions of such differential equations, since $\theta = \pi/2$ for all λ satisfies the equation. Physically, the result is obvious from spherical symmetry.

With $\theta = \pi/2$, Eqs. (12.4.4) and (12.4.5) become

$$p_\phi \equiv r^2\dot{\phi} = \text{constant} \equiv l, \quad (12.4.7)$$

$$-p_t \equiv \left(1 - \frac{2M}{r}\right)\dot{t} = \text{constant} \equiv E. \quad (12.4.8)$$

These are simply the constants of the motion corresponding to the ignorable coordinates ϕ and t in Eq. (12.4.1) (cf. Section 5.2). To understand their physical significance, consider a measurement of the particle's energy made by a static observer in the equatorial plane. This locally measured energy is the time component of the 4-momentum as measured in the observer's local orthonormal frame—that is, the projection of the 4-momentum along the time basis vector:

$$E_{\text{local}} \equiv p^{\hat{t}} = -p_{\hat{t}} = -\vec{\mathbf{p}} \cdot \vec{\mathbf{e}}_{\hat{t}} = -\vec{\mathbf{p}} \cdot \left(1 - \frac{2M}{r}\right)^{-1/2} \vec{\mathbf{e}}_t$$

$$= -\left(1 - \frac{2M}{r}\right)^{-1/2} p_t,$$

that is,

$$E = \left(1 - \frac{2M}{r}\right)^{1/2} E_{\text{local}}. \tag{12.4.9}$$

For $r \rightarrow \infty$, $E_{\text{local}} \rightarrow E$, so the conserved quantity E is called the "energy-at-infinity." It is related to E_{local} by a redshift factor.

Exercise   12.2   For an alternative derivation of the redshift formula, use the fact that E is constant along the photon's path to show that

$$\frac{\nu_{\text{em}}}{\nu_{\text{rec}}} = \left(1 - \frac{2M}{r_{\text{em}}}\right)^{-1/2} \tag{12.4.10}$$

for a static emitter at $r = r_{\text{em}}$ and a receiver at $r \rightarrow \infty$. Explain why the event horizon for a Schwarzschild black hole is sometimes called the "surface of infinite redshift."

The physical interpretation of l follows from considering the locally measured value of $v^{\hat{\phi}}$, the tangential velocity component:

$$v^{\hat{\phi}} = \frac{p^{\hat{\phi}}}{p^{\hat{t}}} = \frac{p_{\hat{\phi}}}{p^{\hat{t}}} = \frac{\vec{\mathbf{p}} \cdot \vec{\mathbf{e}}_{\hat{\phi}}}{E_{\text{local}}} = \frac{\vec{\mathbf{p}} \cdot \vec{\mathbf{e}}_{\phi}/r}{E_{\text{local}}} = \frac{p_{\phi}/r}{E_{\text{local}}},$$

and so

$$l = E_{\text{local}} r v^{\hat{\phi}}. \tag{12.4.11}$$

Comparing with the Newtonian expression $mv^{\hat{\phi}}r$, we see that l is the conserved angular momentum of the particle.

We now consider separately the cases $m \neq 0$ and $m = 0$. For particles of nonzero rest mass, it is convenient to renormalize E and l to quantities expressed per unit mass. Define

$$\tilde{E} = \frac{E}{m}, \qquad \tilde{l} = \frac{l}{m}. \tag{12.4.12}$$

Then, recalling that $\lambda = \tau/m$, we find from Eqs. (12.4.6)–(12.4.8):

$$\left(\frac{dr}{d\tau}\right)^2 = \tilde{E}^2 - \left(1 - \frac{2M}{r}\right)\left(1 + \frac{\tilde{l}^2}{r^2}\right), \tag{12.4.13}$$

$$\frac{d\phi}{d\tau} = \frac{\tilde{l}}{r^2}, \tag{12.4.14}$$

$$\frac{dt}{d\tau} = \frac{\tilde{E}}{1 - 2M/r}. \tag{12.4.15}$$

Equation (12.4.13) can be solved for $r = r(\tau)$ (in general, an elliptic integral); then Eq. (12.4.14) gives $\phi(\tau)$ and Eq. (12.4.15) gives $t(\tau)$.

It is interesting to consider orbits just outside the event horizon. The locally measured value of $v^{\hat{r}}$, the radial velocity component, is given by

$$v^{\hat{r}} = \frac{p^{\hat{r}}}{p^{\hat{t}}} = \frac{p_{\hat{r}}}{p^{\hat{t}}} = \frac{\vec{p} \cdot \vec{e}_{\hat{r}}}{E_{\text{local}}} = \frac{p_r(1 - 2M/r)^{1/2}}{E_{\text{local}}} = \frac{p^r}{E}, \tag{12.4.16}$$

from Eqs. (12.3.4) and (12.4.9). Recalling $p^r \equiv m \, dr/d\tau$ and Eq. (12.4.13), we get

$$v^{\hat{r}} = \frac{dr}{\tilde{E} \, d\tau} = \left[1 - \frac{1}{\tilde{E}^2}\left(1 - \frac{2M}{r}\right)\left(1 + \frac{\tilde{l}^2}{r^2}\right)\right]^{1/2}. \tag{12.4.17}$$

So as $r \to 2M$, $v^{\hat{r}} \to 1$ and the particle is observed by a local static observer at r to approach the event horizon along a *radial* geodesic at the *speed of light*, independent of $\tilde{l}$.

Exercise 12.3 Show that the same observer at r finds that the tangential velocity of the particle satisfies

$$v^{\hat{\phi}} = \left(1 - \frac{2M}{r}\right)^{1/2}\frac{\tilde{l}}{r\tilde{E}}, \tag{12.4.18}$$

so that $v^{\hat{\phi}} \to 0$ as $r \to 2M$.

Exercise 12.4

(a) Show from Eq. (12.4.17) that a local observer at r finds that the velocity of a radially freely-falling particle released from rest at infinity is given by

$$v^{\hat{r}} = \left(\frac{2m}{r}\right)^{1/2}, \tag{12.4.19}$$

which has precisely the same form as the Newtonian velocity!

(b)   Obtain the same result from Eq. (12.4.9), noting that $E_{local} \equiv \gamma m$.

Exercise   12.5   A particle moves along a geodesic from r and ϕ to $r + dr$ and $\phi + d\phi$ in time dt. A local static observer at (r, ϕ) measures the proper length of the particle's path to have increased by $ds(t, \theta, \phi = \text{const}) = g_{rr}^{1/2} dr (= d\hat{r})$ and $ds(t, r, \theta = \text{const}) = g_{\phi\phi}^{1/2} d\phi (= d\hat{\phi})$ in the r and ϕ directions, respectively, during this time; the proper time for this motion as measured on the observer's clock lasts $[-ds^2(r, \theta, \phi = \text{const})]^{1/2} = (-g_{00})^{1/2} dt (= d\hat{t})$. [Note that $d\hat{t}$ for the *observer* is *not* equal to $d\tau$ appearing, e.g., in Eqs. (12.4.13)–(12.4.15) for the particle!] Use the expressions for these measurements together with Eqs. (12.4.13)–(12.4.15) to rederive Eqs. (12.4.17) and (12.4.18).

The simplest geodesics are those for radial infall, $\phi = \text{constant}$. This occurs if $\tilde{l} = 0$, and Eq. (12.4.13) becomes

$$\frac{dr}{d\tau} = -\left(\tilde{E}^2 - 1 + \frac{2M}{r}\right)^{1/2}. \tag{12.4.20}$$

By considering the limit $r \rightarrow \infty$ of Eq. (12.4.20), we see that there are three cases: (i) $\tilde{E} < 1$, particle falls from rest at $r = R$, say; (ii) $\tilde{E} = 1$, particle falls from rest at infinity; (iii) $\tilde{E} > 1$, particle falls with finite inward velocity from infinity, $v \equiv v_\infty$.

Exercise   12.6
   (a)   Integrate Eq. (12.4.20) for the case $\tilde{E} < 1$, so that $1 - \tilde{E}^2 = 2M/R$, to get ($\tau = 0$ at $r = R$):

$$\tau = \left(\frac{R^3}{8M}\right)^{1/2}\left[2\left(\frac{r}{R} - \frac{r^2}{R^2}\right)^{1/2} + \cos^{-1}\left(\frac{2r}{R} - 1\right)\right]. \tag{12.4.21}$$

   (b)   Introduce the "cycloid parameter" η by

$$r = \frac{R}{2}(1 + \cos\eta), \tag{12.4.22}$$

and show that

$$\tau = \left(\frac{R^3}{8M}\right)^{1/2}(\eta + \sin\eta). \tag{12.4.23}$$

   (c)   Integrate Eq. (12.4.15) for t in terms of η to get ($t = 0$ at $r = R$):

$$\frac{t}{2M} = \ln\left|\frac{(R/2M - 1)^{1/2} + \tan(\eta/2)}{(R/2M - 1)^{1/2} - \tan(\eta/2)}\right| + \left(\frac{R}{2M} - 1\right)^{1/2}\left[\eta + \frac{R}{4M}(\eta + \sin\eta)\right]. \tag{12.4.24}$$

Note the following important results for radial infall: from Eq. (12.4.21), the *proper time* to fall from rest at $r = R > 2M$ to $r = 2M$ is *finite*. In fact, the proper time to fall to $r = 0$ is $\pi(R^3/8M)^{1/2}$, also finite. However, from Eqs. (12.4.23) and (12.4.24), the *coordinate time* (proper time for an observer at infinity) to fall to $r = 2M$ is *infinite* [at $r = 2M$, $\tan(\eta/2) = (R/2M - 1)^{1/2}$]. These results are displayed in Figure 12.1.

Exercise 12.7

(a) Find $\tau(r)$ and $t(r)$ for radial infall when $\tilde{E} = 1$.

(b) Find $\tau(r)$, $r(\eta)$, $\tau(\eta)$, and $t(\eta)$ when $\tilde{E} > 1$. You can get these from Eqs. (12.4.21)–(12.4.24) by defining R such that $2M/R = \tilde{E}^2 - 1$ and changing the sign of R in these equations. Show that $2M/R = v_\infty^2/(1 - v_\infty^2)$.

Answer: See Lightman et al. (1975), p. 407.

Turn now to nonradial motion. The elliptic integrals resulting from Eqs. (12.4.13)–(12.4.15) are not particularly informative, but we can get a general picture of the orbits by considering an "effective potential,"

$$V(r) \equiv \left(1 - \frac{2M}{r}\right)\left(1 + \frac{\tilde{l}^2}{r^2}\right). \qquad (12.4.25)$$

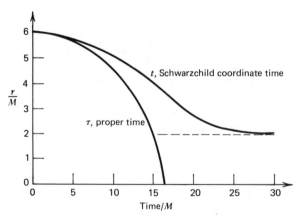

Figure 12.1 Fall from rest toward a Schwarzschild black hole as described (*a*) by a comoving observer (proper time τ) and (*b*) by a distant observer (Schwarzschild coordinate time t). In the one description, the point $r = 0$ is attained, and quickly [see Eq. (12.4.23)]. In the other description, $r = 0$ is never reached and even $r = 2M$ is attained only asymptotically [Eq. (12.4.24)]. [From *Gravitation* by Charles W. Misner, Kip S. Thorne, and John Archibald Wheeler, W. H. Freeman and Company. Copyright © 1973.]

Equation (12.4.13) then becomes

$$\left(\frac{dr}{d\tau}\right)^2 = \tilde{E}^2 - V(r). \tag{12.4.26}$$

For a fixed value of $\tilde{l}$, V is depicted schematically in Figure 12.2. Shown on the diagram are three horizontal lines corresponding to different values of $\tilde{E}^2$. From Eq. (12.4.26) we see that the distance from the horizontal line to V gives $(dr/d\tau)^2$. Consider orbit 1, the horizontal line labeled 1 corresponding to a particle coming in from infinity with energy $\tilde{E}^2$. When the particle reaches the value of r corresponding to point A, $dr/d\tau$ passes through zero and changes sign—the particle returns to infinity. Such an orbit is *unbound*, and A is called a *turning point*. Orbit 2 is a *capture* orbit; the particle plunges into the black hole. Orbit 3 is a *bound* orbit, with two turning points A_1 and A_2. The point B corresponds to a *stable circular* orbit. If the particle is slightly perturbed away from B, the orbit remains close to B. The point C is an *unstable circular* orbit; a particle placed in such an orbit will, upon experiencing the slightest inward radial perturbation, fall toward the black hole and be captured. If it is perturbed outward, it flies off to infinity. Orbits like 1 and 3 exist in the Newtonian case for motion in a central gravitational field; capture orbits are unique to general relativity.

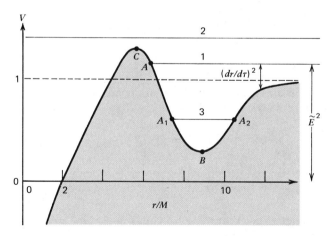

Figure 12.2 Sketch of the effective potential profile for a particle with *nonzero* rest mass orbiting a Schwarzschild black hole of mass M. The three horizontal lines labeled by different values of $\tilde{E}^2$ correspond to an (1) unbound, (2) capture, and (3) bound orbit, respectively. See text for details.

Exercise 12.8 Show that Eq. (12.4.26) reduces to the familiar Newtonian expression for particle motion in a central gravitational field when $2M/r \ll 1$.

Exercise 12.9

(a) Show that $\partial V/\partial r = 0$ when

$$Mr^2 - \tilde{l}^2 r + 3M\tilde{l}^2 = 0, \tag{12.4.27}$$

and hence that there are no maxima or minima of V for $\tilde{l} < 2\sqrt{3}\,M$.

(b) Show that $V_{\max} = 1$ for $\tilde{l} = 4M$.

The variation of V with $\tilde{l}$ is shown in Figure 12.3.

Circular orbits occur when $\partial V/\partial r = 0$ and $dr/d\tau = 0$. Equations (12.4.26) and (12.4.27) give

$$\tilde{l}^2 = \frac{Mr^2}{r - 3M}, \tag{12.4.28}$$

$$\tilde{E}^2 = \frac{(r - 2M)^2}{r(r - 3M)}. \tag{12.4.29}$$

Thus circular orbits exist down to $r = 3M$, the limiting case corresponding to a photon orbit ($\tilde{E} = E/m \rightarrow \infty$). The circular orbits are stable if V is concave up; that is, $\partial^2 V/\partial r^2 > 0$ and unstable if $\partial^2 V/\partial r^2 < 0$ (Why?).

Exercise 12.10 Show the circular Schwarzschild orbits are stable if $r > 6M$, unstable if $r < 6M$.

Exercise 12.11

(a) Show that in Newtonian theory, a distant nonrelativistic test particle can only be captured by a star of mass M and radius R if

$$\tilde{l} < \tilde{l}_{\text{crit}} \simeq (2MR)^{1/2}.$$

(b) Taking into account general relativity, can particles with much larger values of angular momentum be captured by neutron stars? by white dwarfs?

The binding energy per unit mass of a particle in the last stable circular orbit at $r = 6M$ is, from Eq. (12.4.29),

$$\tilde{E}_{\text{binding}} = \frac{m - E}{m} = 1 - \left(\frac{8}{9}\right)^{1/2} = 5.72\%. \tag{12.4.30}$$

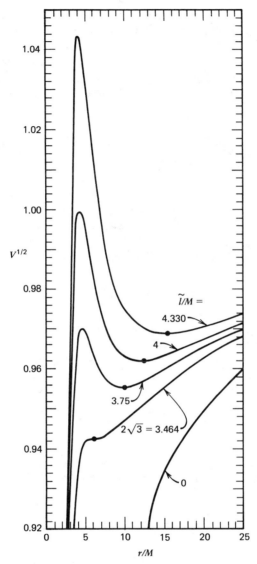

Figure 12.3 The effective potential profile for *nonzero* rest-mass particles of various angular momenta $\tilde{l}$ orbiting a Schwarzschild black hole of mass M. The dots at local minima locate radii of stable circular orbits. Such orbits exist only for $\tilde{l} > 2\sqrt{3}\, M$. [From *Gravitation* by Charles W. Misner, Kip S. Thorne, and John Archibald Wheeler, W. H. Freeman and Company. Copyright © 1973.]

This is the fraction of rest-mass energy released when, say, a particle originally at rest at infinity spirals slowly toward a black hole to the innermost stable circular orbit, and then plunges into the black hole. Thus, the conversion of rest mass to other forms of energy is potentially much more efficient for accretion onto a black hole than for nuclear burning, which releases a maximum of only 0.9% of the rest mass (H → Fe). This high efficiency will be important in our discussion of accretion disks around black holes (Section 14.5). It is the basis for invoking black holes as the energy source in numerous models seeking to explain astronomical observations of huge energy output from compact regions (e.g., Cygnus X-1; quasars; double radio galaxies, etc.).

Exercise   12.12

(a)   Use Eq. (12.4.18) to show that the velocity of a particle in the innermost stable circular orbit as measured by a local static observer is $v^{\hat{\phi}} = \frac{1}{2}$ ($c = 1$).

(b)   Suppose the particle in part (a) is emitting monochromatic light at frequency ν_{em} in its rest frame. Show that the frequency received at infinity varies periodically between

$$\frac{\sqrt{2}}{3}\nu_{em} < \nu_\infty < \sqrt{2}\,\nu_{em}.$$

Hint: Write $\nu_\infty/\nu_{em} = (\nu_\infty/\nu_{stat})(\nu_{stat}/\nu_{em})$, where ν_{stat} is the frequency measured by the local static observer and is related to ν_{em} by the special relativistic Doppler formula.

(c)   Compute the orbital period for the particle as measured by the local static observer and by the observer at infinity.

Hint: Since $d\hat{\phi} = r\,d\phi$, the proper circumference of the orbit is simply $2\pi r$.

Answer: $T_{stat} = 24\pi M$, $T_\infty = T_{stat}/(2/3)^{1/2} = 4.5 \times 10^{-4}$ s $(M/M_\odot)$

Exercise   12.13

(a)   Show that the angular velocity as measured from infinity, $\Omega \equiv d\phi/dt$, has the same form in the Schwarzschild geometry as for circular orbits in Newtonian gravity—namely,

$$\Omega = \left(\frac{M}{r^3}\right)^{1/2}. \tag{12.4.31}$$

(b)   Use this result to confirm the value of T_∞ found in Exercise 12.12.

In our later discussion of accretion onto black holes, we will need to know the capture cross section for particles falling in from infinity. This is simply

$$\sigma_{capt} = \pi b_{max}^2, \tag{12.4.32}$$

where b_{max} is the maximum impact parameter of a particle that is captured. To express b in terms of $\tilde{E}$ and $\tilde{l}$, consider the definition of the impact parameter (cf. Fig. 12.4)

$$b = \lim_{r \to \infty} r \sin \phi. \tag{12.4.33}$$

Now for $r \to \infty$, Eqs. (12.4.13) and (12.4.14) give

$$\frac{1}{r^4} \left(\frac{dr}{d\phi} \right)^2 \simeq \frac{\tilde{E}^2 - 1}{\tilde{l}^2}. \tag{12.4.34}$$

Substituting $r \simeq b/\phi$, we identify

$$\frac{1}{b^2} = \frac{\tilde{E} - 1}{\tilde{l}^2}, \tag{12.4.35}$$

or in terms of the velocity at infinity, $\tilde{E} = (1 - v_\infty^2)^{-1/2}$,

$$\tilde{l} = bv_\infty \left(1 - v_\infty^2 \right)^{-1/2}$$

$$\to bv_\infty \quad \text{for } v_\infty \ll 1. \tag{12.4.36}$$

Consider now a nonrelativistic particle moving towards the black hole ($\tilde{E} \simeq 1$, $v_\infty \ll 1$). From Exercise 12.9, we know that it is captured if $\tilde{l} < 4M$. Thus

$$b_{\max} = \frac{4M}{v_\infty}, \tag{12.4.37}$$

which gives a capture cross section

$$\sigma_{\text{capt}} = \frac{4\pi (2M)^2}{v_\infty^2}. \tag{12.4.38}$$

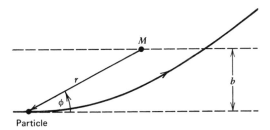

Figure 12.4 Impact parameter b for a particle with trajectory $r = r(\phi)$ about mass M.

This value should be compared with the geometrical capture cross section of a particle by a sphere of radius R in Newtonian theory:

$$\sigma_{\text{Newt}} = \pi R^2 \left(1 + \frac{2M}{v_\infty^2 R} \right). \tag{12.4.39}$$

A black hole thus captures nonrelativistic particles like a Newtonian sphere of radius $R = 8M$.

12.5   Massless Particle Orbits in the Schwarzschild Geometry

For $m = 0$ (e.g., a photon), Eqs. (12.4.6)–(12.4.8) become

$$\frac{dt}{d\lambda} = \frac{E}{1 - 2M/r}, \tag{12.5.1}$$

$$\frac{d\phi}{d\lambda} = \frac{l}{r^2}, \tag{12.5.2}$$

$$\left(\frac{dr}{d\lambda} \right)^2 = E^2 - \frac{l^2}{r^2} \left(1 - \frac{2M}{r} \right). \tag{12.5.3}$$

Now by the Equivalence Principle, we know that the particle's worldline should be independent of its energy. We can see this by introducing a new parameter

$$\lambda_{\text{new}} = l\lambda. \tag{12.5.4}$$

Writing

$$b \equiv \frac{l}{E} \tag{12.5.5}$$

and dropping the subscript "new," we find

$$\frac{dt}{d\lambda} = \frac{1}{b(1 - 2M/r)}, \tag{12.5.6}$$

$$\frac{d\phi}{d\lambda} = \frac{1}{r^2}, \tag{12.5.7}$$

$$\left(\frac{dr}{d\lambda} \right)^2 = \frac{1}{b^2} - \frac{1}{r^2} \left(1 - \frac{2M}{r} \right). \tag{12.5.8}$$

The worldline depends only on the parameter b, which is the particle's *impact parameter*, and not on l or E separately. Taking the limit $m \to 0$ of Eq. (12.4.35), we see that b of Eq. (12.5.5) is the same quantity defined in the previous section for massive particles.

We can understand photon orbits by means of an effective potential

$$V_{\text{phot}} = \frac{1}{r^2}\left(1 - \frac{2M}{r}\right),$$ (12.5.9)

so that Eq. (12.5.8) becomes

$$\left(\frac{dr}{d\lambda}\right)^2 = \frac{1}{b^2} - V_{\text{phot}}(r).$$ (12.5.10)

Clearly the distance from a horizontal line of height $1/b^2$ to V_{phot} gives $(dr/d\lambda)^2$. The quantity V_{phot} has a maximum of $1/(27M^2)$ at $r = 3M$; it is displayed in Figure 12.5. We see that the critical impact parameter separating capture from scattering orbits is given by $1/b^2 = 1/(27M^2)$, or

$$b_c = 3\sqrt{3}\, M.$$ (12.5.11)

The capture cross section for photons from infinity is thus

$$\sigma_{\text{phot}} = \pi b_c^2 = 27\pi M^2.$$ (12.5.12)

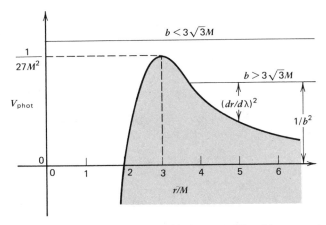

Figure 12.5 Sketch of the effective potential profile for a particle with *zero* rest mass orbiting a Schwarzschild black hole of mass M. If the particle falls from $r = \infty$ with impact parameter $b > 3\sqrt{3}\, M$ it is scattered back out to $r = \infty$. If, however, $b < 3\sqrt{3}\, M$ the particle is captured by the black hole.

To calculate the observed emission from gas near a black hole we must know those propagation directions, as measured by a static observer, for which a photon emitted at radius r can escape to infinity. Referring to Figure 12.5, we see that a photon at $r \geq 3M$ escapes only if (i) $v^{\hat{r}} > 0$, or (ii) $v^{\hat{r}} < 0$ and $b > 3\sqrt{3}\, M$. In terms of the angle ψ between the propagation direction and the radial direction (see Figure 12.6), we have since $|\mathbf{v}| = 1$,

$$v^{\hat{\phi}} = \sin\psi, \qquad v^{\hat{r}} = \cos\psi. \qquad (12.5.13)$$

But Eqs. (12.4.12) and (12.4.18) give, with $b = l/E$,

$$v^{\hat{\phi}} = \frac{b}{r}\left(1 - \frac{2M}{r}\right)^{1/2}. \qquad (12.5.14)$$

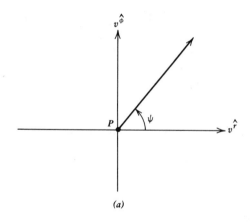

(a)

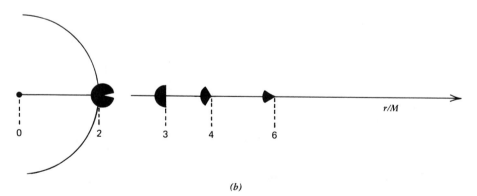

(b)

Figure 12.6 (a) The angle ψ between the propagation direction of a photon and the radial direction at a given point P. (b) Gravitational capture of radiation by a Schwarzschild black hole. Rays emitted from each point into the interior of the *shaded* conical cavity are captured. The indicated capture cavities are those measured in the orthonormal frame of a local static observer.

Thus an inward-moving photon escapes the black hole if

$$\sin \psi > \frac{3\sqrt{3}\,M}{r}\left(1 - \frac{2M}{r}\right)^{1/2}.$$ (12.5.15)

At $r = 6M$, escape requires $\psi < 135°$; at $r = 3M$, $\psi < 90°$ so that all inward-moving photons are captured (i.e., 50% of the radiation from a stationary, isotropic emitter at $r = 3M$ is captured).

Exercise   12.14   Show that an outward-directed photon emitted between $r = 2M$ and $r = 3M$ escapes if

$$\sin \psi < \frac{3\sqrt{3}\,M}{r}\left(1 - \frac{2M}{r}\right)^{1/2}.$$

Only the outward-directed radial photons escape as the source approaches $r = 2M$. See Figure 12.6 for a diagram of these effects.

12.6   Nonsingularity of the Schwarzschild Radius

The metric (12.3.1) appears singular at $r = 2M$; the coefficient of dt^2 goes to zero, while the coefficient of dr^2 becomes infinite. However, we cannot immediately conclude that this behavior represents a true physical singularity. Indeed, the coefficient of $d\phi^2$ becomes zero at $\theta = 0$, but we know this is simply because the polar coordinate system itself is singular there. The coordinate singularity at $\theta = 0$ can be removed by choosing another coordinate system (e.g., stereographic coordinates on the 2-sphere).

We already have a clue that the Schwarzschild radius $r = 2M$ is only a *coordinate singularity*. Recall that a radially infalling particle does not notice anything strange about $r = 2M$; there is nothing special about $r(\tau)$ at this point. However, the coordinate time t becomes infinite as $r \to 2M$. This strongly suggests the presence of a coordinate singularity rather than a physical singularity.

There are many different coordinate transformations that can be used to show explicitly that $r = 2M$ is not a physical singularity. We shall exhibit one—the *Kruskal[6] coordinate system*. It is defined by the transformation

$$u = \left(\frac{r}{2M} - 1\right)^{1/2} e^{r/4M} \cosh\frac{t}{4M},$$ (12.6.1)

$$v = \left(\frac{r}{2M} - 1\right)^{1/2} e^{r/4M} \sinh\frac{t}{4M}.$$ (12.6.2)

[6] Kruskal (1960); Szekeres (1960).

The inverse transformation is

$$\left(\frac{r}{2M} - 1\right)e^{r/2M} = u^2 - v^2, \tag{12.6.3}$$

$$\tanh\frac{t}{4M} = \frac{v}{u}. \tag{12.6.4}$$

The metric (12.3.1) takes the form

$$ds^2 = \frac{32M^3}{r}e^{-r/2M}(-dv^2 + du^2) + r^2\,d\theta^2 + r^2\sin^2\theta\,d\phi^2, \tag{12.6.5}$$

where r is defined implicitly in terms of u and v by Eq. (12.6.3). Clearly the metric (12.6.5) is nonsingular at $r = 2M$. However, $r = 0$ is still a singularity. One can show that it is a real *physical singularity* of the metric, with infinite gravitational field strengths there.

Note that $r = 0$ is, from Eq. (12.6.3), at $v^2 - u^2 = 1$, or $v = \pm(1 + u^2)^{1/2}$. There are two singularities! Note also that $r \geqslant 2M$ is the region $u^2 \geqslant v^2$—that is, $u \geqslant |v|$ or $u \leqslant -|v|$. Two regions correspond to $r \geqslant 2M$!

The original Schwarzschild coordinate system covers only part of the spacetime manifold. Kruskal coordinates give an analytic continuation of the same solution of the field equations to cover the whole spacetime manifold. This situation is depicted in the *Kruskal diagram*, Figure 12.7. Kruskal coordinates have the nice property that radial light rays travel on 45° straight lines [see Eq. (12.6.5) with $ds^2 = 0$]. A Kruskal diagram is a spacetime diagram, with the time coordinate v plotted vertically and the spatial coordinate u plotted horizontally. It can be read like a special relativistic spacetime diagram, because the light cones are at 45° at every point, and particle worldlines must lie inside the light cones.

Region I is "our universe," the original region $r > 2M$. Region II is the "interior of the black hole," the region $r < 2M$. Regions III and IV are the "other universe": region III is asymptotically flat, with $r > 2M$, while region IV corresponds to $r < 2M$.

If one checks the signs of u and v in the various quadrants, one finds that the relationship between Kruskal and Schwarzschild coordinates in the various regions is [cf. Eqs. (12.6.1) and (12.6.2)]

$$u = \pm\left(\frac{r}{2M} - 1\right)^{1/2}e^{r/4M}\cosh\frac{t}{4M}, \tag{12.6.6}$$

$$v = \pm\left(\frac{r}{2M} - 1\right)^{1/2}e^{r/4M}\sinh\frac{t}{4M}, \quad (r \geqslant 2M), \tag{12.6.7}$$

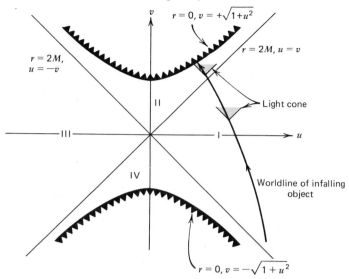

Figure 12.7   Kruskal diagram of the Schwarzschild metric.

and

$$u = \pm \left(1 - \frac{r}{2M}\right)^{1/2} e^{r/4M} \sinh\frac{t}{4M}, \tag{12.6.8}$$

$$v = \pm \left(1 - \frac{r}{2M}\right)^{1/2} e^{r/4M} \cosh\frac{t}{4M}, \quad (r \leqslant 2M). \tag{12.6.9}$$

Here the upper sign refers to "our universe," the lower to the "other universe." Equation (12.6.3) holds everywhere, while the right-hand side of Eq. (12.6.4) becomes u/v for $r \leqslant 2M$. Equation (12.6.4) shows that lines of constant t are straight lines. These relationships are shown in Figure 12.8.

The singularity at the top of the Kruskal diagram at $r = 0$ occurs inside of the black hole. Clearly any timelike worldline at $r \leqslant 2M$ (i.e., in region II) *must* strike the singularity. The singularity at the bottom of the diagram represents a "white hole," from which anything can come spewing out.

It is important to realize that the full analytically continued Schwarzschild metric is merely a mathematical solution of Einstein's equations. For a black hole formed by gravitational collapse, part of spacetime must contain the collapsing matter. We know from Birkhoff's theorem (Section 5.6) that outside the collapsing star the geometry is still described by the Schwarzschild metric. Thus the worldline of a point on the surface of the star will be the boundary of the physically meaningful part of the Kruskal diagram (see Fig. 12.9). The "white hole" and the "other universe" are not present in real black holes.

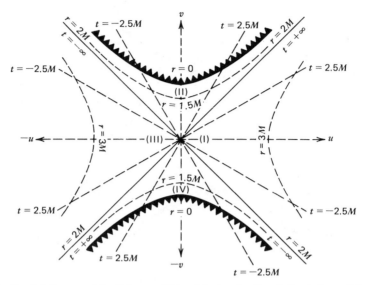

Figure 12.8 Kruskal diagram of the Schwarzschild metric showing the relation between Schwarzschild (t, r) and Kruskal (v, u) coordinates.

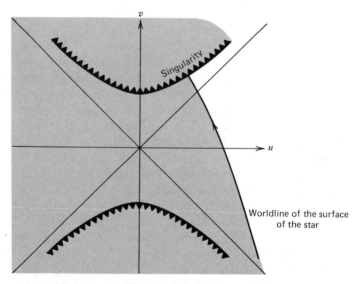

Figure 12.9 Kruskal diagram for gravitational collapse. Only the *unshaded* portion to the right of the surface of the star is physically meaningful. The remainder of the diagram must be replaced by the spacetime geometry of the interior of the star.

The Kruskal diagram makes clear the two key properties of black holes: once an object crosses $r = 2M$, it must strike the singularity at $r = 0$; and once inside $r = 2M$, it cannot send signals back out to infinity. However, there is no local test for this horizon. An observer does not notice anything significantly different in crossing from $r = 2M + \varepsilon$ to $r = 2M - \varepsilon$.

12.7   Kerr Black Holes

The most general stationary black hole metric, with parameters M, J, and Q, is called the Kerr–Newman metric.[7] Special cases are the Kerr metric ($Q = 0$), the Reissner–Nordstrom metric ($J = 0$), and the Schwarzschild metric ($J = 0$, $Q = 0$).

It is usually true that a charged astrophysical object is rapidly neutralized by the surrounding plasma. We shall accordingly simplify our discussion by assuming that charged black holes are not likely to be important astrophysically. All astrophysical objects rotate however, and so we expect black holes formed by gravitational collapse to be rotating in general. Remarkably, when all the radiation of various kinds produced by the collapse has been radiated away, the gravitational field settles down asymptotically to the Kerr metric.

This solution of Einstein's equations, discovered by Kerr in 1963, was not at first recognized to be a black hole solution. Its properties are more transparent in Boyer–Lindquist (1967) coordinates, where

$$ds^2 = -\left(1 - \frac{2Mr}{\Sigma}\right) dt^2 - \frac{4aMr\sin^2\theta}{\Sigma} dt\, d\phi + \frac{\Sigma}{\Delta} dr^2$$

$$+ \Sigma\, d\theta^2 + \left(r^2 + a^2 + \frac{2Mra^2\sin^2\theta}{\Sigma}\right)\sin^2\theta\, d\phi^2. \qquad (12.7.1)$$

Here the black hole is rotating in the ϕ direction, and

$$a \equiv \frac{J}{M}, \qquad \Delta \equiv r^2 - 2Mr + a^2, \qquad \Sigma \equiv r^2 + a^2\cos^2\theta. \qquad (12.7.2)$$

The metric is stationary (independent of t) and axisymmetric about the polar axis (independent of ϕ). Note that a, the angular momentum per unit mass, is measured in cm when expressed in units with $c = G = 1$. Setting $a = 0$ in Eq. (12.7.1) gives the Schwarzschild metric.

Exercise 12.15   The angular momentum of the sun (assuming uniform rotation) is $J = 1.63 \times 10^{48}$ g cm^2 s^{-1}. What is a/M for the sun?

Answer: 0.185.

[7]See, for example, Misner, Thorne, and Wheeler (1973).

The horizon occurs where the metric function Δ vanishes. This occurs first at the larger root of the quadratic equation $\Delta = 0$,

$$r_+ = M + (M^2 - a^2)^{1/2}. \tag{12.7.3}$$

Note that a must be less than M for a black hole to exist. If a exceeded M, one would have a gravitational field with a "naked" singularity (i.e., one not "clothed" by an event horizon). A major unsolved problem in general relativity is Penrose's Cosmic Censorship Conjecture, that gravitational collapse from well-behaved initial conditions never gives rise to a naked singularity. Certainly no mechanism is known to take a Kerr black hole with $a < M$ and spin it up so that a becomes greater than M (see Section 12.8). A black hole with $a \equiv M$ is called a *maximally rotating* black hole.

Static observers were useful for understanding the Schwarzschild metric. We can generalize to rotating black holes by introducing *stationary* observers, who are at fixed r and θ, but rotate with a constant angular velocity

$$\Omega = \frac{d\phi}{dt} = \frac{u^\phi}{u^t}. \tag{12.7.4}$$

The condition $\mathbf{u} \cdot \mathbf{u} = -1$ (i.e., that the observers follow a timelike worldline) is

$$-1 = (u^t)^2 \left[g_{tt} + 2\Omega g_{t\phi} + \Omega^2 g_{\phi\phi} \right], \tag{12.7.5}$$

where Eq. (12.7.4) has been used to eliminate u^ϕ. The quantity in square brackets in Eq. (12.7.5) must therefore be negative. Since $g_{\phi\phi}$ in Eq. (12.7.1) is positive, this is true only if Ω lies between the roots of the quadratic equation obtained by setting the bracketed expression equal to zero. Thus

$$\Omega_{min} < \Omega < \Omega_{max}, \tag{12.7.6}$$

where

$$\Omega_{\substack{max \\ min}} = \frac{-g_{t\phi} \pm \left(g_{t\phi}^2 - g_{tt}g_{\phi\phi} \right)^{1/2}}{g_{\phi\phi}}. \tag{12.7.7}$$

Exercise   12.16   Discuss the restriction (12.7.6) in the weak-field limit.

   Answer: $-c/(r \sin \theta) < \Omega < c/(r \sin \theta)$; stationary observers must rotate about the z-axis with $v < c$.

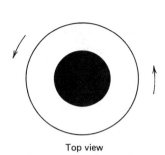

Top view

Side view

Figure 12.10 Ergosphere of a Kerr black hole: the region between the static limit [the flattened outer surface $r = M + (M^2 - a^2\cos^2\theta)^{1/2}$] and the event horizon [inner sphere $r = M + (M^2 - a^2)^{1/2}$].

Note that $\Omega_{\min} = 0$ when $g_{tt} = 0$; that is, when $r^2 - 2Mr + a^2\cos^2\theta = 0$. This occurs at

$$r_0 = M + (M^2 - a^2\cos^2\theta)^{1/2}. \tag{12.7.8}$$

Observers between r_+ and r_0 must have $\Omega > 0$; no static observers ($\Omega = 0$) exist within $r_+ < r < r_0$. The surface $r = r_0$ is therefore called the static limit. It is also called the "boundary of the ergosphere," for reasons that will become clearer later.

Note that the horizon r_+ and the static limit r_0 are distinct for $a \neq 0$. This is depicted in Figure 12.10. From Eq. (12.7.7), we see that stationary observers fail to exist when $g_{t\phi}^2 - g_{tt}g_{\phi\phi} < 0$—that is, when $\Delta < 0$; this can occur only when $r < r_+$. This is the generalization of the fact that static observers do not exist inside the horizon for a Schwarzschild black hole.

The Kerr metric can be analytically continued inside $r = r_+$ in a manner similar to the Kruskal continuation of the Schwarzschild metric. However, this interior solution is not physically meaningful for two reasons. First, part of it must be replaced by the interior of the collapsing object that formed the black hole. More important, there is no Birkhoff's theorem for rotating collapse. The Kerr metric is not the exterior metric *during* the collapse; it is only the *asymptotic* form of the metric when all the dynamics has ceased. Its mathematical continuation inside r_+ is essentially irrelevant. We shall therefore restrict our discussion to the region $r \geq r_+$, adopting the viewpoint that anything entering inside that region becomes causally disconnected from the rest of the universe.

A complete description of the geodesics of a Kerr black hole is quite complicated because of the absence of spherical symmetry. There is, however, a "hidden" symmetry that can be exploited to solve for the geodesics analytically.[8]

[8] Carter (1968); see also Misner, Thorne, and Wheeler (1973).

To get some insight into the effects of rotation on the geodesics, we shall restrict ourselves to considering test particle motion in the equatorial plane. This can be done in a straightforward way without invoking the hidden symmetry.

Setting $\theta = \pi/2$ in Eq. (12.7.1), we obtain the Lagrangian

$$2L = -\left(1 - \frac{2M}{r}\right)\dot{t}^2 - \frac{4aM}{r}\dot{t}\dot{\phi} + \frac{r^2}{\Delta}\dot{r}^2 + \left(r^2 + a^2 + \frac{2Ma^2}{r}\right)\dot{\phi}^2,$$

(12.7.9)

where $\dot{t} = dt/d\lambda$, and so on. Corresponding to the ignorable coordinates t and ϕ, we obtain two first integrals:

$$p_t \equiv \frac{\partial L}{\partial \dot{t}} = \text{constant} = -E,$$

(12.7.10)

$$p_\phi \equiv \frac{\partial L}{\partial \dot{\phi}} = \text{constant} \equiv l.$$

(12.7.11)

Evaluating the derivatives from Eq. (12.7.9) and solving the two equations for $\dot{t}$ and $\dot{\phi}$, we obtain

$$\dot{t} = \frac{(r^3 + a^2 r + 2Ma^2)E - 2aMl}{r\Delta},$$

(12.7.12)

$$\dot{\phi} = \frac{(r - 2M)l + 2aME}{r\Delta}.$$

(12.7.13)

We get a third integral of the motion as usual be setting $g_{\alpha\beta}p^\alpha p^\beta = -m^2$—that is, $L = -m^2/2$. Substituting Eqs. (12.7.12) and (12.7.13), we obtain after some simplification

$$r^3\left(\frac{dr}{d\lambda}\right)^2 = R(E, l, r),$$

(12.7.14)

where

$$R \equiv E^2(r^3 + a^2 r + 2Ma^2) - 4aMEl - (r - 2M)l^2 - m^2 r\Delta. \quad (12.7.15)$$

We can regard R as an effective potential for radial motion in the equatorial plane.[9] For example, circular orbits occur where $dr/d\lambda$ remains zero (perpetual

[9]Some authors define the effective potential as that value of E which makes $R = 0$, as we did in the Schwarzschild case, Eq. (12.4.26).

turning point). This requires

$$R = 0, \qquad \frac{\partial R}{\partial r} = 0. \tag{12.7.16}$$

After considerable algebra, Eqs. (12.7.16) can be solved for E and l to give

$$\tilde{E} = \frac{r^2 - 2Mr \pm a\sqrt{Mr}}{r(r^2 - 3Mr \pm 2a\sqrt{Mr})^{1/2}}, \tag{12.7.17}$$

$$\tilde{l} = \pm \frac{\sqrt{Mr}(r^2 \mp 2a\sqrt{Mr} + a^2)}{r(r^2 - 3Mr \pm 2a\sqrt{Mr})^{1/2}}. \tag{12.7.18}$$

Here the upper sign refers to *corotating* or direct orbits (i.e., orbital angular momentum of particle parallel to black hole spin angular momentum), the lower sign to *counterrotating* orbits. These formulas generalize Eqs. (12.4.28)–(12.4.29) for the Schwarzschild metric.

Exercise 12.17 Show that Kepler's Third Law takes the form

$$\Omega = \pm \frac{M^{1/2}}{r^{3/2} \pm aM^{1/2}} \tag{12.7.19}$$

for circular equatorial orbits in the Kerr metric. Here $\Omega \equiv d\phi/dt = \dot{\phi}/\dot{t}$.

Circular orbits exist from $r = \infty$ all the way down to the limiting circular photon orbit, when the denominator of Eq. (12.7.17) vanishes. Solving the resulting cubic equation in $r^{1/2}$, we find for the photon orbit[10]

$$r_{\rm ph} = 2M\{1 + \cos[\tfrac{2}{3}\cos^{-1}(\mp a/M)]\}. \tag{12.7.20}$$

For $a = 0$, $r_{\rm ph} = 3M$, while for $a = M$, $r_{\rm ph} = M$ (direct) or $4M$ (retrograde).

For $r > r_{\rm ph}$, not all circular orbits are bound. An unbound circular orbit has $E/m > 1$. Given an infinitesimal outward perturbation, a particle in such an orbit will escape to infinity on an asymptotically hyperbolic trajectory. Bound circular orbits exist for $r > r_{\rm mb}$, where $r_{\rm mb}$ is the radius of the marginally bound circular orbit with $E/m = 1$:

$$r_{\rm mb} = 2M \mp a + 2M^{1/2}(M \mp a)^{1/2}. \tag{12.7.21}$$

Note also that $r_{\rm mb}$ is the minimum periastron of all parabolic ($E/m = 1$) orbits.

[10] Bardeen et al. (1972).

In astrophysical problems, particle infall from infinity is very nearby parabolic, since $v_\infty \ll c$. Any parabolic trajectory which penetrates to $r < r_{mb}$ must plunge directly into the black hole. For $a = 0$, $r_{mb} = 4M$; for $a = M$, $r_{mb} = M$(direct) or $5.83M$(retrograde).

Even the bound circular orbits are not all stable. Stability requires that

$$\frac{\partial^2 R}{\partial r^2} \leqslant 0. \tag{12.7.22}$$

From Eq. (12.7.15), we get

$$1 - (\tilde{E})^2 \geqslant \frac{2}{3}\frac{M}{r}. \tag{12.7.23}$$

Substituting Eq. (12.7.17), we get a quartic equation in $r^{1/2}$ for the limiting case of equality. The solution for r_{ms}, the radius of the marginally stable circular orbit, is given by Bardeen et al. (1972):

$$r_{ms} = M\{3 + Z_2 \mp [(3 - Z_1)(3 + Z_1 + 2Z_2)]^{1/2}\},$$

$$Z_1 \equiv 1 + \left(1 - \frac{a^2}{M^2}\right)^{1/3}\left[\left(1 + \frac{a}{M}\right)^{1/3} + \left(1 - \frac{a}{M}\right)^{1/3}\right],$$

$$Z_2 \equiv \left(3\frac{a^2}{M^2} + Z_1^2\right)^{1/2}. \tag{12.7.24}$$

For $a = 0$, $r_{ms} = 6M$; for $a = M$, $r_{ms} = M$(direct) or $9M$(retrograde). A quantity of great interest for the potential efficiency of a black hole accretion disk as an energy source is the binding energy of the marginally stable circular orbit. If we eliminate r from Eq. (12.7.17) using Eq. (12.7.23), we find

$$\frac{a}{M} = \mp\frac{4\sqrt{2}(1 - \tilde{E}^2)^{1/2} - 2\tilde{E}}{3\sqrt{3}(1 - \tilde{E}^2)}. \tag{12.7.25}$$

The quantity $\tilde{E}$ *decreases* from $\sqrt{8/9}$ ($a = 0$) to $\sqrt{1/3}$ ($a = M$) for *direct* orbits, while it *increases* from $\sqrt{8/9}$ to $\sqrt{25/27}$ for *retrograde* orbits. The maximum binding energy $1 - \tilde{E}$ for a maximally rotating black hole is $1 - 1/\sqrt{3}$, or 42.3% of the rest-mass energy! This is the amount of energy that is released by matter spiralling in toward the black hole through a succession of almost circular equatorial orbits. Negligible energy is released during the final plunge from r_{ms} into the black hole.

Note that the Boyer–Lindquist coordinate system collapses r_{ms}, r_{mb}, r_{ph}, and r_+ into $r = M$ as $a \to M$. This is simply an artifact of the coordinates as discussed by Bardeen et al. (1972): the radii actually correspond to distinct spacetime regions.

An exceedingly interesting property of rotating black holes is that there exist negative energy test particle trajectories. If we solve Eq. (12.7.14) for E, we find

$$ E = \frac{2aMl + (l^2 r^2 \Delta + m^2 r \Delta + r^3 \dot{r}^2)^{1/2}}{r^3 + a^2 r + 2Ma^2}. \qquad (12.7.26) $$

(The sign of the square root is determined by letting $r \to \infty$.) To have $E < 0$, we require a retrograde orbit ($l < 0$), with

$$ l^2 r^2 \Delta + m^2 r \Delta + r^3 \dot{r}^2 < 4a^2 M^2 l^2. \qquad (12.7.27) $$

The boundary of the region of negative energy orbits is found by making the left-hand side of inequality (12.7.27) as small as possible. Thus let $m \to 0$ (highly relativistic particle) and $\dot{r} \to 0$. We then find that the boundary is at $r = 2M = r_0$ ($\theta = \pi/2$). One can in fact show that the static limit r_0 is the boundary of the region containing negative energy orbits for *all* values of θ. A particle can only be injected into such an orbit *inside* the static limit; it then plunges into the black hole.

Penrose (1969) exploited this property of Kerr black holes in a remarkable thought experiment to demonstrate that rotating black holes are potentially vast storehouses of energy. Imagine sending a particle in from infinity with an energy E_{in}. The trajectory is carefully chosen so that it penetrates inside the static limit. The particle is then "instructed" (or preprogrammed) to split into two. One piece goes into a negative energy trajectory and down the black hole, with energy $E_{down} < 0$. The other comes back out to infinity with energy E_{out}. Energy conservation gives

$$ E_{in} = E_{out} + E_{down}, \quad \text{i.e.,} \quad E_{out} > E_{in}! \qquad (12.7.28) $$

Even though some rest-mass has been lost down the hole, there is a net gain of energy at infinity. This energy is extracted from the rotational energy of the hole, which slows down slightly when the retrograde negative energy particle is captured.

The region $r_+ < r < r_0$, where energy extraction is possible, is called the *ergosphere*, from the Greek word for work.

Unfortunately, the original Penrose process is not likely to be important astrophysically. Bardeen et al. (1972) have shown that the breakup of the two particles inside the ergosphere must happen with a relative velocity of at least $\frac{1}{2}c$; it is hard to imagine astrophysical processes that produce such large relative velocities.

Energy amplification also occurs when waves (electromagnetic or gravitational) of suitable frequency are scattered by a rotating black hole. Part of the wave is absorbed, but the part that scatters can, under the right conditions, have more energy than the incident wave. Whether such "superradiant scattering" is important astrophysically or not is an open question. Superradiant scattering has been invoked to design a "black hole bomb," and for advanced civilizations to solve their energy crisis.[11] Incidentally, it has been shown that rotating black holes are dynamically stable objects, in the sense that they cannot spontaneously explode in a burst of energy.[12]

Exercise   12.18   Consider a particle with $\tilde{l} = 0$ released from rest far from a Kerr black hole. Show that the particle "corotates with the geometry" as it spirals toward the hole along a conical surface of constant θ. In other words, show that the particle acquires an angular velocity $d\phi/dt = \omega(r, \theta)$ as viewed from infinity, where

$$\omega(r, \theta) = \frac{2aMr}{(r^2 + a^2)^2 - \Delta a^2 \sin^2\theta}.$$

Note: Observers at fixed r and θ with zero angular momentum also "corotate with the geometry" with angular velocity $\omega(r, \theta)$. Such observers define the so-called "locally nonrotating frame" (LNRF) (see Bardeen et al., 1972); according to such observers, the released particle described above appears to move *radially* locally.

The procedure for determining the emission angles leading to photon capture and escape from a radiating source near a Kerr hole has been outlined by Bardeen (1973). Capture must be accounted for whenever one wishes to determine the actual radiation observed at infinity that originates from a local source near the hole. For a rotating hole, photons emitted with $v^\phi > 0$ in the LNRF preferentially escape to infinity. In general applications, the calculation of the escape angles must be performed numerically.[13]

12.8   The Area Theorem and Black Hole Evaporation

Hawking[14] proved a remarkable theorem about black holes: in any interaction, the surface area of a black hole can never decrease. If several black holes are present, it is the sum of the surface areas that can never decrease.

We can compute the surface area of a Kerr black hole quite easily from the metric (12.7.1). Setting t = constant, $r = r_+$ = constant, and using Eq. (12.7.3), we

[11] Press and Teukolsky (1972).

[12] Press and Teukolsky (1973), Teukolsky and Press (1974); but see Section 12.8.

[13] See Cunningham and Bardeen (1972) and Shapiro (1974).

[14] See Hawking and Ellis (1973).

find for the metric on the surface

$$ds^2 = \left(r_+^2 + a^2\cos^2\theta\right) d\theta^2 + \frac{(2Mr_+)^2}{r_+^2 + a^2\cos^2\theta}\sin^2\theta \, d\phi^2. \qquad (12.8.1)$$

The area of the horizon is

$$A = \int\int \sqrt{g}\, d\theta\, d\phi$$

$$= \int\int 2Mr_+ \sin\theta\, d\theta\, d\phi$$

$$= 8\pi M\left[M + (M^2 - a^2)^{1/2}\right], \qquad (12.8.2)$$

where g is the determinant of the metric coefficients appearing in Eq. (12.8.1). Note that for $a = 0$, $A = 4\pi(2M)^2$, as we would expect.

Exercise   12.19   Use Hawking's area theorem to find the minimum mass M_2 of a Schwarzschild black hole that results from the collision of two Kerr black holes of equal mass M and opposite angular momentum parameter a. Show that if $|a| \to M$, 50% of the rest mass is allowed to be radiated away. Show that no other combinations of masses and angular momenta lead to higher possible efficiencies. Show that if $a = 0$, the maximum efficiency is 29%.

   Note: The *actual* amount of radiation generated by such a collision is amenable to numerical computation. The result is not yet known for the general case, but for $a = 0$ it is $\sim 0.1\%$ (Smarr, 1979a). See also Chapter 16.

We can use the area theorem to show that one cannot make a naked singularity by adding particles to a maximally rotating black hole in an effort to spin it up. From Eq. (12.8.2), we find that $\delta A > 0$ implies

$$\left[2M(M^2 - a^2)^{1/2} + 2M^2 - a^2\right]\delta M > Ma\, \delta a. \qquad (12.8.3)$$

As $a \to M$, this becomes

$$M\,\delta M > a\,\delta a. \qquad (12.8.4)$$

Thus M^2 always remains greater than a^2, and the horizon is not destroyed [cf. Eq. (12.7.3)]. The capture cross section for particles that increase a/M goes to zero as $a \to M$.

   The law of increase of area looks very much like the second law of thermodynamics for the increase of entropy. Bekenstein (1973) tried to develop a thermodynamics of black hole interactions. However, in classical general relativity there is no equilibrium state involving black holes. If a black hole is placed in a radiation bath, it continually absorbs radiation without ever coming to equilibrium.

This situation was changed by Hawking's remarkable discovery[15] that when quantum effects are taken into account, black holes radiate with a thermal spectrum. The expectation number of particles of a given species emitted in a mode with frequency ω is

$$\langle N \rangle = \frac{\Gamma}{\exp(\hbar\omega/kT) \mp 1}, \tag{12.8.5}$$

where Γ is the absorption coefficient for that mode incident on the hole. The absorption coefficient Γ is a slowly varying function of ω depending on the kind of particle emitted and is close to unity for wavelengths much less than M; we shall simply take it to be unity. The temperature of the black hole is inversely proportional to its mass:

$$T = \frac{\hbar}{8\pi kM} \simeq 10^{-7}\,\mathrm{K}\left(\frac{M_\odot}{M}\right). \tag{12.8.6}$$

Note that the "Planck mass" and "Planck radius" are

$$\hbar^{1/2} = 2.2 \times 10^{-5}\,\mathrm{g} = 1.6 \times 10^{-33}\,\mathrm{cm} \tag{12.8.7}$$

in units with $c = G = 1$.

Exercise   12.20   Verify the numerical relations in Eqs. (12.8.6) and (12.8.7).

We are giving Hawking's formulas for the case of a Schwarzschild black hole. They can be easily generalized to include charge and rotation. Dimensionally, T is obtained by setting a thermal wavelength[16] $\hbar c/kT$ equal to the Schwarzschild radius. Because of the thermal nature of the spectrum, mainly massless particles are produced (photons, neutrinos, and gravitons). To create a significant amount of particles of mass m, one requires $kT \sim mc^2$; that is, the Schwarzschild radius must be of order the Compton wavelength $\lambda_c \sim \hbar/mc$ of the particle.

We can now compute the entropy of a black hole: since the area is (restoring the c's and G's)

$$A = 4\pi\left(\frac{2GM}{c^2}\right)^2, \tag{12.8.8}$$

---

[15] Hawking (1974), (1975).
[16] A thermal wavelength is roughly the average separation between photons in equilibrium at temperature T.

we have

$$d(Mc^2) = \frac{c^6}{G^2}\frac{1}{32\pi M}dA \equiv T\,dS \qquad (12.8.9)$$

provided we identify the entropy S with

$$S = \frac{kc^3}{G\hbar}\frac{1}{4}A. \qquad (12.8.10)$$

The ratio of a macroscopic quantity (A) to a microscopic quantity $(\hbar)$ ensures that black holes have large entropy. This is in accord with the "no hair" ideas about many different internal states of a black hole corresponding to the same external gravitational field, and that information is lost from the outside world once black holes form.[17]

Note that during black hole "evaporation" (emission of thermal quanta), M decreases by energy conservation and thus so does A (and S). This violates Hawking's area theorem. However, one of the postulates of the area theorem is that matter obeys the "strong" energy condition, which requires that a local observer always measures positive energy densities and that there are no spacelike energy fluxes. Black hole evaporation can be understood as pair creation in the gravitational field of the black hole, one member of the pair going down the black hole and the other coming out to infinity. In pair creation the pair of particles materializes with a spacelike separation—effectively there is a spacelike energy flux.

The area theorem of classical general relativity gets replaced by a generalized second law of thermodynamics: in any interactions, the *sum* of the entropies of all black holes plus the entropy of matter outside black holes never decreases. Thus black holes do not simply have laws analogous to those of thermodynamics; they actually fit very naturally into an extended framework of thermodynamics.

We can get a qualitative understanding of the Hawking process by first considering the case of pair production in a strong electric field E (cf. Fig. 12.11). In quantum mechanics, the vacuum is continuously undergoing fluctuations, where a pair of "virtual" particles is created and then annihilated. The electric field tends to separate the charges. If the field is strong enough, the particles tunnel through the quantum barrier and materialize as real particles. The critical field strength is achieved when the work done in separating them by a Compton wavelength equals the energy necessary to create the particles:

$$eE\lambda_c \sim 2mc^2. \qquad (12.8.11)$$

[17]Compare Bekenstein (1975).

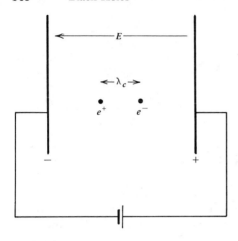

Figure 12.11   Pair production in a strong electric field.

In the black hole case, the tidal gravitational force across a distance λ_c is of order

$$\frac{GmM}{r^3}\lambda_c. \tag{12.8.12}$$

The work done is the product of this with λ_c. Setting $r \sim GM/c^2$, since the maximum field strength is near the horizon, and equating the work to $2mc^2$ gives

$$\lambda_c \sim \frac{GM}{c^2}. \tag{12.8.13}$$

Thus particles are created when their Compton wavelength is of order the Schwarzschild radius, as we mentioned earlier. [The argument has to be modified for massless particles, which have no barrier to penetrate. Their production rate is controlled by the phase space available. If we equate the mean separation of blackbody photons, $\hbar c/kT$, to the only length scale associated with the black hole, GM/c^2, we get a dimensional estimate of the black hole temperature, Eq. (12.8.6).]

The rate of energy loss from an evaporating black hole is given by the blackbody formula

$$\frac{dE}{dt} \sim \text{area} \times T^4 \sim M^2 \times M^{-4} \sim M^{-2}. \tag{12.8.14}$$

The associated timescale is

$$\tau \sim \frac{E}{dE/dt} \sim M^3. \tag{12.8.15}$$

To get the dimensions correct (with $c = G = 1$), we must restore a factor of $\hbar$:

$$\tau \sim \frac{M^3}{\hbar} \sim 10^{10} \text{ yr} \left(\frac{M}{10^{15} \text{ g}}\right)^3. \tag{12.8.16}$$

For solar mass black holes, Hawking evaporation is completely unimportant, as is clear from Eqs. (12.8.6) and (12.8.16). Only when $M \lesssim 10^{15}$ g is the timescale shorter than the age of the Universe. Presumably such "mini" black holes could only have been formed from density fluctuations during the Big Bang. Their Schwarzschild radius is about a fermi. If a spectrum of mini black holes were in fact formed in the early Universe, those with $M \ll 10^{15}$ g would long since have exploded. Those with $M \sim 10^{15}$ g would just now be exploding, with

$$\frac{dE}{dt} \sim 10^{20} \text{ erg s}^{-1} \left(\frac{10^{15} \text{ g}}{M}\right)^2 \tag{12.8.17}$$

[cf. Eq. (12.8.14)], producing quanta with energy

$$\hbar\omega \sim 100 \text{ MeV} \left(\frac{10^{15} \text{ g}}{M}\right) \tag{12.8.18}$$

[cf. Eq. (12.8.6)]. Rees (1977a) and Blandford (1977) have discussed the possibility of detecting such events. Page (1976) has computed detailed predictions of the energy spectrum emitted by 10^{15} g black holes.

The observed energy density of γ-rays at around 100 MeV is[18] $\sim 10^{-38}$ g cm^{-3}. The calculations of Page suggest that about 10% of the energy of an exploding black hole is emitted in the form of photons (as opposed to neutrinos, gravitons or particles with mass). Thus the density of 10^{15} g black holes must be less than 10^{-37} g cm^{-3}, which is about 10^{-8} times the critical density to close the universe.[19]

Exercise   12.21

(a) Compute the entropy of a $1M_\odot$ black hole in units of k, Boltzmann's constant.
Answer: $S = 1.0 \times 10^{77}\, k$.

(b) Estimate the entropy of the sun. Assume it consists of completely ionized hydrogen, with a mean density of 1 g cm^{-3} and a mean temperature of 10^6 K.
Answer: $S \sim 2 \times 10^{58}\, k$.

(c) Estimate the entropy of a $1M_\odot$ iron white dwarf and a $1M_\odot$ neutron star. Take the mean temperature to be 10^8 K, and the mean densities to be 10^6 g cm^{-3} and 10^{14} g cm^{-3}, respectively. Note that the expression (11.8.1) for C_v for a degenerate ideal gas is also equal to S, since $C_v = T\, dS/dT$. [Bekenstein (1975) has discussed the very large entropy of black holes from an information-theoretic viewpoint.]

---

[18]Fichtel et al. (1975).
[19]See also Chapline (1975); Page and Hawking (1976); Carr (1976).

13

Compact X-Ray Sources

13.1 Discovery and Identification

A new era in astronomy was ushered in on June 18, 1962, when Giacconi et al.[1] launched an Aerobee rocket carrying three Geiger counters aloft from White Sands Missile Range, New Mexico. They soon discovered that our Galaxy contains discrete X-ray sources. For example, they discovered the object Scorpius X-1, the brightest source in the sky in the energy band 1–10 keV. (Its name, abbreviated Sco X-1, denotes that it was the first X-ray source discovered in the constellation Scorpio.) Subsequent rocket and balloon flights soon followed, confirming the Aerobee observations, refining the source positions, and identifying new sources.

By the end of the decade about 20 X-ray sources had been identified. One of the strongest sources, Cygnus X-1, was found to vary in time.[2] Many of the sources were situated near the Galactic plane, establishing the bulk of the sources as Galactic objects.[3]

In 1966, an optical counterpart was located[4] at the position of Sco X-1; the optical source is an old 12th–13th magnitude star. The following year, an amusing theoretical model (for that time!) was proposed by Shklovskii (1967) for Sco X-1: the X-ray emission originates from high-temperature gas flowing onto a neutron star from a close binary companion. The companion is an ordinary, cool, dwarf star responsible for the observed optical radiation. This model was a modification of the original idea of Hayakawa and Matsuoka (1964) that gas accretion in close binaries might be a source of X-ray emission. (They had in mind unusual conditions in otherwise normal binary systems.)

Cameron and Mock (1967) criticized the model of Shklovskii, arguing that the observed soft X-rays from Sco X-1 would more likely originate from accretion onto a white dwarf than onto a neutron star. Some details of their proposal were sketched for the case of spherical accretion onto a white dwarf in a binary system.

---

[1] Giacconi, Gursky, Paolini, and Rossi (1962).
[2] Byram et al. (1966); Overbeck and Tananbaum (1968).
[3] Morrison (1967).
[4] Sandage et al. (1966).

Following these developments, Prendergast and Burbidge (1968) argued that gas flowing onto a compact star from a binary companion would have too much orbital angular momentum to flow radially. Instead, the gas would form a thin accretion *disk* around the compact star, with approximately Keplerian circular velocities and a small inward drift velocity. Details were worked out for accretion onto a white dwarf.

Now as early as 1965, Zel'dovich and Guseynov[5] had pointed out that the discovery of X-rays or γ-rays from a *single-line spectroscopic binary* would provide strong evidence for the presence of either a black hole or a neutron star. (A "spectroscopic" binary means that the binary orbital motion is inferred from the Doppler shift of spectral lines. "Single-line" means that spectral lines from only one object are visible.) Clearly a black hole or neutron star could not produce optical spectral lines, and so only one set of lines would be seen.

The possibility that accretion onto black holes would be an efficient way of converting gravitational potential energy into radiation had been foreseen earlier by Salpeter (1964) and Zel'dovich (1964). At the time, however, they were considering accretion onto supermassive black holes ($M \gg 10^8 M_\odot$) to explain the enormous luminosities of the newly discovered quasars.

By 1969, Trimble and Thorne concluded[6] that few, if any, of the single-line spectroscopic binaries known at that time contained unseen companions so massive that they could only be black holes or neutron stars. Furthermore, none coincided with any of the published X-ray source positions at that time.

So in spite of the expanding theoretical effort to explain the newly discovered Galactic X-ray sources, there was no compelling observational evidence at that time that these sources had anything to do with close binary systems or compact objects. But clearly the idea had been planted. The discovery of pulsars in 1967 (Chapter 10) made the notion of neutron stars more credible.

On December 12, 1970, the first astronomy satellite, *Uhuru*, was launched by NASA off the coast of Kenya.[7] This satellite, entirely devoted to X-ray observations in the 2–20 keV band, revolutionized our understanding of cosmic X-ray sources and compact objects. Before its demise in March of 1973 (much later than its design life), it had identified over 300 discrete X-ray sources.[8] Most significant was its positive detection of X-ray emission from *binary stellar systems*,[9] and its discovery of *binary X-ray pulsars*.[10] Most of the Galactic X-ray sources are probably compact objects accreting gas from a nearby, normal companion star.

---

[5] Zel'dovich and Guseynov (1965).

[6] Trimble and Thorne (1969).

[7] "Uhuru" means "freedom" in Swahili; the launch occurred on the anniversary of Kenya's independence.

[8] See, for example, The Fourth UHURU Catalog of X-ray Sources, Forman et al. (1978).

[9] About 100 optical companions have been identified; Bradt and McClintock (1983).

[10] As of 1982, approximately 19 of the X-ray sources now known have been identified as binary X-ray pulsars.

This profound interpretation of the observational data follows from these facts:

1. The variability of the X-ray emission on short timescales implies a small emitting region.
2. Many of the sources are positively confirmed to be in binary systems, with optical primaries orbiting optically invisible secondaries.
3. Mass accretion onto a compact object, especially a neutron star or a black hole, is an extremely efficient means of converting released gravitational potential energy into X-ray radiation.

In general, the list of possible Galactic X-ray source candidates includes all three kinds of compact objects: white dwarfs, neutron stars, and black holes. But in special cases, the specific nature of the compact object can be identified with high probability. Consider, for example, the source Hercules X-1, which exhibits very short, regular pulsations with a 1.24 s period, or the source SMC X-1 in the Small Magellanic Cloud (0.71 s period). The constancy of the period suggests regular motion in a gravitational field. Such motion—rotation, pulsation or orbital motion—is characterized by periods

$$P \gtrsim (G\bar{\rho})^{-1/2}, \qquad (13.1.1)$$

where $\bar{\rho}$ is the mean density in the volume containing the motion (cf. Section 10.2). So for these sources the observed period requires $\bar{\rho} \gtrsim 10^6$ g cm^{-3}, implying the presence of a compact object. Now white dwarfs are unlikely, for although such short periods are possible for massive, rapidly rotating *cold* dwarfs, hot dwarfs with surface temperatures in the keV range would have bloated atmospheres and longer periods. Orbital motion, whether modulating some emission mechanism directly or exciting short-period pulsations, would decay very quickly due to gravitational radiation. Rotating black holes are ruled out, since there are no stationary nonaxisymmetric black holes (cf. Chapter 12), which would be required to produce the regular variability. One is therefore inevitably led to identifying these (and possibly all) pulsating X-ray sources with neutron stars.

Presumably, we see pulses because the X-ray beam pattern is misaligned with the rotation axis of the accreting, magnetized neutron star. The measurement of a spectral feature in Her X-1, which can be interpreted as a cyclotron line, implies a magnetic field strength $B \sim 5 \times 10^{12}$ G near the surface (cf. Exercise 13.2). Such strong fields are consistent with pulsar magnetic fields (cf. Chapter 10), and are not found on the surfaces of white dwarfs.

On the other hand, consideration of the nonperiodic, rapidly varying binary X-ray source Cygnus X-1 leads to a dramatically different conclusion, as we shall document in Section 13.5. The high mass inferred for this source from the combined optical and X-ray data (certainly $M/M_\odot \gtrsim 3$ and probably $9 \lesssim$

$M/M_\odot \lesssim 15$) rules out the possibility of a white dwarf or neutron star. It is this conclusion, together with the compact nature of the source, that leads to the preliminary identification of Cyg X-1 as a black hole. If correct, this finding will undoubtedly be one of the most remarkable discoveries in the history of science.

In the wake of the triumphant Uhuru experiment, approximately 10 additional satellites solely or partially devoted to X-ray observations were launched during the 1970s. They were known by such names as Copernicus, Ariel-5, ANS, SAS-3, OSO-7, and OSO-8, COS-b, HEAO-1, and HEAO-2 (the "Einstein Observatory"), and Hakucho. These satellites served to confirm and reanalyze Uhuru observations, obtain spectral information of selected sources, pinpoint the position of some sources, and expand the list of known sources by detecting fainter, more distant objects. In addition, entire new classes of X-ray sources were discovered during the decade. Most notable of these were the 30 or so X-ray *burst sources*, the first few of which were discovered in the ANS data by Grindlay et al. (1976). These Galactic sources emit nonperiodic bursts of X-rays on timescales of minutes to days. At least nine of them lie in globular clusters. Although no eclipses have been observed for any of these sources, it is generally believed that most of these sources are neutron stars in low-mass binary systems.[11]

In the next few sections we will review some of the features of the Galactic X-ray sources. In no way will our discussion be complete; instead, our intent will be to convey the "flavor" of this relatively new branch of astronomy and how it has already had a revolutionary impact on the theory of compact objects.[12] To appreciate the observations summarized below, we urge the reader to review Appendix A, which highlights the properties of *ordinary*, luminous stars. Such stars form the companions to compact stars in binary X-ray systems.

In Chapter 14, we will discuss the physics of accretion onto compact objects.

13.2 General Characteristics of the Galactic X-Ray Sources

The distribution of all 339 X-ray sources listed in the 4U Catalog is shown in Figure 13.1, which is a plot of source positions in Galactic coordinates. Dramatically clear is the concentration of strong sources along the Milky Way, especially at low Galactic latitude. (Galactic plane corresponds to Galactic latitutde $\equiv 0°$; Galactic center is at Galactic longitude $\equiv 0°$.) The sources at high Galactic latitude are much weaker and many of them can, in fact, be identified with specific extragalactic objects. Exceptions include Her X-1 and Sco X-1, which are

---

[11]See Section 13.6.

[12]More detailed reviews can be found in Gursky and Ruffini (1975), Giacconi and Ruffini (1978) (observations of compact X-ray sources); Bahcall (1978a) and Rappaport and Joss (1981) (X-ray pulsars); Lewin and Joss (1981) (globular cluster sources and burst sources). A recent catalog of the X-ray sky is the Fourth Uhuru Catalog (Forman et al., 1978).

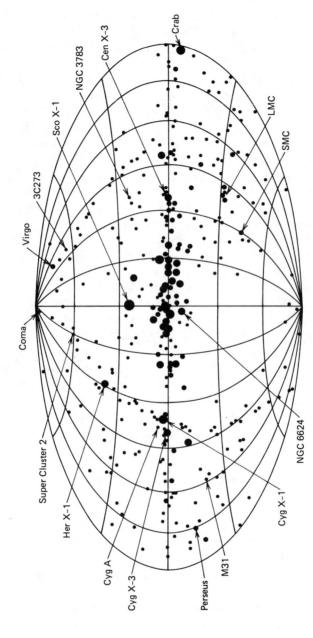

Figure 13.1 The X-ray sky. The sources are displayed in Galactic coordinates. The Galactic plane corresponds to 0° latitude; the Galactic center lies along 0° latitude, 0° longitude. The size of the symbols representing the sources is proportional to the logarithm of the peak source intensity. [Reprinted courtesy of Forman et al. and *The Astrophysical Journal*, published by the University of Chicago Press; © 1978 The American Astronomical Society.]

both Galactic. We are mainly concerned with the *strong* sources here, which constitute the majority of the identified Galactic sources.

Typically, the identified sources reside at distances ranging from 500 pc to 10 kpc. The spread in luminosities is in the range 10^{33}–10^{38} erg s^{-1} from about 2–6 keV.

At least two very different kinds of stellar systems contain known binary X-ray sources. One group of sources with identified optical components is associated with a late O or early B supergiant, a star that is very massive and luminous (cf. Appendix A). These stars are relatively rare; they have lifetimes of $\lesssim 10^7$ yr. Such young, Population I systems are associated with spiral arms and regions of active star formation near the Galactic plane. Examples of Pop I X-ray sources with B0 supergiant companions include Cyg X-1, Cen X-3, and 2U 0900 – 40.

Another group of sources is associated with stars with much later spectral type. Such stars are more like the Sun in luminosity, temperature, and mass. Such low-mass Pop II stars have longer lifetimes than massive O and B stars, so they are typically older and more common. Examples of Pop II X-ray sources include Her X-1, Sco X-1, Cyg X-2, and Cyg X-3.[13]

Most of the 50 X-ray sources that reside within 30 degrees of the Galactic center seem to represent a distinct class of high luminosity, old Pop II sources. Contained in this group are 11 known sources located in globular clusters (at least 9 of which product X-ray bursts). They are commonly referred to as the "Galactic bulge sources" due to their concentration near the Galactic center and will be described below.

X-ray emission from all the sources is found to be highly variable on timescales ranging from milliseconds to years. The observed variability is both random and periodic. The possible black hole candidate, Cyg X-1, is highly variable on timescales down to milliseconds.

13.3   Binary X-Ray Pulsars

Binary X-ray sources displaying *periodic* variations are called *binary X-ray pulsars* (e.g., Her X-1). They are distinguished from aperiodically varying sources such as Cyg X-1.

Much of our information about the Galactic X-ray sources has been derived from studies of binary X-ray pulsars. Knowledge of pulse profiles, pulse periods, and orbital parameters from such systems provides key information regarding the physics of accreting compact stars, especially neutron stars.

---

[13]Actually, the binary nature of Sco X-1, the first such source discovered, was not really substantiated until 1975 [Gottlieb et al. (1975); Cowley and Crampton (1975)].

Table 13.1
Binary X-Ray Pulsars[a]

Source	Pulse Period (s)	Discovery of Pulsation
SMC X-1	0.714	Lucke et al. (1976)
Her X-1	1.24	Tananbaum et al. (1972a)
4U 0115 + 63	3.61	Cominsky et al. (1978)
Cen X-3	4.84	Giacconi et al. (1971)
4U 1626 − 67	7.68	Rappaport et al. (1977)
LMC X-4	13.5	Kelley et al. (1983)
2S 1417 − 62	17.6	Kelley et al. (1981)
OAO 1653 − 40	38.2	White and Pravdo (1979)
A 0535 + 26	104	Rosenberg et al. (1975)
GX1 + 4	122	Lewin et al. (1971); White et al. (1976a)
4U 1320 − 61 } A 1239 − 59 }	191	Huckle et al. (1977)
GX304 − 1	272	McClintock et al. (1977), Huckle et al. (1977)
4U 0900 − 40	283	McClintock et al. (1976)
4U 1145 − 61	292	White et al. (1978)
1E 1145.1 − 6141	297	White et al. (1978), Lamb et al. (1980)
A 1118 − 61	405	Ives, Sanford, and Bell-Burnell (1975)
4U 1538 − 52	529	Davison (1977), Becker et al. (1977)
GX301 − 2	696	White et al. (1976a)
4U 0352 + 30	835	White et al. (1976b)

[a]After Rappaport and Joss (1983).

(a) Pulse Profiles and Periods

The pulse periods of the 19 known binary X-ray pulsars range over three decades from 0.7 s to 835 s; see Table 13.1. The pulse profiles are characterized by[14] (i) large "duty" cycles of $\geq 50\%$ (cf. 3% for typical radio pulsars); (ii) amplitude variations ranging from 25–90%; (iii) a range from symmetric to highly asymmetri shapes; and (iv) no obvious trend in profile morphology as a function of period. For several of the sources (e.g., 4U 1626 − 67, 4U 0900 − 40) the pulse profiles vary significantly with energy, while for others (e.g., Her X-1, Cen X-3) the basic pulse profile is maintained over a large span of energies. Characteristic pulse profiles for 14 of the X-ray pulsars are shown in Figure 13.2.

To date, no theoretical calculations can reproduce the observed pulse shapes shown in Figure 13.2. However, it is generally accepted that the pulse profiles result from the misalignment of the X-ray beam pattern with the rotation axis of

[14]Rappaport and Joss (1981).

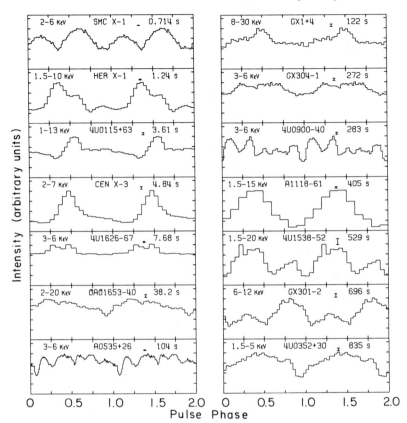

Figure 13.2 Sample pulse profiles for 14 X-ray pulsars thought to be in binary systems. In each case, the data are folded modulo the pulse period (indicated in seconds) and plotted against the pulse phase for two complete cycles. [From Rappaport and Joss (1981). Copyright © 1981 by D. Reidel Publishing Company, Dordrecht, Holland.]

an accreting, magnetized neutron star. This beam, collimated along, for example, the axis of the star's dipole magnetic field, is then viewed in different directions as the star rotates.[15] The X-ray pulse is then determined by the complicated transfer of X-ray photons from the neutron star surface through regions of magnetized, accreting plasma (see Fig. 13.3).

The pulse period histories of a number of X-ray pulsars have been reliably charted for almost a decade. When the periods of the best-measured eight pulsars are plotted vs. time, it is apparent that "spin-up" occurs in at least six (Fig. 13.4). Even for these six sources, the change in pulse period is not always monotonic on short timescales, but the trend toward a secular decrease in pulse period is evident. This secular behavior can be explained in terms of torques exerted on a

[15]Pringle and Rees (1972); Davidson and Ostriker (1973); Lamb, Pethick, and Pines (1973).

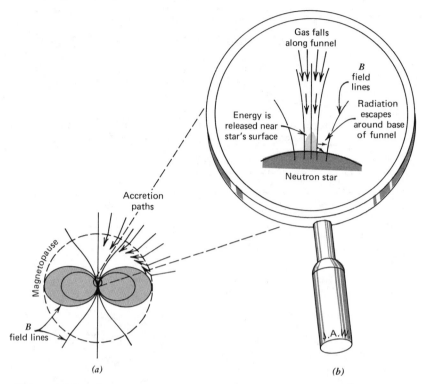

Figure 13.3 (*a*) Schematic dipole magnetosphere around a neutron star that is accreting material. Infalling gas is excluded from the toroidal region whose cross section is *shaded*. (*b*) Enlargement of the base of an accretion funnel, near a magnetic pole of the neutron star. [After Davidson and Ostriker (1973). Reprinted courtesy of the authors and *The Astrophysical Journal*, published by the University of Chicago Press; © 1973 The American Astronomical Society.]

neutron star by the accreting matter, as we will discuss in Section 15.2. In fact, the data provide further evidence in support of neutron stars in binary X-ray pulsars as opposed to white dwarfs.

(b)   Orbits and Masses

Measurements of the pulse arrival times from X-ray pulsars have been employed very successfully to determine the orbits of several of the systems. In fact, there are six sources for which sufficient data—X-ray and optical—exist to estimate the masses of the compact star: Her X-1, Cen X-3, SMC X-1, LMC X-4, 4U 0900 − 40, and 4U 1538 − 52. The method was discussed in Section 9.4 and the results summarized in Figure 9.6.

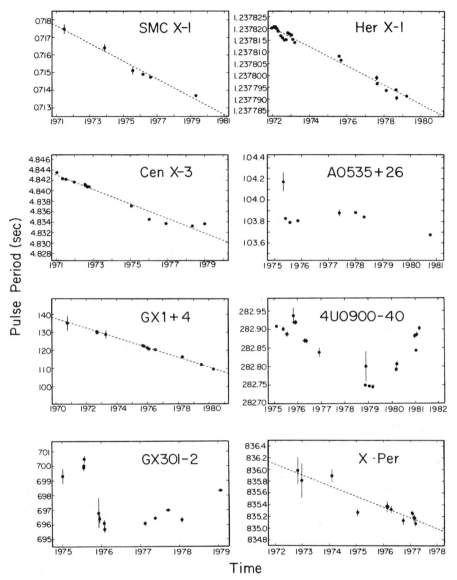

Figure 13.4 Pulse period histories for eight binary X-ray pulsars. The *heavy dots* are individual measurements of pulse period; the vertical bars represent the 1σ uncertainties in the period determination. The data are from the *Uhuru*, Copernicus, Ariel 5, SAS-3, OSO-8, and HEAO-1 satellites, the Apollo–Soyuz Test Project, and a balloon and sounding rocket experiment. The *dashed* lines are minimum chi-squared fits of a straight line to the data points. [From Rappaport and Joss (1983), in *Accretion Driven Stellar X-Ray Sources*, edited by W. H. G. Lewin and E. P. J. van den Heuvel, Cambridge University Press.]

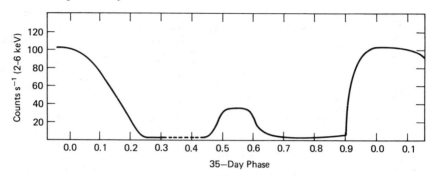

Figure 13.5   Uhuru satellite observations of the Hercules X-1 35-day cycle. Plotted is the envelope of the X-ray light curve of Her X-1 covering the entire 35-day cycle. No observations were available for the portion of the light curve represented with broken lines. Note that 1 Uhuru count s^{-1} = 1.7 × 10^{-11} erg cm^{-2} s^{-1}. [Reprinted courtesy of Jones and Forman and *The Astrophysical Journal*, published by the University of Chicago Press; © 1976 The American Astronomical Society.]

13.4   Hercules X-1: A Prototype Binary X-Ray Pulsar

Since its discovery[16] in the Uhuru data in 1972 Her X-1 has become the most widely studied binary X-ray pulsar. Many of its X-ray properties are similar to those of Cen X-3, which was the first pulsating binary X-ray source to be discovered.[17] In both cases, the observations are interpreted in terms of a mass-transfer binary system in which a rotating neutron star is accreting matter from a companion.

Her X-1 is distinguished by the fact that its X-ray data exhibit *three* periodicities: a pulsation period of 1.24 s, an orbital period of 1.7 days, and an approximate 35-day "on-off" cycle. The estimated mass of the neutron star is $0.4 \leqslant M_x/M_\odot \leqslant 2.2$ and the mass of its companion, HZ Herculis, is $1.4 \leqslant M_{opt}/M_\odot \leqslant 2.8$.[18] The companion exhibits optical pulsations of 1.24 s period; they have been interpreted as the result of the pulsed X-ray emission from Her X-1 being absorbed, thermalized, and re-emitted in the atmosphere of HZ Herculis. The low mass of HZ Herculis is consistent with its spectral type—late A or early F.

This curious 35-day cycle, unique to Her X-1, consists of a 12-day high-intensity state followed by a 23-day low state (see Fig. 13.5). In some theoretical models[19] this 35-day intensity modulation is explained by the *precession* of an accretion disk about the X-ray source. The precession causes the X-ray source to be shadowed from the view of the earth during the extended low state. This model

[16]Tananbaum et al. (1972a).
[17]Giacconi et al. (1971); Schreier et al. (1972).
[18]Bahcall and Chester (1977).
[19]For example, Katz (1973), Roberts (1974), and Petterson (1975).

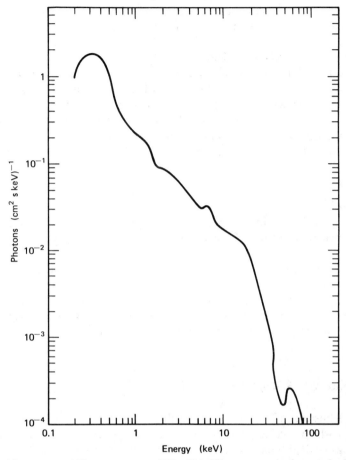

Figure 13.6 Time-averaged X-ray spectrum of Her X-1. The spectrum is characterized by an excess of soft ($E \leq 1$ keV) X-rays, a nearly power-law continuum for $2 \leq E \leq 20$ keV, an iron emission line at $E \approx 7$ keV, a break in the spectrum for $E \geq 20$ keV and structure at $E \approx 50$ keV that may be a cyclotron line feature. [After Holt and McCray (1982). Reproduced, with permission, from the *Annual Review of Astronomy and Astrophysics*, Vol. 20. © 1982 by Annual Reviews, Inc.]

is favored because 1.7-day *optical* modulations, which apparently result from X-ray heating, persist throughout the 35-day cycle.

The average Her X-1 X-ray spectrum from 1–40 keV is shown in Figure 13.6. The continuum is relatively flat below ~ 24 keV [$dL(E)/dE$ ~ constant, where L is the luminosity and E the energy], but falls abruptly above this energy. If we adopt an approximate though reasonable distance of ~ 4 kpc[20] the total luminosity from Her X-1 is ~ 10^{37} erg s^{-1} from 2–24 keV.

[20] Davidsen and Henry (1972).

Exercise 13.1 Show that the above luminosity is far too high to be furnished by the loss of *rotational* kinetic energy of a neutron star.

There exists some recent evidence[21] of a spectral line in the pulsed, hard X-ray spectrum of Her X-1 near 58 keV. The most plausible explanation for this line is that it represents an electron cyclotron absorption or emission feature, which would imply a surface magnetic field of $B \sim (4-6) \times 10^{12}$ G. If correct, this observation would strengthen the identification of Her X-1 as a neutron star. A comparable feature has been observed[22] in the pulsed high-energy spectrum of the 3.6 s source 4U 0115 − 63 at 20 keV, indicating a field of $B \sim 2 \times 10^{12}$ G for that source.

Exercise 13.2

(a) Assuming that the line observed at 58 keV for Her X-1 results from emission when an electron makes a transition from the first excited Landau level to the ground state, estimate B.

Hint: $E_{n,q} = \mu_B B(2n + 1) + q^2/2m_e$, where μ_B = Bohr magneton, $n = 0, 1, 2, \ldots$, and q = momentum along B.

(b) Assuming that the observed line width of ≤ 12 keV results from thermal broadening, estimate T, the gas temperature of the line-emission region. Assume a "typical" viewing angle of the source with respect to the magnetic field.

13.5 Cygnus X-1: A Black Hole Candidate

Cygnus X-1 has played a central role in the study of compact X-ray sources. It is the source with the widest range of time variability; it was the first source to be located in an optical binary system; it is the source most likely to be a black hole. It was the rapid X-ray variability that triggered the search in the Uhuru data for other variable sources, leading to the discovery of binary X-ray pulsars.[23]

Cygnus X-1 is variable on all timescales varying from months and years down to milliseconds. The most dramatic variability is the 1-ms bursts,[24] which set a maximum size for the X-ray source of $R \lesssim ct \sim 300$ km and establish the object to be highly compact (see Fig. 13.7).

The optical star in the Cyg X-1 system is the 5.6-day period single-line spectrocopic binary known as HDE 226868, a 9th magnitude supergiant star. The identification of HDE 226868 with Cyg X-1 was originally established by the sudden appearance in March 1971 of a radio source in the direction of Cyg X-1,

---

[21] Trümper et al. (1978).
[22] Wheaton et al. (1979).
[23] For a comprehensive discussion of Cyg X-1 and a list of references, see Oda (1977).
[24] Rothschild et al. (1974).

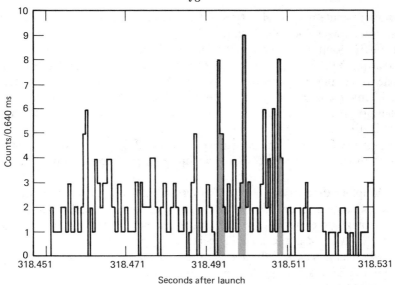

Figure 13.7 Time variability of Cygnus X-1 exhibiting several "bursts" of millisecond duration. Shown here are rocket flight observations near 318 s after launch containing eighty milliseconds of exposure to Cyg X-1. The count rates are binned every 0.640 ms. Bursts with more than 12 counts per 1.28 ms are shaded. [Reprinted courtesy of Rothschild et al. and *The Astrophysical Journal*, published by the University of Chicago Press; © 1974 The American Astronomical Society.]

and a simultaneous variation in the X-ray intensity.[25] The small error box for the radio source ($\lesssim 1''$) contained the optical star HDE 226868, a binary with an invisible companion. Later, a 5.6-day modulation in the X-ray intensity was detected, confirming the identification.[26]

Unfortunately, the X-ray data for Cyg X-1 do not furnish as much information to determine the orbital parameters as they do for the binary X-ray pulsars. No regular X-ray pulsations have been observed, and no definite eclipse has been detected.

Nevertheless, a lower limit to the mass of Cyg X-1 of greater than $3M_\odot$ can be obtained in two independent ways.[27] Both methods require a reliable distance determination. This is obtained from the interstellar absorption observed in the spectrum of HDE 226868. If one knew the absolute magnitude of the star, then comparison with the apparent magnitude (corrected for absorption) would give the distance [cf. Eq. (A.4)]. Although an OB-supergiant like HDE 226868 gener-

[25] Hjellming (1973); Tananbaum et al. (1972b).
[26] Holt et al. (1976).
[27] See Bahcall (1978c) for a careful discussion and references.

ally has an absolute magnitude $M_v \sim -6$, there are stars with very similar spectra of much lower mass and luminosity.[28]

Instead of assuming a luminosity for HDE 226868, one can survey a large number of stars in the same direction and compare absorption with distance for each one assuming the spectrum does in fact give the intrinsic luminosity. Although any *individual* distance determination might be wrong, by having a large sample one gets a reasonably accurate calibration of absorption versus distance in that direction. Two independent studies,[29] each with about 50 stars within 1° of HDE 226868, gave distances of order 2.5 kpc for HDE 226868, with absolute minimum distances of 2 kpc in order to produce the observed absorption.

The Doppler curve of the spectrum of HDE 226868 gives information about the orbital elements, as in Section 9.4. Gies and Bolton (1982) obtain: $P = 5.60$ days, with negligible error for our purposes, $a_1 \sin i = (5.82 \pm 0.08) \times 10^6$ km, $f = (0.252 \pm 0.010) M_\odot$, and a small eccentricity, probably less than 0.02.

Now whereas a *typical* OB supergiant has a mass greater than $20 M_\odot$, stellar structure calculations[30] show that one requires the mass to be *at least* $8.5 M_\odot$ to have the right luminosity at a distance $d \sim 2$ kpc. Since

$$f = \frac{(M_2 \sin i)^3}{(M_1 + M_2)^2}, \tag{13.5.1}$$

where 1 refers to the optical star and 2 to the X-ray source, we get a minimum value for M_2 by setting $\sin i = 1$. This gives

$$M_2 \gtrsim 3.3 M_\odot. \tag{13.5.2}$$

More compelling is an argument[31] that sets a lower limit using only the absence of a prominent X-ray eclipse and assuming *nothing* about the mass of the optical star. The absence of eclipses implies (see Fig. 13.8) that

$$\cos i > \frac{R}{a}, \tag{13.5.3}$$

where R is the radius of the optical star (taken to be spherical) and a is the separation of the two objects. Using Eq. (9.4.5), we can write this as

$$\cos i \geqslant \frac{R}{a_1 \sin i} \frac{M_2 \sin i}{M_1 + M_2}. \tag{13.5.4}$$

---

[28] Compare Trimble, Rose, and Weber (1973).
[29] Margon et al. (1973); Bregman et al. (1973).
[30] For example, Kippenhahn (1969).
[31] Paczynski (1974); compare also Mauder (1973) and Bahcall (1978c).

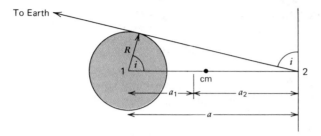

Figure 13.8 No eclipse condition for Cyg X-1.

Hence

$$M_2 \sin i \cos^2 i \geqslant \frac{fR^2}{(a_1 \sin i)^2}. \qquad (13.5.5)$$

Now $\sin i \cos^2 i$ has a maximum value of $2/3\sqrt{3}$ when $\sin i = 1/\sqrt{3}$, and so

$$M_2 \geqslant \frac{3\sqrt{3} fR^2}{2(a_1 \sin i)^2}. \qquad (13.5.6)$$

The quantity R^2 can be found from the luminosity and effective temperature:

$$R^2 = \frac{L}{4\pi\sigma T_e^4}. \qquad (13.5.7)$$

Combining Eqs. (A.4), (A.5), and (A.8), we have

$$R^2 = \frac{L_\odot}{4\pi\sigma T_e^4} \left(\frac{d}{1 \text{ kpc}} \right)^2 10^{4+0.4(4.72-m_v-BC+A_v)}. \qquad (13.5.8)$$

Here m_v is the apparent magnitude in the V (visual) energy band, BC is the bolometric correction, and A_v is the absorption in the V band. Bregman et al. (1973) found $V(\equiv m_v) = 8.87$, and a color index $B - V = 0.81$. Now the precise spectral classification[32] of HDE 226868 implies[33] that the intrinsic color corrected for absorption is $(B - V)_0 = -0.30$. The "color excess," due to the preferential absorption of short wavelength optical radiation by interstellar gas, is

$$E(B - V) \equiv B - V - (B - V)_0 = 1.11. \qquad (13.5.9)$$

---

[32] It is classified as an O9.7 Iab star by Walborn (1973).
[33] See, for example, Harris (1963).

The total absorption in the V band, A_v, is then given by

$$A_v = 3.0E(B - V) = 3.3. \qquad (13.5.10)$$

The factor of 3.0 in Eq. (13.5.10) is an empirical result of detailed studies of interstellar absorption[34] and represents a fairly reliable lower limit,[35] which is what we require.

The quantities BC and T_e for supergiant stars are rather uncertain, as they are obtained largely from theoretical calculations and, until the work of Code et al. (1976), with little direct observational check. Conti (1978) incorporates the results of Code et al. and gives for our case $T_e = 30,000$ K, $BC = -2.9$. Fortunately, since $-BC$ decreases as T_e decreases, the radius of the star R in Eq. (13.5.8) is not particularly sensitive to uncertainties in T_e and BC. Conti's values give

$$R^2 = (6.62 \times 10^6 \text{ km})^2 \left(\frac{d}{1 \text{ kpc}} \right)^2, \qquad (13.5.11)$$

and hence [36] from Eq. (13.5.6),

$$M_2 \geqslant 3.4 M_\odot \left(\frac{d}{2 \text{ kpc}} \right)^2. \qquad (13.5.12)$$

Exercise 13.3 From model atmosphere calculations of Auer and Mihalas (1972), Mauder (1973) has estimated that the surface gravity of HDE 226868 is

$$g \equiv \frac{GM}{R^2} = 1.6 \times 10^3 \text{ cm s}^{-2}. \qquad (13.5.13)$$

Use f and R from Eqs. (13.5.1) and (13.5.11) to set a lower limit on M_2.
   Answer: $M_2 \geqslant 5.7 M_\odot$.

Exercise 13.4 (based on Paczynski, 1974) Another restriction one may wish to place on the binary system is that the optical star not overflow its Roche lobe (cf. Section 13.7). In other words,

$$R \leqslant R_{\text{Roche}}. \qquad (13.5.14)$$

---

[34]Spitzer (1978), Ch. 7.
[35]Seaton (1979).
[36]Paczynski's (1974) values for T_e and BC would have given a coefficient of 3.7 in Eq. (13.5.12).

A good approximation in the mass range we are interested in is[37]

$$R_{Roche} = a\left[0.38 + 0.2\log\left(\frac{M_1}{M_2}\right)\right],\qquad(13.5.15)$$

where a is given by Eq. (9.4.5).

(a) Show that the minimum M_2 is given by finding the minimum of

$$M_2 = \frac{(1+q)^2 f}{\sin^3 i},\qquad(13.5.16)$$

subject to the constraints (13.5.14) and (13.5.4):

$$f_1 \equiv \frac{1+q}{\sin i}(0.38 + 0.2\log q) \geqslant x,\qquad(13.5.17)$$

$$f_2 \equiv \frac{1+q}{\sin i}\cos i \geqslant x,\qquad(13.5.18)$$

where

$$q \equiv \frac{M_1}{M_2}, \qquad x \equiv \frac{R}{a_1 \sin i}.\qquad(13.5.19)$$

(b) Varying q and $\sin i$, show that the minimum occurs when equality holds in the expressions (13.5.17) and (13.5.18). Evaluate the minimum for $d = 2$ kpc and $d = 2.5$ kpc.

(c) Show that this also gives a minimum to M_1, and evaluate M_1 numerically.

Answer: $M_2 \geqslant 6.5 M_\odot$ ($d = 2$ kpc); $M_2 \geqslant 9.5 M_\odot$ ($d = 2.5$ kpc)

We can summarize the situation as follows: the lower limit (13.5.12) of $3.4 M_\odot$ is very solid. Adopting the more reasonable value of $d \sim 2.5$ kpc increases this to $5.3 M_\odot$. Various other less rigorous but more realisitic arguments give even bigger masses.

Now even the most rapidly rotating, stable white dwarf has a maximum mass of only about $2.5 M_\odot$ [cf. Eq. (7.4.41)]. The maximum mass of a stable neutron star is of order $3 M_\odot$ (Chapter 9)—a fact that depends only on the validity of general relativity, not on an assumed equation of state. Likely maximum neutron star masses are closer to $2 M_\odot$ for currently fashionable stiff equations of state. We are therefore led to a profound conclusion: Cyg X-1 is most likely a black hole! The argument is summarized in Box 13.1.

The case for a black hole is strengthened by additional constraints implied by observed optical light variations. Avni and Bahcall (1975a, b) conclude that the

[37]Paczynski (1971).

mass of Cyg X-1 is *likely* to be in the range $9-15 M_\odot$. This value is well above *any* upper limit we have derived for white dwarfs and neutron stars.

Devil's advocates have proposed alternative models for Cyg X-1 that do not involve black holes. One such model[38] requires a *third* body in the system, so that an X-ray emitting neutron star orbits one or both members of a massive binary system of normal stars. Another model[39] postulates an exotic energy source other than gas accretion onto a compact star.

A third body would perturb the optical light curve from HDE 226868. Observational upper limits[40] on such perturbations place severe constraints on triple system models, but they remain possible in principle.[41]

Box 13.1

The Case for Cygnus X-1 as a Black Hole

1. Cyg X-1 is a compact object (variability timescale $\Delta t \sim 1-10$ ms $\Rightarrow$ emission zone size $R \leq c\,\Delta t \leq 10^8$ cm).

2. Cyg X-1 is the unseen companion of HDE 226868 (simultaneous radio and X-ray intensity transition + 5.6-day period for binary and for soft X-ray variability).

3. Cyg X-1 has a mass $M_x > 3.4 M_\odot$ (optical mass function + spectral type + absence of eclipses + distance).

Further support for the black hole model comes from the details of the X-ray spectrum. The source undergoes frequent transitions from a "low" state, where it spends about 90% of its time, to a "high" state.[42] During the low state, the source emits primarily a hard X-ray component, characterized by a single power-law spectrum over the range 1–250 keV:

$$\frac{1}{4\pi d^2}\frac{dL}{dE} = AE^{-\alpha}. \tag{13.5.20}$$

---

[38] Bahcall et al. (1974); Fabian et al. (1974).

[39] Bahcall et al. (1973).

[40] Abt et al. (1977).

[41] When asked to discuss the Cyg X-1 situation, E. E. Salpeter concluded, "A black hole in Cyg X-1 is the most *conservative* hypothesis."

[42] For a summary of the observations, see Dolan et al. (1979).

Here L is the luminosity, E the energy, d is the distance, and A and α are constants. The constant α is in the range 0.5–1 for integration times longer than a few seconds.

The high state sets in on a timescale of days and lasts for about a month. A new, intense, very soft spectral component appears in the 3–6 keV band. The spectral intensity above 10 keV decreases somewhat, retaining the power-law form (13.5.20) with $\alpha = 1.2$ (Fig. 13.9). The total luminosity in the high state increases by a factor of about 2 to over 6×10^{37} erg s^{-1}.

The observed X-ray spectrum from Cyg X-1 is reasonably well modeled by emission from a gaseous accretion disk around a central black hole (cf. Sections 14.5 and 14.6). A great triumph of such models is that they can reproduce both the soft and hard components of the spectrum in either the "high" or the "low" state. No models to date, however, can reliably explain the peculiar transitions between the two states. The long time scales characterizing the transitions suggest an association with changes in the mass transfer rate between primary and secondary.

Exercise 13.5 Use the power-law fits to the soft and hard X-ray data for Cyg X-1 as shown in Figure 13.9 to calculate:

   (a) the soft X-ray (1–20 keV) luminosity L in erg s^{-1} when Cyg X-1 is in the high and low states, respectively;

   (b) the hard X-ray (20–250 keV) luminosity in the high and low states;

   (c) the total luminosity from 1–250 keV in both states. Assume a distance of 2.5 kpc.

   Note: Preliminary γ-ray balloon observations suggest that the hard X-ray power-law fall-off may continue at least up to 3 MeV in the low state; Mandrou et al. (1978).

There are other promising black hole candidates besides Cyg X-1. Circinus X-1 and OAO 1653 $-$ 40 are the best of these at present. Cir X-1 is an Uhuru binary X-ray source with rapid time variability, much like Cyg X-1.[43] The faintness of the red companion has made it impossible so far to observe spectral lines and determine orbital parameters. The eclipsing X-ray source OAO 1653 $-$ 40, detected by the Copernicus satellite, has been identified with the single-line spectroscopic binary V861 Scorpii. Orbital parameters suggest a large mass $7–11M_\odot$ for the X-ray source.[44]

13.6 Galactic Bulge Sources: Bursters

Most of the 50 or so X-ray sources within 30° of the direction of the Galactic center share the following traits[45]: (i) soft X-ray spectra; (ii) X-ray variability on

---

[43] For a review and discussion of Cir X-1 observations, see Dower et al. (1982).
[44] Polidan et al. (1978).
[45] See Lewin and Clark (1980) for a full discussion.

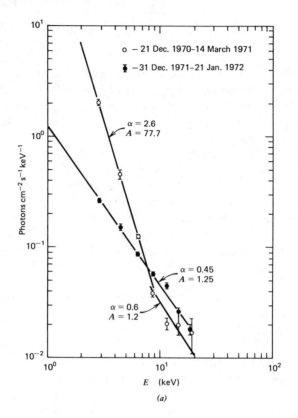

(a)

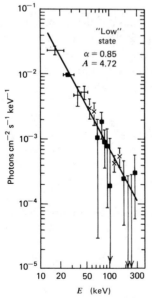

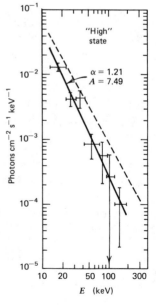

(b)

390

timescales of minutes to days; (iii) $L_x \lesssim 10^{36}$ erg s^{-1}; (iv) no periodic pulsations; (v) no eclipses. Roughly half the sources produce X-ray bursts. Together, they are referred to as Galactic bulge sources, because they are distributed like the older stellar population that comprises the Galactic bulge (Population II). At least 11 sources with the same traits (i)–(v) reside in (old) globular star clusters, and of these at least 9 emit X-ray bursts.[46] The similarity suggests that bulge and cluster sources constitute a class of old Population II X-ray sources, distinct from the binary X-ray sources associated with massive, young Population I stars. More generally, the bulge and cluster sources may be identified with the older "spheroid" component of the Galaxy, while the X-ray sources orbiting massive young stars may be identified with the younger "disk" component.

Since their discovery in 1975 by Grindlay et al.,[47] over 25 burst sources have been detected. The most common bursts, called [48] "Type I," come in intervals of hours to days, and generally their spectra soften as the burst decays. Representative pulse profiles are shown in Figure 13.10.

More peculiar are the so-called "Type II" bursts that arise from the Rapid Burster (MXB 1730 − 335; MXB stands for MIT X-Ray Burst Source). These bursts occur in rapid succession on timescales of seconds to minutes and show very little spectral softening. The energy in each burst is roughly proportional to the waiting time to the next burst. To complicate matters, the Rapid Burster also produces Type I bursts at intervals of 3–4 hours.

Many models have been proposed to explain the bursts. All invoke gas accretion onto a compact object. The compact object serves at least one of two

---

[46]Grindlay (1981). Recent analysis of the Einstein Observatory data by Grindlay indicates that a second class of X-ray sources with lower luminosities ($10^{34} \lesssim L_x \lesssim 10^{36}$ erg s^{-1}) resides in globular clusters. These sources are not concentrated near the cluster centers like the high-luminosity sources. A possible interpretation is that the low-luminosity sources are low-mass binary systems containing white dwarfs and the high-luminosity sources are higher-mass binaries containing neutron stars.

[47]Grindlay et al. (1976), but also see Belian, Conner, and Evans (1976).

[48]Hoffman et al. (1978).

Figure 13.9 (*a*) The average soft X-ray ($2 \leq E \leq 20$ keV) photon spectra for Cyg X-1 before and after the 1971 March and April transition from a "high" to "low" state. Power law fits to the corresponding luminosity spectra [see Eq. (13.5.20) with E in keV, d in cm, and dL/dE in keV s^{-1} keV^{-1}] are indicated. [After Tananbaum et al. (1972b). Reprinted courtesy of the authors and *The Astrophysical Journal*, published by the University of Chicago Press; © 1972 The American Astronomical Society.] (*b*) The average hard X-ray ($15 \leq E \leq 250$ keV) photon spectra for Cyg X-1. The "low"-state (as defined at lower energies) spectrum is plotted on the left and the "high"-state spectrum on the right. The *dashed* line on the right is the power law that fits the low-state spectrum shown on the left. Power law fits to the corresponding luminosity spectra (in the same units as in Fig. 13.9*b*) are indicated. [After Dolan et al. (1979). Reprinted courtesy of the authors and *The Astrophysical Journal*, published by the University of Chicago Press; © 1979 The American Astronomical Society.]

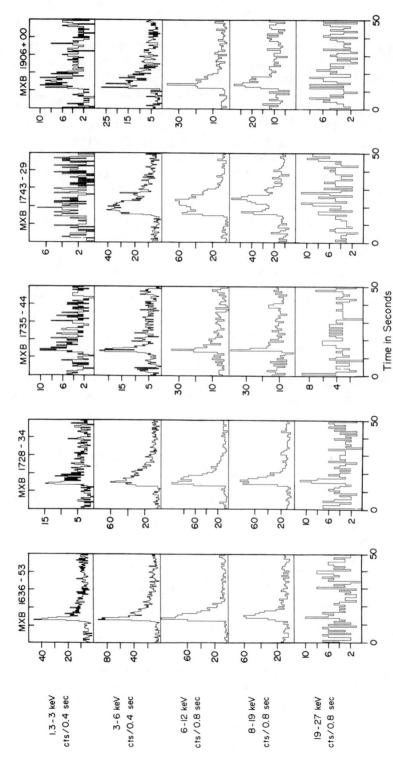

Figure 13.10 Profiles of Type-I X-ray bursts from five different sources in five different energy channels. The burst profiles are reasonably distinctive for each particular source. Note that in all cases the gradual decay (tail) persists longer at lower energies than at higher energies. [From Lewin and Joss (1977). Reprinted by permission from *Nature*, Vol. 270, No. 5634, p. 214. Copyright © 1977 Macmillan Journals Limited.]

functions: the deep gravitational potential well provides an energy store for efficient conversion into radiation; and the small size is consistent with the observed rapid time variability.

The suggested models fall into two categories[49]: (i) those involving instabilities in the accretion flow; and (ii) those invoking thermonuclear flashes in the surface layers of an accreting neutron star. It is currently fashionable to identify all Galactic bulge sources as compact objects of around a solar mass, probably neutron stars, accreting matter from low-mass binary companions. Type I bursts are possibly the result of thermonuclear flashes, while Type II bursts from the Rapid Burster may result from instabilities in the accretion flow.

The absence of X-ray pulsations and eclipses casts some doubt on the identification of bursters as accreting neutron stars. Pulsations are expected from such objects provided they possess magnetic fields sufficiently strong to funnel the accretion flow and misaligned with the rotation axis, as in the case of the binary X-ray pulsars associated with Population I stars. Possibly the bursters are *older* neutron stars: their magnetic fields may have decayed away[50] or may have aligned themselves with the rotation axis. It is also possible that the channeling of gas by the strong magnetic fields in young neutron stars may suppress the thermonuclear flash phenomenon, thus explaining why binary X-ray pulsars are not bursters.

The absence of eclipses[51] in all the well-studied sources is more troublesome. Some accretion disk models,[52] however, permit an "X-ray shadow" in the orbital plane of a low-mass binary, thereby eliminating otherwise detectable eclipses.

Because of their very high frequency of occurrence and association with the oldest stellar population, the globular cluster X-ray sources are possibly quite a distinct class.[53] Of the 150 or so known Galactic X-ray sources, 7% are in globular clusters. However, the total mass of stars in globular clusters is only about 0.1% of the Galaxy mass.

The deep potential well of a centrally condensed cluster would be able to retain gas ejected from old stars. This motivated in part the idea that a supermassive central black hole ($100-1000M_\odot$) might accrete the gas and power the observed source.[54] This idea was also influenced by the theoretical result that the cores of globular clusters undergo secular "collapse" to a state of high stellar density, which might lead to star collisions, coalescenses, and the ultimate formation of a supermassive black hole.[55]

The tendency toward energy equipartition in star clusters (via gravitational scattering) means that the massive stars acquire lower velocities and so are

---

[49] See, for example, Lewin and Joss (1981) for a more detailed review.
[50] See Exercise 10.3.
[51] But see discussion at the end of Section 13.6.
[52] For example, Milgrom (1978).
[53] Katz (1975).
[54] Bahcall and Ostriker (1975); Silk and Arons (1975).
[55] See, for example, Spitzer (1975) or Lightman and Shapiro (1978) for an extensive discussion.

confined to the centermost parts of the cluster. Data[56] from the Einstein X-ray observatory determined the X-ray positions of eight globular cluster sources with an accuracy of $1-3''$. The results, which showed several sources more than $1''$ away from the centers, argue against supermassive black holes, and are consistent with objects of mass around $2 M_{\odot}$.[57] The issue is not yet fully resolved.

Very recently, absorption dips with a period of 50 ± 0.5 min have been observed in the burster 4U $1916-05$.[58] In addition, a 4.3-hr modulation in the optical flux from the burst source[59] MXB $1735-44$ and a 4-hr modulation from the burst source[60] MXB $1636-53$ have been detected. Although preliminary, these periodic variations may be the first direct evidence for the binary nature of X-ray bursters.

13.7   The Standard Model: Accretion in a Close Binary System

(a)   Simple Estimates

We have summarized above some of the observational evidence that the Galactic X-ray sources are compact objects accreting gas in binary systems. In this picture, matter falling onto a compact star releases gravitational potential energy, which heats the gas and generates radiation. It is remarkable that the most naive description of this process demonstrates just how natural and efficient a mechanism this is for generating the observed X-ray emission.

Consider gas accreting steadily at a rate $\dot{M}$ onto a neutron star of mass M_x and radius R_x. Neglect magnetic fields and assume that the flow is spherical, with the gas moving at the freefall velocity v_{ff} until it reaches the "hard" stellar surface. The gas will be decelerated abruptly at the surface and its infall kinetic energy will be converted into heat and radiation. In steady state, the emergent luminosity L_x will be

$$L_x = \tfrac{1}{2}\dot{M}v_{\text{ff}}^2 = \frac{G\dot{M}M_x}{R_x}. \tag{13.7.1}$$

Thus the *efficiency* of radiative emission, in units of the available rest-mass energy incident on the neutron star, is

$$\varepsilon = \frac{L_x}{\dot{M}c^2} = \frac{GM_x}{R_x c^2}. \tag{13.7.2}$$

So for typical neutron stars, the efficiency is quite high: $\varepsilon \sim 0.1$.

---

[56] Grindlay (1981).
[57] Lightman et al. (1980).
[58] Walter, White, and Swank (1981); White and Swank (1982); Walter et al. (1982).
[59] McClintock and Petro (1981).
[60] Pedersen et al. (1981).

Now suppose that the bulk of the radiation is emitted thermally as blackbody radiation from the surface at temperature T_{bb}:

$$L_x = 4\pi R_x^2 \sigma T_{bb}^4. \qquad (13.7.3)$$

For observed soft X-ray source luminosities of 10^{37} erg s^{-1}, Eq. (13.7.3) gives $T_{bb} \sim 10^7$ K for typical neutron star radii. Thus accreting neutron stars with luminosities of 10^{37} erg s^{-1} are natural emitters of *X-ray* radiation (10^7 K $\sim$ 1 keV).

From Eq. (13.7.1), we see that the average accretion rate required to generate this luminosity is $\dot{M} \sim 10^{-9} M_\odot$ yr^{-1}. Such rates are not difficult to attain in close binary systems (see Section 13.7b).

Exercise   13.6   Repeat the above naive calculations for spherical accretion onto a white dwarf and comment on the differences.

Finally, let us consider why the *steady* X-ray luminosities of the observed Galactic sources generally fall in the neighborhood of 10^{37} erg s^{-1} and never exceed a few times 10^{38} erg s^{-1}.[61] Continuing with our simple model, let us compute the upward force exerted on the infalling matter, taken to be ionized hydrogen, by the radiation flux from the surface. Assume the force is due to Thomson scattering off the electrons, which then communicate the force to the protons by their electrostatic coupling. Suppose further that the mean momentum of an emitted photon is p. Since the Thomson cross section is forward–backward symmetric, a photon of momentum p deposits on the average a momentum p to the electron per collision. The energy of the typical photon is pc, and if all the photons are moving radially, then the number of photons crossing unit area in unit time at radius r is $L_x/(4\pi r^2 pc)$. We get the number of collisions per electron per unit time by multiplying by $\sigma_T = 0.66 \times 10^{-24}$ cm^2, the Thomson cross section. The force per electron is just the rate at which momentum is deposited per unit time, so we multiply by p to obtain

$$F_x = \frac{L_x \sigma_T}{4\pi r^2 c}. \qquad (13.7.4)$$

Note that this expression holds even if the photons are not all streaming radially: only the component p_r of the momentum is transferred radially per collision, but since the energy flux $L_x/4\pi r^2$ is radial the number of photons crossing unit area in unit time is $L_x/4\pi r^2 p_r c$.

---

[61]Of those Galactic sources whose distances can be reliably estimated, the X-ray pulsar in the Small Magellanic Cloud, SMC X-1, has the highest mean X-ray luminosity: $L_x \approx 5 \times 10^{38}$ erg s^{-1}; Primini et al. (1976).

In order for accretion to occur, gravity must exceed the photon force (13.7.4). The gravitational force per electron is communicated via the proton:

$$F_{\text{grav}} = \frac{GM_x m_p}{r^2}.$$ (13.7.5)

Since (13.7.4) and (13.7.5) both scale is $1/r^2$, there is a critical luminosity, the *Eddington limit*,[62] above which radiation pressure exceeds gravity:

$$L_{\text{Edd}} = \frac{4\pi c G M_x m_p}{\sigma_T} = 1.3 \times 10^{38} \left(\frac{M_x}{M_\odot}\right) \text{ erg s}^{-1}.$$ (13.7.6)

Exercise 13.7   Show that the same limit (13.7.6) applies for the luminosity of massive stars supported in hydrostatic equilibrium by radiation pressure. Assume that the optical depth is sufficiently large that the radiation flux may be calculated in the diffusion approximation, Eq. (4.1.1), and that Thomson scattering is the dominant opacity source.

In spite of the oversimplification in the above argument (i.e., spherical symmetry, steady state, Thomson scattering, pure ionized hydrogen, Newtonian gravity[63]), it is interesting that 36 Uhuru sources with known or estimated distances exhibit a 2–10 keV luminosity cut-off at $\sim 10^{37.7}$ erg s^{-1}, in agreement with Eq. (13.7.6) for a $1M_\odot$ star.[64] This lends further support to the model of gas accretion onto compact stars for Galactic X-ray sources. Smaller luminosities merely mean that the source is "starved" for fuel.

We shall return to the accretion mechanism in more detail in the next two chapters, and shall consider white dwarfs and black holes as well as neutron stars.

Similar naive considerations facilitate the identification of the burst sources. The tails of observed burst spectra fit a blackbody curve reasonably well. If the measured flux is F_x and the temperature T_{bb}, then since $F_x = L_x/4\pi d^2$ (assuming isotropic emission and a distance d), Eq. (13.7.3) gives

$$R_x = d\left(\frac{F_x}{\sigma T_{\text{bb}}^4}\right)^{1/2}.$$ (13.7.7)

---

[62] Eddington (1926).

[63] Note that in a strong gravitational field, F_{grav} is larger by a relativistic correction factor $(1 - 2GM_x/rc^2)^{-1/2}$. Accordingly, the *local* Eddington limit L_{Edd} is larger by the same factor. In the case of spherical accretion, radiation pressure will still dominate gravity at large r whenever L exceeds the value given by Eq. (13.7.6), so this value is still the maximum observable luminosity in steady state. Alternatively, for radiation from a thin atmosphere heated from below the surface of a hot star, the critical luminosity measured locally is larger than Eq. (13.7.6) by $(1 - 2GM_x/R_x c^2)^{-1/2}$, where R_x is the (Schwarzschild) stellar radius. This implies a maximum observable luminosity at *infinity* that is smaller than Eq. (13.7.6) by a factor $(1 - 2GM_x/R_x c^2)^{1/2}$ (cf. Exercise 11.10).

[64] Margon and Ostriker (1973).

For one particular burster, Swank et al. (1977) assumed a distance of 10 kpc and derived a radius of about 100 km during the first 10 s of the burst and about 15 km later in the burst. Van Paradijs (1978) studied 10 SAS-3 bursters with good spectral data. He assumed that the peak luminosity provided a "standard candle" for bursters (i.e., it is assumed to be the same for all sources and so $d \propto F_{\max}^{-1/2}$). Then Eq. (13.7.7) implies that

$$R_x \propto \left(\frac{F_x}{F_{\max}} \right)^{1/2} \frac{1}{T_{bb}^2}. \qquad (13.7.8)$$

For each source, the combination of observed quantities in the expression (13.7.8) did not change significantly during a burst. This strongly suggests that one is observing a cooling surface of constant size R_x in the tail of the burst. Moreover, the quantity in expression (13.7.8) was nearly the same from source to source, with a standard deviation of about 20%. This implies that all compact objects giving rise to bursts have roughly the same size.

Assuming that the peak luminosity is the Eddington luminosity, Eq. (13.7.6), for a mass of $1.4M_\odot$, Van Paradijs found a mean value of $R_x \sim 7$ km with a scatter of 20%. Relativistic corrections and modifications to the blackbody spectra due to scattering may alter the above conclusions somewhat.[65] However, even with a generous allowance for errors in the above numbers, they strongly suggest that neutron stars are involved, and not white dwarfs or supermassive black holes with $M \gtrsim 100M_\odot$.

Exercise 13.8

(a) Define the "observed blackbody radius" of a burster to be

$$R_\infty \equiv d \left(\frac{F_\infty}{\sigma T_\infty^4} \right)^{1/2}. \qquad (13.7.9)$$

Use Exercise (11.10) and Eq. (13.7.3) to show that ($c = G = 1$)

$$R_\infty = R \left(1 - \frac{2M}{R} \right)^{-1/2}, \qquad (13.7.10)$$

where R is the actual radius in Schwarzschild coordinates.

(b) Now suppose that $2M < R < 3M$. Then only photons within an angle ψ of the radial direction escape to infinity, where ψ is given in Exercise (12.14). Show that for isotropic emission, the fraction of flux that escapes is $\sin^2 \psi$, and hence that

$$R_\infty = 3\sqrt{3}\, M, \qquad 2M < R < 3M. \qquad (13.7.11)$$

---

[65] See the discussion above and also Goldman (1979) and Van Paradijs (1979), (1980); note that $2GM/Rc^2 \approx 0.6(M/1.4M_\odot)(R/7 \text{ km})^{-1}$.

Are neutron stars with radii in this range allowed?

   Hint: See Exercise 5.5.

It is also interesting that the ratio α of the time-averaged energy in the persistent (though variable) X-ray flux to that emitted in typical Type I bursts is $\gtrsim 100$. This value is consistent with the idea that the Type I bursts are produced by thermonuclear flashes of, for example, He or heavier elements on the surfaces of *neutron stars*. From Eq. (13.7.2), we see that the gravitational potential energy liberated at the neutron star surface and released as persistent emission should be $\varepsilon m_B c^2 \sim 100$ MeV per accreted nucleon. The energy released by nuclear reactions involving He and heavier elements and released as burst emission is $\sim 0.1\% \, m_B c^2 \sim 1$ MeV per accreted nucleon. Thus thermonuclear flash models predict[66]

$$\alpha \simeq 100 \frac{M}{M_\odot} \left(\frac{R}{10 \text{ km}} \right)^{-1}. \qquad (13.7.12)$$

Exercise 13.9   The average power in Type II bursts from the Rapid Burster is about 130 times that in Type I bursts from this source. What does this suggest for accretion driven instabilities vs. thermonuclear flashes for the origin of Type II bursts? (cf. Hoffman et al. 1978).

To complicate matters, however, we note that some Type I burst sources have values of α much less than 100 (e.g., 4U 1608 − 52, which produced two bursts less than 10 min apart with $\alpha \lesssim 2.5$). Plausible explanations include anisotropic emission of the accretion driven luminosity, but the situation is unresolved.

(b)   The Binary System

We have outlined some of the direct observational evidence for the binary nature of the compact X-ray sources. This evidence includes periodic eclipses of the X-ray source by its companion, periodic Doppler shifts of the optical spectral

---

[66] Joss (1977); Maraschi and Cavaliere (1977); Lamb and Lamb (1978).

Figure 13.11   Two possible modes of mass transfer in a binary X-ray source. The *solid curve* shows the "Roche lobe" beyond which the normal primary will begin to shed gas onto its compact companion. In (*a*) the primary lies *inside* the Roche lobe but loses mass via a stellar wind. The orbiting compact star is an obstacle in the wind stream and a bow-shaped shock front is formed around it by the action of its gravitational field. Some of the shocked material is captured by the compact star. In (*b*) the corotating primary has expanded to, and begins to *overflow*, its Roche lobe. The gas flowing through the inner Lagrange point L_1 possesses significant angular momentum. Some of this gas is captured by the compact star and flows toward it in an accretion disk.

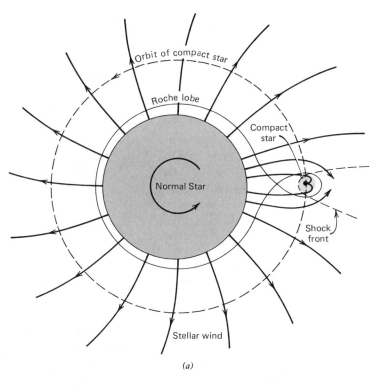

(a)

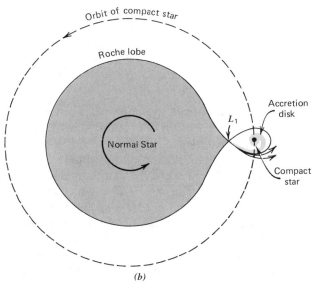

(b)

399

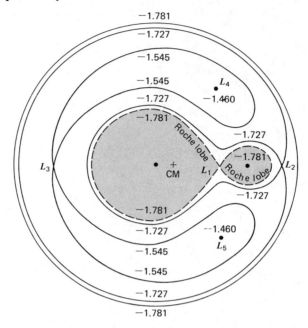

Figure 13.12   The equipotentials ϕ_{gc} = constant for the Newtonian gravitational + centrifugal potential in the orbital plane of the binary star system with a circular orbit. For the case shown here the stars have a mass ratio $M_N : M_c = 10 : 1$, where the normal star M_N is on the left and the compact star M_c is on the right. The equipotentials are labeled by their values of ϕ_{gc} measured in units of $G(M_N + M_c)/a$, where a is the separation of the centers of mass of the two stars. The innermost equipotential shown is the "Roche lobe" of each star. Inside each Roche lobe, but outside the stellar surface, the potential ϕ_{gc} is dominated by the "Coulomb" $(1/r)$ field of the star, so the equipotentials are nearly spheres. The potential ϕ_{gc} has local stationary points ($\nabla\phi_{gc} = 0$), called "Lagrange points," at the locations marked L_J. [From Novikov and Thorne (1973).]

lines of the companion and of the pulsation period of the X-ray source, and in some cases heating of one face of the companion star by the X-ray source. Binary periods are typically of the order of days.

The all-important transfer of mass from the primary to the compact secondary can take place in two different ways: by *Roche* or *tidal lobe* overflow, or by a *stellar wind* (see Fig. 13.11). In the case of Roche or tidal lobe overflow, material flows slowly over the gravitational potential saddle point between the two stars and is quickly captured by the compact object. In the frame corotating with the binary, the equipotential surfaces are spherical close to each star. They gradually increase in size and deform until they touch at the "inner Lagrangian" saddle point L_1 (see Fig. 13.12). The first common equipotential is called the "Roche lobe" if the spin of the primary is synchronized with the binary rotation and the "tidal lobe" if the primary is not rotating significantly. It is usually assumed that the primary expands to "fill its Roche lobe," thus facilitating the spill of material across L_1. Mass loss rates of 3×10^{-4}–$3 \times 10^{-8} M_\odot$ yr^{-1} are possible for such a process.[67] In the overflow scenario, the captured plasma will posses sufficient angular momentum to form an accretion disk around the compact star.

For the case of accretion from a stellar wind, a small fraction ($\lesssim 0.1\%$) of the plasma ejected from the primary is gravitationally captured by the compact star. The captured gas has some angular momentum with respect to the compact star, but it may not be sufficient to form an accretion disk. In this case the accretion is more or less spherical. Theory and observations of stellar winds suggest that appreciable winds are generated by sufficiently massive, upper main sequence stars; for such stars, mass-loss rates of 10^{-7}–$10^{-6} M_\odot$ yr^{-1} are reasonable.[68]

(c)   Origin and Evolution

The origin and evolution of compact X-ray sources is ultimately related to the evolution of close binary systems, a subject of intense research even prior to the discovery of X-ray binaries.[69] Specific evolutionary scenarios for X-ray binaries have been constructed.[70]

The story begins with a massive close binary system. The more massive star evolves faster and expands to fill its Roche lobe after $\sim 10^6$–10^7 yr, prior to the onset of helium ignition in its interior. Soon ($\sim 10^4$ yr) most of the mass is transferred to the *other* component. Thus, the *less* massive component is now the more evolved star and its evolution determines the final nature of the system. For small values of its mass, it presumably becomes a white dwarf. For larger values, it probably undergoes gravitational collapse after another $\sim 10^6$ yr, leading to a supernova explosion and to formation of a neutron star or a black hole. The more massive star continues to evolve and if the binary has not been disrupted by the supernova, this star will reach its Roche lobe in $10^6 \sim 10^7$ yr. At this point, it will begin transferring matter back onto the compact companion, leading to X-ray production.

Exercise   13.10   Consider a binary star system in which one member undergoes a supernova explosion and ejects a considerable fraction of its mass. Assume that the mass is lost in a time much shorter than the orbital period. Calculate the condition on the fraction of mass lost necessary for the preservation of the binary system, assuming a circular orbit. What is the criterion if the orbit is elliptical?

Answer: $\Delta M/M < \frac{1}{2}$ for circular orbit, where M = total system mass.

[67]Davidson and Ostriker (1973); Van den Heuvel (1974).
[68]Conti (1978).
[69]See Paczynski (1971) for a review.
[70]See, for example, Van den Heuvel and Heise (1972).

In the above picture, the crucial factor is the *mass* of the system. In low-mass systems, which one might expect to produce white dwarfs or neutron stars, the evolution timescales are long, and the optical stars will appear as members of an "older" stellar population (e.g., Sco X-1, Cyg X-2, Cyg X-3, Her X-1). In high-mass systems, evolution timescales are short and the stars will belong to a "younger" stellar population. Because the loss of $\sim \frac{1}{2}$ of the total mass would disrupt the binary, the remaining mass after an explosion would also be high and this identification would be maintained (e.g., Cyg X-1, Cen X-3, 3U 0900 − 40, SMC X-1).

Of the well-studied X-ray binaries, Her X-1 (eccentricity $e < 0.002$), Cen X-3 ($e = 0.0008$), and SMC X-1 ($e < 0.0007$) have nearly circular orbits[71] as does Cyg X-1 ($e < 0.02$).[72] Only 4U 0900 − 40 ($e = 0.092$), GX 301 − 2 ($e = 0.44$), and 4U 0115 + 63 ($e = 0.340$) have measurably eccentric orbits as does the binary radio pulsar ($e = 0.617$).[73] The evolutionary history of binary eccentricities is still not understood theoretically.

In contrast to the above scenario, the available evidence argues *against* explaining the cluster X-ray sources in terms of the evolution of primordial binary systems. Globular clusters are ancient—about 10^{10} years old—and most of their high- and moderate-mass stars have already evolved. Also, there is no observational evidence as yet for the presence of normal binary stars in clusters.

If cluster sources are indeed stellar-mass compact objects, and not supermassive black holes, then it is likely that the X-ray sources were formed under conditions peculiar to the dense cores of centrally condensed star clusters. One possibility is that short-lived massive stars in the cluster gave rise to compact objects near the center. These compact objects may then have undergone close, dissipative encounters with low-mass normal stars, leading to capture and the formation of binary systems. If the capture hypothesis is true, the evolution of the globular cluster sources will probably differ from the other galactic bulge sources. If the latter are low-mass binaries containing neutron stars, they are presumably primordial binaries.

[71] Rappaport and Joss (1981).
[72] Gies and Bolton (1982).
[73] Taylor and Weisberg (1982).

14
Accretion onto
Black Holes

14.1 Introduction

The process by which compact stars gravitationally capture ambient matter is called *accretion*. We have already seen in Chapter 13 that the accretion of gas onto compact stars of mass $M \sim M_\odot$ is the likely source of energy in the observed binary X-ray sources. The same process may also be at work on a much larger scale in quasars and active galactic nuclei where rapidly varying emission at high luminosity is observed from relatively compact regions. Here, accretion onto supermassive black holes with $M \gtrsim 10^6$–$10^9 M_\odot$ may power the system. Isolated, compact stars with $M \sim M_\odot$ may also be accreting gas as they wander through the interstellar medium of our Galaxy. The number of such objects could be considerable.[1]

What makes all of these possibilities worth investigating is the fact that, in falling through the steep gravitational potential of, say, a neutron star or a black hole, roughly 10% of the accreted rest-mass energy may be converted into radiation. Thus accretion is a process that can be considerably more efficient as a cosmic energy source than many other commonly invoked mechanisms in astrophysics (e.g., nuclear fusion).

Calculations of the accretion flow onto a compact star and the emitted radiation pattern are, in general, very difficult. Consider what is involved. Suppose that the effective mean free path for gas particle collisions is sufficiently short that the flow is *hydrodynamical* in nature. First one must determine the flow *geometry*; in general, if the gas possesses intrinsic *angular momentum*, the flow will be two- or three-dimensional, depending upon the flow symmetry. In simple cases, the flow may be spherical (such is the case, e.g., when there is no mean motion of gas far from a stationary compact star) or disk-like (as in axisymmetric flow of gas with intrinsic angular momentum). Such flow patterns simplify the

---

[1]See Section 1.3 for a crude "count."

analysis enormously. Second, one must enumerate the dominant *heating* and *cooling* mechanisms that characterize the accreting plasma. If the gas is optically thick to the emitted radiation (i.e., photons are scattered or absorbed by the gas before escaping to infinity), the *net* heating and cooling rates will themselves depend on the radiation field, which is to be determined self-consistently. Third, the possible role of *magnetic fields* in the plasma must be assessed. Magnetic fields may already thread the plasma far from the compact star, or they may originate from the star itself (or both!). In either case, the contributions of the magnetic field to the stresses and to the heating and cooling rates in the inflowing plasma must be evaluated. Fourth, the effect of radiation pressure in holding back the flow must be properly accounted for. Fifth, one must understand the flow *boundary conditions* both at large distances where the gas "joins on" to the ambient medium, and at the stellar surface, where the gas merges smoothly into the star. So in the general accretion case, one must solve the time-dependent, multidimensional, relativistic, magneto-hydrodynamic equations with coupled radiative transfer!

It is thus not surprising that the problem of gas accretion onto compact stars has been solved in only a few idealized cases. Even with the simplest assumptions—for example, steady-state flow with spherical or thin-disk geometry—the analysis is made complicated by the requirement of *self-consistency*. For example, the radiation field to be calculated is determined, in general, by heating and cooling processes that are used to determine gas temperatures, which in turn determine the radiation field!

In spite of the difficulties, progress has been achieved in recent years in understanding accretion flows and applying that understanding to compact X-ray stars. Indeed, as we have already seen in Section 13.7, even the most naive assumptions regarding the outgoing luminosity and spectrum yield results in qualitative agreement with much of the X-ray data.

In the next few sections we shall discuss some very *idealized* accretion models that illustrate some of the physics associated with such a process. We shall focus initially on accretion onto black holes, since in this case complications in the flow pattern at the stellar surface are reduced somewhat by the absence of magnetic fields[2] and the presence of an event horizon ("vacuum cleaner" boundary condition). In Chapter 15, we will return to consider accretion onto neutron stars and white dwarfs. For all cases we shall consider, the gravitational field of the central compact star dominates the field of the inflowing material. Accordingly, we shall ignore the self-gravity of the accreting gas as well as the very slow increase in the mass of the central object due to accretion.[3]

---

[2] Recall the "no hair" theorem for black holes, Section 12.1.

[3] Accretion onto compact objects is discussed in a number of books and review articles, including Zel'dovich and Novikov (1971) [pre-Uhuru], Novikov and Thorne (1973), Lightman, Shapiro, and Rees (1978), Pines (1980b), Lamb and Pines (1979), and Kylafis et al. (1980). We refer the reader to these discussions for further details and for references to the relevant literature.

An introduction to radiative transfer is given in Appendix I. Readers unfamiliar with basic radiative transfer theory are encouraged to read this section before reading the rest of this chapter.

14.2   Collisionless Spherical Accretion

In this section we consider the accretion of a *collisionless* gas of identical particles of mass m onto a central star of mass M and radius R. This situation was first examined by Zel'dovich and Novikov (1971). Our general discussion will be Newtonian, but the accretion rate derived for a central Schwarzschild black hole will be relativistically correct.

All information about the accreting gas is contained in the particle distribution function, $f(\mathbf{r}, \mathbf{v}, t)$, defined by

$$f(\mathbf{r}, \mathbf{v}, t)\, d^3r\, d^3v = \begin{array}{l}\text{the number of particles in the phase}\\ \text{space volume element } d^3r\, d^3v \text{ centered}\\ \text{about } \mathbf{r} \text{ and } \mathbf{v}, \text{ at time } t.\end{array} \quad (14.2.1)$$

The quantity f is also called the "phase-space density" since d^3v and d^3p are proportional for nonrelativistic particles. The particle number density in coordinate space is given by

$$n(\mathbf{r}, t) = \int_{\text{all } \mathbf{v}} f(\mathbf{r}, \mathbf{v}, t)\, d^3v, \quad (14.2.2)$$

the velocity dispersion (i.e., "temperature") by

$$\langle v^2(\mathbf{r}, t) \rangle = \frac{1}{n(\mathbf{r}, t)} \int_{\text{all } \mathbf{v}} v^2 f(\mathbf{r}, \mathbf{v}, t)\, d^3v, \quad (14.2.3)$$

and so on.

The distribution function f for a collisionless gas is determined by the *collisionless Boltzmann equation* or *Vlasov equation*:

$$\frac{D}{Dt} f(\mathbf{r}, \mathbf{v}, t) \equiv \frac{\partial f}{\partial t} + \mathbf{v} \cdot \nabla_{\mathbf{r}} f + \dot{\mathbf{v}} \cdot \nabla_{\mathbf{v}} f = 0, \quad (14.2.4)$$

where $\mathbf{v} \equiv \dot{\mathbf{r}}$ is the particle velocity along the coordinates $\mathbf{r}$, and $\dot{\mathbf{v}}$ is the acceleration. Equation (14.2.4) is simply a continuity equation for the flow of particles in six-dimensional phase space.[4] Equation (14.2.4) is a statement of *Liouville's theorem*: the distribution function is conserved along the trajectory of

---

[4]See, for example, Reif (1965), Chapter 13, for a derivation and discussion.

each particle. For the problems we are interested in, $\dot{\mathbf{v}} = -\nabla\Phi$, where Φ is the gravitational potential:

$$\Phi = \frac{-GM}{r} + \Phi_{\text{self}}, \tag{14.2.5}$$

where

$$\nabla^2\Phi_{\text{self}} = 4\pi G\rho, \tag{14.2.6}$$

and where the mass density $\rho \equiv mn$ is obtained from Eq. (14.2.2). Henceforth we shall ignore the self-gravity of the particles and thus set $\Phi = -GM/r$, where r is the distance to the central star.

For stationary flow in which the distribution function is independent of time, f must be a function only of the dynamical constants of motion.[5] For spherical systems, there are just two constants: the energy E and (magnitude of) the angular momentum, J, which, per unit mass, are given by

$$E = \frac{1}{2}v^2 + \Phi(r) = \frac{1}{2}v_r^2 + \frac{1}{2}\frac{J^2}{r^2} - \frac{GM}{r}, \tag{14.2.7}$$

$$J = rv_t. \tag{14.2.8}$$

Here v_r and v_t are the radial and transverse particle velocities. Given E and J, the particle trajectory is completely determined in a spherical potential, hence $f = f(E, J)$. If, in addition, the velocity distribution is everywhere isotropic, then f simplifies further since it is independent of J. In this case, $f = f(E)$, where, by Eq. (14.2.7), $E \geqslant \Phi(r)$ [i.e., there are no particles with $E < \Phi(r)$]. For stationary, spherical distributions with isotropic velocities, Eq. (14.2.2) reduces to

$$n(r) = 4\pi\int v^2 f \, dv = 4\pi\int_{E=\Phi}^{\infty} [2(E - \Phi)]^{1/2} f(E) \, dE, \tag{14.2.9}$$

and Eq. (14.2.3) reduces to

$$\langle v^2(r)\rangle = \frac{4\pi}{n(r)}\int_{E=\Phi}^{\infty} [2(E - \Phi)]^{3/2} f(E) \, dE. \tag{14.2.10}$$

Exercise 14.1 Prove that when $f = f(E)$, $\langle v^2(r)\rangle = 3\langle v_r^2(r)\rangle$, and $\langle v_r^2(r)\rangle = \langle v_t^2(r)\rangle/2$.

Consider the accretion of noninteracting particles onto the star. Particles moving with angular momentum less than a critical value, $J_{\text{min}}(E)$, will be

---

[5] This result is often quoted as "Jeans' theorem"; Jeans (1919).

captured by the star as they approach pericenter. For nonrelativistic (NR) particles orbiting a Newtonian star of radius R we have

$$J_{\min}(E) \equiv \left[2\left(E + \frac{GM}{R} \right) \right]^{1/2} R \quad \text{(NR particles; Newtonian star).}\quad (14.2.11)$$

Exercise   14.2   Verify Eq. (14.2.11).

For nonrelativistic particles orbiting a black hole (or very compact neutron star), we have

$$J_{\min}(E) = \frac{4GM}{c} \quad \text{(NR particles; black hole).} \quad (14.2.12)$$

This result follows from Exercise 12.9, since the condition for capture is $\tilde{E}^2 > V_{\max}$ and $\tilde{E}^2 \simeq 1$ for NR particles. The quantity $J_{\min}(E)$ thus defines a "loss-cone" in velocity space within which particles are consumed by the central star near pericenter. Because of the existence of a loss-cone $J < J_{\min}(E)$, it is convenient to determine the particle density as a function of E and J. Equations (14.2.7) and (14.2.8) allow us to rewrite the velocity element d^3v as

$$d^3v = 2\pi v_t \, dv_t \, dv_r = \frac{4\pi J \, dJ \, dE}{r^2 |v_r|}, \quad (14.2.13)$$

where the "extra" factor of 2 arises from the fact that for a given E, v_r can be positive or negative. Now we can determine $N^-(r, E, J)$, the number of particles per unit r, E, and J with inward-directed radial velocity:

$$N^-(r, E, J) \, dr \, dE \, dJ = \tfrac{1}{2} f(E, J) \, d^3r \, d^3v$$

$$= 8\pi^2 \frac{J}{|v_r|} f \, dr \, dE \, dJ. \quad (14.2.14)$$

Accordingly, the *total* capture rate for particles onto the central mass is

$$\dot{N}_{\text{tot}} = \int_{\Phi(r)}^{\infty} dE \int_{J=0}^{J_{\min}(E)} dJ \, |v_r| N^-(r, E, J) \Bigg|_{r=R} = 8\pi^2 \int_{\Phi(R)}^{\infty} dE \int_0^{J_{\min}(E)} dJ \, fJ,$$

$$(14.2.15)$$

and the corresponding mass accretion rate is $\dot{M}_{\text{tot}} = m\dot{N}_{\text{tot}}$.

Let us now consider a concrete example by analyzing the capture of *unbound* nonrelativistic particles with energies $E > 0$. We shall assume that at all times there is an infinite bath of such particles surrounding the star and that the particle distribution is isotropic and monoenergetic. Far from the star the particle density is uniform and equal to n_∞ and the particle speed is $v_\infty \ll c$. Thus

$$f = f(E) = n_\infty \frac{\delta(E - E_\infty)}{4\pi(2E_\infty)^{1/2}}, \qquad (14.2.16)$$

where

$$E_\infty = \tfrac{1}{2}v_\infty^2, \qquad v_\infty \ll c. \qquad (14.2.17)$$

The normalization of the δ-function in Eq. (14.2.16) can be checked using Eq. (14.2.9), since $\Phi = 0$ at $r = \infty$. At large distances, $f \to n_\infty \delta(v - v_\infty)/4\pi v_\infty^2$, but Eq. (14.2.16) applies *everywhere* by Liouville's theorem. From Eq. (14.2.15) we immediately calculate the unbound capture rate:

$$\dot{N}(E > 0) = 8\pi^2 \int_0^\infty dE\, f(E) \int_0^{J_{\min}} dJ\, J = 4\pi^2 \int_0^\infty dE\, f(E) J_{\min}^2(E). \qquad (14.2.18)$$

This result applies even in the case of black hole accretion since for steady consumption of unbound particles, the first integrand in Eq. (14.2.15) can be evaluated at any arbitrarily large r. Thus, Eqs. (14.2.11), (14.2.12), and (14.2.16)–(14.2.18) give

$$\dot{M}(E > 0) = m\dot{N}(E > 0)$$

$$= \begin{cases} 2\pi GM^2 \rho_\infty v_\infty^{-1} \dfrac{R}{M}\left(1 + \dfrac{v_\infty^2 R}{2MG}\right) & (14.2.19\ ) \\ \qquad \text{(NR particles; Newtonian star)}, \\ 16\pi(GM)^2 \rho_\infty v_\infty^{-1} c^{-2} & \\ \qquad \text{(NR particles; black hole)}, & (14.2.20\ ) \end{cases}$$

where $\rho_\infty = mn_\infty$.

Exercise 14.3   Rederive Eqs. (14.2.19) and (14.2.20) by noting that the mass per unit time crossing inward through a sphere of large radius $r \gg R$ with particles in capture orbits is $\dot{M}(E > 0) = (\rho_\infty v_\infty/4\pi)4\pi r^2 \Delta\Omega_c$, where the solid angle for capture is $\Delta\Omega_c = \sigma_c/r^2$ and where the capture cross section σ_c is given by Eqs. (12.4.38) or (12.4.39). Thus $\dot{M}(E > 0) = \rho_\infty v_\infty \sigma_c$.

Exercise 14.4

(a) Calculate the (mass) accretion rate of a Schwarzschild black hole immersed in an isotropic bath of photons with energy density ε_∞^r far from the hole.

Hint: Use the method of the previous exercise and Eq. (12.5.12) for σ_c.

Answer: $\dot{M} = 27\pi G^2 M^2 \varepsilon_\infty^r / c^5$

(b)   What is the minimum mass of a black hole that can double its mass in less than the age of the Universe $t = 10^{10}$ yr when placed in the $3K$ cosmic microwave (blackbody) background? Take the microwave background temperature to be independent of time during this period.

Note that Eq. (14.2.20) must be employed instead of Eq. (14.2.19) whenever $Rc^2/GM < 8$ as, for example, in the case of very compact neutron stars. Evaluating Eq. (14.2.20) for conditions appropriate to the ionized component of the interstellar medium in our Galaxy, we obtain a black hole accretion rate

$$\dot{M}(E > 0) = 1.56 \times 10^{-23} \left(\frac{\rho_\infty}{10^{-24} \text{ g cm}^{-3}} \right) \left(\frac{M}{M_\odot} \right)^2 \left(\frac{v_\infty}{10 \text{ km s}^{-1}} \right)^{-1} M_\odot \text{ yr}^{-1}.$$

$$(14.2.21)$$

It is straightforward to derive the unbound particle velocity dispersion and density as functions of the distance from the star for the distribution function given by Eq. (14.2.16). We shall require $r \gg R$ so that we can ignore the removal of particles by collision with the stellar surface and relativistic corrections near a black hole event horizon. Accordingly, Eq. (14.2.9), evaluated for $E > 0$, yields

$$n_{E>0}(r) = n_\infty \left(1 + \frac{2GM}{v_\infty^2 r} \right)^{1/2}, \qquad (14.2.22)$$

while Eq. (14.2.10) yields

$$\langle v^2(r) \rangle_{E>0} \equiv v^2(r) = v_\infty^2 \left(1 + \frac{2GM}{v_\infty^2 r} \right). \qquad (14.2.23)$$

Equation (14.2.23) can also be obtained from Eqs. (14.2.7) and (14.2.17). It can be used to define the particle "temperature" according to

$$T_{E>0}(r) = T_\infty \left(1 + \frac{2GM}{v_\infty^2 r} \right), \qquad \tfrac{3}{2}kT \equiv \tfrac{1}{2}mv^2. \qquad (14.2.24)$$

Now Eqs. (14.2.22)–(14.2.24) illustrate a very general feature of spherical accretion flows onto a central mass. Define the "accretion" or "capture" radius r_a to be that radius at which the kinetic energy of a particle is equal to its potential energy:

$$r_a \equiv \frac{2GM}{v_\infty^2}. \qquad (14.2.25)$$

Then clearly, for $r \gg r_a$, the density and temperature profiles vary little from their asymptotic values at infinity. However, for $r \ll r_a$ the gravitational potential of the central mass influences the distribution, focusing particles together and increasing their density and temperature. The same qualitative behavior also arises in hydrodynamic accretion (cf. Section 14.3), but there the density enhancement is much greater than in Eq. (14.2.22) because of collisions.

Exercise 14.5   Consider the accretion of unbound particles obeying a Maxwell–Boltzmann velocity distribution far from the central star, with density n_∞ and *rms* velocity v_∞. Calculate $\dot{M}(E > 0)$, $n(r)$, $\langle v^2(r) \rangle$, and $T(r)$ for such a distribution; examine both the Newtonian and black hole cases (but for $r \gg R$ only).

Answer:

$$\dot{M}_{MB}(E > 0) = 8(24\pi)^{1/2} G^2 M^2 \rho_\infty v_\infty^{-1} c^{-2} \text{ (NR particles; black hole).} \quad (14.2.26)$$

As emphasized by Zel'dovich and Novikov (1971), the above equations for $\dot{M}$, $n(r)$, and so on, only apply to the *unbound* particles that approach the central star from infinity with $E > 0$. The number of *bound* particles is unlimited, if collisions are neglected completely. In this case, the distribution of bound particles, which move in elliptic orbits about the central star, is an initial condition that can be specified arbitrarily and independently of Eq. (14.2.16). In the strict absence of collisions, the bound particles in capture orbits are never replenished by the unbound particle distribution. Only with collisions can a particle with $E > 0$ occasionally scatter into a bound orbit with $E < 0$. Accordingly, any contribution of bound particles to $\dot{M}$ is transient, and vanishes by the time steady state is achieved.

More interesting, perhaps, is the case in which collisions *do* occur. When the mean free path for collisions is much shorter than the characteristic length scale, that is, $\lambda_c \ll r$ (large collision cross sections), then the flow is *hydrodynamical* and can be determined by analyzing the *fluid* equations of motion. We shall discuss this important regime in detail in the next section. When, on the other hand, we have $\infty > \lambda_c \gg r$ (so that collision cross sections are small but finite), the situation is described by the *collisional Boltzmann equation*:

$$\frac{Df}{Dt} = \left(\frac{\partial f}{\partial t} \right)_c. \quad (14.2.27)$$

Here the right-hand side of Eq. (14.2.27) is nonzero, because collisions can alter the distribution in phase space of particles.

In general, Eq. (14.2.27) is extremely difficult to solve. One simplification arises when the inequality $\lambda_c \gg r$ is strong: the distribution function, to lowest order, satisfies $Df/Dt \approx 0$, with the collisional terms serving as first-order *perturbations*. In this regime the characteristic "dynamical crossing time" or orbital period of a particle, $t_d \sim r/v$, is much shorter than the characteristic collision time or "relaxation time," $t_r \sim \lambda_c/v$; that is, $t_d \ll t_r$. Thus on *dynamical* time-scales f must still satisfy the Vlasov equation (14.2.4) in first approximation; to high accuracy it still obeys Liouville's theorem and, for spherical systems, it must be of the form $f = f(E, J)$. However, f must also undergo slow, secular changes on *relaxation* timescales, due to collisions. Indeed for $t \gg t_r$, the collisions will in fact determine the *form* of the distribution function, choosing from among the many possible solutions to the Vlasov equation those that satisfy the collisional constraints. The situation is reminiscent of dynamical equilibrium in evolving stars, which satisfy hydrostatic equilibrium to high accuracy, but which undergo secular expansion or contraction on thermonuclear timescales.

A further simplification arises when, in addition, the dominant collisional effects arise from *small-angle scattering*. In such a situation, the net deflection of particles results from the cumulative effect of repeated, small-angle collisions rather than from a few large-angle ($\sim 90°$) encounters. In this case the right-hand side of Eq. (14.2.27) may be treated in a (second-order) perturbation expansion called the *Fokker–Planck equation*. Such an expansion is suitable for example, when the collisions are Coulombic in nature (i.e., when the force between particles varies as $1/r^2$) as in Coulomb scattering between charged particles in a plasma or gravitational scattering between masses in a star cluster.

One expects that, typically, the flow of plasma onto a compact star is hydrodynamical in nature[6] with $\lambda_c \ll r$. On the other hand, the dynamical behavior of stars in a globular star cluster is governed by the Fokker–Planck equation. In this situation λ_c due to star-star gravitational "encounters" (i.e., scattering) satisfies $\infty > \lambda_c \gg r$ or $t_d \ll t_r$. The dynamical evolution of globular star clusters has been the subject of much theoretical interest recently. This interest results partly from the development of new numerical methods to solve the Fokker–Planck equation, the availability of good observational data for globular clusters, and the discovery of X-ray emission from globular clusters (cf. Chapter 13).[7] We note, finally, that the dynamical behavior of normal galaxies is governed by the collisionless Boltzmann equation since $t_d \ll t_H \ll t_r$, where $t_H \sim 1/H \sim 10^{10}$ yr $(H/100 \text{ km s}^{-1} \text{ Mpc}^{-1})^{-1}$ is the age of the Universe and H is Hubble's constant.

Exercise 14.6   Calculate the rate at which unbound stars with a Maxwell–Boltzmann distribution would be consumed by a massive black hole M at the center of a very dense, active galactic nucleus. Assume that the star density is $n_\infty = 10^7$ pc^{-3} and the velocity dispersion is $\langle v_\infty^2 \rangle^{1/2} = 250$ km s^{-1} far from the hole. Assume further that stars are consumed following tidal disruption by the hole at a tidal-breakup radius $r_D = R(M/m)^{1/3}$, where m is the mass and R is the radius of a star.

   Hint: $J_{\min}(E) = [2(E + GM/r_D)]^{1/2} r_D$.

[6]But see Begelman (1977) for scenarios in which $\lambda_c > r$.
[7]For a general discussion of the dynamical evolution of globular star clusters, the reader is referred to the review articles of Spitzer (1975) and Lightman and Shapiro (1978) and references therein.

14.3 Hydrodynamic Spherical Accretion

For typical gas dynamical conditions found in the interstellar medium and in the matter exchanged between binary stars, it is expected that the accretion flow onto compact objects will be *hydrodynamical* in nature. Sometimes collisions alone will *not* be adequate to couple particles together effectively, as is the case for Coulomb collisions in interstellar plasma accreting onto a solar-mass compact star. However, the presence of macroscopically weak magnetic fields or the existence of two-stream plasma instabilities or other plasma collective effects is usually sufficient to keep the effective particle mean free path small (i.e., $\lambda_{\text{eff}} \ll r$) and thereby ensure hydrodynamical flow. Such is the situation observed, for example, in the solar wind where magnetic fields and plasma instabilities provide strong coupling between the charged particle constituents and maintain hydrodynamical outflow.[8]

Consider then the steady, spherical accretion of ambient gas onto a stationary, nonrotating black hole of mass M in the fluid limit. Assume the gas flow to be adiabatic in first approximation, treating entropy loss due to radiation as a small perturbation. The gas will then be characterized by an adiabatic index, Γ, and everywhere the pressure P will be related to the rest-mass density ρ according to

$$P = K\rho^{\Gamma} \qquad K, \Gamma \text{ constant.} \qquad (14.3.1)$$

The sound speed a is given by $a \equiv (dP/d\rho)^{1/2} = (\Gamma P/\rho)^{1/2}$ everywhere. Assume that the gas is at rest at infinity, where the density is ρ_{∞}, the pressure is P_{∞} and the sound speed is a_{∞}.

The basic characteristics of the flow are reasonably well described by employing Newtonian gravity, especially at large distances from the hole, $r \gg GM/c^2$. The distinguishing feature of accretion onto a black hole, in contrast to accretion onto an uncollapsed star with a hard surface, is that the black hole imposes some unique regularity conditions on the flow at small radii near $r = 2GM/c^2$. These must be evaluated by using relativistic gravity.[9] The black hole regularity conditions serve to determine unambiguously the mass accretion rate $\dot{M}$. For accretion onto a star with a surface, the accretion rate depends on the boundary conditions near the surface. The black hole solution is but one possible solution for the flow above a star with a surface. As we shall see below, this solution gives the maximum accretion rate.

In addition to Eq. (14.3.1), which replaces the entropy equation, the flow is completely governed by the continuity equation,

$$\nabla \cdot \rho\mathbf{u} = \frac{1}{r^2}\frac{d}{dr}(r^2\rho u) = 0, \qquad (14.3.2)$$

---

[8] Flow in a stellar wind is essentially "time-reversed" accretion; see below.
[9] See Appendix G, where the relativistic treatment is contrasted with the Newtonian treatment.

and the Euler equation,

$$u\frac{du}{dr} = -\frac{1}{\rho}\frac{dP}{dr} - \frac{GM}{r^2}, \qquad (14.3.3)$$

[cf. Eqs. (6.1.1) and (6.1.2)]. The equations have been written above for steady-state, spherical flow; the *inward* radial velocity is denoted by $u > 0$.

Equation (14.3.2) can be integrated immediately to yield an equation for $\dot{M}$, which, as we shall see, enters as a kind of eigenvalue:

$$4\pi r^2 \rho u = \dot{M} = \text{constant} \quad \text{(independent of } r\text{).} \qquad (14.3.4)$$

Equation (14.3.3) can also be integrated, using Eq. (14.3.1), to yield the familiar Bernoulli equation,

$$\frac{1}{2}u^2 + \frac{1}{\Gamma - 1}a^2 - \frac{GM}{r} = \text{constant} = \frac{1}{\Gamma - 1}a_\infty^2, \qquad (14.3.5)$$

where we have used the boundary conditions at infinity to evaluate the constant in Eq. (14.3.5). The flow is determined once $\dot{M}$ and the distributions $P(r)$ and $u(r)$ are known.

As shown by Bondi (1952), different values of $\dot{M}$ lead to physically distinct classes of solutions for the same boundary conditions at infinity. Here we shall only be interested in that unique solution for which the velocity u rises *monotonically* from 0 at $r = \infty$ to free-fall velocity at small radii, $u \rightarrow (2GM/r)^{1/2}$ as $r \rightarrow 0$. In fact, the relativistic equations at $r = 2GM/c^2$ *demand* that we choose this solution to avoid singularities in the flow outside the event horizon (cf. Appendix G). To calculate the required accretion rate $\dot{M}$ we rewrite Eq. (14.3.2) in the form

$$\frac{\rho'}{\rho} + \frac{u'}{u} + \frac{2}{r} = 0 \qquad (14.3.6)$$

(prime ' denotes d/dr), and Eq. (14.3.3) in the form

$$uu' + a^2\frac{\rho'}{\rho} + \frac{GM}{r^2} = 0. \qquad (14.3.7)$$

Solving Eqs. (14.3.6) and (14.3.7) for u' and ρ', we get

$$u' = \frac{D_1}{D}, \qquad \rho' = -\frac{D_2}{D}, \qquad (14.3.8)$$

where

$$D_1 = \frac{2a^2/r - GM/r^2}{\rho},$$ (14.3.9)

$$D_2 = \frac{2u^2/r - GM/r^2}{u},$$ (14.3.10)

and

$$D = \frac{u^2 - a^2}{u\rho}.$$ (14.3.11)

Equation (14.3.8) shows that to guarantee the smooth, monotonic increase of u with decreasing r and simultaneously avoid singularities in the flow, the solution must pass through a "critical point," where

$$D_1 = D_2 = D = 0 \quad \text{at } r \equiv r_s.$$ (14.3.12)

From Eqs. (14.3.9)–(14.3.12), we find that at the critical radius

$$u_s^2 = a_s^2 = \frac{1}{2}\frac{GM}{r_s},$$ (14.3.13)

so that the critical radius corresponds to the *transonic radius* at which the flow speed equals the sound speed. By combining Eq. (14.3.13) with Eq. (14.3.5) we can relate a_s, u_s, and r_s to the known sound speed at infinity:

$$a_s^2 = u_s^2 = \left(\frac{2}{5 - 3\Gamma}\right)a_\infty^2, \qquad r_s = \left(\frac{5 - 3\Gamma}{4}\right)\frac{GM}{a_\infty^2}.$$ (14.3.14)

Thus at the transonic radius the gravitational potential GM/r_s is comparable to the internal ambient thermal energy per unit mass, a_∞^2.

Now we can calculate the accretion rate from Eq. (14.3.4) using

$$\rho = \rho_\infty \left(\frac{a}{a_\infty}\right)^{2/(\Gamma - 1)}.$$ (14.3.15)

We obtain

$$\dot{M} = 4\pi\rho_\infty u_s r_s^2 \left(\frac{a_s}{a_\infty}\right)^{2/(\Gamma - 1)} = 4\pi\lambda_s \left(\frac{GM}{a_\infty^2}\right)^2 \rho_\infty a_\infty,$$ (14.3.16)

where the nondimensional accretion eigenvalue λ_s for the transonic solution is given by

$$\lambda_s = \left(\frac{1}{2}\right)^{(\Gamma+1)/2(\Gamma-1)}\left(\frac{5-3\Gamma}{4}\right)^{-(5-3\Gamma)/2(\Gamma-1)}. \qquad (14.3.17)$$

Values of λ_s as a function of Γ are given in Table 14.1.

Exercise 14.7 Verify that for $\Gamma = 1$, $\lambda_s = e^{3/2}/4$, and that for $\Gamma = \frac{5}{3}$, $\lambda_s = \frac{1}{4}$.

We can rewrite the transonic accretion rate (14.3.16) in the form

$$\dot{M} = 4\pi\lambda_s(GM)^2\rho_\infty a_\infty^{-1}c^{-2}\frac{c^2}{a_\infty^2}. \qquad (14.3.18)$$

Immediately we see by direct comparison with Eq. (14.2.20) that the approximate equality between the sound speed a_∞ and the mean particle speed v_∞ implies that the hydrodynamic accretion rate is larger than the collisionless accretion rate by the large factor $(c/a_\infty)^2$ ($\sim 10^9$ for typical ionized interstellar gas with $a_\infty \sim 10$ km s^{-1}). The physical reason for the disparity is simple: the presence of collisions between particles restricts tangential motion and funnels particles effectively in the radial direction for efficient capture. For an ideal Maxwell–Boltzmann gas of mean molecular weight μ we have

$$P = \frac{\rho kT}{\mu m_u}, \qquad a^2 = \frac{\Gamma kT}{\mu m_u}, \qquad T = T_\infty\left(\frac{\rho}{\rho_\infty}\right)^{\Gamma-1}. \qquad (14.3.19)$$

Employing Eq. (14.3.19) to evaluate $\dot{M}$ for the ionized component of the interstel-

Table 14.1

Values of the Accretion Eigenvalue λ_s

Γ	λ_s
1	1.120
$\frac{4}{3}$	0.707
$\frac{7}{5}$	0.625
$\frac{3}{2}$	0.500
$\frac{5}{3}$	0.250

lar medium (assumed to be pure hydrogen: $\mu = \frac{1}{2}$), yields, for $\Gamma = \frac{5}{3}$,

$$\dot{M} = 8.77 \times 10^{-16} \left(\frac{M}{M_\odot}\right)^2 \left(\frac{\rho_\infty}{10^{-24} \text{ g cm}^{-3}}\right) \left(\frac{a_\infty}{10 \text{ km s}^{-1}}\right)^{-3} M_\odot \text{ yr}^{-1}$$

$$= 1.20 \times 10^{10} \left(\frac{M}{M_\odot}\right)^2 \left(\frac{\rho_\infty}{10^{-24} \text{ g cm}^{-3}}\right) \left(\frac{T_\infty}{10^4 \text{ K}}\right)^{-3/2} \text{ g s}^{-1}. \qquad (14.3.20)$$

Exercise 14.8 Prove that the *transonic* accretion rate is also the *maximum* possible steady accretion rate for a given Γ (Bondi, 1952).

Hint: (a) Write the Bernoulli equation (14.3.5) in the nondimensional form

$$f(v) = \lambda^{-2(\Gamma-1)/(\Gamma+1)} g(x),$$

where

$$f(v) = v^{-2(\Gamma-1)/(\Gamma+1)} \left(\frac{1}{2} v^2 + \frac{1}{\Gamma-1}\right),$$

$$g(x) = x^{4(\Gamma-1)/(\Gamma+1)} \left(\frac{1}{x} + \frac{1}{\Gamma-1}\right),$$

and where $x \equiv r/(GM/a_\infty^2)$, $v \equiv u/a$, $z \equiv \rho/\rho_\infty = (a/a_\infty)^{2/(\Gamma-1)}$. Use Eq. (14.3.4) to eliminate z.

(b) Show that f and g both pass through minimum values, f_s and g_s, respectively, where

$$f_s = \frac{1}{2} \frac{\Gamma+1}{\Gamma-1} \qquad \text{at } v = v_s = 1,$$

$$g_s = \frac{1}{4} \frac{\Gamma+1}{\Gamma-1} \left[\frac{1}{4}(5-3\Gamma)\right]^{-(5-3\Gamma)/(\Gamma+1)} \qquad \text{at } x = x_s = \frac{1}{4}(5-3\Gamma).$$

(c) Argue from parts (a) and (b) that the largest value that λ can attain is then λ_s, where

$$\lambda_s = \left(\frac{g_s}{f_s}\right)^{(\Gamma+1)/2(\Gamma-1)} = \left(\frac{1}{2}\right)^{(\Gamma+1)/2(\Gamma-1)} \left(\frac{5-3\Gamma}{4}\right)^{-(5-3\Gamma)/2(\Gamma-1)}.$$

Exercise 14.9 Prove that there are no steady transonic polytropic solutions with $\Gamma > \frac{5}{3}$.

Hint: Equations (14.3.8) must give *real* values for u' and ρ' when the right-hand sides are evaluated using l'Hôpital's rule at $r = r_s$. Note that by Eq. (14.3.6) l'Hôpital's rule need only be applied to one of the equations.

The transonic flow profiles are straightforward to deduce. Far outside the transonic radius (i.e., $r \gg r_s$), the gravitational potential of the central mass is barely felt and the temperature and density remain close to their asymptotic values at infinity:

$$\rho \approx \rho_\infty, \qquad T \approx T_\infty, \qquad a \approx a_\infty, \qquad \frac{r}{r_s} \gg 1. \qquad (14.3.21)$$

From Eqs. (14.3.4), (14.3.16), and (14.3.21) we see that the flow speed decreases with increasing r according to

$$\frac{u}{a_\infty} \approx \lambda_s \left(\frac{GM}{a_\infty^2}\right)^2 r^{-2}, \qquad \frac{r}{r_s} \gg 1. \qquad (14.3.22)$$

Well inside the transonic radius, the flow is significantly influenced by the gravitational field of the central mass. For $r/r_s \ll 1$, the term GM/r dominates over the term $a^2/(\Gamma - 1)$ in Eq. (14.3.5). The flow velocity approaches the free-fall speed, and deceleration due to gas pressure becomes negligible:

$$u \approx \left(\frac{2Gm}{r}\right)^{1/2}, \qquad \frac{r}{r_s} \ll 1 \quad \left(1 \leqslant \Gamma < \frac{5}{3}\right). \qquad (14.3.23)$$

According to Eqs. (14.3.4), (14.3.16), and (14.3.23), the density enhancement increases with decreasing r as

$$\frac{\rho}{\rho_\infty} \approx \frac{\lambda_s}{2^{1/2}} \left(\frac{GM}{a_\infty^2}\right)^{3/2} r^{-3/2}, \qquad \frac{r}{r_s} \ll 1 \quad \left(1 \leqslant \Gamma < \frac{5}{3}\right). \qquad (14.3.24)$$

This enhancement is significantly larger than the enhancement found in the collisionless case, Eq. (14.2.22), even though the characteristic particle speeds are the same. The reason is again due to the ability of collisions to "channel" the flow effectively in the radial direction in the hydrodynamical limit. Evaluating Eq. (14.3.24) at $r = 2GM/c^2$ indicates that at the horizon the density enhancement in the hydrodynamical case is larger than in the collisionless gas by roughly the factor (c^2/a_∞^2). The temperature rise at small r is obtained by combining Eqs. (14.3.19) and (14.3.24):

$$\frac{T}{T_\infty} \approx \left[\frac{\lambda_s}{2^{1/2}} \left(\frac{GM}{a_\infty^2}\right)^{3/2}\right]^{\Gamma-1} r^{-3(\Gamma-1)/2}, \qquad \frac{r}{r_s} \ll 1 \quad \left(1 \leqslant \Gamma < \frac{5}{3}\right).$$

$$(14.3.25)$$

Comparing Eqs. (14.3.23)–(14.3.25) with Eqs. (G.34)–(G.36), we see that the former are relativistically "correct," provided we identify r with the Schwarzschild radial coordinate, u with the radial 4-velocity, and ρ with the proper rest-mass density.

Equations (14.3.23)–(14.3.25) are modified somewhat for the special case $\Gamma = \frac{5}{3}$, for which, in the *Newtonian* approximation, $r_s = 0$. In this case, we find that for $r/(GM/a_\infty^2) \ll 1$, the critical solution satisfies

$$a \approx u \approx \left(\frac{GM}{2r}\right)^{1/2}, \qquad \frac{r}{GM/a_\infty^2} \ll 1 \quad \left(\Gamma = \frac{5}{3}\right). \qquad (14.3.26)$$

The corresponding density and temperature profiles at small r are then given by Eqs. (14.3.4), (14.3.16), (14.3.17), and (14.3.19):

$$\left.\begin{array}{l} \dfrac{\rho}{\rho_\infty} \approx \dfrac{2^{1/2}}{4}\left(\dfrac{GM}{a_\infty^2}\right)^{3/2} r^{-3/2} \\[4mm] \dfrac{T}{T_\infty} \approx \dfrac{1}{2}\dfrac{GM}{a_\infty^2} r^{-1} \end{array}\right\}, \qquad \frac{r}{GM/a_\infty^2} \ll 1 \quad \left(\Gamma = \frac{5}{3}\right). \quad (14.3.27)$$

Equations (14.3.26) and (14.3.27) are altered by multiplicative numerical factors of order unity when we employ the *relativistic* flow equations. In Appendix G we show that the radial 4-velocity $u_h = 0.782$ at the horizon $r = 2GM/c^2$, and we calculate the corresponding density and temperature enhancements there [cf. Eq. (G.39)]. We also work out explicit numerical values for these ratios for the case of nonrelativistic baryons accreting from typical interstellar gas.

Exercise 14.10 Consider transonic accretion with $\Gamma = 1$ (i.e., isothermal flow).
(a) How is Eq. (14.3.5) modified?
(b) Explain why Eqs. (14.3.21)–(14.3.25) are still valid for this case.

The boundary conditions at infinity (i.e., $u = 0$, $a = a_\infty$, $\rho = \rho_\infty$) do not uniquely specify the accretion solution to the Newtonian Eqs. (14.3.2) and (14.3.3) [or Eqs. (14.3.4) and (14.3.5)]. Indeed, as Bondi pointed out, there exists a *second* class of accretion solutions satisfying the same boundary conditions. This class is characterized by *subsonic* motion everywhere. A different subsonic solution exists for each value of λ in the range of $0 \leqslant \lambda < \lambda_s$; see Figure 14.1. Define r_s according to Eq. (14.3.14). The subsonic solutions again satisfy Eq. (14.3.21) for $r/r_s \gg 1$. For $r \ll r_s$, however, the flow differs significantly from transonic flow. In the subsonic cases, the second term on the left-hand side of Eq. (14.3.5) dominates the first term as r decreases, yielding for $1 < \Gamma < \frac{5}{3}$

$$\frac{1}{\Gamma - 1}\left[\left(\frac{a}{a_\infty}\right)^2 - 1\right] \approx \frac{GM}{a_\infty^2 r}, \qquad r \ll r_s \quad \left(1 < \Gamma < \frac{5}{3}; \text{subsonic}\right).$$

$$(14.3.28)$$

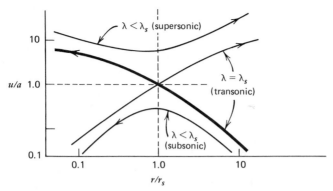

Figure 14.1 Sketch of some possible solutions to the Bondi equations for stationary, adiabatic, spherical flow about a gravitating point mass. The Mach number ($\equiv u/a$) profile is plotted for three values of the nondimensional parameter λ. Arrows pointing to the left denote accretion flows and to the right denote wind flows. The *heavy solid* line corresponds to transonic accretion.

Equation (14.3.28) is nothing more than hydrostatic equilibrium for a polytropic equation of state.

Exercise   14.11   Derive Eq. (14.3.28) directly by integrating the equation of hydrostatic equilibrium for a polytropic equation of state [i.e., Eq. (14.3.1)].

Thus the flow is "choked off" by back pressure in the subsonic regime. This back-pressure is generated by the large density gradient that builds up as matter accumulates in the vicinity of the central mass. From Eq. (14.3.28) we find

$$\frac{\rho}{\rho_\infty} = \left(\frac{a}{a_\infty}\right)^{2/(\Gamma-1)}$$

$$\approx \left[1 + \frac{(\Gamma-1)GM}{a_\infty^2 r}\right]^{1/(\Gamma-1)}$$

$$\approx \left[\frac{(\Gamma-1)GM}{a_\infty^2 r}\right]^{1/(\Gamma-1)}, \qquad r \ll r_s \quad \left(1 < \Gamma < \frac{5}{3}; \text{ subsonic}\right).$$

$$(14.3.29)$$

Exercise   14.12   Evaluate the subsonic density profile for $\Gamma = 1$ and $\Gamma = \frac{5}{3}$ for small r.

The density enhancement given by Eq. (14.3.29) for subsonic flow is clearly larger than the enhancement for transonic flow [Eq. (14.3.24)]. This general result is true for all $1 \leqslant \Gamma \leqslant \frac{5}{3}$.

Exercise   14.13   Evaluate the subsonic flow velocity for small r.

   Answer:

$$\frac{u}{a_\infty} \approx \frac{\lambda}{(\Gamma - 1)^{1/(\Gamma - 1)}} \left(\frac{GM}{a_\infty^2 r} \right)^{(2\Gamma - 3)/(\Gamma - 1)} \qquad (\Gamma \neq 1, \text{ subsonic}). \quad (14.3.30)$$

In the extreme limit corresponding to $\lambda = 0 = u$ (that is, no accretion) the subsonic regime reduces to a hydrostatic "extended atmosphere."

In general, boundary conditions at the surface of the accreting star will determine which regime—transonic or subsonic—is applicable in a given physical situation. For a star with a hard surface (e.g., a white dwarf or neutron star), steady-state subsonic flow is allowed. For black holes, the flow must be transonic (cf. Appendix G).

We note in passing that additional types of flow, corresponding to different boundary conditions, are described by the Bondi equations (14.3.4) and (14.3.5). In one, the motion corresponds to steady *outflow* from a central star. The inner regions of gas at $r < r_s$ move *outward* with subsonic velocities, cross a sonic point at $r = r_s$ and move supersonically toward infinity outside r_s. The outflow rate is given by Eq. (14.3.16) with $\lambda = \lambda_s$ (see Fig. 14.1). This solution may apply to stars that generate a stellar wind.[10] A second type of outflow corresponds to supersonic flow everywhere, with $\lambda > \lambda_s$ (see Fig. 14.1). Such flow can occur physically only as a result of some nonhydrodynamical mechanism of particle acceleration, not embodied in our equations.[11]

Finally, we note that the Bondi solution is applicable, with some modification, to the case of accretion onto a *moving* hole in a uniform medium. If the hole is moving very subsonically in the ambient medium, with velocity $V \ll a_\infty$, then the modifications are very small. If the hole is moving with arbitrary (nonrelativistic) velocity $V > a_\infty$, the effects of the hole's motion are more substantial. In order of magnitude the quantity a_∞^2, which appears in the Bondi solution for the accretion rate, density and temperature profile, and so on, must everywhere be replaced by the factor $(V^2 + a_\infty^2)$. Thus the accretion rate becomes

$$\dot{M} = 4\pi \tilde{\lambda} (GM)^2 \left(a_\infty^2 + V^2 \right)^{-3/2} \rho_\infty, \qquad (14.3.31)$$

where $\tilde{\lambda}$ is a constant of order unity. Equation (14.3.31) was originally suggested, up to the factor $\tilde{\lambda}$, by Bondi (1952). He thereby combined in a single formula the spherical accretion rate appropriate for subsonic flow far from the gravitating mass with the accretion rate estimated by Hoyle and Lyttleton (1939) and Bondi and Hoyle (1944) for supersonic flow.

[10] Parker (1965).
[11] Zel'dovich and Novikov (1971).

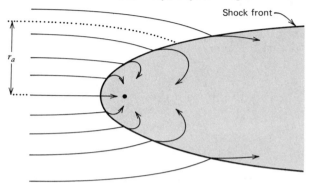

Figure 14.2 Streamlines of gas particles incident supersonically onto a gravitating point mass, as viewed in the rest frame of the mass.

If $V > a_\infty$, one expects a shock wave to form behind the hole (see Fig. 14.2).[12] After compression in the shock wave the gas temperature increases to $kT \sim m_B V^2$. For any V, gas particles within a distance $r_a \sim GM/(V^2 + a_\infty^2)$ from the black hole will be captured. Outside r_a, the directed kinetic energy of a gas particle relative to the hole ($\frac{1}{2}V^2$ per unit mass) will be greater than the gravitational potential energy (GM/r), so the hole's gravitational pull will be negligible. Gas pressure serves to symmetrize the flow inside r_a so that for $r \ll r_a$ the motion of the gas becomes quasiradial. In this inner region the Bondi solution is recovered. These qualitative expectations have been verified in detailed two-dimensional hydrodynamical calculations by Hunt (1971).

Accretion by a black hole moving supersonically through an ambient medium is likely to characterize a binary system in which the normal primary star lies inside its Roche lobe but emits a stellar wind. In this case, only the gas inside an accretion cylinder of radius $r_a = \xi 2GM/V_{rel}^2$, where M is the black hole mass, V_{rel} is the relative velocity between the hole and wind, and ξ is a constant of order unity, will be captured by the hole [cf. Section 13.7(b) and Fig. 13.11]. For this case the accretion rate will be given by Eq. (14.3.31) with $V^2 \rightarrow V_{rel}^2$.

Exercise   14.14

(a)   Estimate the rate of gas accretion onto the compact secondary M_x (the black hole) in Cyg X-1. Assume that accretion is from the stellar wind of the primary with the wind velocity constant and comparable to the escape velocity from the primary surface,

$$V_w = \eta \left(\frac{2GM_*}{R_*} \right)^{1/2}, \qquad \eta = \frac{V_w}{V_{esc}}.$$

Here $M_* \gg M_x$ and R_* are the mass and radius of the primary. Show that the accretion

---

[12]Salpeter (1964).

rate is given by (Shapiro and Lightman, 1976a)

$$\dot{M} \approx \tfrac{1}{4}\left(\frac{r_a}{a}\right)^2 \frac{V_{\text{rel}}}{V_w}\dot{M}_w$$

$$\approx \left(7 \times 10^{-9}M_\odot \text{ yr}^{-1}\right)\xi^2\eta^{-4}\left(\frac{M_x}{10M_\odot}\right)^2\left(\frac{M_*}{30M_\odot}\right)^{-8/3}$$

$$\times \left(\frac{R_*}{20R_\odot}\right)^2\left(\frac{P}{5^d.6}\right)^{-4/3}\left(\frac{\dot{M}_w}{10^{-6}M_\odot \text{ yr}^{-1}}\right). \qquad (14.3.32)$$

In Eq. (14.3.32), $\dot{M}_w$ is the steady-state, spherical wind ejection rate ($\dot{M}_w = 4\pi\rho r^2 V_w = $ constant), a is the separation between the primary and secondary, and P is the period of the orbit, assumed circular.

Hint: $V_{\text{rel}}^2 \approx V_x^2 + V_w^2$; where V_x is the velocity of the secondary relative to the primary. Note that the normalization of factors in Eq. (14.3.32) has been chosen to fit the observations of the Cyg X-1 binary system (cf. Section 13.5).

(b) Show that the normalization of parameters in Eq. (14.3.32) is consistent with the observed X-ray luminosity of $\sim 5 \times 10^{37}$ erg s^{-1}, assuming around 10% efficiency of conversion of rest mass to radiation.

(c) Calculate the fraction $\dot{M}/\dot{M}_w$ of emitted gas that actually accretes onto the compact secondary.

14.4   Radiation from Spherical Accretion onto Black Holes

In this section we estimate the amount of radiation emitted during steady, spherical accretion onto a stationary black hole. In general, the calculation is nontrivial, for it requires a self-consistent treatment of the hydrodynamical equations of motion coupled to the equations of radiative transfer. Moreover, the results depend strongly on the physical (boundary) conditions assumed to exist far from the black hole ($r \gg r_s$). In the following discussion we shall consider a simple case in which the hydrodynamical flow is, to lowest order, adiabatic, and the radiation losses represent a small perturbation. Thus, we can employ the adiabatic Bondi equations to determine the flow, and use the resulting flow profile to calculate the radiative emission rate. Provided this radiation rate is less than the rate of accretion of *thermal* energy, the flow will be nearly adiabatic and our analysis will be entirely self-consistent. When we are done, we will have demonstrated an important point: *spherical* accretion onto a black hole is *not necessarily* an efficient mechanism for converting rest-mass energy into radiation. This important result is, of course, very "regime" dependent. It *does* apply, however, to at least one astrophysically relevant situation: the accretion of interstellar gas onto a stellar mass black hole.[13]

---

[13]Shvartsman (1971); Shapiro (1973a,b) and (1974).

For definiteness, let us assume that a Schwarzschild black hole is at rest in a uniform ionized gas of pure hydrogen. We shall assume that the gas is at rest at infinity with density n_∞ and temperature T_∞ and accretes steadily onto the black hole. This situation will correspond to accretion in a typical interstellar "H II" (i.e., ionized hydrogen) region where $n_\infty \sim 1$ cm^{-3} and $T_\infty \sim 10^4$ K.

In the adiabatic limit, the flow will be governed by the Bondi equations, with one important exception: the adiabatic index $\Gamma = \Gamma(T)$ will vary with temperature T, hence with distance r from the hole. Specifically, as r decreases and kT increases above $m_e c^2$, the electrons become relativistic, causing Γ to fall below $\frac{5}{3}$. Basically, there are two distinct adiabatic flow regimes distinguished by $kT \ll m_e c^2$ and $kT \gg m_e c^2$ (but always $kT \ll m_p c^2$). In both regimes, Γ is essentially constant. For adiabatic changes, we have (recalling that here $\rho = mn$ denotes rest-mass density)

$$d\left(\frac{\varepsilon}{\rho}\right) = -P d\left(\frac{1}{\rho}\right). \tag{14.4.1}$$

Here ε is the total energy density, $\varepsilon = \rho c^2 + \varepsilon'$. Using Eq. (14.3.1) and integrating, we obtain for the thermal energy density in either regime

$$\varepsilon' = \frac{P}{\Gamma - 1}, \qquad \Gamma = \text{constant.} \tag{14.4.2}$$

In the regime $kT \ll m_e c^2$ (i.e., $T \ll 6 \times 10^9$ K), the hydrogen plasma is nonrelativistic and so

$$\varepsilon' = 3nkT, \qquad P = 2nkT. \tag{14.4.3}$$

The protons and electrons contribute equally since n = number density of protons = number density of electrons. The total energy density is

$$\varepsilon = n(m_p + m_e)c^2 + 3nkT. \tag{14.4.4}$$

Comparing Eqs. (14.4.2) and (14.4.3) gives the familiar result

$$\Gamma = \tfrac{5}{3}, \qquad kT \ll m_e c^2. \tag{14.4.5}$$

Nearer the black hole, we will find $m_e c^2 \ll kT \ll m_p c^2$ (i.e., 6×10^9 K $\ll T \ll 1 \times 10^{13}$ K). The electrons become relativistic while the protons remain nonrelativistic. Thus

$$\varepsilon' = \left(3 + \tfrac{3}{2}\right)nkT, \qquad P = 2nkT,$$

$$\varepsilon = \tfrac{9}{2}nkT + nm_p c^2, \qquad \Gamma = \tfrac{13}{9}, \qquad m_e c^2 \ll kT \ll m_p c^2. \tag{14.4.6}$$

Here we have used $\varepsilon = \varepsilon' = 3nkT$ for relativistic electrons.

Exercise 14.15 Show that for a fully ionized mixture of hydrogen and helium one obtains

$$\Gamma = \frac{\frac{13}{9}X + \frac{7}{3}(1 - X)}{X + \frac{5}{3}(1 - X)}$$

in place of Eq. (14.4.6). Here X is the fractional abundance by number of hydrogen ions, hence $1 - X$ is the fractional abundance of helium ions.

Exercise 14.16 Show that for relativistic electrons *and* protons (i.e., $kT \gg m_p c^2$), $\Gamma = \frac{4}{3}$. Explain why this is the same value of Γ obtained for a pure photon gas.

Since ε and P match continuously at $kT = \frac{2}{3}m_e c^2$, it is reasonable to define an "effective adiabatic index" Γ^* according to

$$\Gamma^* = \frac{5}{3}, \qquad \frac{kT}{m_e c^2} \leqslant \frac{2}{3}$$

$$= \frac{13}{9}, \qquad \frac{m_p}{m_e} \gg \frac{kT}{m_e c^2} > \frac{2}{3}. \qquad (14.4.7)$$

Of course the transition from the nonrelativistic to relativistic regime is not abrupt but smooth and gradual; however, Eq. (14.4.7) will suffice for our purposes.

It is now straightforward to obtain the flow solution. According to the discussion in Section 14.3, the accretion rate is determined by the gas parameters at large distances from the black hole where the electrons are still nonrelativistic. In this region we have $\Gamma^* = \frac{5}{3}$ and

$$a_\infty = \left[2 \times \frac{5}{3} \frac{kT_\infty}{m_p}\right]^{1/2}. \qquad (14.4.8)$$

The accretion rate $\dot{M}$ is given by Eq. (14.3.16), which is evaluated in Eq. (14.3.20) for the case considered here ($\lambda_s = \frac{1}{4}$). For large $r \gg r_a \equiv GM/a_\infty^2$, Eqs. (14.3.21) and (14.3.22) describe the flow parameters. For small $r \ll r_a$, there are two regions, depending upon whether or not the electrons are relativistic. Define r_* to be the transition radius at which $kT \equiv kT_* = \frac{2}{3}m_e c^2$. The flow parameters are then given by Eqs. (14.3.26) and (14.3.27) for the nonrelativistic domain $r_* < r < r_a$, where the velocity is subsonic.[14]

---

[14]As shown in Appendix G, when general relativity is taken into account, the flow for $\Gamma \equiv \frac{5}{3}$ becomes *transonic* at the radius $r_s = 3GM/4a_\infty c > 0$. The corresponding accretion rate, determined at r_s, is still given by Eq. (14.3.16) and the gas parameters remain comparable everywhere to their Newtonian values. For $\Gamma \equiv \Gamma^*$, the accretion rate is given by the *same* value provided $r_s > r_*$—that is, provided $\frac{10}{3}(m_e/m_p)c/a_\infty > 1$ or $T_\infty < 1 \times 10^7$ K [cf. Eqs. (14.4.9) and (14.3.19)]. The fact that Γ changes at r_* is irrelevant in determining $\dot{M}$ because the change cannot be "communicated" upstream against the supersonic flow. Alternatively, Eq. (14.3.16) will also give $\dot{M}$ rather accurately whenever $1 \gg \frac{5}{3} - \Gamma^* > \frac{9}{30}(a_\infty/c)^2 m_p/m_e$ for all $r > r_*$. This condition ensures that, although Γ^* is close to $\frac{5}{3}$ upstream, it deviates sufficiently to allow $r_s > r_*$ [cf. Eq. (14.3.14)].

Exercise 14.17

(a) Show that r_* is given by

$$r_* = \frac{9}{40} \frac{GM}{c^2} \frac{m_p}{m_e}.$$ (14.4.9)

(b) Evaluate the radii $r_h = 2GM/c^2$, r_*, and r_a numerically in terms of $M/M_\odot$ and $T_\infty/10^4$ K.

In the relativistic electron domain, $2GM/c^2 < r \ll r_*$, the flow is supersonic, so that Eqs. (14.3.23) and (14.3.24) apply for u and ρ, respectively. The temperature T is given by the general relationship $T \propto \rho^{\Gamma^* - 1}$, which may be combined with Eqs. (14.3.24) and (14.4.7) to yield

$$\frac{T}{T_*} \approx \left(\frac{r_*}{r} \right)^{2/3}.$$ (14.4.10)

Exercise 14.18

(a) Evaluate T at the event horizon and compare it with the temperature there for constant $\Gamma = \frac{5}{3}$ adiabatic flow [i.e., Eq. (G.46)].
Answer:

$$T_h \approx \frac{2}{3} \left(\frac{9}{80} \right)^{2/3} \left(\frac{m_e}{m_p} \right)^{1/3} \frac{m_p c^2}{k}$$

$$\approx 1.4 \times 10^{11} \text{ K} < T\left(\Gamma = \tfrac{5}{3} \right)$$ (14.4.11)

(b) Evaluate $\rho(r)/\rho_\infty = n(r)/n_\infty$ at the event horizon.
Answer:

$$n(r)/n_\infty \approx 3.7 \times 10^{11} \left(T_\infty/10^4 \right)^{-3/2}.$$ (14.4.12)

Note that a careful integration of the relativistic Euler equations using Eq. (14.4.7) gives $T = 1.0 \times 10^{11}$ K and $\rho/\rho_\infty = 3.9 \times 10^{11}(T_\infty/10^4 \text{ K})^{-3/2}$ at the horizon.[15] Replacing Eq. (14.4.7) by the exact, temperature-dependent continuous function $\Gamma = \Gamma(T)$ gives, instead, $T = 0.76 \times 10^{11}$ K and the same value of ρ/ρ_∞ at the horizon.[16]

[15]Shapiro (1973a).
[16]Brinkmann (1980).

Now that we know the flow parameters as a function of position we can calculate the emission. Most of the radiation will originate from the region just outside the event horizon, where the gas temperature and density reach their maximum values. For the temperature and densities of interest, the dominant emission mechanism will be *thermal bremsstrahlung* or *free-free emission*. Relativistic bremsstrahlung will be generated via the inelastic scattering of relativistic thermal electrons off (nonrelativistic) ions and off other electrons. The emissivity (emission rate per unit volume) in the extreme relativistic limit is given by

$$\Lambda_{\text{ff}} = (\Lambda_{ei} + \Lambda_{ee}) \text{ erg cm}^{-3} \text{ s}^{-1}, \tag{14.4.13}$$

where

$$\Lambda_{ei} = 12\alpha Z^2 r_0^2 n_e n_i ckT \left[\frac{3}{2} + \ln\left(\frac{2kT}{m_e c^2}\right) - 0.577 \right],$$

$$\Lambda_{ee} = 24\alpha r_0^2 n_e^2 ckT \left[\frac{5}{4} + \ln\left(\frac{2kT}{m_e c^2}\right) - 0.577 \right]. \tag{14.4.14}$$

In Eq. (14.4.14), $\alpha = e^2/\hbar c$ is the fine structure constant and $r_0 = e^2/m_e c^2$ is the classical radius of the electron.[17] For the pure hydrogen gas considered here, $Z = 1$ and $n_e = n_i \equiv n$.

Exercise 14.19
(a) Explain why $\Lambda_{ee}/\Lambda_{ei} \approx 2$ in the extreme relativistic limit for a pure hydrogen gas.
(b) What is the ratio $\Lambda_{ee}/\Lambda_{ei}$, approximately, in a nonrelativistic hydrogen gas?

Answer: $\dfrac{\Lambda_{ee}}{\Lambda_{ei}} \sim \dfrac{\langle v^2 \rangle}{c^2} \sim \dfrac{kT}{m_e c^2}$ (nonrelativistic gas).

Ignoring special relativistic effects due to the motion of the infalling gas and general relativistic effects arising from the strong gravitational field of the black hole, the total emitted luminosity of the gas is

$$L_{\text{ff}} \approx \int_{r_h}^{\infty} \Lambda_{\text{ff}} 4\pi r^2 \, dr \sim \Lambda_{\text{ff}} \left(\frac{4}{3}\pi r^3 \right) \Big|_{r_h}$$

$$\sim 8 \times 10^{20} \left(\frac{n_\infty}{1 \text{ cm}^{-3}}\right)^2 \left(\frac{T_\infty}{10^4 \text{ K}}\right)^{-3} \left(\frac{M}{M_\odot}\right)^3 \text{ erg s}^{-1}. \tag{14.4.15}$$

[17]Maxon (1972). The interested reader can assemble Eq. (14.4.14) approximately from the equations and discussion given in Jackson (1975), Section 15.3.

In obtaining Eq. (14.4.15) we have used the fluid parameters at the horizon derived approximately in Exercise 14.18. Because the integrand rises sharply as r decreases, the dominant contribution to the emission originates from the region just outside the event horizon. A careful relativistic integration of the emitted radiation yields[18]

$$L_{ff} = 1.2 \times 10^{21} \left(\frac{n_\infty}{1 \text{ cm}^{-3}} \right)^2 \left(\frac{T_\infty}{10^4 \text{ K}} \right)^{-3} \left(\frac{M}{M_\odot} \right)^3 \text{ erg s}^{-1}. \quad (14.4.16)$$

This equation accounts properly for the relativistic flow, Doppler and gravitational redshifts, photon recapture by the black hole, and so on. The corresponding spectrum is moderately flat below energy

$$h\nu \sim kT_h \sim 10 \text{ MeV}, \quad (14.4.17)$$

and falls off exponentially above this energy. Thus, the emission consists largely of very hard X-rays and γ-rays. For characteristic interstellar densities, the accreting material remains "optically thin" to these emergent photons and they travel outward toward infinity unimpeded by scattering or absorption.

Exercise 14.20 The dominant opacity source for typical $E \sim 10$ MeV photons is Compton scattering off relativistic electrons, for which the cross section is $\sigma(E) \sim \sigma_T(m_e c^2/E)$, where σ_T is the Thomson cross section. Show that the corresponding scattering optical depth $\tau = \int_{r_h}^\infty \sigma n_e \, dr$ is much less than unity for typical interstellar densities (i.e., $n_\infty \sim 1$ cm^{-3}). Interpret your result.

What is remarkable about Eq. (14.4.16) is that it represents a very low efficiency for the conversion of rest-mass energy into radiation:

$$\varepsilon = \frac{L_{ff}}{\dot{M}c^2} \sim 6 \times 10^{-11} \left(\frac{n_\infty}{1 \text{ cm}^{-3}} \right) \left(\frac{T_\infty}{10^4 \text{ K}} \right)^{-3/2} \left(\frac{M}{M_\odot} \right). \quad (14.4.18)$$

Even if the hole rotates with its maximal allowed angular momentum, the efficiency of spherical accretion only increases by a meagre 15% above the value in Eq. (14.4.18).[19]

The presence of tangled magnetic fields, leading to synchrotron radiation in addition to bremsstrahlung, can increase the efficiency somewhat.[20] In general, however, the spherical accretion of interstellar gas by stellar mass black holes tends to be an *inefficient* radiation mechanism. This fact is in contrast to spherical

---

[18]Shapiro (1973a), (1974).

[19]Shapiro (1974).

[20]Shvartsman (1971); Novikov and Thorne (1973); Shapiro (1973b); Meszaros (1975); Ipser and Price (1977); see also Exercise 14.21.

accretion onto a neutron star for which $\varepsilon \sim 0.1$, or to disk accretion onto a black hole, for which $\varepsilon = 0.05 - 0.42$, depending on the value of the hole's angular momentum a/M and on the direction of the gas rotation relative to the hole (cf. Section 14.5). With different gas boundary conditions in lieu of typical interstellar parameters or with turbulence, the efficiency may be quite different. However, reliable relativistic calculations do not now exist for such regimes.

Exercise 14.21 Assume that the accreting gas possesses an equipartition magnetic field, where

$$\frac{B^2}{8\pi} \sim \frac{GM\rho}{r}. \tag{14.4.19}$$

Such a field might arise if everywhere the growth of a frozen-in, tangled field threading the infalling plasma were limited by the *magnetic reconnection* of oppositely directed, radial field lines (Shvarstman, 1971; Shapiro, 1973b). Reconnection timescales are given by $t_{\text{recon}} \sim r/v_A$, where $v_A \sim (B^2/8\pi\rho)^{1/2}$ is the Alfvén velocity or magnetic "sound speed." The amplification of the B field is limited by the condition $t_{\text{recon}} \sim t_{\text{free-fall}}$, where $t_{\text{free-fall}} \sim r/(GM/r)^{1/2}$ is the infall time. This condition gives Eq. (14.4.19).

(a) Estimate the synchrotron luminosity, L_{syn} resulting from relativistic electrons gyrating around equipartition field lines.

Hint: $\Lambda_{\text{syn}} = \dfrac{16}{3}\dfrac{e^2}{c}\left(\dfrac{eB}{m_e c}\right)^2\left(\dfrac{kT}{m_e c^2}\right)^2 n_e$ erg cm^{-3} s^{-1}

Answer: $L_{\text{syn}} \sim 10^{27}(n_\infty/1 \text{ cm}^{-3})^2(T_\infty/10^4 \text{ K})^{-3}(M/M_\odot)^3$ erg s^{-1}

Note: A general relativistic calculation along these lines gives luminosities lower by more than a factor of 10.

(b) Estimate the characteristic frequency of the emitted radiation.

Hint: $h\nu \sim \hbar\omega_H \sim \hbar(eB/m_e c)\gamma^2$, where ω_H is the synchrotron frequency and where $\gamma \equiv E_e/m_e c^2$ satisfies $\langle\gamma^2\rangle = 12kT/m_e c^2$ for a relativistic, thermal electron gas.

Answer: $\nu \sim 10^{13}$ Hz or $\lambda \sim 30$ μm (infrared).

14.5 Disk Accretion onto a Compact Star

Accretion onto a compact star in a binary system may be far from spherical, since the accreted gas possesses angular momentum. Consider the case when the compact object is a black hole. Whenever the angular momentum per unit mass, $\tilde{l}$, exceeds $\sim r_I c$, where r_I is the innermost stable circular orbit (cf. Chapter 12), centrifugal forces will become significant before the gas plunges through the event horizon. In this case, the gas will be thrown into circular orbits about the hole, moving inward only after viscous stresses in the gas have transported away the excess angular momentum. The corresponding viscous heating and increased time spent outside the horizon result in a significant increase in the luminosity over the values computed in the previous section for spherical accretion. In fact, the total energy radiated by a unit mass as it drifts inward through the disk toward the hole must be equal to the gravitational binding energy of the unit mass at the inner edge of the disk at $r = r_I$ (assuming little additional energy is radiated later as the mass plunges rapidly from $r = r_I$ to the event horizon). Thus the radiation efficiency for disk accretion onto a black hole is guaranteed to be at least[21]

$$\varepsilon = \tilde{E}_I = 0.057 \quad \text{for a nonrotating hole,}$$
$$= 0.42 \quad \text{for a maximally rotating hole and prograde disk.}$$

$$(14.5.1)$$

For disk accretion onto a white dwarf or neutron star, the inner edge of the disk r_I might be close to the star's surface.[22] In this case, maximum disk efficiencies are comparable to

$$\varepsilon \sim \frac{1}{2}\frac{GM}{Rc^2} \sim 10^{-4} \quad \text{for a white dwarf,}$$
$$\sim 10^{-1} \quad \text{for a neutron star.}$$

$$(14.5.2)$$

Thus, in order of magnitude, the total *disk* luminosity must be

$$L \sim 10^{-4}\dot{M}c^2 \sim \left(10^{34}\ \text{erg s}^{-1}\right)\left(\frac{\dot{M}}{10^{-9}M_\odot\ \text{yr}^{-1}}\right) \quad \text{for a white dwarf,}$$

$$\sim 10^{-1}\dot{M}c^2 \sim \left(10^{37}\ \text{erg s}^{-1}\right)\left(\frac{\dot{M}}{10^{-9}M_\odot\ \text{yr}^{-1}}\right) \quad \begin{array}{l}\text{for a neutron star or}\\ \text{black hole.}\end{array}$$

$$(14.5.3)$$

For accretion onto a neutron star or white dwarf, there is, in addition to the disk luminosity, a luminosity of comparable magnitude emitted when the fluid collides with the stellar surface (or surface B-fields), decelerates, and dissipates its kinetic energy. Consequently, the order of magnitude of the *total* luminosity does not depend on the nature of the flow—disk or spherical—in the case of neutron star or white dwarf accretion. In either scenario *all* of the gravitational binding energy at the surface is released as radiation as gas falls through the star's potential field and comes to rest on the stellar surface. The total luminosity is

---

[21]See Eq. (12.4.30) and discussion following Eq. (12.7.25).
[22]Here we are assuming that only very weak magnetic fields extend from their surfaces; see Chapter 15 for alternative scenarios.

thus *always* equal to $L = G\dot{M}M/R$ (ignoring general relativistic corrections).[23] For accretion onto a black hole, however, the situation can depend quite sensitively on the flow geometry. For disk accretion, at least, the efficiency is invariably high.

Disk accretion will always occur in a binary system when the companion supplies gas to the secondary via Roche-lobe overflow. In this case the intrinsic angular momentum of the gas $\tilde{l}$ greatly exceeds $r_l c$, as is obvious from Figure 13.11*b*. When the companion supplies gas by a spherical wind, the situation is not so obvious (cf. Fig. 13.11*a*). Whether or not $\tilde{l}$ exceeds $r_l c$ depends sensitively on the wind and binary system parameters.[24]

Exercise 14.22 Return to Exercise 14.14 for accretion onto a compact secondary via a spherical wind. For the same system:

(a) Show that the net angular momentum per unit mass carried by the accreted gas is

$$\tilde{l} = \frac{1}{2} V_x a \left(\frac{r_a}{a} \right)^2.$$
(14.5.4)

(b) Estimate the ratio of $\tilde{l}$ from spherical wind accretion to $\tilde{l}$ from Roche lobe overflow.

Answer: $(r_a/a)^2 \sim 10^{-3} - 10^{-4}$

(c) Assume that the outer edge of the disk forms at the radius r_D at which Eq. (14.5.4) equals the angular momentum per unit mass of a gas element in circular orbit about the hole: $\tilde{l} = (GMr_D)^{1/2}$. Calculate r_D/r_I for the parameters appropriate to the Cyg X-1 binary system. Comment.

Answer:

$$\frac{r_D}{r_I} \approx 160 \xi^4 \eta^{-8} \left[1 + \left(\frac{V_x}{V_w} \right)^2 \right]^{-4} \left(\frac{M_x}{10 M_\odot} \right)^2 \left(\frac{M_*}{30 M_\odot} \right)^{-4} \left(\frac{P}{5^d.6} \right)^{-2} \left(\frac{R_*}{20 R_\odot} \right)^4$$

Disk accretion may also be likely for the flow of turbulent gas onto a supermassive black hole in a dense galactic nucleus or quasar or at the center of the Galaxy.

In this section we will discuss the structure of a *Keplerian accretion disk* around a central point mass. Historically, the significance of angular momentum for accretion in a binary system was first emphasized by Prendergast.[25] He constructed models for disk accretion onto white dwarfs in binary systems. Subse-

---

[23] In general relativity the luminosity from the steady accretion of matter falling from rest at infinity onto a spherical star is $L = [1 - (1 - 2GM/Rc^2)^{1/2}]\dot{M}c^2$, assuming the matter comes to rest at the surface [cf. Eq. (12.4.9)].

[24] Illarionov and Sunyaev (1975); Shapiro and Lightman (1976a); see Exercise 14.22.

[25] See Prendergast and Burbidge (1968).

quently, Shakura (1972), Pringle and Rees (1972), Shakura and Sunyaev (1973), and Novikov and Thorne (1973) constructed Newtonian accretion disk models for flow onto neutron stars and black holes. In addition, Novikov and Thorne (1973) considered the effects of general relativity on the inner parts of the accreting disk. Lynden-Bell (1969) was the first to propose that galaxy cores might contain supermassive black holes surrounded by gaseous accretion disks.[26]

We begin with a qualitative overview of disk accretion. Consider what happens as gas is transferred from the surface of a normal star to a compact companion. If the process occurs by Roche lobe overflow, gas is transferred through the critical Lagrange point L_1, after which gravitational and Coriolis forces combine to keep it in a roughly circular orbit about the compact star (cf. Fig. 13.12). Some of the gas dumped earlier orbits the compact star in an accretion disk. The incoming gas interacts with the gas in the disk via viscous stresses. Some of the incoming gas acquires angular momentum from the disk via viscous torques and subsequently gets ejected from the disk region; part of this gas falls back onto the normal star and part is thrown out of the entire system through the Lagrange point L_2. If the process occurs by accretion from a stellar wind, the outer boundary of the disk is likely to be much closer to the compact star. Outside this boundary the accretion will be almost spherical, changing character at that radius at which the intrinsic angular momentum per unit mass of accreted gas from the wind equals the angular momentum of an element in circular, Keplerian orbit about the compact star (cf. Fig. 13.11b and Exercise 14.22).

However the disk is formed, most of the gravitational energy is released and most of the radiation originates from the *innermost* parts of the disk closest to the compact star. Gas deposition and angular momentum removal occur predominantly in the outer parts. Because we wish to examine the emitted radiation, we shall henceforth focus our attention on the inner regions of the disk.

While orbiting the compact star in a nearly circular fashion, each gas element in the disk acquires a small, inward radial motion as viscous torques remove angular momentum. Gas thus drifts toward the compact star in a tightly wound spiral trajectory as its angular momentum is transported outward from the inner to the outer regions of the disk. Simultaneously, the viscous stresses, acting in response to the shearing orbital motion of the gas, generate "frictional" heat. Most of this heat is then radiated away from the top and bottom faces of the disk in steady state.

Let us now examine quantitatively the structure of the inner regions of a Newtonian accretion disk about a compact star of mass M. Assume that gas is deposited into the disk at a constant rate $\dot{M}$ and accretes onto the compact star at the same rate. Ignore any tidal gravitational forces exerted on the disk by the

---

[26]Reviews and references to recent work on disk accretion are given in Lightman, Shapiro and Rees (1978) and Pringle (1981). The discussion in Section 14.5 below will follow closely the treatment of Novikov and Thorne (1973) for Newtonian disk accretion.

normal companion. Assume that the central plane of the disk lies in the equatorial plane of the compact star, defined by $z = 0$.

In Keplerian circular motion, each element possesses specific angular momentum $\tilde{l}$, where

$$\tilde{l} = (GMr)^{1/2}. \tag{14.5.5}$$

For small r, near the stellar surface, this value of $\tilde{l}$ is far less than $\tilde{l}$ at the outer edge of the disk, r_D. So in steady state the rate at which angular momentum must be removed from the disk by passing gas must be

$$\dot{J} = \dot{M}\tilde{l}(r_D) = \dot{M}(GMr_D)^{1/2}. \tag{14.5.6}$$

Define $2h$ to be the disk thickness and Σ to be the surface density of the disk at radius r. Then we may write

$$\Sigma \equiv \int_{-h}^{h} \rho \, dz \approx 2h\rho, \tag{14.5.7}$$

where z measures vertical height perpendicular to the disk midplane and where ρ is evaluated at $z = 0$ on the right-hand side of Eq. (14.5.7). Here and below we shall average z-dependent quantities in the vertical direction, replacing integrals of products by products of averages. Let v_r be the magnitude of the *inward* radial velocity of the gas ($v_r > 0$) and let v_ϕ and Ω be the Keplerian orbital velocity and angular velocity, respectively[27]:

$$v_\phi = r\Omega = \left(\frac{GM}{r}\right)^{1/2}. \tag{14.5.8}$$

For relevant viscosities, v_r will always be much smaller than v_ϕ.

We shall assume that the disk is thin—that is, that h everywhere satisfies

$$h(r) \ll r. \tag{14.5.9}$$

We shall find that Eq. (14.5.9) requires the temperature to be "cool" in that $kT \ll GMm_p/r$. Such low temperatures will be achieved provided the heat generated by viscous stresses is radiated away efficiently (and not stored) by the disk. Thus, in contrast to spherical accretion, thin disk accretion *must* be highly nonadiabatic [cf. Eq. (14.3.27), i.e., $kT \sim GMm_p/r$ for spherical accretion with

---

[27]In Chapters 14 and 15 all spatial vector and tensor components refer to the "usual" orthonormal basis vectors (e.g., $\mathbf{v} = v_r\mathbf{e}_{\hat{r}} + v_\phi\mathbf{e}_{\hat{\phi}} + v_z\mathbf{e}_{\hat{z}}$ in cylindrical geometry). However, we shall henceforth follow the convention of omitting the caret $\wedge$ in component subscripts. No confusion should arise in what follows from this simplification.

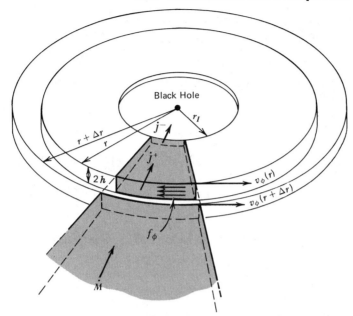

Figure 14.3   Slice of a thin, Keplerian accretion disk around a central black hole.

$\Gamma = \frac{5}{3}$.] In the limit in which Eq. (14.5.9) is valid, the coupled, two-dimensional axisymmetric flow equations decouple into separate equations for the radial and vertical directions. This approximate decoupling simplifies the hydrodynamic equations considerably.

Define f_ϕ, to be the viscous stress (force per unit area) exerted in the ϕ direction by fluid elements at r on neighboring elements at $r + dr$ (cf. Fig. 14.3). The stress is related to the stress tensor according to

$$f_\phi = -t_{r\phi},\tag{14.5.10}$$

where for a Keplerian disk

$$t_{r\phi} = -\frac{3}{2}\eta\Omega = -\frac{3}{2}\eta\left(\frac{GM}{r^3}\right)^{1/2}.\tag{14.5.11}$$

Here η is the coefficient of dynamic viscosity (g cm^{-1} s^{-1}). For a brief discussion of the hydrodynamic equations of motion for a viscous fluid, see Appendix H, where Eqs. (14.5.10) and (14.5.11) are derived. We shall postpone a discussion of the origin of the viscosity to later. Finally, define F to be the total radiation flux emitted from the upper (or lower) face of the disk.

In steady state, the disk structure is determined by solving simultaneously four conservation equations (conservation of mass, angular momentum, energy, and vertical momentum). In addition, a viscosity law for η (e.g., expressed in terms of local thermodynamic parameters) must be specified as well as a law describing the transport of radiation from the center to the surface. We discuss these equations immediately below.

(a) Rest-Mass Conservation

Integrating the steady-state continuity equation for the mass flow, $\nabla \cdot (\rho \mathbf{v}) = 0$, yields

$$\dot{M} = 2\pi r \Sigma v_r = \text{constant.} \tag{14.5.12}$$

Equation (14.5.12) says that the rate of mass flowing inward through a cylinder of radius r is independent of r.

(b) Angular Momentum Conservation

Let $\dot{J}^+ = \dot{M}(GMr)^{1/2}$ be the inward rate of the angular momentum transport across radius r in the disk due to inflowing gas (cf. Fig. 14.3). Let $\dot{J}^-$ be the rate at which angular momentum is consumed by the compact star. Since the specific angular momentum deposited in the compact star cannot exceed the value $\tilde{l}(r_I)$ at the inner edge of the disk, we may write

$$\dot{J}^- = \beta \dot{M}(GMr_I)^{1/2}, \qquad \text{where } |\beta| \leqslant 1. \tag{14.5.13}$$

(For example, for accretion onto a black hole, $\beta \approx 1$.)[28] Angular momentum conservation requires that the *net* rate of change of angular momentum within r equal the torque exerted by the viscous stress. Accordingly,

$$\text{torque} = \left(\text{force along } \mathbf{e}_\phi/\text{area}\right) \times (\text{area}) \times (r) = \dot{J}^+ - \dot{J}^-,$$

or

$$(f_\phi)(2\pi r \cdot 2h)(r) = \dot{M}\left[(GMr)^{1/2} - \beta(GMr_I)^{1/2}\right]. \tag{14.5.14}$$

Note that the required stress f_ϕ is determined uniquely in steady state by M and $\dot{M}$.

---

[28] Note that the condition $\beta = 1$ for a black hole is only an approximation. It is equivalent to a zero stress boundary condition at r_I [cf. Eq. (14.5.14)] and results in $\Sigma \to \infty$ and $\rho \to \infty$ at this radius [cf. Eq. (14.5.37)]. Setting $\beta = 1$ requires that within r_I the gas spirals into the hole rapidly without radiating. Although not strictly valid, such an idealization is probably justified; see Stoeger (1980).

(c)   Energy Conservation

From Eq. (H.6), we know that entropy (heat) is generated by viscosity at the rate

$$\dot{Q} \equiv \rho T \dot{s} = \tfrac{1}{2}\frac{t^2}{\eta} \approx \frac{(t_{\phi r})^2}{\eta} = \frac{-f_\phi t_{\phi r}}{\eta}. \qquad (14.5.15)$$

Using Eqs. (14.5.11) and (14.5.14) in Eq. (14.5.15) we find

$$2h\dot{Q} = \frac{3\dot{M}}{4\pi r^2}\frac{GM}{r}\left[1 - \beta\left(\frac{r_I}{r}\right)^{1/2}\right]. \qquad (14.5.16)$$

Let us assume that this heat is not stored, but is totally radiated away. Then Eq. (14.5.16) gives directly the integrated flux emitted from the top and bottom faces of the disk at r:

$$F(r) = \frac{1}{2} \times 2h\dot{Q} = \frac{3\dot{M}}{8\pi r^2}\frac{GM}{r}\left[1 - \beta\left(\frac{r_I}{r}\right)^{1/2}\right]. \qquad (14.5.17)$$

(Since the disk is thin, the emission is directed primarily in the vertical rather than the radial direction.)

The significant feature of Eq. (14.5.17) is that the radial distribution of the emitted flux is completely *independent* of the viscosity law, which we have yet to specify.

The total luminosity of the disk is

$$L = \int_{r_I}^{\infty} 2F \times 2\pi r\, dr = \left(\frac{3}{2} - \beta\right)\frac{GM\dot{M}}{r_I}. \qquad (14.5.18)$$

This luminosity is exactly what one expects from energy conservation. The *Newtonian* gravitational binding energy per gram at r_I is $E_B = GM/2r_I$ and the rotational kinetic energy per gram extracted from the compact star is (Exercise 14.23) $E_R = (1 - \beta)GM/r_I$. Accordingly, Eq. (14.5.18) correctly gives $L = (E_B + E_R)\dot{M}$.

Exercise   14.23   Verify the expression quoted for E_R using Eq. (14.5.13).
   Hint: Calculate the work done by the torque exerted by the star on the disk at r_I.

Exercise   14.24   Calculate the radiation efficiency of a thin, *Newtonian* accretion disk that extends inward to $r_I = 6GM/c^2$ around a central black hole. Compare your result to the efficiency of a relativistic disk around a nonrotating black hole.
   Answer: $\varepsilon = \frac{1}{12}$; 8.3% vs. 5.7%

(d) Vertical Momentum Conservation

Since there is no net motion of the gas in the vertical direction, momentum conservation along $\mathbf{e}_z$ reduces to a hydrostatic equilibrium condition. Equating the component of the gravitational force of the compact star along $\mathbf{e}_z$ to the vertical pressure gradient in the disk gives

$$\frac{1}{\rho}\frac{dP}{dz} = -\frac{GM}{r^2}\frac{z}{r} \quad (z \ll r). \tag{14.5.19}$$

Replacing the differentials in Eq. (14.5.19) by finite differences (that is, setting $\Delta P \approx P$, where P is the pressure evaluated at $z = 0$, and setting $\Delta z \approx h$) yields for the disk half-thickness

$$h \approx \left(\frac{P}{\rho}\right)^{1/2}\left(\frac{r^3}{GM}\right)^{1/2} \approx \frac{c_s}{\Omega}. \tag{14.5.20}$$

In Eq. (14.5.20), c_s is the sound speed in the disk midplane.

Exercise 14.25 Consider the thin disk requirement, Eq. (14.5.9), which, together with Eq. (14.5.20) demands

$$\frac{h}{r} \approx \frac{c_s}{v_\phi} \ll 1. \tag{14.5.21}$$

What constraint does Eq. (14.5.21) impose on the disk interior temperature T in regions that are gas pressure dominated [cf. the first term in Eq. (14.5.29)]?

(e) The Viscosity Law

No piece of physics is more poorly understood in the theory of disk accretion then the nature of the viscosity. The source of viscosity is likely to be small-scale turbulence in the gas-dynamical flow. Random magnetic fields, which are dragged in with the accreting plasma, sheared by the disk differential velocity and then reconnect at the boundaries between chaotic cells, should also contribute significantly to the viscosity.[29] Fortunately, the integrated flux profile given by Eq. (14.5.17) and the total luminosity are independent of the uncertain viscosity law. Unfortunately, details of the disk structure and the emitted radiation spectrum depend on the viscosity.

To make progress, we shall follow Shakura and Sunyaev (1973) and construct a plausible turbulent viscosity–shear-stress relation based on dynamical considerations. In turbulent motion, the coefficient of dynamic viscosity η is given by

$$\eta \approx \rho v_{\text{turb}} l_{\text{turb}}, \tag{14.5.22}$$

[29]See, for example, Novikov and Thorne (1973); Eardley and Lightman (1975); Pringle (1981).

where v_{turb} is the velocity of turbulent cells relative to the mean gas motion and l_{turb} is the size of the largest turbulent cells.[30] Shocks will dissipate turbulent kinetic energy into heat whenever the motion is supersonic, so we require $v_{\text{turb}} \lesssim c_s$. The cell sizes are bounded by the disk thickness, so $l_{\text{turb}} \lesssim h$. Using Eqs. (14.5.10), (14.5.11), and (14.5.20), we then have a bound on the stress:

$$f_\phi = -t_{\phi r} \lesssim (\rho c_s h)\Omega \approx \rho c_s^2 \approx P. \qquad (14.5.23)$$

We may then write, in general,

$$f_\phi = \alpha P, \qquad (14.5.24)$$

where we have introduced a nondimensional viscosity parameter, α, which we cannot compute in any detail, but which satisfies

$$\alpha \lesssim 1. \qquad (14.5.25)$$

Models constructed using Eq. (14.5.24) are called "α-disks." Such models typically leave α as a free, constant parameter in the disk structure equations. Interestingly, α can be calibrated empirically from the time-dependent spectra of the observed outbursts in mass-transferring binaries containing dwarf novae. Comparing the observations from such "cataclysmic variables" with the time-dependent flux expected from a thin accretion disk in the system gives[31] α in the range 0.1–1. This value is consistent with the estimates of Eardley and Lightman (1975) who consider, from first principles, viscosity arising from shear amplification and reconnection of a chaotic magnetic field. Their viscosity law gives $0.01 \lesssim \alpha \lesssim 1$.[32]

(f)   Opacity

For typical accretion parameters involving compact stars with $M \sim M_\odot$, the dominant source of photon *absorption* in the disk is nonrelativistic thermal bremsstrahlung or "free-free" transitions. There may be comparable (but somewhat smaller) contributions from "bound-bound" line transitions and "bound-

---

[30]See, for example, Landau and Lifshitz (1959), Section 31.

[31]Lynden-Bell and Pringle (1974); Bath and Pringle (1981). It is not clear at present whether or not α remains at this level or falls to a smaller value during the quiescent period between outbursts.

[32]An entirely different approach to employing magnetic fields to transfer away angular momentum has been made by Blandford (1976) and by Lovelace (1976); see also Blandford and Znajek (1977). They find that if the disk contains an *ordered* magnetic field with a large perpendicular component, both energy *and* angular momentum can be transported away via a magnetized relativistic wind. The "pulse-like" electrodynamical model can give rise to twin "jets" of relativistic plasma moving perpendicular to the disk. Under extreme conditions, the magnetic torque can replace the viscous torque considered in Eq. (14.5.24).

free" ionization transitions. The frequency-averaged, Rosseland mean absorption opacity is thus (cf. Appendix I)

$$\bar{\kappa}_{abs} \simeq \bar{\kappa}_{ff} \simeq 0.64 \times 10^{23} \left(\rho \left[\text{g cm}^{-3} \right] \right) \left(T[\text{K}] \right)^{-7/2} \text{cm}^2 \text{ g}^{-1}. \quad (14.5.26)$$

The major source of photon *scattering* is Thomson scattering, for which

$$\bar{\kappa}_{scatt} \simeq \bar{\kappa}_{es} = 0.40 \text{ cm}^2 \text{ g}^{-1}. \quad (14.5.27)$$

Typically, absorption dominates scattering in the cool, outer regions of the disk at large radii, while scattering dominates absorption in the hot, inner regions. Roughly, the total Rosseland mean opacity satisfies [cf. Eq. (I.30)]

$$\frac{1}{\bar{\kappa}(\rho, T)} \approx \frac{1}{\bar{\kappa}_{scatt}} + \frac{1}{\bar{\kappa}_{abs}}. \quad (14.5.28)$$

(g)   Pressure

The total pressure of the disk material is, essentially, the sum of thermal gas and radiation pressure. For ionized hydrogen this yields

$$P(\rho, T) \simeq \frac{2\rho kT}{m_p} + P_{rad}, \quad (14.5.29)$$

where, in local thermodynamic equilibrium, we have

$$P_{rad} \simeq \tfrac{1}{3} aT^4. \quad (14.5.30)$$

Typically, gas pressure dominates radiation pressure throughout the disk except (depending upon parameters) in the innermost regions where the temperature is high.

(h)   Radiative Transport

For typical parameters, it is by radiation, rather than by conduction or convection, that the heat generated internally by viscosity is transported vertically through the disk before being radiated at the surface. There are, in general, several different regimes of radiative transport that may apply in different regions of the disk and (depending upon the accretion rate, mass of the compact star, etc.) in different disk models.

If the total optical thickness, τ, of the disk (measured in the vertical direction) exceeds unity, photons are transported to the surface via *diffusion* [Eq. (I.34)]

$$F(r, z) = -\frac{c}{3} \frac{d(aT^4)}{d\tau}, \qquad \tau > 1. \quad (14.5.31)$$

Here $F(r, z)$ is the vertical photon flux and we have assumed local thermodynamic equilibrium between the photon gas and the matter.[33] The quantity τ is the optical depth computed from the *total* Rosseland-mean opacity:

$$\tau = \int_0^h \bar{\kappa}\rho \, dz \approx \bar{\kappa}(\rho, T)\Sigma \qquad (14.5.32)$$

(cf. Appendix I). Replacing differentials by finite differences, we may integrate Eq. (14.5.31) approximately to yield for the surface flux $F(r, z = h) = F(r)$,

$$F(r) \approx \frac{acT^4}{\tau} \approx \frac{acT^4}{\bar{\kappa}\Sigma}, \qquad \tau(r) > 1. \qquad (14.5.33)$$

In Eq. (14.5.33) the quantities Σ and F are given by Eqs. (14.5.7) and (14.5.17), respectively, and T and ρ are to be evaluated near the disk center at $z = 0$. This optically thick regime applies in the distant outer regions of virtually all disk models, as well as in the inner regions of all but the hottest disks.

If τ is less than unity, than the disk becomes "optically thin" to outgoing photons. In this case the photons can escape freely from their point of emission without ever undergoing absorption or scattering. In this case Eq. (14.5.33), appropriate only for diffusion, must be replaced by

$$F(r) \approx \int_0^h \Lambda(\rho, T) \, dz \approx h\Lambda(\rho, T), \qquad \tau(r) < 1, \qquad (14.5.34)$$

where $\Lambda(\rho, T)$ is the average photon emissivity (erg s^{-1} cm^{-3}) in the disk. Typically, Λ is due to thermal bremsstrahlung (free-free emission) and/or to Comptonization, which is discussed in Appendix I.

Exercise   14.26   Show (roughly) that Eqs. (14.5.33) and (14.5.34) are equivalent when $\tau \sim 1$ whenever the opacity is dominated by a true thermal absorption process.

Hint: Use Kirchhoff's Law.

(i)   Solution: Structure of the "Standard" Disk Model

Equations (14.5.7), (14.5.12), (14.5.14), (14.5.17), (14.5.20), (14.5.24), (14.5.28), (14.5.29), and (14.5.33) [or (14.5.34)] give nine algebraic relations for the nine quantities $\rho(r), h(r), \Sigma(r), v_r(r), P(r), T(r), f_\phi(r), \bar{\kappa}(r)$, and $F(r)$ as functions of $r, \dot{M}$, and M. The algebraic solution is somewhat tedious, but has been obtained by Shakura and Sunyaev (1973) and by Novikov and Thorne (1973). They find

---

[33]Actually, Eq. (14.5.31) is not likely to apply in the deep interior of the disk where turbulent transport is probably important. Radiative diffusion *does* apply in the disk outer atmosphere, where the emitted spectrum is formed, whenever $\tau > 1$.

that, for fixed values of M and $\dot{M}$, the disk can conveniently be divided into *three* distinct regions, depending on r. These regions are:

1. An *outer* region, at large r, in which gas pressure dominates radiation pressure and in which the opacity is controlled by free-free absorption;
2. A *middle* region, at smaller r, in which gas pressure dominates radiation pressure but the opacity is mainly due to electron scattering; and
3. An *inner* region, at very small r, in which radiation pressure dominates gas pressure and again, scattering dominates absorption in the opacity.

(For some choices of $\dot{M}$, the inner and middle regions may not exist at all.) Typically, the transition from the outer to middle region occurs when $\bar{\kappa}_{ff} \sim \bar{\kappa}_{es}$, which occurs at

$$\frac{r_{om}}{r_s} = 4 \times 10^3 \left(\frac{\dot{M}_{17}}{M/M_\odot} \right)^{2/3} \mathcal{G}^{2/3}. \qquad (14.5.35)$$

Here[34] $r_s \equiv GM/c^2$, $\mathcal{G}$ (≈ 1) is defined in Exercise 14.28, and $\dot{M}_{17}$ is in units of 10^{17} g s^{-1}. (Recall that 10^{17} g s$^{-1} \sim 10^{-9} M_\odot$ yr^{-1}, a value for $\dot{M}$ that gives a total luminosity of $L = \varepsilon \dot{M} c^2 \sim 10^{37}$ erg s^{-1} for an efficiency $\varepsilon \sim 10\%$; this luminosity is typical of strong Galactic X-ray sources.) The transition from the middle to inner region occurs when $P_g \sim P_{rad}$, which occurs at[35]

$$\frac{r_{mi}}{r_s} = 80 \alpha^{2/21} \left(\frac{M}{M_\odot} \right)^{-2/3} \dot{M}_{17}^{16/21} \mathcal{G}^{16/21}. \qquad (14.5.36)$$

It is clear from Eq. (14.5.17) that the emission from the disk peaks at small radii $r \sim 10 r_s$. Thus, for stellar mass black holes, most of the radiation (and virtually all of the X-rays) originates from the "inner region" of the disk.

Exercise 14.27 Consider a typical binary X-ray source with $M/M_\odot = \dot{M}_{17} = 1 = \alpha$.

(a) Assuming the compact star is a black hole, calculate the fraction of the total disk luminosity emitted by the inner, middle, and outer disk regions (let the Roche lobe $r_R \sim 3 \times 10^{11}$ cm).

(b) Estimate the ratio of the total disk luminosity for a white dwarf to that of a black hole or neutron star.

Exercise 14.28 Solve the Newtonian accretion disk structure equations for the *inner* region around a black hole. Assume that $P = P_{rad} = \frac{1}{3} a T^4$ and $\bar{\kappa} = \kappa_{es}$. Show (cf. Shakura

---

[34]The definition of r_s employed here should not be confused with its earlier definition as the transonic radius.
[35]Equation (14.5.36) can be verified with the results of Exercise (14.28).

and Sunyaev, 1973; Novikov and Thorne, 1973):

$$F = \left(5 \times 10^{26} \text{ erg cm}^{-2} \text{ s}^{-1}\right)\left(M^{-2}\dot{M}_{17}\right)r^{-3}\mathcal{G},$$

$$\Sigma = \left(7 \text{ g cm}^{-2}\right)\left(\alpha^{-1}M\dot{M}_{17}^{-1}\right)r^{3/2}\mathcal{G}^{-1},$$

$$h = \left(1 \times 10^{5} \text{ cm}\right)\left(\dot{M}_{17}\right)\mathcal{G},$$

$$\rho = \left(3 \times 10^{-5} \text{ g cm}^{-3}\right)\left(\alpha^{-1}M\dot{M}_{17}^{-2}\right)r^{3/2}\mathcal{G}^{-2},$$

$$T = \left(5 \times 10^{7} \text{ K}\right)\left(\alpha M\right)^{-1/4}r^{-3/8},$$

$$\tau_{\text{es}} = 3\left(\alpha^{-1}M\dot{M}_{17}^{-1}\right)r^{3/2}\mathcal{G}^{-1}, \tag{14.5.37}$$

where

$$\mathcal{G} \equiv 1 - \left(\frac{6}{r}\right)^{1/2}.$$

Here M is measured in units of $M_\odot$ and r in units of r_s.

(j)   Solution: Spectrum of the "Standard" Disk Model

The temperature T appearing in the disk structure equations [e.g., Eq. (14.5.37)] refers to the typical *interior* temperature at $z \approx 0$. In general $T > T_s$, where T_s is the characteristic temperature of the *surface* at which the emergent photon spectrum is formed. The inequality follows from the radiative diffusion equation (14.5.31), which induces a temperature gradient in the disk to drive the photon flux. The quantities T and T_s become comparable whenever the disk becomes optically thin to absorption, in which case Eq. (14.5.34) applies. A careful analysis of the temperature and density profile at the surface of the disk is thus required in general to determine the photon spectral profile.

Whenever the disk is optically thick *and* absorption dominates scattering, the local emission will be blackbody. In such regions the spectrum will be described by a Planck function at a temperature T_s equal to the local effective temperature. Accordingly,

$$T_s(r) = \left[\frac{4F(r)}{a}\right]^{1/4} \approx \left(5 \times 10^{7} \text{ K}\right)\left(\frac{M}{M_\odot}\right)^{-1/2}\dot{M}_{17}^{1/4}\left(\frac{r}{r_s}\right)^{-3/4}\mathcal{G}^{1/4}. \tag{14.5.38}$$

This temperature characterizes the matter at optical depth $\bar{\tau}_{\text{ff}} \sim 1$ below the surface. It is at this depth that a typical escaping photon is created before emerging from the surface. Escaping photons of a particular frequency ν are created at $\tau_\nu^{\text{ff}} \sim 1$, where

$$\tau_\nu^{\text{ff}} \sim \kappa_\nu^{\text{ff}} \rho \, \Delta z \sim 1, \tag{14.5.39}$$

and where we tacitly assume that the surface is homogeneous and isothermal. Photons created at larger values of τ_ν^{ff} are absorbed prior to escaping from the surface. Accordingly, the intensity of the emerging radiation I_ν is given by adding up the emission in the layers from $z = 0$ to $z = \Delta z$ below the surface:

$$I_\nu \sim j_\nu^{\text{ff}} \Delta z \sim \frac{j_\nu^{\text{ff}}}{\kappa_\nu^{\text{ff}} \rho} = B_\nu(T_s), \tag{14.5.40}$$

where j_ν^{ff} is the free-free emissivity, B_ν is the Planck function, and where the ratio $j_\nu^{\text{ff}}/\kappa_\nu^{\text{ff}} \rho$ has been evaluated using Kirchhoff's law. Equation (14.5.40) is derived more carefully in Appendix I [cf. Eq. (I.19)]. The flux crossing outward through the surface is then related to the intensity I_ν by

$$F_\nu = \int_0^{\pi/2} I_\nu \cos \theta \, d\Omega \sim 2\pi B_\nu(T_s),$$

$$F = \int_0^\infty F_\nu \, d\nu \sim aT_s^4 \qquad (\bar{\kappa}_{\text{ff}} \gg \kappa_{\text{es}}), \tag{14.5.41}$$

which gives the familiar blackbody result.

It is a major triumph of accretion disk theory that even the most naive assumptions yield effective surface temperatures in the soft X-ray domain for reasonable parameters of binary systems containing compact objects. This result further motivates the identification of the observed binary X-ray sources with accretion onto compact stars.

From the discussion in Section 14.5(i), we note that absorption dominates scattering only in the *outer* region of the accretion disk. Thus Eq. (14.5.38) applies only when $r > r_{\text{om}}$ [cf. Eq. (14.5.35)]. Rewriting Eq. (14.5.38) according to

$$T_s \sim (1 \times 10^5 \text{ K}) \dot{M}_{17}^{-4} \left(\frac{r}{r_{\text{om}}} \right)^{-3/4} \mathcal{g}^{1/4}, \tag{14.5.42}$$

we see that the outer region actually emits blackbody radiation with temperature $T_s \lesssim 10^5$ K. Closer in, the temperature exceeds somewhat the value given by Eq. (14.5.38), as we now proceed to show.

Consider the modification to the spectrum in the middle and inner regions of the disk, where electron scattering dominates absorption for typical photons. Again assume for simplicity that the surface is homogeneous and isothermal. Following emission, photons may undergo many (nearly) elastic scatterings before escaping from the surface. These scatterings will cause a typical photon to "random-walk" in a zig-zag path through the disk prior to reaching the surface. Let Δz^* be the *vertical* depth below the surface at which an escaping photon of frequency ν was created by free-free emission. Let Δs be the *total* path length traversed by the photon in its zig-zag path prior to escape. Accordingly,

$$\tau_\nu^{ff} \sim 1 \sim \kappa_\nu^{ff} \rho \, \Delta s. \tag{14.5.43}$$

Adding up the emission from the region $0 \leqslant z \leqslant \Delta z^*$ again gives the emergent intensity:

$$I_\nu \sim j_\nu \Delta z^*. \tag{14.5.44}$$

The quantity Δz^* is less than Δz found above when we ignored scattering. Now the zig-zag wandering of the photon enhances the probability for absorption prior to escape and thus reduces the depth at which the emission contributes to the emergent flux. In particular, if $N_{\nu s}$ is the total number of scatterings prior to escape, then

$$N_{\nu s} = \frac{\Delta s}{\lambda_{es}}, \qquad \lambda_{es} \sim \frac{1}{\kappa_{es}\rho}, \tag{14.5.45}$$

where λ_{es} is the mean free path for scattering. Since scattering induces a random-walk photon motion, the *net* distance traversed in the *vertical* direction is then $N_{\nu s}^{1/2} \lambda_{es}$. So

$$\Delta z^* = N_{\nu s}^{1/2} \lambda_{es}. \tag{14.5.46}$$

Combining Eqs. (14.5.43), (14.5.45), and (14.5.46), we obtain

$$N_{\nu s} \sim \frac{\kappa_{es}}{\kappa_\nu^{ff}} \tag{14.5.47}$$

and

$$\Delta z^* \sim \frac{1}{\left(\kappa_{es}\kappa_\nu^{ff}\right)^{1/2}\rho} \sim \Delta z \left(\frac{\kappa_\nu^{ff}}{\kappa_{es}}\right)^{1/2}. \tag{14.5.48}$$

Substituting Eq. (14.5.48) into Eq. (14.5.44) gives the intensity,

$$I_\nu \sim \frac{j_\nu^{ff}}{\kappa_\nu^{ff}\rho} \left(\frac{\kappa_\nu^{ff}}{\kappa_{es}}\right)^{1/2} \sim B_\nu(T_s)\left(\frac{\kappa_\nu^{ff}}{\kappa_{es}}\right)^{1/2}, \tag{14.5.49}$$

which differs from Eq. (14.5.40).

Now the opacity coefficient for free-free absorption in ionized hydrogen is given by

$$\kappa_\nu^{\text{ff}} \sim \left(1.5 \times 10^{25} \text{ cm}^2 \text{ g}^{-1}\right)\rho T^{-7/2} g_{\text{ff}} \frac{1 - e^{-x}}{x^3}, \qquad (14.5.50)$$

[Eq. (I.43)], where g_{ff} is a slowly varying "Gaunt factor" of order unit and where $x \equiv h\nu/kT$. For sufficiently high ν, scattering will indeed dominate absorption. Accordingly

$$F_\nu = \int_0^{\pi/2} I_\nu \cos\theta\, d\Omega \sim 2\pi B_\nu(T_s)\left(\frac{\kappa_\nu^{\text{ff}}}{\kappa_{\text{es}}}\right)^{1/2}, \qquad (14.5.51)$$

or

$$F_\nu \propto \frac{x^{3/2}\exp(-x/2)}{(e^x - 1)^{1/2}}, \qquad \left(\kappa_\nu^{\text{ff}} \ll \kappa_{\text{es}}\right), \qquad (14.5.52)$$

which is a "modified" blackbody spectral distribution. Integrating over all ν gives the total flux, assuming scattering dominates absorption even at low ν satisfying $x \ll 1$:

$$F \sim \left(6.2 \times 10^{19} \text{ erg cm}^{-2} \text{ s}^{-1}\right)\rho^{1/2} T_s^{9/4}, \qquad \left(\bar\kappa_{\text{ff}} \ll \kappa_{\text{es}}\right).[36] \quad (14.5.53)$$

Crudely, the integration of Eq. (14.5.51) may be approximated by the relation

$$T_{\text{eff}} \approx T_s\left(\frac{\bar\kappa_{\text{ff}}}{\kappa_{\text{es}}}\right)^{1/8}, \qquad \left(\kappa_{\text{es}} \gg \bar\kappa_{\text{ff}}\right), \qquad (14.5.54)$$

where we have used the identity $F \equiv \sigma T_{\text{eff}}^4$ and where we have employed the Rosseland mean opacity in the ratio in parentheses. Equation (14.5.54) demonstrates that the effect of scattering is to *increase* the mean energy of the emergent photons, $\sim kT_s$, above the value it would have been if the radiation occurred in thermodynamic equilibrium. Note that Eq. (14.5.54) can also be derived (up to a numerical factor near unity) directly from Eqs. (I.38) and (I.39), provided we identify $\tau = \tau_{\text{es}}$ in these equations with the optical depth at which the emergent photons originate. In this case

$$\tau_{\text{es}} \approx \kappa_{\text{es}}\rho\,\Delta z^* \approx \left(\frac{\kappa_{\text{es}}}{\kappa_{\text{ff}}}\right)^{1/2} \qquad (14.5.55)$$

by Eq. (14.5.48). The result, Eq. (14.5.54), follows.

---

[36] The condition $\bar\kappa_{\text{ff}} \lesssim \kappa_{\text{es}}$ for the Rosseland mean opacities is *roughly* equivalent to the requirement that $\kappa_\nu^{\text{ff}} \lesssim \kappa_{\text{es}}$ for all $x \gtrsim 6$ [cf. Eqs. (I.43) and (I.44)].

Exercise   14.29 Argue that the above analysis of the intensity from an isothermal, homogeneous, optically thick slab with scattering and free-free absorption can be conveniently summarized as follows:

$$I_\nu \approx B_\nu \left(\frac{\kappa_\nu^{ff}}{\kappa_\nu^{ff} + \kappa_{es}} \right)^{1/2}, \qquad \tau^* > 1, \tag{14.5.56}$$

for *arbitrary* frequency ν, provided the *effective optical depth for absorption*, τ^*, exceeds unity. Here

$$\tau^* \equiv \begin{cases} (\bar{\tau}_{ff} \tau_{es})^{1/2}, & \bar{\kappa}_{ff} < \kappa_{es} \\ \bar{\tau}_{ff}, & \bar{\kappa}_{ff} > \kappa_{es}. \end{cases} \tag{14.5.57}$$

If $\tau^* < 1$, photons are never reabsorbed, so they cannot thermalize. Consequently, the emission is optically thin and

$$I_\nu \approx j_\nu(\rho, T)h, \qquad \tau^* < 1. \tag{14.5.58}$$

If the surface regions are neither homogeneous nor isothermal, as we would expect for a relativistic atmosphere in hydrostatic equilibrium, then the spectrum acquires a slightly different shape (cf. Shakura and Sunyaev, 1973). Ignoring such deviations and assuming that the surface density is comparable to the central density, we can employ Eq. (14.5.53) together with Eqs. (14.5.17) and (14.5.37) to estimate T_s in the optically thick inner regions:

$$T_s = (2 \times 10^9 \text{ K}) \alpha^{2/9} \left(\frac{M}{M_\odot} \right)^{-10/9} \dot{M}_{17}^{8/9} \left(\frac{r}{r_s} \right)^{-17/9} \mathcal{g}^{8/9}. \tag{14.5.59}$$

The key feature about this "modified" blackbody temperature is that it is appreciably *higher* than the effective temperature, Eq. (14.5.38), in the inner region. Consequently, the photons emitted from this region have higher energy (i.e., they are "harder") than if the disk radiated as a true blackbody. Indeed, since blackbody emission is thermodynamically the *most* efficient radiation mechanism, any other radiation process must occur at higher temperature to generate the same flux.

The very innermost regions of the disk that, depending on parameters, are optically thin (i.e., $\tau^* < 1$ from $z = 0$ to $z = h$) radiate via free-free emission and Comptonization. Comptonization is the mechanism by which photons can exchange energy with electrons by virtue of second-order (i.e., $v^2/c^2 \sim kT/m_e c^2$)

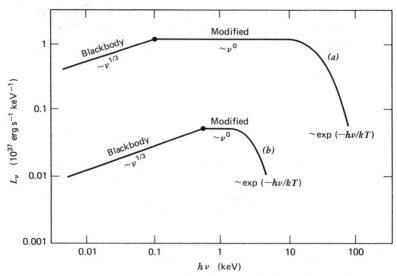

Figure 14.4 Sketch of the power spectra emitted by two model disks around black holes, as calculated by Shakura and Sunyaev (1973) without taking account of relativistic corrections or capture of radiation by the hole. In both models the hole was assumed to be nonrotating, so the inner edge of the disk was at $r_I = 6M$. Model (a) corresponds to $\alpha \sim 10^{-3}$, $M = M_\odot$, $\dot{M} = 10^{-8} M_\odot$ yr^{-1}, $L \approx L_{\rm Edd} \approx 10^{38}$ erg s^{-1}; model (b) corresponds to $\alpha \sim 10^{-2}$ to 1 (spectrum insensitive to α), $M = M_\odot$, $\dot{M} = 10^{-6} M_\odot$ yr^{-1}, $L \approx 10^{36}$ erg s^{-1}. The portion of the spectrum marked "blackbody" is generated primarily by the outer, cold region of the disk where electron scattering is unimportant. The portion marked "modified" is generated by the middle and inner regions where electron scattering is the dominant source of opacity. The temperature of the exponential tail is the surface temperature of the innermost region. [From Novikov and Thorne (1973).]

Doppler shifts of photon energy following Thomson scattering.[37] These regions can have temperatures somewhat hotter than Eq. (14.5.59), reaching[38] maximum values $T_{\max} \leqslant 4 \times 10^8$ K for $\alpha \sim 1$, $M \sim 6 M_\odot$, and $\dot{M}_{17} \gtrsim 10$.

By determining the local flux and spectrum from each region of the disk and then adding the contributions together, a composite emission spectrum can be determined. The results of such a calculation are shown in Figure 14.4 for accretion onto a solar-mass black hole. The ability of disk accretion to generate a copious soft X-ray flux is evident from the figure.

---

[37]For a brief summary of Comptonization, see Appendix I, Section I.3. For a general discussion see, for example, Kompaneets (1957), Peebles (1971), or Rybicki and Lightman (1979). For a discussion of Comptonization of soft X-rays produced by free-free emission see, for example, Illarionov and Sunyaev (1972) and Felten and Rees (1972). For applications to accretion disks see Shakura and Sunyaev (1973) and Shapiro et al. (1976). For a discussion of time-dependent effects, see Lightman and Rybicki (1979) and Payne (1980).

[38]Lightman and Shapiro (1975).

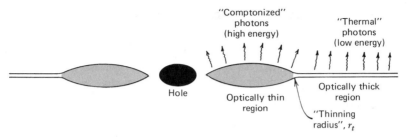

Figure 14.5   Schematic diagram of the X-ray emitting region of a "hot" accretion disk. Typical dimensions are: radius of hole ~ 12 km, thinning radius r_t ~ 30 to 300 km, radius of entire X-ray emitting region ~ 300 km. [Reprinted courtesy of Thorne and Price and *The Astrophysical Journal*, published by the University of Chicago Press; © 1975 The American Astronomical Society.]

14.6   Other Disk Models

The above accretion disk model is, conceptually, the simplest solution to the hydrodynamic equations for thin disk accretion. However, it is not the only solution. Moreover, the inner region of this "standard" disk, where radiation pressure dominates gas pressure, has been demonstrated to be secularly unstable to clumping into rings.[39] Furthermore, the "standard" disk model proposed to explain black hole accretion in a close binary system *cannot* produce the observed hard X-rays (~ 100 keV) from Cyg X-1 (cf. Section 13.5 and Fig. 13.9).

It was originally suggested by Thorne and Price (1975) that *if* the inner portion of an accretion disk is optically thin and at high temperature (~ 10^9 K), instead of optically thick and at low temperature, as in the "standard" model, *then* the observed hard component of the Cyg X-1 spectrum near 100 keV could be explained. They argued that the secular instability present in the inner region of the standard, "cool disk" could swell this optically thick, radiation-pressure dominated region into a hotter, gas-pressure dominated, optically thin region (see Fig. 14.5). A self-consistent "hot disk" model with these properties has been constructed by Shapiro, Lightman, and Eardley (1976). When applied to Cyg X-1 the model yields quite naturally the required thermal emission temperature of 10^9 K. In this "two-temperature" model, the electrons are at 10^9 K throughout the inner region while the ions are 3–300 times hotter. The hard X-ray spectrum above ~ 8 keV is produced by the Comptonization of soft X-ray photons generated in optically thick portions of the disk; the spectral curve is a power-law from 8 to ~ 500 keV, above which it decays exponentially. The power depends on

---

[39]Lightman and Eardley (1974). Abramowicz (1981) has shown that, by virtue of general relativity, accretion disks—"standard" or otherwise—are all thermally and secularly stable *close* to their inner edges. However, the stabilizing influence of general relativity cannot reach far out into the inner region which extends out to radius $r_{mi} \gg r_s$, where relativistic effects are unimportant.

the nondimensional Comptonization parameter y, where

$$y \equiv \frac{4kT_e}{m_e c^2} \tau_{es}^2, \tag{14.6.1}$$

and where, typically, $y \approx 1$ throughout the inner region. The intensity is then given by (cf. Appendix I, Section I.3)

$$I_\nu \propto x^{m+3} \exp(-x), \tag{14.6.2}$$

$$m = -\frac{3}{2} - \left(\frac{9}{4} + \frac{4}{y}\right)^{1/2}, \tag{14.6.3}$$

so that, typically, $I_\nu \propto x^{-1} \exp(-x)$ from the disk. This result compares favorably with the observed hard X-ray spectrum from Cyg X-1 (cf. Section 13.5 and Figs. 13.9 and I.1). However, the observational status of the spectrum above ~ 200 keV is still unclear. The lack of an observed high-energy cut-off out to $\gtrsim 100$ keV indicates that the hard X-ray emitting region has an electron temperature $\gtrsim 10^9$ K, but to date the observations only establish this as a lower limit.[40]

Other "hot disk" models have been proposed that can generate hard X-ray emission. Some of these models are motivated by the Cyg X-1 observations; others by the observations of hard X-ray (typically power-law) spectra of quasars and active galactic nuclei. The "hot corona" model, first proposed by Ostriker (1976), Liang and Price (1977), and Bisnovatyi-Kogan and Blinnikov (1977), consists of a hot corona sandwiching an optically thick disk. The high temperature of the corona is produced by acoustic and Alfvén waves and magnetic dissipation in the upper disk atmosphere. Hard X-rays are emitted from the corona, where they are again generated via the Comptonization of soft-photons by hot, thermal electrons.[41]

Still another class of self-consistent "hot disk" models has been proposed by Pringle, Rees, and Pacholczyk (1973) and Payne and Eardley (1977). They assume that the disk is optically thin to emission and absorption, supported by gas pressure *and* dominated by pure bremsstrahlung cooling. For appropriate model parameters ($\dot{M}_{17} \lesssim 0.1 M/M_\odot$), the spectrum has a bremsstrahlung shape, that is $dL/dE \propto \exp(-x)$, with an emission temperature $\lesssim 10$ keV. Thus the model cannot produce power-law profiles above 10 keV. Both the two-temperature and bremsstrahlung optically thin models may be subject to thermal instabilities,

---

[40] For a recent summary of the hard X-ray observations of Cyg X-1 and the status of the Comptonization model, see, for example, Liang (1980).

[41] See Liang and Thompson (1980) for a comparison of the optically thin two-temperature model and corona disk model.

which lead to the formation of condensations.[42] Even in this case, however, the disk may be stable in a time-averaged sense, but subject to large fluctuations.[43]

Exercise   14.30

(a)   Solve the coupled disk structure equations for the optically thin, bremsstrahlung model. Assume $P \approx P_g$, $\alpha \approx 1$, and $\Lambda = \Lambda_{ff} = 1.43 \times 10^{-27}(\rho/m_p)^2 T^{1/2} g_{ff}$ [$g_{ff} \sim 1$; cf. Eq. (I.46)]. Show (Payne and Eardley, 1977):

$$T = (1.1 \times 10^8 \text{ K})\left(\frac{\dot{M}_{17}}{M_3}\right)^{1/2} R^{-3/4} \mathcal{J}_R^{1/2},$$

$$\rho = (2.7 \times 10^{-8} \text{ g cm}^{-3})\left(\frac{\dot{M}_{17}}{M_3}\right)^{1/4} M_3^{-1} R^{-15/8} \mathcal{J}_R^{1/4},$$

$$\tau_{es} = (0.24)\left(\frac{\dot{M}_{17}}{M_3}\right)^{1/2} R^{-3/4} \mathcal{J}_R^{1/2},$$

$$\frac{h}{r} = (1.5 \times 10^{-2})\left(\frac{\dot{M}_{17}}{M_3}\right)^{1/4} R^{1/8} \mathcal{J}_R^{1/4},$$

where

$$R \equiv \frac{r}{10GM/c^2}, \qquad \mathcal{J}_R \equiv \frac{1 - (R_I/R)^{1/2}}{1 - R_I^{1/2}},$$

$$M_3 \equiv \frac{M}{10^3 M_\odot}, \quad \text{and} \quad \dot{M}_{17} = \frac{\dot{M}}{10^{17} \text{ g s}^{-1}}.$$

Note that the structure depends on $\dot{M}_{17}$ and M_3 only through the ratio $\dot{M}_{17}/M_3 \sim L/(10^{-4} L_{Edd})$, where L is the total disk luminosity and L_{Edd} is the Eddington limiting luminosity for the hole [cf. Eq. (13.7.6)].

(b)   Compare the emission temperature of the optically thin, bremsstrahlung disk and "standard" accretion disk at the radius of maximum flux ($R \sim 1$). Graph your results as a function of M ($1 \leqslant M/M_\odot \leqslant 10^9$) for $L/L_{Edd} = 0.1$. Comment on the relative applicability of the two disk models vis-à-vis the soft X-ray emission from Galactic black holes in binary systems ($M \sim M_\odot$) and from supermassive black holes in dense galactic nuclei and quasars ($M/M_\odot \sim 10^8$).[44]

---

[42] Pringle (1976).

[43] Shakura and Sunyaev (1976).

[44] For a detailed discussion of the emission spectra from different kinds of accretion disks around massive black holes, see Eardley et al. (1978).

15

Accretion onto Neutron Stars and White Dwarfs

In the first three sections of this chapter we describe several aspects of accretion peculiar to neutron stars. This discussion is motivated by the general belief that the pulsating X-ray sources are magnetic neutron stars that are accreting gas from a binary companion (cf. Chapter 13). As we shall see, models for accretion onto neutron stars are subject to some severe observational constraints. In particular, they must explain the period changes observed for many of the X-ray pulsars. In the final section of the chapter we shall summarize (albeit briefly) some highlights of accretion onto white dwarfs.

15.1 Accretion onto Neutron Stars: The Magnetosphere

Our theoretical and observational understanding of neutron star accretion has expanded considerably during the past decade or so. However, the true picture is very far from being fully resolved. One crucial, but complicating, feature of neutron star accretion is the presence of a strong magnetic field extending from the stellar surface outward into the stellar magnetosphere. This field can control the manner in which gas flows onto the stellar surface, the torques exerted on the spinning star, the pulse shape and spectrum of the emitted radiation, and so on. We therefore first discuss the role played by the neutron star magnetic field during accretion.[1]

---

[1] The structure of the magnetosphere of a rotating neutron star and its influence on the flow of accreting gas was first discussed in pioneering papers by Pringle and Rees (1972), Davidson and Ostriker (1973), and Lamb, Pethick, and Pines (1973). For a recent review, see Lamb (1979a).

Far from the neutron star the magnetic field has little influence on the flow, so at large radii the spherical or disk flow solutions presented in the previous chapter remain valid. Close to the stellar surface, however, the field is likely to control the plasma flow completely.

Consider the influence of a dipole magnetic field outside the star of magnitude

$$B \approx \frac{\mu}{r^3} = (10^{12} \text{ G}) \mu_{30} R_6^{-3} \left(\frac{R}{r} \right)^3. \tag{15.1.1}$$

In Eq. (15.1.1), μ is the magnetic moment and μ_{30} is in units of 10^{30} G cm^3, R is the stellar radius, and R_6 is in units of 10^6 cm. Assume first that, far from the star, the flow is spherically symmetric. Then intuitively we expect that the field will begin to dominate the flow when the magnetic energy density becomes comparable to the total kinetic energy density of the accreting gas. This condition specifies the characteristic radius of the magnetospheric boundary, r_A, called the *Alfvén radius*:

$$\frac{B^2(r_A)}{8\pi} \simeq \frac{1}{2} \rho(r_A) v^2(r_A). \tag{15.1.2}$$

Assuming steady, transonic flow at nearly free-fall velocity we may write from Eqs. (14.3.23) and (14.3.4)

$$v(r) \approx v_{\text{ff}} = \left(\frac{2GM}{r} \right)^{1/2},$$

$$\rho(r) \approx \rho_{\text{ff}} = \frac{\dot{M}}{4\pi v_{\text{ff}} r^2}. \tag{15.1.3}$$

Substituting Eq. (15.1.1), which, strictly speaking, holds only in the equatorial plane, and Eq. (15.1.3) into Eq. (15.1.2) gives

$$r_A = \left(\frac{\mu^4}{2GM\dot{M}^2} \right)^{1/7} = 3.2 \times 10^8 \dot{M}_{17}^{-2/7} \mu_{30}^{4/7} \left(\frac{M}{M_\odot} \right)^{-1/7} \text{ cm}. \tag{15.1.4}$$

Equation (15.1.4) may be rewritten in terms of L, the total accretion luminosity, given by

$$L = \dot{M} \left(\frac{GM}{R} \right). \tag{15.1.5}$$

Substituting Eq. (15.1.5) into Eq. (15.1.4) yields

$$r_A = \left(\frac{\mu^4 GM}{2L^2 R^2} \right)^{1/7} = 3.5 \times 10^8 L_{37}^{-2/7} \mu_{30}^{4/7} \left(\frac{M}{M_\odot} \right)^{1/7} R_6^{-2/7} \text{ cm}. \tag{15.1.6}$$

where L_{37} is the luminosity in units of 10^{37} ergs s^{-1}. Typically, therefore, $r_A \gg R$ for a neutron star.

The magnetic field does not penetrate the plasma much beyond r_A. At the Alfvén surface, currents will be induced in the infalling plasma by the magnetic field. These surface currents screen the stellar field outside r_A and enhance it inside r_A.[2]

Exercise 15.1 Consider a neutral beam of cold plasma, initially moving with velocity v in a field-free region, incident normally to a plane surface, beyond which there is a uniform B field parallel to the plane (Lamb, Pethick, and Pines, 1973). Neglect collisions between particles.

(a) Describe the motion of the electrons and ions as they penetrate the plane surface. In particular, argue that a current will be established parallel to the surface, and estimate the penetration depth of the ions.

(b) Estimate the ion and electron contributions to the surface current density $\mathcal{J}$ and the corresponding jump in the magnetic field that the current generates.

Answer: $\Delta B \sim 4\pi(\mathcal{J}_i + \mathcal{J}_e)/c \simeq 4\pi\rho v^2/B$

(c) Argue from part (b) and Eq. (15.1.2) that these screening currents induced at the Alfvén surface will cancel the stellar field outside r_A.

How the infalling plasma crosses or is channeled by the field lines to the stellar surface is an extremely difficult problem to solve. It is thought that a standing shock develops in the neighborhood of the Alfvén surface, which acts as a blunt obstacle in the path of the accretion flow. Gas temperatures immediately behind the shock, which serves to convert bulk kinetic to random thermal energy, are roughly given by

$$kT \sim m_i v^2 \sim \frac{GMm_i}{r_A} \tag{15.1.7}$$

or

$$T \sim 10^{10} \text{ K} \left(\frac{M}{M_\odot}\right)\left(\frac{r_A}{10^8 \text{ cm}}\right)^{-1}. \tag{15.1.8}$$

A first guess regarding the flow inside r_A might be that plasma could reach the stellar surface by tunneling down the magnetic poles in a narrow accretion column.[3] Imagine that this region is delineated by those field lines that would, in the absence of accretion, have penetrated out beyond r_A (cf. Fig. 13.3). Now the

---

[2] Technically, the surface where screening currents flow is called the "magnetopause." We shall not distinguish between the magnetopause and Alfvén surface, which in general are distinct, depending on the matter flow; see Arons and Lea (1980).

[3] Pringle and Rees (1972); Davidson and Ostriker (1973); Lamb, Pethick, and Pines (1973).

field lines for a dipole field are defined by $(\sin^2\theta)/r = $ constant, so the base of the last undistorted field line which would close inside r_A lies at an angle θ_c given by

$$\sin^2\theta_c = \frac{R}{r_A} \approx 3 \times 10^{-3}\left[R_6^{9/7}L_{37}^{2/7}\mu_{30}^{-4/7}\left(\frac{M}{M_\odot}\right)^{-1/7}\right], \qquad (15.1.9)$$

where we have used Eq. (15.1.6) to evaluate r_A. Thus, at the stellar surface near either pole, the cross-sectional area of the accretion column will be approximately

$$A \approx \pi R^2 \sin^2\theta_c \approx 10^{10} \text{ cm}^2, \qquad (15.1.10)$$

which is a small fraction of the total spherical surface area.

Given the concentration of plasma near the polar caps, the emitted radiation pattern will exhibit a strongly anisotropic angular pattern that will depend on the detailed nature of the accretion flow. For an oblique rotator (i.e., a star in which the magnetic field and rotation axes are not aligned), one has a plausible mechanism for the generation of pulsed X-ray radiation.

Exercise   15.2   Use Eq. (15.1.10) to estimate the effective blackbody temperature of the radiation emitted from one of the magnetic polar caps, assuming $L_{37} \sim 1$.

Recently, however, Arons and Lea (1976, 1980) and Elsner and Lamb (1976, 1977) have argued that, under a wide range of conditions, a Rayleigh–Taylor "interchange" instability at the magnetospheric boundary is likely to be the most important process by which plasma enters the magnetosphere. This instability allows, in principle, blobs or filaments of plasma to penetrate the magnetosphere "between field lines" and fall all the way to the stellar surface without becoming "threaded." The material still lands predominantly near the magnetic poles, but the effective area of the heated polar caps will exceed that given by Eq. (15.1.10) by an appreciable factor. These issues are not fully resolved.

15.2   Disk Accretion and Period Changes in Pulsating X-ray Sources

Turn now to the case in which matter far from the star flows inward in a Keplerian accretion disk. Determining the location of the Alfvén radius is much more subtle for disk accretion. The most detailed model of disk accretion at present is that developed by Ghosh and Lamb (1978, 1979a, b), who take into account the appreciable magnetic coupling between the star and the accreting plasma in the disk. They treat the special, but nontrivial, case of axisymmetric disk accretion by an aligned rotator. The flow is analyzed in an approximate fashion, utilizing the hydromagnetic equations in two dimensions. The "slippage" of the stellar field lines through the disk plasma is modeled by an "effective" conductivity. This conductivity accounts for the penetration of the inner part of the disk via plasma instabilities (specifically, the Kelvin–Helmholtz instability), turbulent diffusion and magnetic reconnection. However, the *magnitude* of the conductivity is determined solely by the condition of steady flow.

Ghosh and Lamb find that field penetration produces a broad transition zone joining the unperturbed disk flow far from the star to the magnetospheric flow near the star. The characteristic radius r_0, which divides the outer part of the transition zone where the velocity is Keplerian, from the inner part where it is no longer Keplerian, can be estimated in a straightforward way.

Start with the magnetohydrodynamic Euler equation (7.1.1). Using Eq. (7.1.2) and setting all time derivatives equal to zero in steady state, we find

$$\rho(\mathbf{v}\cdot\nabla)\mathbf{v} = -\nabla P - \rho\nabla\Phi + \frac{1}{4\pi}(\nabla\times\mathbf{B})\times\mathbf{B}. \qquad (15.2.1)$$

Using a vector identity, we rewrite the left-hand side as

$$\rho(\mathbf{v}\cdot\nabla)\mathbf{v} = \rho\left[\tfrac{1}{2}\nabla v^2 + (\nabla\times\mathbf{v})\times\mathbf{v}\right]. \qquad (15.2.2)$$

We are going to derive an equation for conservation of angular momentum along the z axis. This is equivalent to considering the ϕ component of the equation of force balance, Eq. (15.2.1). All the gradient terms give zero by axisymmetry, and in cylindrical coordinates we have

$$\left[(\nabla\times\mathbf{B})\times\mathbf{B}\right]_\phi = \frac{1}{r}B_r\left(rB_\phi\right)_{,r} + B_z B_{\phi,z}. \qquad (15.2.3)$$

Using the Maxwell equation

$$0 = \nabla\cdot\mathbf{B} = \frac{1}{r}\left(rB_r\right)_{,r} + B_{z,z}, \qquad (15.2.4)$$

Eq. (15.2.3) becomes

$$\left[(\nabla\times\mathbf{B})\times\mathbf{B}\right]_\phi = \frac{1}{r^2}\left(r^2 B_r B_\phi\right)_{,r} + \left(B_z B_\phi\right)_{,z}. \qquad (15.2.5)$$

An expression analogous to Eq. (15.2.3) holds for the last term in Eq. (15.2.2) with $\mathbf{B}$ replaced by $\mathbf{v}$, except that we may set $v_z = 0$. Thus Eq. (15.2.1) gives

$$\rho v_r r\left(rv_\phi\right)_{,r} = \frac{1}{4\pi}\left[\left(r^2 B_r B_\phi\right)_{,r} + r^2\left(B_z B_\phi\right)_{,z}\right]. \qquad (15.2.6)$$

Now consider the magnetic field configuration in the disk. In the *absence* of the disk, the dipole field lines are poloidal (i.e., no ϕ component) and, near the equatorial plane, dominated by the B_z component. In the presence of the disk, the lines are initially "frozen-in" to the orbiting plasma, and hence are sheared in the ϕ direction. This shearing generates a sizeable B_ϕ component. Assume that the field lines, though sheared, remain continuous across the disk surfaces. Then B_ϕ must be equal in magnitude but opposite in direction above and below the disk plane. As a result, the magnetic field in the disk will vary much more rapidly with z than with r (e.g., B_ϕ goes from $+B_\phi$ to $-B_\phi$ as z goes from $-h$ to h, where $h \ll r$ is the disk thickness). In addition, we expect $B_r \ll B_\phi, B_z$, since the disk lies close to the equatorial plane and the gas moves in nearly circular orbits. Thus we can neglect the first term on the right-hand side of Eq. (15.2.6).

Recall that the equation of continuity, $\nabla \cdot (\rho \mathbf{v}) = 0$, implies

$$r\rho v_r = \text{constant}. \tag{15.2.7}$$

Integrating Eq. (15.2.7) with respect to z from $-h$ to h and recalling Eq. (14.5.7) gives Eq. (14.5.12):

$$4\pi r h v_r \rho = \dot{M}. \tag{15.2.8}$$

Thus if we integrate Eq. (15.2.6) from $-h$ to h and neglect the variation of v_ϕ with z, we get

$$\dot{M}(rv_\phi)_{,r} = r^2 B_z B_\phi. \tag{15.2.9}$$

Following Ghosh and Lamb, we can evaluate Eq. (15.2.9) in an approximate fashion to obtain the transition radius r_0. Assume that the transition region has a finite radial thickness $\delta \ll r$ over which the torque exerted by the magnetic field on the disk plasma brakes its azimuthal motion. Equation (15.2.9) then gives r_0 implicitly via

$$\dot{M}\left(\frac{r}{\delta}\right) v_\phi \approx r^2 B_z B_\phi. \tag{15.2.10}$$

Now one expects that where the stellar magnetic field begins to control the azimuthal motion, one has $B_\phi \approx B_z$, where B_z is comparable to the unperturbed dipole field in the equatorial plane. Using this approximation, together with the relation $v_\phi \approx v_{ff}$ for Keplerian orbital motion, and assuming that the inequality $h \lesssim \delta \lesssim r$ delineates the possible range of the width of the transition layer (Ghosh and Lamb actually find $\delta \sim 4h$ at r_0), we may solve Eq. (15.2.10) for r_0:

$$\left(\frac{h}{r}\right)_0^{2/7} r_A \lesssim r_0 \lesssim r_A. \tag{15.2.11}$$

Here r_A is the Alfvén radius defined by Eq. (15.1.4) for radial accretion. Near r_0 the unperturbed disk will have a half-thickness h appropriate for the "middle region" of a standard accretion disk. This implies

$$\frac{h}{r} \approx 9 \times 10^{-3} r^{1/20} \alpha^{-1/10} \left(\frac{M}{M_\odot}\right)^{-3/10} (\dot{M}_{17})^{1/5} \mathcal{G}^{1/5}, \qquad (15.2.12)$$

where on the right-hand side r is in units of GM/c^2. [Equation (15.2.12) can be derived from the results of Section 14.5.][4] Inserting Eq. (15.2.12), with $\alpha \approx 1 \approx \dot{M}_{17} \approx (M/M_\odot)$, into Eq. (15.2.11) gives, finally,

$$0.3 r_A \lesssim r_0 \lesssim r_A \quad \text{(roughly)}. \qquad (15.2.13)$$

Somewhat more carefully, Ghosh and Lamb find $\delta \approx 0.03 r_0$ and

$$r_0 \approx 0.5 r_A. \qquad (15.2.14)$$

The significance of Eq. (15.2.14) and of the disk accretion model to which it applies is that they imply a particularly simple relation between the spin-up rate of an accreting neutron star, $-\dot{P}$, and the product $PL^{3/7}$ for a given star, where P is the stellar rotation period. To appreciate the result one must recall that one of the very few precise observational probes of pulsating X-ray sources is the rate of change of the pulsation period (cf. Section 13.3 and Fig. 13.4). In contrast to radio pulsars, which are observed to be spinning down, the X-ray pulsars exhibit a secular *spin-up*. We thus expect that this spin-up rate reflects the manner in which angular momentum is being deposited on the accreting star by inflowing plasma that, in turn, depends upon the flow pattern outside the magnetosphere.

Consider, then, angular momentum transport via disk accretion onto a rotating neutron star with an aligned magnetic field. Imagine the star-plus-magnetosphere to be enclosed by a surface S_0 that lies just outside r_0. The rate of change of the angular momentum of the star-plus-magnetosphere is given by

$$\frac{d}{dt}(I\Omega_s) = \dot{M}\tilde{l}(r_0) + N, \qquad (15.2.15)$$

where I is the moment of inertia of the star-plus magnetosphere (and is essentially equal to the moment of inertia of the star alone), Ω_s is the angular velocity of the star, $\tilde{l}$ is the specific angular momentum of the accreting plasma, and N is the magnetic and viscous torque acting just outside S_0. Writing

$$\frac{dI}{dt} = \frac{dI}{dM}\dot{M}, \qquad \frac{d\Omega_s}{dt} = -\Omega_s \frac{\dot{P}}{P}, \qquad (15.2.16)$$

---

[4] For example, Novikov and Thorne (1973), Eq. (5.9.8).

we obtain[5]

$$\frac{\dot{P}}{P} = \frac{\dot{M}}{M}\left[\frac{M}{I}\frac{dI}{dM} - \frac{\tilde{l}(r_0)}{\tilde{l}_s}\right] - \frac{N}{I\Omega_s}, \qquad (15.2.17)$$

where $\tilde{l}_s \equiv I\Omega_s/M$. Typical nuclear equations of state give

$$\frac{M}{I}\frac{dI}{dM} \simeq 1, \qquad (15.2.18)$$

with the exception of the very lightest neutron stars, and so this term can be neglected in comparison with the second term on the right-hand side of Eq. (15.2.17). For Keplerian orbital motion,

$$\tilde{l}(r_0) = (GMr_0)^{1/2}, \qquad (15.2.19)$$

and so typically $\tilde{l}(r_0) > \tilde{l}_s$.

Suppose we neglect viscous and magnetic torques by setting $N = 0$ and determine the resulting spin-up rate due to the second term in Eq. (15.2.17) alone. Combining Eqs. (15.1.5), (15.1.6), (15.2.14), (15.2.17), and (15.2.18) yields

$$-\dot{P} \approx 5.8 \times 10^{-5}\left[\mu_{30}^{2/7}R_6^{6/7}\left(\frac{M}{M_\odot}\right)^{-3/7}I_{45}^{-1}\right]\left(PL_{37}^{3/7}\right)^2 \text{ s yr}^{-1}. \quad (15.2.20)$$

We thus obtain the result (at least in the limiting case $N = 0$) that the value of $\dot{P}$ for a star of given mass (hence radius) and magnetic moment is a function only of $PL^{3/7}$. In fact, Ghosh and Lamb have shown that this conclusion remains valid *even* when torques arising from the magnetic coupling outside r_0 become important (viscous torques in the disk are never as important, typically). Such magnetic torques become particularly important for "fast rotators." By definition, fast rotators have large values of the nondimensional "fastness parameter" ω_s, defined by

$$\omega_s \equiv \frac{\Omega_s}{\Omega_K(r_0)}, \qquad (15.2.21)$$

where $\Omega_K(r_0)$ is the Keplerian angular frequency at radius r_0.[6] For example, Her X-1, for which $\omega_s \approx 0.4$, is probably a "fast rotator", while GX 1 + 4, for which $\omega_s \approx 3 \times 10^{-3}$, is probably a "slow rotator." Now for slow rotators, the magnetic coupling serves to *enhance* the *spin-up* torque by a modest amount, about 40%;

[5]Lamb, Pethick, and Pines (1973).
[6]Elsner and Lamb (1977).

Eq. (15.2.20) thus remains roughly reliable, both in magnitude and sign. For "fast rotators," the magnetic torques in the outer part of the disk oppose the torques associated with the angular momentum in the matter flow and magnetic field at the disk inner edge, thereby *decreasing* the spin-up rate. In fact, for sufficiently high stellar angular velocities—or, equivalently, for sufficiently low accretion rates—the star will even *spin down*, while accretion and X-ray emission continues.

Exercise 15.3 Why is a "low accretion rate" equivalent to a "high stellar angular velocity" as remarked above?

Answer:

$$\omega_s = 1.2 P^{-1} \dot{M}_{17}^{-3/7} \mu_{30}^{6/7} \left(\frac{M}{M_\odot} \right)^{-5/7}. \tag{15.2.22}$$

Ghosh and Lamb thus explain the existence of a large number of long-period sources ($P \geqslant 10^2$ s) with short spin-up times (~ 50–100 yr) as a consequence of recurrent "low" states when the accretion rate falls and the star is then acted upon by large spin-down torques.

The model of Ghosh and Lamb provides a quantitative picture of the earlier notion[7] that one expects Ω_s to stabilize at a value $\sim \Omega_K(r_0)$, thereby driving ω_s toward unity. Their model also supports the idea that steady accretion is not possible for $\Omega_s \gg \Omega_K(r_0)$. In this case, the centrifugal force on the plasma at r_0 is too great to allow for corotation; the net radial force (gravitation + centrifugal + magnetic) is large and outward and the inflow velocity rapidly falls to zero. Some authors[8] have proposed that for sufficiently fast rotators, there will be a spin-down torque exerted on the star-plus magnetosphere due to mass ejection; this mechanism has been called the "propellor" effect for obvious reasons.

Perhaps the strongest evidence in support of the picture of accretion onto a neutron star for the pulsating X-ray sources, as well as the general correctness of current magnetospheric models, comes from a comparison of the observed pulsation period changes with the theoretical predictions. The disk accretion model of Ghosh and Lamb predicts that an ensemble of pulsating X-ray sources would all reside on the same curve $-\dot{P} = f(PL^{3/7})$ [cf. Eq. (15.2.20)] provided they all had the same mass M and the same magnetic moment μ. (Recall that I and R are uniquely determined by M for a given equation of state.) Although all pulsating X-ray sources are not expected to have identical values of M and μ, the correlation between $\dot{P}$ and $PL^{3/7}$ should still exist if the variation in M and μ were not too large from source to source. In Figure 15.1, the observed values of $-\dot{P}$ are plotted against $PL^{3/7}$ for nine sources. Also plotted are the theoretical spin-up curves for neutron stars with $\mu_{30} = 0.48$ and $M/M_\odot = 0.5$, 1.3, and 1.9 (assuming

---

[7]Pringle and Rees (1972); Davidson and Ostriker (1973); and Lamb, Pethick, and Pines (1973).
[8]Davidson and Ostriker (1973); Illarionov and Sunyaev (1975).

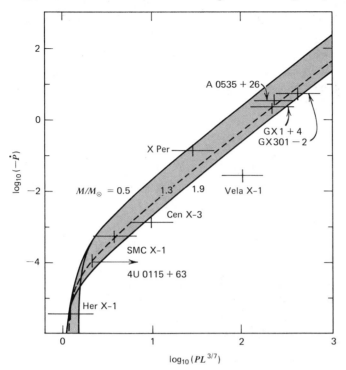

Figure 15.1   The theoretical relation between the spin-up rate, $-\dot{P}$, and the quantity $PL^{3/7}$, superposed on the data (pre-1979) for nine pulsating X-ray sources. The units of $-\dot{P}$, P, and L are s yr^{-1}, s, and 10^{37} erg s^{-1}, respectively. Shown is the effect of varying the neutron star mass, assuming a stellar magnetic moment $\mu_{30} = 0.48$ and the TI equation of state for all stars. Each curve is labeled with the corresponding value of $M/M_{\odot}$. The *shaded* area indicates the region spanned by the theoretical curves for $0.5 \leqslant M/M_{\odot} \leqslant 1.9$. The *dashed* line is the theoretical curve for $M/M_{\odot} = 1.3$, which gives a rough best fit to the data. Because the curves corresponding to different stellar masses cross, the upper boundary of the *shaded* region is defined by the envelope of the curves (*light solid line*). [From Ghosh and Lamb (1979b). Reprinted courtesy of the authors and *The Astrophysical Journal*, published by the University of Chicago Press; © 1979 The American Astronomical Society.]

the "stiff" *TI* equation of state of Pandharipande and Smith (1975a); cf. Section 8.5 and Table 8.2). Note that the linear portion of the theoretical curves at large values of $PL^{3/7}$ (corresponding to small values of ω_s by Exercise 15.3) is described approximately by Eq. (15.2.20). The downturn in the theoretical curves at small values of $PL^{3/7}$ results from the magnetic coupling between the disk and star near r_0. The key conclusion of this plot is that except for Vela X-1 all the sources lie in the shaded region spanned by curves corresponding to neutron star masses in the range $0.5 \leqslant M/M_{\odot} \leqslant 1.9$. Agreement with the indicated Vela X-1 observations can be obtained if either (i) this source has a much larger magnetic moment than the others (e.g., $\mu_{30} \approx 86$) or, more likely, (ii) if the source is not accreting from a disk but, rather, is accreting spherically from a stellar wind (cf. Exercise 14.14 and associated discussion).[9]

Exercise 15.4

(a)   Determine the spin-up rate for wind accretion analogous to Eq. (15.2.20) for disk accretion (Ghosh and Lamb, 1979b).

Answer:

$$-\dot{P} \approx 3.8 \times 10^{-5} R_6 \left(\frac{M}{M_\odot} \right)^{-1} I_{45}^{-1} \left(\frac{\tilde{l}_a}{10^{17} \text{ cm}^2 \text{ s}^{-1}} \right) P^2 L_{37} \quad \text{s yr}^{-1}.$$

Note that $\dot{P}$ is independent of μ (why?).

(b)   Use the results of Exercise 14.22 for $\tilde{l}_a$, the specific angular momentum of the accreted gas, to estimate $-\dot{P}$.

The apparent agreement between the theoretical predictions for disk accretion and the observations of eight of the nine accurately measured sources provides support for the magnetic neutron star-disk accretion picture.[10]

15.3   Emission From Accreting Neutron Stars

Provided $R < r_A$, the dominant emission originates well inside the Alfvén surface, near the stellar surface. A number of physical processes and radiative mechanisms contribute to the emission. The expected radiative spectrum and beam pattern depend in a complicated way upon the flow geometry and upon the manner in which the incident plasma is decelerated near the neutron star surface. Although there have been a number of studies of these issues,[11] an unambiguous prediction for the radiation spectrum has yet to emerge. The total luminosity, of course, *is*

---

[9]From 1975 through 1978 the pulse period of Vela X-1 ($\equiv$ 4U 0900 − 40) decreased at an average range of $-\dot{P}/P \approx 1.5 \times 10^{-4}$ yr^{-1} (except for a period increase in late 1975), but it has turned to an *increase* since 1979 at an average rate of $\dot{P}/P \approx 3 \times 10^{-4}$ yr (see Fig. 13.4). Such fluctuations can be understood if Vela X-1 is a stellar wind-fed X-ray pulsar, with the wind parameters undergoing modest variations with time. Alternatively, fluctuations in the accretion rate might explain the observations if Vela X-1 is a disk-fed "fast" rotator with high μ. Future X-ray observations showing the correlation between the period and luminosity of the source could resolve the issue (Ghosh and Lamb, 1979b).

[10]See, however, Arons and Lea (1980) for an alternative spin-up model for the low-luminosity sources involving spherical, wind-fed accretion outside the magnetosphere and entry inside via the Rayleigh–Taylor instability.

[11]See Lightman, Shapiro, and Rees (1978) for a review and references.

unambiguously determined by M, R, and $\dot{M}$ according to

$$L = \frac{GM}{R}\dot{M}. \tag{15.3.1}$$

Complicating the spectral analysis, again, are the surface magnetic fields. They constrain the plasma to flow along field lines, cause the electrons to lose their transverse energy very rapidly via cyclotron emission, and lead to anisotropic thermal velocities, emissivities, opacities, and, ultimately, X-ray intensities. However, the magnetic fields do not provide the only complication. Indeed, the range of plausible emission spectra is rather large even for *spherical* accretion onto an *unmagnetized* neutron star, depending upon the behavior of the plasma near the surface.

Consider, for example, the case in which the incident ion beam gradually decelerates in a comparatively extended layer of the atmosphere of the neutron star via ordinary Coulomb collisions with electrons. This scenario was first proposed by Zel'dovich and Shakura (1969).[12] If the infalling plasma is stopped by Coulomb encounters with stellar material, most of the energy is deposited several photon mean free paths below the surface and is therefore thermalized before escaping. The photon energy is then comparable to the effective blackbody temperature at the surface, which for spherical flow gives

$$T \gtrsim T_{bb} = \left(\frac{L}{4\pi R^2 \sigma}\right)^{1/4} \approx \left\{\begin{array}{c} 1 \times 10^7 \text{ K} \\ \text{or} \\ 1 \text{ keV} \end{array}\right\} \times R_6^{-1/2} L_{37}^{1/4}. \tag{15.3.2}$$

Processes tending to increase T above the minimum value, T_{bb}, include Thomson scattering of photons by electrons in the "lower" atmosphere or photosphere [cf. Eq. (14.5.54)], Comptonization of thermal photons by hot electrons ($\lesssim 10^8$ K) in the "upper" atmosphere, and (when a magnetic field is present), the funneling of matter to the polar caps, in which case the emitting region has an area *less* than $4\pi R^2$ [cf. Eq. (15.1.10)]. Nevertheless, the general character of the radiation when Coulomb collisions alone are responsible for the deceleration is soft, thermal X-ray emission.

Consider, however, the alternative possibility that plasma collective effects, which are observed, incidentally, in the solar wind, serve to stop the incident plasma stream more suddenly than Coulomb collisions. The "two-stream" instabilities, in which strong electric fields arising from fluctuations in the charge density can decelerate one plasma beam moving through another, provide an example of such a collective mode. In the extreme case in which the collisionless instability is particularly strong and the resulting deceleration particularly rapid, a standing shock front may form just above the stellar surface. Such a scenario

---

[12]See also Alme and Wilson (1973).

was first explored by Shapiro and Salpeter (1975).[13] In this case the shock serves to convert the kinetic energy of the incident stream into thermal energy, so that

$$T \lesssim T_{ff} \approx \frac{\frac{1}{2}mv_{ff}^2}{k}, \qquad (15.3.3)$$

where $v_{ff} \approx (2GM/R)^{1/2}$ is the free-fall velocity near the surface.

It is easy to see that Coulomb collisions are incapable of bringing about such a "randomization" of kinetic energy near a neutron star. Consider the cross section for ion–ion Coulomb collisions in the stream. Up to logarithmic factors, it may be estimated from

$$\sigma_c \approx \pi r_{eff}^2, \qquad (15.3.4)$$

where r_{eff} is the separation at which the ion–ion potential energy is comparable to the kinetic energy. This is the separation at which the ions experience a roughly 90° deflection from their unperturbed trajectories. Thus,

$$\frac{e^2}{r_{eff}} \approx \frac{1}{2}m_p v^2, \qquad (15.3.5)$$

where we have assumed $Z \sim 1$ and where $v \approx v_{ff}$ is the velocity of the incident ion beam as well as the ion thermal velocity in the post-shocked region. Now the mean free path for an ion–ion Coulomb collision is just

$$\lambda_c \approx \frac{1}{n_i \sigma_c} \approx \frac{m_p}{\rho \sigma_c}, \qquad (15.3.6)$$

where n_i is the ion density and ρ is the mass density given by

$$\rho \approx \frac{\dot{M}}{4\pi R^2 v_{ff}} = \frac{2L}{4\pi R^2 v_{ff}^3}, \qquad (15.3.7)$$

and we have used Eqs. (15.1.3) and (15.1.5). Combining Eqs. (15.3.4)–(15.3.7) yields

$$\frac{\lambda_c}{R} \approx \frac{m_p^3 R v_{ff}^7}{2Le^4} \approx 1 \times 10^6 \left(\frac{M}{M_\odot} \right)^{7/2} R_6^{-5/2} L_{37}^{-1}. \qquad (15.3.8)$$

---

[13]See also Tuchman and Yahel (1977) and Klein, Stockman, and Chevalier (1980).

Thus, Eq. (15.3.8) shows that the distance λ_c required to thermalize the incident stream exceeds R by a very large factor. Logarithmic corrections[14] due to distant encounters, which result in cumulative, small-angle scatterings, reduce Eq. (15.3.8) by only a factor $\gtrsim 10$. Accordingly, a thin shock front above the star is possible only if other *collisionless* mechanisms are at work.

Given such a collisionless mechanism, Eq. (15.3.3) predicts very high gas temperatures near the surface,

$$T \lesssim T_{ff} \simeq \left\{ \begin{array}{c} 1 \times 10^{12} \text{ K} \\ \text{or} \\ 100 \text{ MeV} \end{array} \right\} \times \frac{M}{M_\odot} R_6^{-1}. \qquad (15.3.9)$$

In this situation the shocked electrons emit γ-rays in addition to soft X-rays. Roughly half of the total incident kinetic energy goes into thermal energy of relativistic electrons immediately behind the shock front (the other half goes into ion thermal energy). These relativistic electrons emit 10–100 MeV γ-rays, about half of which are directed downwards toward the star and are thermalized, while the other half is emitted outwards and escapes unperturbed in the γ-ray band. Thus, roughly $\frac{1}{2} \times \frac{1}{2} \approx \frac{1}{4}$ of the total observed luminosity would be in the form of a copious γ-ray flux; the rest would be in the form of soft X-rays.

Exercise   15.5   The process by which the shocked electrons emit γ-rays in the above scenario is relativistic Comptonization. Show that photons of mean energy E_i that scatter once off relativistic electrons of energy $E_e \gg m_e c^2$ acquire a mean energy $E_f = \frac{4}{3}(E_e/m_e c^2)^2 E_i$ (assume that both photon and electron distributions are isotropic). Use this result to show why soft X-rays from the stellar surface can emerge from the shock front with γ-ray energies.

Whether or not sufficiently strong collisionless processes exist to generate thin shock fronts, high gas temperatures and hard photons is not known at the present time. Observationally, there are no compelling reasons to suggest that steadily accreting neutron stars emit copius γ-ray photons. But again, difficulties in detecting photons in the $\leqslant$ 10–100 MeV band (balloon or satellite measurements are required) prevent definitive conclusions.[15]

---

[14] See Jackson (1975), Section 13.8, or Spitzer (1962), Chapter 5.

[15] Cosmic γ-ray bursts in the energy range 0.2–1.5 MeV were reported in 1973 by Klebesadel, Strong, and Olson (1973) and since confirmed by numerous subsequent observations; see Klebesadel et al. (1982) for a review and references. The short timescales associated with the bursts (0.1 s $\lesssim \Delta t \lesssim$ 100 s) have motivated compact star (esp., neutron star) models for these sources. See Ruderman (1975) and Lamb (1982) for a summary of these and other models for the observed γ-ray bursts.

15.4   White Dwarf Accretion

Space limitations prevent us from discussing the theory of white dwarf accretion in any detail.[16] Nevertheless, white dwarf accretion is important for several reasons: (i) It is generally believed that the soft and hard X-ray emission observed from known dwarf binary systems like Am Her, DQ Her and SS Cyg represents radiation from accretion onto *magnetic* white dwarfs.[17] AM Her, and sources comparable to it, show strong circular polarization ($\gtrsim 10\%$) at visual wavelengths: strong surface fields of magnitude $B \leq 10^7 – 10^8$ G are inferred for these sources. DQ Her, and sources like it, undergo nova outbursts and show coherent, small amplitude pulsations (period = 71 s for DQ Her). DQ Her sources exhibit clear optical evidence of a large accretion disk, which suggests weaker magnetic fields than those inferred for AM Her. (The presence of significant magnetic fields may disrupt the disk and lead to radial inflow near the star.) SS Cyg and comparable sources (like U Gem and AK Per) are cataclysmic variables characterized by frequent outbursts, like AM Her objects, but show strong optical evidence of a disk. (ii) Many of the identified low-mass X-ray binaries such as Cyg X-2 and Sco X-1, as well as some of the galactic "bulge" sources (cf. Chapter 13), may also be moderate luminosity *nonmagnetic* white dwarfs.

Accretion onto white dwarfs differs from accretion onto neutron stars in several ways. If the infalling gas has typical cosmic abundances (e.g., ~ 70% H by mass, 25% He, $\leq$ few percent heavier elements), then approximately 8 MeV per nucleon of *nuclear* energy will be released below the stellar surface in nuclear reactions. Averaged over time, this extra energy will, in the case of white dwarf accretion, overwhelm the gravitational potential energy release from infall. However, the nuclear energy may be released in explosive outbursts so that for most of the time the quasisteady gravitational output dominates.

Next we note that the temperature of the white dwarf photosphere, which is optically thick and emits blackbody radiation, cannot be high enough to generate a reasonable luminosity in the X-ray band. According to Eq. (15.3.2) with $R_6 \approx 500$, we have $T_{bb} \approx 44$ eV, so that the photospheric emission is predominantly hard UV radiation.

However, accretion onto a white dwarf is likely to result in a standing shock front above the surface. Unlike the situation for neutron star accretion, Coulomb collisions *alone* are capable of maintaining a thin shock front above the surface [$\lambda_c / R \ll 1$ by Eq. (15.3.8)]. Equation (15.3.9) then indicates that the electron temperature behind the front can approach $T_{ff} \lesssim 100$ keV. Consequently, hard X-rays, generated via optically thin bremsstrahlung, can be emitted from the

---

[16] Interested readers should consult the references cited in Section 14.1 and later in this section for further discussion and additional references.

[17] See Kylafis et al. (1980) for a review.

post-shock region above the photosphere.[18] In addition, if a magnetic field is present, cyclotron emission will also occur in this region.

According to detailed calculations by Lamb and Masters (1979), roughly half of this bremsstrahlung radiation is emitted outward from the surface and is in the hard X-ray band (10–100 keV); roughly half of the (high harmonic) cyclotron radiation is emitted outward and is in the UV band. The other halves of the bremsstrahlung and cyclotron fluxes are emitted inward and are either reflected or absorbed by the surface. The resulting thermalized blackbody flux from the surface consists of a UV or very soft X-ray component $L_{bb} \approx L_{cyc} + L_{brem}$, where L_{bb}, L_{cyc}, and L_{brem} are the blackbody, cyclotron, and bremsstrahlung luminosities, respectively. Lamb and Masters predict that magnetic dwarfs can be very powerful UV sources ($\lesssim 10$ eV) and that the UV flux may greatly dominate the X-ray flux.

The spectra resulting from accretion onto nonmagnetic or weakly magnetic degenerate dwarfs are distinguished from the strong field dwarfs by the absence of UV cyclotron radiation (i.e., $L_{cyc} \ll L_{brem}$ for $B \ll 10^6$ G; Kylafis and Lamb, 1979). If the total luminosity exceeds a maximum value $L_{max} \approx 10^{36}$ erg s^{-1} ($\ll L_{Edd}$), the cool infalling matter *above* the shock will be sufficiently opaque to scatter and degrade (via Comptonization) most of the X-rays. As a result a pronounced correlation between the X-ray spectral temperature and luminosity will exist for nonmagnetic dwarfs. Interestingly, observations of Cyg X-2 by the Copernicus X-ray satellite show such spectral changes in luminosity. If the theory is correct, the observations suggest[19] that Cyg X-2 is an accreting degenerate dwarf with a mass of $\sim 0.4\ M_\odot$ at a distance of 250 ± 50 pc. Lamb (1979b) has also argued that many of the low luminosity, hitherto unidentified "Galactic bulge" sources (cf. Chapter 13) are likely candidates for accreting nonmagnetic white dwarfs because they too exhibit the expected spectral temperature X-ray luminosity correlation. In fact, he estimates that the total number of such objects in our Galaxy may be $\sim 10^4$, which can be compared to an estimated number of magnetic accreting dwarfs of $\sim 10^6$ and roughly 100 strong ($L \gtrsim 10^{36}$ erg/s) accreting neutron star X-ray sources.

---

[18]Hoshi (1973); Aizu (1973); Fabian, Pringle, and Rees (1976); Katz (1977).
[19]Branduardi et al. (1980).

16

Gravitational Radiation

16.1 What is a Gravitational Wave?

General relativity describes gravitational waves as ripples in the curvature of spacetime, which propagate with the speed of light. Like water waves on the ocean, the concept of a gravitational wave requires the idealization of a smooth, unperturbed background on which the waves propagate. Unlike water waves, however, gravitational waves are not motions of a material medium; they are ripples in the fabric of spacetime itself.

Once the waves leave their source ("near zone"), they are generally in regions where their wavelengths λ are very small compared to the radius of curvature R of the background spacetime through which they propagate.

Exercise 16.1

(a) Show that λ for self-gravitating sources oscillating on a dynamical timescale varies from a few kilometers to a few astronomical units as the mean density of the source varies from black hole densities to stellar densities

Hint: Recall Eq. (10.2.2).

(b) The strength of the background gravitational field is measured by the tidal gravitational field [cf. Eq. (5.1.21)]

$$F_{\text{tid}} \sim \frac{M}{L^3}, \quad (c = G = 1). \tag{16.1.1}$$

Here M is the mass of the source of gravitation and L the characteristic size. Equivalently, the radius of curvature of the background field is

$$R \sim (F_{\text{tid}})^{-1/2}. \tag{16.1.2}$$

Show that $R \sim 10^{10}$ lt. yr between galaxies, $R \sim 10^7$ lt. yr in a galaxy, and $R \sim 0.1$ lt. yr in the solar system.

The fact that R is large in the solar system is merely a restatement of the fact that gravitation is weak there. One can introduce nearly Minkowski coordinates in the solar system so that

$$g_{\mu\nu} = \eta_{\mu\nu} + h_{\mu\nu}, \qquad |h_{\mu\nu}| \ll 1. \tag{16.1.3}$$

Here $h_{\mu\nu}$ contains quasistatic contributions from the sun, planets and so on (cf. Chapter 5), plus any possible gravitational waves from astrophysical sources. Henceforth we ignore the quasistatic contributions and focus only on the wave contributions to $h_{\mu\nu}$. We shall generally only summarize results, with little derivation.[1]

Gravitational waves are completely described by two dimensionless amplitudes, h_+ and $h_\times$, say. Take the propagation direction to be the z-direction. Then h_+ and $h_\times$ are functions only of $t - z/c$. If we introduce *polarization tensors* $\underset{\sim}{e}^+$ and $\underset{\sim}{e}^\times$ such that

$$e_{xx}^+ = -e_{yy}^+ = 1, \qquad e_{xy}^\times = e_{yx}^\times = 1, \quad \text{all other components zero,} \tag{16.1.4}$$

then we can write the gravitational wave as

$$h_{jk}^{TT} = h_+ e_{jk}^+ + h_\times e_{jk}^\times. \tag{16.1.5}$$

This is a symmetric spatial tensor that is traceless and transverse to the propagation direction (no z-component). This h_{jk}^{TT} is the analog of the vector potential of electrodynamics in the Lorentz gauge. There one has in vacuum

$$A_0 = 0, \qquad A_{i,i} = 0, \qquad \Box A_i = 0, \tag{16.1.6}$$

where the last equation comes from Maxwell's field equations in this gauge. Here we have

$$h_{0\mu}^{TT} = 0, \qquad h_{jk,k}^{TT} = 0, \qquad \Box h_{jk}^{TT} = 0, \tag{16.1.7}$$

together with the trace-free condition. The last equation is Einstein's field equation in this gauge. The coordinate system in which Eqs. (16.1.7) are valid is called the *TT gauge* in general relativity, for *transverse-traceless gauge*.

When an electromagnetic wave hits a charged particle, it produces an acceleration that is transverse to the wave's propagation direction and is proportional to e/m, the particle's charge-to-mass ratio. Similarly, when a gravitational wave hits a free particle, it imparts a transverse acceleration. However, the "gravitational charge" of a particle (its response to a gravitational force) is equal to its inertial

---

[1] Full details of the theory of gravitational waves are given, for example, by Misner, Thorne, and Wheeler (1973).

mass m (Equivalence Principle). Thus in general relativity all particles have the same gravitational "e/m." Therefore, all free particles at the same location experience the same transverse acceleration. Since local inertial frames are determined by freely moving particles, the local inertial frames themselves undergo this same acceleration. The acceleration is *locally* undetectable. On the other hand, the acceleration is different at different locations in spacetime, and this provides a means for detecting the wave. (We have merely restated the fact that the "true" gravitational field is the tidal field, which can only be measured by nonlocal comparisons.)

Consider the local inertial frame of a fiducial free test particle. In this coordinate system let ξ_j be the vector separation of a second test particle from the fiducial particle. Then a passing gravitational wave will produce a tiny relative acceleration of the particles given by

$$\ddot{\xi}_j = \tfrac{1}{2}\ddot{h}^{TT}_{jk}\xi_k, \qquad (16.1.8)$$

which in turn will produce a tiny change

$$\delta\xi_j = \tfrac{1}{2}h^{TT}_{jk}\xi_k \qquad (16.1.9)$$

in their separation. In order of magnitude, the *relative strain* produced by a gravitational wave is

$$\frac{\delta\xi}{\xi} \sim h, \qquad (16.1.10)$$

a very useful result for rough estimates.

Note that the relative acceleration (16.1.8) is transverse both in the sense that it is orthogonal to the propagation direction and that it vanishes if ξ_j is parallel to the propagation direction.

Exercise   16.2

(a)   Consider a plane wave propagating in the z-direction with $h_+ \neq 0$, $h_\times = 0$. Show that the relative acceleration field (16.1.8) is divergence-free, and hence can be represented by "lines of force" analogous to a vacuum electric field.

(b)   Sketch the lines of force and show that they are quadrupole-shaped, with the spacing decreasing with distance from the origin.

When a gravitational wave hits an object with internal forces, the various pieces of the object cannot move as free test particles. Instead, the object oscillates in accord with its standard equations of motion, the oscillations being driven by a gravitational-wave driving force

$$F_j = \tfrac{1}{2} m \ddot{h}_{jk}^{TT} \xi_k \qquad (16.1.11)$$

acting on each element of mass m. Here ξ_k is the displacement of the mass element from the center of mass.

Actually, expressions (16.1.8)–(16.1.11) are only correct if the separation distance $|\xi|$ is small compared to the wavelength λ of the waves. If this is not so, retardation effects change the linear dependence on ξ_j into a sinusoidal one,[2] roughly as $\sin(2\pi|\xi|/\lambda)$.

Since gravitational waves can exert forces and do work, they must carry energy and momentum. The energy and momentum densities cannot be localized at a point (no gravitational forces at a point!), but *can* be localized within a few wavelengths. The stress-energy tensor for waves propagating in the z-direction has nonzero components

$$T^{00} = \frac{T^{0z}}{c} = \frac{T^{zz}}{c^2} = \frac{1}{16\pi}\frac{c^2}{G}\langle \left(\dot{h}_+ \right)^2 + \left(\dot{h}_\times \right)^2 \rangle, \qquad (16.1.12)$$

where the angle brackets denote an average over several wavelengths. Here T^{00} is the energy density, T^{0z} the energy flux ($= c^2 \times$ momentum density), and T^{zz} is the momentum flux.

16.2   The Generation of Gravitational Waves

In electromagnetism, the leading-order multipole radiation from a nonrelativistic charge distribution is dipole radiation. The Lorentz-gauge vector potential in the wave zone is

$$A_j(t, \mathbf{x}) = \frac{1}{cr}\dot{d}_j\left(t - \frac{r}{c} \right), \qquad (16.2.1)$$

where $r \equiv |\mathbf{x}|$ and $\mathbf{d}$ is the electric dipole moment. The $1/r$ electric and magnetic fields computed from Eq. (16.2.1) depend only on the components of $\mathbf{d}$ *transverse* to the propagation direction $\mathbf{n} \equiv \mathbf{x}/r$, so we can replace d_j in Eq. (16.2.1) by its transverse part,

$$d_j^T \equiv P_{jk} d_k, \qquad (16.2.2)$$

where P_{jk} is the projection tensor,

$$P_{jk} \equiv \delta_{jk} - n_j n_k. \qquad (16.2.3)$$

---

[2] See Exercise 37.6 of Misner, Thorne, and Wheeler (1973).

Substituting the **E** and **B** fields obtained from Eq. (16.2.1) into the Poynting vector, we obtain the angular distribution of the energy flux,

$$\frac{d^2E}{dt\,d\Omega} = \frac{1}{4\pi c^3}\ddot{d}_j^T\ddot{d}_j^T \tag{16.2.4a}$$

$$= \frac{1}{4\pi c^3}\left[\ddot{d}_j\ddot{d}_j - \left(n_j\ddot{d}_j\right)^2\right]. \tag{16.2.4b}$$

The quantity d_j is to be evaluated at the retarded time $t - r/c$. Choosing the z axis along **n**, we can easily integrate Eq. (16.2.4b) over solid angles to obtain

$$L_{\text{em}} \equiv \frac{dE}{dt} = \frac{2}{3c^3}\ddot{d}_j\ddot{d}_j. \tag{16.2.5}$$

Writing $d_j = ex_j$ for a point charge, we see that this is simply the Larmor formula.

Now on dimensional grounds we might expect the leading-order gravitational radiation from a source with low internal velocities also to be dipole:

$$L_{\text{GW}} \propto \frac{G}{c^3}\ddot{d}_j\ddot{d}_j, \tag{16.2.6}$$

where the gravitational dipole moment is

$$d_j = \sum_A m_A x_j^A \tag{16.2.7}$$

and we have let $e^2 \to Gm^2$ in Eq. (16.2.5). However, Eq. (16.2.7) gives

$$\ddot{d}_j = \sum_A m_A \ddot{x}_j^A = \sum_A \dot{p}_j^A, \tag{16.2.8}$$

where $\mathbf{p}^A$ is the momentum of the Ath particle. Since the total momentum of the system is conserved, $\ddot{d}_j = 0$. There is no dipole radiation in general relativity.

The next-order multipole radiation is magnetic dipole and electric quadrupole. The "mass magnetic dipole moment" is

$$\boldsymbol{\mu} \equiv \frac{1}{c}\sum_A \mathbf{x}^A \times \left(m_A\dot{\mathbf{x}}^A\right) = \frac{1}{c}\sum_A \mathbf{j}^A, \tag{16.2.9}$$

where $\mathbf{j}^A$ is the angular momentum of the Ath particle. By conservation of angular momentum, $\dot{\boldsymbol{\mu}} = 0$: there is no magnetic dipole radiation in general relativity. The lowest-order radiation is electric *quadrupole*.

The analog of Eq. (16.2.1) is

$$h_{jk}^{TT}(t,\mathbf{x}) = \frac{2}{r}\frac{G}{c^4}\ddot{I}_{jk}^{TT}\left(t-\frac{r}{c}\right), \tag{16.2.10}$$

where I_{jk} is the mass quadrupole moment

$$I_{jk} \equiv \sum_A m_A\left[x_j^A x_k^A - \frac{1}{3}\delta_{jk}(x^A)^2\right]. \tag{16.2.11}$$

We follow Misner, Thorne, and Wheeler (1973) in using the "slash" notation to distinguish our definition of quadrupole moment from others in the literature. The superscript "TT" means take the transverse traceless part:

$$I_{jk}^{TT} \equiv P_{jl}P_{km}I_{lm} - \tfrac{1}{2}P_{jk}(P_{lm}I_{lm}). \tag{16.2.12}$$

Note that in order of magnitude

$$h \sim \frac{r_{\text{Sch}}}{r}\frac{v^2}{c^2}, \tag{16.2.13}$$

where r_{Sch} is the "Schwarzschild radius" GM/c^2 associated with the mass in quadrupole motion and v is the characteristic velocity.

The energy flux is given by the stress-energy tensor

$$T_{0r} = \frac{1}{32\pi}\frac{c^4}{G}\langle h_{jk,0}^{TT}h_{jk,r}^{TT}\rangle \tag{16.2.14}$$

[cf. Eq. (16.1.12)]. Substituting Eq. (16.2.10), we obtain

$$\frac{d^2E}{dt\,d\Omega} = \frac{1}{8\pi}\frac{G}{c^5}\langle \dddot{I}_{jk}^{TT}\dddot{I}_{jk}^{TT}\rangle \tag{16.2.15a}$$

$$= \frac{1}{8\pi}\frac{G}{c^5}\left\langle \dddot{I}_{jk}\dddot{I}_{jk} - 2n_i\dddot{I}_{ij}\dddot{I}_{jk}n_k + \frac{1}{2}\left(n_jn_k\dddot{I}_{jk}\right)^2\right\rangle, \tag{16.2.15b}$$

[cf. Eqs. (16.2.4a, b)]. Integrating over $\mathbf{n}$, we get

$$L_{\text{GW}} \equiv \frac{dE}{dt} = \frac{1}{5}\frac{G}{c^5}\langle \dddot{I}_{jk}\dddot{I}_{jk}\rangle. \tag{16.2.16}$$

The analogous electromagnetic formula has a coefficient of $\frac{1}{20}$ instead of $\frac{1}{5}$ because the waves are vector, not tensor fields.

The quadrupole formula (16.2.16), with the definition (16.2.11), is valid for *slow-motion* sources ($v \ll c$), with *weak internal gravity* (Newtonian potential $\phi \ll c^2$). References to other situations for which the radiation can be calculated are given by Thorne (1980).

Note that the vanishing of L_{GW} for spherically symmetric sources is a general result, by Birkhoff's theorem (Chapter 5).

The radiation-reaction force corresponding to the energy loss (16.2.16) can be written as the gradient of a Newtonian-type potential:

$$\mathbf{F}^{(\text{react})} = -m\nabla\Phi^{(\text{react})}, \qquad \Phi^{(\text{react})} = \frac{1}{5}\frac{G}{c^5}\mathcal{I}_{jk}^{(5)}x_j x_k. \qquad (16.2.17)$$

Here the bracketed superscript "5" denotes five time derivatives. We can easily verify this:

$$\frac{dE}{dt} = \sum_A \mathbf{v}_A \cdot \mathbf{F}_A^{(\text{react})}$$

$$= -\sum_A m_A v_{Aj} \frac{2}{5}\frac{G}{c^5}\mathcal{I}_{jk}^{(5)}x_k^A$$

$$= -\frac{1}{5}\frac{G}{c^5}\mathcal{I}_{jk}^{(5)}\frac{d}{dt}\sum_A m_A x_j^A x_k^A$$

$$= -\frac{1}{5}\frac{G}{c^5}\mathcal{I}_{jk}^{(5)}\mathcal{I}_{jk}^{(1)}, \qquad (16.2.18)$$

where in the last line we have used the fact that $\mathcal{I}_{jk}\delta_{jk} = 0$. Averaging over several cycles (for a periodic source) or over a time long compared with the dynamical timescale (for bounded source motion) allows us to integrate by parts twice and change $\mathcal{I}^{(5)}\mathcal{I}^{(1)}$ into $\mathcal{I}^{(3)}\mathcal{I}^{(3)}$, thus recovering Eq. (16.2.16).

Exercise 16.3 The angular momentum carried off by gravitational waves is given by

$$\frac{dJ_i}{dt} = \sum_A \varepsilon_{ijk} x_j^A F_k^{A(\text{react})}. \qquad (16.2.19)$$

Show that this gives

$$\frac{dJ_i}{dt} = -\frac{2}{5}\frac{G}{c^5}\varepsilon_{ijk}\langle\ddot{\mathcal{I}}_{jm}\dddot{\mathcal{I}}_{km}\rangle. \qquad (16.2.20)$$

Note that no angular momentum is carried off if the source is axisymmetric, a result that is quite general.

16.3   Order-of-Magnitude Estimates

We gave an estimate of the amplitude h of a gravitational wave in Eq. (16.2.13). To estimate dE/dt, note that

$$\dddot{\bar{I}}_{jk} \sim \frac{MR^2}{T^3} \sim \frac{Mv^3}{R}, \tag{16.3.1}$$

where M, R, T, and v are the characteristic mass, size, timescale, and velocity of the source, respectively. Thus from Eq. (16.2.16)

$$\frac{dE}{dt} \sim \frac{G}{c^5}\left(\frac{M}{R}\right)^2 v^6 \sim L_0 \left(\frac{r_{\mathrm{Sch}}}{R}\right)^2 \left(\frac{v}{c}\right)^6, \tag{16.3.2}$$

where

$$L_0 \equiv \frac{c^5}{G} = 3.6 \times 10^{59} \text{ erg s}^{-1}. \tag{16.3.3}$$

Exercise   16.4   A steel rod of mass 10^5 kg and length 20 m rotates at its breakup speed (30 rad s^{-1}). Estimate L_{GW}.
   Answer: $\sim 10^{23}$ erg s^{-1}

Exercise   16.5   Show that

$$L_{\mathrm{GW}} \sim L_{\mathrm{int}} \frac{L_{\mathrm{int}}}{L_0}, \tag{16.3.4}$$

where $L_{\mathrm{int}} \sim Mv^2/T$ is the internal power in quadrupole motions.

From Eq. (16.3.2), we see that one obtains maximal gravitational radiation luminosities when $r_{\mathrm{Sch}} \sim R$ and $v \sim c$. *Compact objects are therefore important potential sources of gravitational waves.*
   Astrophysical systems are generally gravitationally bound, so that by the virial theorem

$$\text{Kinetic energy} \sim \frac{MR^2}{T^2} \sim |\text{Potential energy}| \sim \frac{GM^2}{R}, \tag{16.3.5}$$

or

$$T \sim \left(\frac{R^3}{GM}\right)^{1/2}. \tag{16.3.6}$$

Equation (16.3.6) is the relation between timescale and mean density we have seen many times before. Eliminating v/c from Eq. (16.3.2), we get

$$L_{\mathrm{GW}} \sim L_0 \left(\frac{r_{\mathrm{Sch}}}{R} \right)^5. \tag{16.3.7}$$

Exercise 16.6 Show that the gravitational wave energy radiated by a nonspherical self-gravitating system is

$$\Delta E \sim L_{\mathrm{GW}} T \sim Mc^2 \left(\frac{r_{\mathrm{Sch}}}{R} \right)^{7/2}. \tag{16.3.8}$$

Equations (16.3.7) and (16.3.8) again underscore the importance of compact objects as potential sources of gravitational waves.

We can parametrize the efficiency of gravitational wave emission by writing

$$\Delta E = \varepsilon Mc^2, \qquad \varepsilon \sim \left(\frac{r_{\mathrm{Sch}}}{R} \right)^{7/2}. \tag{16.3.9}$$

Then Eqs. (16.2.13), (16.3.5), and (16.3.8) give

$$h \sim \varepsilon^{2/7} \frac{r_{\mathrm{Sch}}}{r}$$

$$\sim 3 \times 10^{-18} \left(\frac{\varepsilon}{0.1} \right)^{2/7} \frac{(M/M_\odot)}{(r/10 \text{ kpc})}. \tag{16.3.10}$$

In Eq. (16.3.10) we have scaled r to the distance to the center of the Galaxy, and ε to the optimistic value of 10%. If we use the total mass of the system in making estimates, then ε incorporates the degree of nonsphericity as well as the degree of compaction.

The above estimates refer to the energy radiated in one dynamical time T. If the system evolves for longer, then ε and ΔE will be correspondingly larger, but h will be the same.

Exercise 16.7 Show that the damping time from gravitational radiation, $\tau \sim E/(dE/dt)$, is roughly

$$\tau \sim \frac{R}{c} \left(\frac{R}{r_{\mathrm{Sch}}} \right)^3. \tag{16.3.11}$$

Estimate τ for nonspherical white dwarfs and neutron stars. (Similar considerations were used in Sections 7.4 and 9.6 to argue that white dwarfs and neutron stars with $T/|W| \gtrsim 0.14$ would be secularly unstable due to gravitational radiation.)

Answer: White dwarfs, $\tau \lesssim 10^3$ yr; neutron stars, $\tau \lesssim 300$ s.

What are the chances of detecting gravitational waves from an astrophysical source? Consider a highly nonspherical supernova explosion at the center of our Galaxy, leading to the formation of a neutron star or possibly a black hole. Taking $\varepsilon \sim 10\%$ and $M \sim M_\odot$, we find from Eqs. (16.1.10) and (16.3.10) that the displacement of the end of a 1.5-m bar on the earth is only

$$\Delta \xi \sim 5 \times 10^{-16} \text{ cm}, \qquad (16.3.12)$$

or $\frac{1}{200}$ of a fermi! Even if our estimate is not too optimistic, recall[3] that the event rate is expected to be only about one every 30 years! To achieve a more "tolerable" event rate of one per month say, one must go out to the Virgo cluster of galaxies ($\sim$ 20 Mpc) before the supernova rate becomes sufficiently high. From Eq. (16.3.10), we see that this implies $h \lesssim 10^{-21}$.

Joseph Weber has pioneered the development of gravitational wave astronomy.[4] A "Weber bar" is made from some high-Q material such as aluminum, sapphire, or niobium. (Here Q is the "quality factor" of the resonant bar.) The bar is isolated from all possible terrestrial disturbances. An impinging gravitational wave of the right frequency sets the fundamental mode into oscillation, and a suitable sensor detects the displacement of the bar. (Weber originally used piezoelectric crystals bonded to the surface of the bar). The signal is then amplified electronically and scanned for evidence of a burst of gravitational waves. By comparing signals from two widely separated bars, one can have greater confidence that any detected "event" is of nonterrestrial origin.

Weber's original bars were resonant at kHz frequencies, corresponding to solar mass-type compact object events [cf. Eq. (16.3.6)]. Sensitivities of the "first generation" of bars built by Weber and others reached about $h \sim 10^{-16}$. Although Weber reported seeing numerous events at this sensitivity, no other group could confirm his results and it is generally agreed that his events were not caused by gravitational waves.

"Second generation" bars now coming on line are cooled to liquid helium temperatures and should reach sensitivites of $h \sim 10^{-19}$. To achieve $h \sim 10^{-21}$ will require a "third generation," possibly cooled to millidegrees!

Another promising line of development involves laser interferometers. A passing gravitational wave changes the relative path length of the two arms of the interferometer and the experimentalist searches for the resulting fringe shifts. Forward (1978) built a prototype that achieved $h \sim 10^{-15}$. Detectors of greater sensitivity are currently under construction at several places, and it is not inconceivable that $h \sim 10^{-22}$ might be reached. Interferometers have an advantage over resonant systems in that they record the entire waveform $h(t)$,

---

[3] Equation (1.3.27).
[4] For example, Weber (1960).

whereas a resonant system essentially measures only the Fourier component of $h(t)$ at the resonant frequency.

A third method of detection involves very accurate Doppler tracking of spacecraft from the earth. Here one is interested in lower-frequency gravitational waves, in the range 10^{-2} to 10^{-4} Hz. (These limits are set by the 100 s or so necessary for accurate readout of an atomic clock, and the earth's rotation period which prevents continuous tracking from a single site.) Such low-frequency waves might be produced by catastrophic events involving supermassive compact objects.[5]

16.4   Gravitational Radiation from Binary Systems

Consider first the case of two point masses M_1 and M_2 in a *circular* orbit of radius a about each other. If a_1 and a_2 are their respective distances from the center of mass, then

$$M_1 a_1 = M_2 a_2 = \mu a, \qquad (16.4.1)$$

where

$$\mu \equiv \frac{M_1 M_2}{M_1 + M_2}, \qquad (16.4.2)$$

is the reduced mass.

If the z-axis is the axis of rotation, and ϕ is the azimuthal angle from the x-axis to the line joining the masses, then

$$\bar{t}_{xx} = \left(M_1 a_1^2 + M_2 a_2^2 \right)\cos^2\phi + \text{constant.} \qquad (16.4.3)$$

Here we use the fact that the term $M_i a_i^2 / 3$, $i = 1, 2$, is constant for each particle. Equation (16.4.3) can be rewritten as

$$\bar{t}_{xx} = \tfrac{1}{2}\mu a^2 \cos 2\phi + \text{constant.} \qquad (16.4.4)$$

Similarly,

$$\bar{t}_{yy} = -\tfrac{1}{2}\mu a^2 \cos 2\phi + \text{constant,}$$

$$\bar{t}_{xy} = \bar{t}_{yx} = \tfrac{1}{2}\mu a^2 \sin 2\phi. \qquad (16.4.5)$$

---

[5] Further details on gravitational wave detection are given in Thorne (1980) and references therein.

Since $\phi = \Omega t$, where Ω is the orbital angular velocity, we find

$$
\begin{aligned}
L_{\text{GW}} &= \frac{1}{5}\frac{G}{c^5}\langle \dddot{\mathcal{f}}_{jk}\dddot{\mathcal{f}}_{jk}\rangle \\
&= \frac{1}{5}\frac{G}{c^5}(2\Omega)^6\left(\frac{1}{2}\mu a^2\right)\langle \sin^2 2\Omega t + \sin^2 2\Omega t + 2\cos^2 2\Omega t\rangle \\
&= \frac{32}{5}\frac{G^4}{c^5}\frac{M^3\mu^2}{a^5}.
\end{aligned}
\tag{16.4.6}
$$

Here we have used Kepler III,

$$
\Omega^2 = \frac{GM}{a^3}, \qquad M \equiv M_1 + M_2. \tag{16.4.7}
$$

The loss of energy leads to a decrease in the separation a and hence a decrease in the orbital period $P \equiv 2\pi/\Omega$. Since the energy is

$$
\begin{aligned}
E &= \left(\frac{1}{2}M_1 a_1^2 + \frac{1}{2}M_2 a_2^2\right)\Omega^2 - \frac{GM_1 M_2}{a} \\
&= -\frac{1}{2}\frac{G\mu M}{a},
\end{aligned}
\tag{16.4.8}
$$

we have

$$
\begin{aligned}
\frac{1}{P}\frac{dP}{dt} &= \frac{3}{2}\frac{1}{a}\frac{da}{dt} \\
&= -\frac{3}{2}\frac{1}{E}\frac{dE}{dt} \\
&= -\frac{96}{5}\frac{G^3}{c^5}\frac{M^2\mu}{a^4}.
\end{aligned}
\tag{16.4.9}
$$

Exercise 16.8 Assuming that the expressions above remain valid as $a \to 0$, show that the time t_0 until $a \to 0$ is

$$
t_0 = \frac{5}{256}\frac{c^5}{G^3}\frac{a_{\text{now}}^4}{M^2\mu}. \tag{16.4.10}
$$

Exercise 16.9 Show that in order of magnitude, t_0 [or $P/(dP/dt)$] is given by

$$
\frac{t_0}{P} \sim 10^5\left(\frac{P}{1\,s}\right)^{5/3} \tag{16.4.11}
$$

for $M_1 \sim M_2 \sim M_\odot$. [This result was used in Eq. (10.2.5) to rule out binary neutron star models for pulsars.]

Exercise   16.10   Using Eq. (16.2.20), show that for circular orbits

$$\frac{dJ}{dt} = -\frac{32}{5} \frac{G^{7/2}}{c^5} \frac{\mu^2 M^{5/2}}{a^{7/2}}. \tag{16.4.12}$$

Show that

$$\frac{dE}{dt} = \Omega \frac{dJ}{dt}, \tag{16.4.13}$$

and hence that a circular orbit remains circular.

If the two masses are in an *elliptic* orbit with eccentricity e, then[6]

$$\frac{dE}{dt} = \frac{dE}{dt}\bigg|_{e=0} f(e), \tag{16.4.14}$$

$$\frac{dJ}{dt} = \frac{dJ}{dt}\bigg|_{e=0} g(e), \tag{16.4.15}$$

$$f(e) \equiv \left(1 + \tfrac{73}{24}e^2 + \tfrac{37}{96}e^4\right)\left(1 - e^2\right)^{-7/2}, \tag{16.4.16}$$

$$g(e) \equiv \left(1 + \tfrac{7}{8}e^2\right)\left(1 - e^2\right)^{-2}. \tag{16.4.17}$$

Here dE/dt and dJ/dt have been averaged over an orbit. Since Eqs. (16.4.7) and (16.4.8) are valid for elliptic orbits, Eq. (16.4.9) becomes

$$\frac{1}{P} \frac{dP}{dt} = -\frac{96}{5} \frac{G^3}{c^5} \frac{M^2\mu}{a^4} f(e). \tag{16.4.18}$$

Exercise   16.11   Using the elliptic orbit relation

$$e^2 = 1 + \frac{2EJ^2}{G^2\mu^3 M^2}, \tag{16.4.19}$$

derive a result for de/dt due to gravitational radiation, and show that $de/dt < 0$: gravitational radiation reaction tends to *circularize* an elliptic orbit.

[6]Peters (1964); Lightman et al. (1975).

16.5   The Binary Pulsar PSR 1913 + 16

The strongest evidence to date for the existence of gravitational waves comes from studies of the orbit of the first binary pulsar to be discovered. In view of its importance, we will discuss this object in some detail.

The binary pulsar, PSR 1913 + 16, was discovered by Hulse and Taylor[7] in 1974. They quickly realized that apparent changes in the pulsar frequency could be explained by the Doppler effect due to orbital motion about an unseen companion with a period of about 8 hours. The presence of a high-precision clock (pulsar) moving with velocity ~ 300 km s^{-1} through the gravitational field of the companion caused a flurry of activity in the relativity community. Nature had provided a testing ground for various relativistic effects. In practice, these effects are sought for by studying the *arrival times* of pulses from the pulsar.[8] We shall give below a simplified treatment based on the *period* of the pulses.

Let the pulsar mass be M_1 and the companion mass M_2. If we assume they are spherical, then in lowest order (Newtonian gravity) they move in elliptic orbits about their common center of mass. Choose the orbit to lie in the x-y plane, with the origin at the center of mass. Let the inclination of the orbital plane to the line of sight be i (see Fig. 16.1). Choose the x axis along the "line of nodes," the line through the origin along the intersection of the orbital plane with the plane perpendicular to the line of sight. Let ω be the angular distance of periastron from the node, measured in the orbital plane. Then the position of the pulsar at any instant is given by

$$x = r_1 \cos \psi, \qquad y = r_1 \sin \psi, \qquad (16.5.1)$$

where

$$\psi = \omega + \phi, \qquad r_1 = \frac{a_1(1 - e^2)}{1 + e \cos \phi}. \qquad (16.5.2)$$

The angle ϕ (polar coordinate measured from periastron) is called the "true anomaly" in celestial mechanics.

The ratio of received to emitted period of the pulsar can be written

$$\frac{(\delta t)_{\text{rec}}}{(\delta t)_{\text{em}}} = \frac{(\delta t)_{\text{rec}}}{(\delta t)_{\text{stat}}} \frac{(\delta t)_{\text{stat}}}{(\delta t)_{\text{em}}}, \qquad (16.5.3)$$

where "stat" denotes an observer at the position of the pulsar who is stationary with respect to the center of mass. Assume for the moment that the receiver (on

---

[7]Hulse and Taylor (1975).
[8]Blandford and Teukolsky (1976); Epstein (1977).

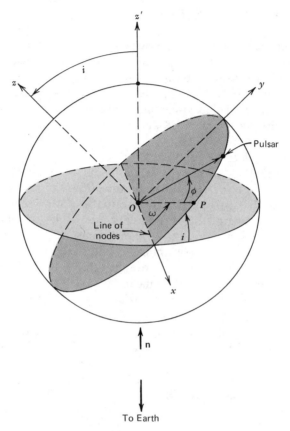

Figure 16.1 The orbit of the binary pulsar. The point O lies at the center of mass of the binary system; P lies at pulsar periastron. The "line of nodes" Ox passes through O along the intersection of the orbital plane with the plane perpendicular to the line of sight (see text).

the earth) is also static with respect to the center of mass. Then if r is the distance between M_1 and M_2,

$$\frac{(\delta t)_{\text{rec}}}{(\delta t)_{\text{stat}}} = \left(1 - \frac{GM_2}{rc^2}\right)^{-1}, \tag{16.5.4}$$

by the gravitational redshift formula (5.4.2). The Doppler formula gives

$$\frac{(\delta t)_{\text{stat}}}{(\delta t)_{\text{em}}} = \left(1 - \frac{v_1^2}{c^2}\right)^{-1/2}\left(1 + \frac{\mathbf{v}_1 \cdot \mathbf{n}}{c}\right), \tag{16.5.5}$$

where $\mathbf{n}$ is a unit vector pointing from the earth to the emitter. Thus Eq. (16.5.3) becomes, to $\mathcal{O}(v^2)$ and $\mathcal{O}(M/r)$,

$$\frac{(\delta t)_{\text{rec}}}{(\delta t)_{\text{em}}} = 1 + \frac{\mathbf{v}_1 \cdot \mathbf{n}}{c} + \frac{1}{2}\frac{v_1^2}{c^2} + \frac{GM_2}{rc^2}. \tag{16.5.6}$$

Now

$$\mathbf{n} = \mathbf{e}_{z'} = \cos i\, \mathbf{e}_z + \sin i\, \mathbf{e}_y \tag{16.5.7}$$

(cf. Fig. 16.1). Thus Eq. (16.5.1) gives

$$\mathbf{v}_1 \cdot \mathbf{n} = \left(\dot{r}_1 \sin \psi + r_1 \dot{\psi} \cos \psi \right)\sin i. \tag{16.5.8}$$

Using Eq. (16.5.2) and Kepler II in the form

$$\dot{\phi} = \frac{2\pi}{P(1 - e^2)^{3/2}}(1 + e\cos\phi)^2, \tag{16.5.9}$$

we find after some simple algebra that

$$\mathbf{v}_1 \cdot \mathbf{n} = K\left[\cos(\omega + \phi) + e\cos\omega\right], \tag{16.5.10}$$

$$K \equiv \frac{2\pi a_1 \sin i}{P(1 - e^2)^{1/2}}. \tag{16.5.11}$$

So far the analysis is exactly the same as for a single-line spectroscopic binary, with the important difference that $(\delta t)_{\text{em}}$, not being from a spectral line, is not known. Thus any *constant* term on the right-hand side of Eq. (16.5.6) is not measurable: it is simply absorbed in $(\delta t)_{\text{em}}$. In particular, a uniform velocity between the solar system center of mass and the pulsar system center of mass is not measurable. (The earth's orbital motion leads to a Doppler effect that must be subtracted out using the known velocity of the earth in the solar system.)

From the first-order Doppler term the following parameters can be found: e and P through Eq. (16.5.9), which gives $\phi(t)$ when integrated, and K and ω from the two independent time-varying terms proportional to $\cos\phi$ and $\sin\phi$ in Eq. (16.5.10). From K one gets $a_1\sin i$, and from P and $a_1\sin i$ one gets the mass function

$$f \equiv \frac{(M_2 \sin i)^3}{(M_1 + M_2)^2} = \frac{(a_1 \sin i)^3}{G}\left(\frac{2\pi}{P}\right)^2 \tag{16.5.12}$$

in the usual way.

Because of the high precision of pulsar timing, the transverse Doppler shift and gravitational redshift terms in Eq. (16.5.6) can also be measured. We find

$$v_1^2 = \dot{r}_1^2 + r_1^2 \dot{\psi}^2$$

$$= \left(\frac{2\pi}{P}\right)^2 \frac{a_1^2}{1 - e^2} (1 + 2e\cos\phi + e^2), \qquad (16.5.13)$$

and

$$\frac{GM_2}{r} = \frac{GM_2^2}{(M_1 + M_2)r_1}. \qquad (16.5.14)$$

Thus, since by Kepler III

$$\left(\frac{2\pi}{P}\right)^2 = \frac{GM_2^3}{(M_1 + M_2)^2 a_1^3}, \qquad (16.5.15)$$

we get

$$\frac{1}{2} v_1^2 + \frac{GM_2}{r} = \beta \cos\phi + \text{constant}, \qquad (16.5.16)$$

where

$$\beta \equiv \frac{GM_2^2(M_1 + 2M_2)e}{(M_1 + M_2)^2 a_1(1 - e^2)}. \qquad (16.5.17)$$

That only one new measurable quantity would arise from the second-order Doppler shift and gravitational redshift could have been foreseen from the virial theorem. However, we now notice that the time dependence in Eq. (16.5.16) is exactly the same as that of the first-order term $K\cos\omega\cos\phi$ in Eq. (16.5.10). For pure elliptic motion, β is unmeasurable.[9]

Fortunately, we are saved by general relativity. The orbit is not exactly an ellipse. There is a periastron advance given by[10]

$$\dot{\omega} = \frac{6\pi GM_2}{a_1(1 - e^2)Pc^2}. \qquad (16.5.18)$$

---

[9] Blandford and Teukolsky (1975); Brumberg et al. (1975).
[10] This expression, first derived by Robertson (1938), is valid whether or not M_1 is a test mass compared with M_2, as is the case for Mercury and the Sun.

The measured value of $\dot{\omega}$ is about 4.2° per year for the binary pulsar (compare 43″ per century for the planet Mercury!). Thus if we let $\omega \rightarrow \omega_0 + \dot{\omega}t$ in Eq. (16.5.10), there are now four independent time-varying trigonometric combinations of ϕ and $\dot{\omega}t$. Thus on a timescale of years one can separate K, ω_0, $\dot{\omega}$, and β. In particular, $\dot{\omega}$ and β involve two *different* combinations of the four parameters M_1, M_2, a_1, and $\sin i$ from the mass function and $a_1 \sin i$. Thus measurement of $\dot{\omega}$ and β allows a complete solution for the parameters of the binary system.

Now recall Eq. (16.4.18) for $\dot{P}$ from a binary system. With all the system parameters known, we can *predict* a value for $\dot{P}$. If this agrees with the measured value, we have confirmed the existence of gravitational waves!

At this stage, the reader might be worried. After all, we can imagine other effects, such as mass loss from or accretion onto the system, which can lengthen or shorten the orbital period. Similarly, the periastron advance need not all be due to general relativity. If the companion is nonspherical, either because it is rotating or because the pulsar has raised a tidal bulge, then its quadrupole gravitational field can also produce a periastron advance.

Suppose, however, that $\dot{P}$ is measured, and that it agrees with the predicted value assuming that general relativity is the only source of $\dot{P}$ and $\dot{\omega}$. It would be a miracle if general relativity were wrong in its prediction of $\dot{P}$, but there was another source of $\dot{\omega}$ and/or $\dot{P}$ that conspired to give *exactly* the measured value. Since physicists cannot invoke miracles, we would conclude instead that the predictions of general relativity are confirmed, and that we are seeing the effects of gravitational radiation.

As the timing accuracy improves, it becomes possible to measure further relativistic effects. One of these is the time-delay of signals as they cross the orbit on the way to the earth, an effect already measured[11] with spacecraft in the solar system. In addition, there are various post-Newtonian periodic deviations from elliptic motion, not yet verified in the solar system. In general relativity each of these terms contains a known different combination of M_1, M_2, a_1, and $\sin i$. As more and more of these terms are measured (and it appears as if we are on the verge of being able to do so) and if they agree with the general relativistic predictions, then the case for general relativity and gravitational waves will become increasingly watertight.

The binary pulsar has been monitored by Taylor and his collaborators with ever-increasing accuracy since they discovered it in 1974. The latest values for the system parameters are given in Table 16.1. The quantity γ is related to β by

$$\gamma \equiv \frac{\beta P (1 - e^2)}{2 \pi c^2} . \qquad (16.5.19)$$

The time-delay effect and the post-Newtonian orbital effects are just barely

---

[11]Shapiro (1964).

Table 16.1

Parameters for the Binary Pulsar[a]

Parameter	Value
P_p (s)	0.0590299952709(20)
$\dot{P}_p$ (10^{-18})	8.628(20)
$\ddot{P}_p$ (10^{-30} s^{-1})	$-58(1200)$
$(a_1 \sin i)/c$ (s)	2.34186(24)
e	0.617139(5)
P (s)	27906.98161(3)
ω (deg)	178.8656(15)
T_0 (Julian Day No.)	2442321.4332092(15)
$\dot{\omega}$ (deg yr^{-1})	4.2261(7)
γ (s)	0.00438(24)
$\dot{P}$ (10^{-12})	$-2.30(22)$

[a] From Taylor and Weisberg (1982). P_p is the pulsar period, P the orbital period. T_0 is the epoch, or origin of time for the measurements. Numbers in parentheses are estimated uncertainties in the last digits of a value.

detectable at present. No parameter corresponding to these effects is given in Table 16.1.

Note that Eq. (16.5.18) can be rewritten as [using Eq. (16.5.12)]

$$\dot{\omega} = \frac{6\pi G M_2 \sin i}{a_1 \sin i (1 - e^2) P c^2}$$

$$= \frac{3 G^{2/3} (M_1 + M_2)^{2/3}}{(1 - e^2) c^2} \left(\frac{2\pi}{P} \right)^{5/3}. \qquad (16.5.20)$$

Using the measured value of $\dot{\omega}$ from Table 16.1, together with the values of P and e, we find

$$M_1 + M_2 = 2.8278(7) M_\odot. \qquad (16.5.21)$$

Here one must be careful *not* to use separately the values of G and $M_\odot$ in cgs units.[12] The gravitational constant is only known to less than four significant digits. The quantity $G M_\odot / c^3 = 4.925490 \times 10^{-6}$ s is known much more accurately, and so $M_1 + M_2$ can be quoted much more accurately in solar masses than in grams!

[12] The values we derive from Table 16.1 differ slightly from those quoted by Taylor and Weisberg (1982) for this reason.

We now rewrite the less accurately determined parameters γ, $\sin i$, and $\dot{P}$ by inserting the numerical values for P, e, $a_1 \sin i$, and $M_1 + M_2$. Equation (16.5.19) becomes

$$\gamma = \frac{G^{2/3}M_2(M_1 + 2M_2)e}{(M_1 + M_2)^{4/3}c^2}\left(\frac{P}{2\pi}\right)^{1/3}$$

$$= (0.0007344 \text{ s})M_2(2.8278 + M_2), \qquad (16.5.22)$$

with M_2 measured in solar masses. Equation (16.5.15) gives

$$\sin i = \left(\frac{2\pi}{P}\right)^{2/3}\frac{(M_1 + M_2)^{2/3}a_1\sin i}{G^{1/3}M_2}$$

$$= \frac{1.019}{M_2}, \qquad (16.5.23)$$

while Eq. (16.4.18) gives

$$\dot{P} = \frac{-192\pi}{5}\frac{G^{5/3}M_1M_2f(e)}{c^5(M_1 + M_2)^{1/3}}\left(\frac{2\pi}{P}\right)^{5/3}$$

$$= -1.202 \times 10^{-12}M_2(2.8278 - M_2). \qquad (16.5.24)$$

Note that in order of magnitude, $\Delta P \sim 10^{-4}$ s yr^{-1}. Considering that the binary pulsar is about 5 kpc away, it is astounding that the timing accuracy allows one to contemplate measuring such a small effect!

Taking $\gamma = (0.00438 \pm 0.00024)$ s, we find from Eq. (16.5.22) that $M_2 = (1.41 \pm 0.06)M_\odot$, and hence from Eq. (16.5.21) M_1 has the same value. Then Eq. (16.5.24) predicts $\dot{P} = -2.40 \times 10^{-12}$, in excellent agreement with the measured value of $(-2.30 \pm 0.22) \times 10^{-12}$.

So far general relativity appears to be all that is needed to explain the binary pulsar system, and the quadrupole formula for gravitational wave emission is confirmed to within the measurement errors of $\sim 10\%$. This error should decrease considerably as the orbital period is monitored for a longer time. The situation is summarized in Box 16.1.

This demonstration of the existence of gravitational waves, if confirmed, has the significance of Hertz's experiment to confirm Maxwell's prediction of electromagnetic waves. Unlike the Hertz experiment, the experimental conditions are not under our control. We are fortunate that Nature has supplied us with the right laboratory!

Box 16.1

The Binary Pulsar PSR 1913 + 16—Gravitational Waves Do Exist!

1.   The binary system consists of a pulsar in an elliptic orbit about an unseen companion. The pulsar has an orbital velocity of about 300 km s^{-1}. Variations in the arrival times of pulses at the earth provide information about the orbit.

2.   The system is characterized by four key parameters: the masses M_1 and M_2; a_1, the semimajor axis of the pulsar orbit about the center of mass (whereby $M_2 a_2 = M_1 a_1$); and i, the inclination of the orbital plane to the line of sight.

3.   Two combinations of these parameters, $a_1 \sin i$ and the mass function, are measured by $\mathcal{O}(v/c)$ Newtonian effects.

4.   General relativity at present enables *three* more combinations of parameters to be measured: one from the periastron advance, one from the combined second-order Doppler shift and gravitational redshift, and one from the decay of the orbital period due to the emission of gravitational waves.

5.   Consistency of the five relations between the four parameters confirms both the existence of gravitational waves *and* the absence of possible perturbing influences on the system that could complicate the interpretation of the data. In particular, the quadrupole formula for gravitational wave emission in general relativity is confirmed to within the present measurement errors of $\sim 10\%$.

6.   With improved timing accuracy, further relativistic effects become measurable and additional relations between the four parameters can be corroborated.

16.6   Radiation from Spinning Masses: Pulsar Slowdown

An axisymmetric object rotating rigidly about its symmetry axis has no time-varying quadrupole (or higher) moment, and hence does not radiate gravitational waves.

If the principal moments of inertia of an object are I_1, I_2, and I_3, then radiation will be produced if it rotates about the principal axis $\mathbf{e}_3$ and is nonaxisymmetric $(I_1 \neq I_2)$. Alternatively, it can radiate if it is axisymmetric $(I_1 = I_2)$, but the rotation axis is not the symmetry axis $\mathbf{e}_3$. The general case would be a nonsymmetric object rotating about an arbitrary axis.

We shall consider first the case $I_1 \neq I_2$, with rotation about $\mathbf{e}_3$. A possible physical application would be a pulsar where the rigid crust can support a "mountain". A set of coordinates x_i' rotating with the object (body coordinates) is related to an inertial coordinate system x_i with common origin at the center of mass by a rotation matrix:

$$x' = Rx, \tag{16.6.1}$$

where

$$R_{ij} = \begin{bmatrix} \cos\phi & \sin\phi & 0 \\ -\sin\phi & \cos\phi & 0 \\ 0 & 0 & 1 \end{bmatrix}, \tag{16.6.2}$$

and $\phi = \Omega t$, Ω = constant (no applied torques). The inertia tensor in the inertial coordinates has components given by

$$I = R^T I' R, \tag{16.6.3}$$

where I' is a diagonal matrix with diagonal elements I_1, I_2, and I_3. We will use 1, 2, 3 to denote components in the body frame and x, y, z for the inertial frame. Equations (16.6.2) and (16.6.3) give

$$I_{xx} = \cos^2\phi\, I_1 + \sin^2\phi\, I_2 = \tfrac{1}{2}\cos 2\phi(I_1 - I_2) + \text{constant}. \tag{16.6.4}$$

Similarly,

$$I_{xy} = I_{yx} = \tfrac{1}{2}\sin 2\phi(I_1 - I_2),$$

$$I_{yy} = \tfrac{1}{2}\cos 2\phi(I_2 - I_1) + \text{constant},$$

$$I_{zz} = \text{constant}, \qquad I_{xz} = I_{yz} = 0. \tag{16.6.5}$$

Since

$$\text{Tr}\, I' = \text{Tr}\, I = I_1 + I_2 + I_3 = \text{constant}, \tag{16.6.6}$$

we can use I_{ij} instead of $\mathcal{I}_{ij}$ in the energy loss formula, Eq. (16.2.16) (recall that

$\hat{I}_{ij} = -I_{ij} + \frac{1}{3}\delta_{ij}\operatorname{Tr} I)$. Thus

$$\frac{dE}{dt} = -\frac{1}{5}\frac{G}{c^5}\langle \dddot{I}_{xx}^2 + 2\dddot{I}_{xy}^2 + \dddot{I}_{yy}^2 \rangle$$

$$= -\frac{1}{5}\frac{G}{c^5}\frac{1}{4}(2\Omega)^6(I_1 - I_2)^2\langle \cos^2 2\phi + 2\sin^2 2\phi + \cos^2 2\phi \rangle$$

$$= -\frac{32}{5}\frac{G}{c^5}(I_1 - I_2)^2\Omega^6. \tag{16.6.7}$$

If the object can be approximated by a homogeneous ellipsoid with semiaxes a, b, and c, then

$$I_1 = \tfrac{1}{5}M(b^2 + c^2), \qquad I_2 = \tfrac{1}{5}M(a^2 + c^2), \qquad I_3 = \tfrac{1}{5}M(a^2 + b^2).$$

$$\tag{16.6.8}$$

For small asymmetry (i.e., $a \approx b$), we may write

$$\frac{dE}{dt} \approx -\frac{32}{5}\frac{G}{c^5}I_3^2\varepsilon^2\Omega^6, \tag{16.6.9}$$

where the ellipticity ε is defined by

$$\varepsilon \equiv \frac{a - b}{(a + b)/2}. \tag{16.6.10}$$

This is the formula quoted in Eq. (10.5.15).

Now turn to the case of rigid rotation about a nonprincipal axis, but assume $I_1 = I_2$ for simplicity. Choose the fixed direction of the angular momentum vector $\mathbf{J}$ to be along $\mathbf{e}_z$ in the inertial frame. The transformation to the body coordinates is given in terms of the Euler angles by[13]

$$R_{ij} = \begin{bmatrix} \cos\psi\cos\phi - \cos\theta\sin\phi\sin\psi & \cos\psi\sin\phi + \cos\theta\cos\phi\sin\psi & \sin\theta\sin\psi \\ -\sin\psi\cos\phi - \cos\theta\sin\phi\cos\psi & -\sin\psi\sin\phi + \cos\theta\cos\phi\cos\psi & \sin\theta\cos\psi \\ \sin\theta\sin\phi & -\sin\theta\cos\phi & \cos\theta \end{bmatrix}$$

$$\tag{16.6.11}$$

In free precession, the symmetry axis $\mathbf{e}_3$ and the angular velocity vector rotate about $\mathbf{e}_z$ with constant angular velocity $\dot{\phi} = J/I_1$, $\mathbf{e}_3$ maintaining a constant angle

---

[13] See, for example, Goldstein (1981), Eq. (4-46).

θ with respect to $\mathbf{e}_z$. In addition, the angular velocity vector precesses about $\mathbf{e}_3$ with angular velocity $\dot{\psi} = (I_1 - I_3)\dot{\phi}\cos\theta/I_3 = $ constant as seen in the body frame.

Equations (16.6.3) and (16.6.11) give

$$I_{xx} = I_1(\cos^2\phi + \cos^2\theta\sin^2\phi) + I_3\sin^2\theta\sin^2\phi$$

$$= \tfrac{1}{2}(I_1 - I_3)\sin^2\theta\cos^2 2\phi + \text{constant.} \qquad (16.6.12)$$

Similarly

$$I_{xy} = I_{yx} = \tfrac{1}{2}(I_1 - I_3)\sin^2\theta\sin 2\phi,$$

$$I_{xz} = I_{zx} = -(I_1 - I_3)\sin\theta\cos\theta\sin\phi,$$

$$I_{yy} = -\tfrac{1}{2}(I_1 - I_3)\sin^2\theta\cos 2\phi + \text{constant,}$$

$$I_{yz} = I_{zy} = (I_1 - I_3)\sin\theta\cos\theta\cos\phi,$$

$$I_{zz} = I_3 + (I_1 - I_3)\sin^2\theta = \text{constant.} \qquad (16.6.13)$$

Writing $\phi = \Omega t$, $\Omega = \dot{\phi} = $ constant, we find

$$\frac{dE}{dt} = -\frac{1}{5}\frac{G}{c^5}\langle \dddot{I}_{xx}^2 + \dddot{I}_{yy}^2 + 2\dddot{I}_{xy}^2 + 2\dddot{I}_{xz}^2 + 2\dddot{I}_{yz}^2 \rangle$$

$$= -\frac{1}{5}\frac{G}{c^5}(I_1 - I_3)^2 \langle \tfrac{1}{4}\sin^4\theta(2\Omega)^6(2\cos^2 2\phi + 2\sin^2 2\phi)$$

$$+ 2\sin^2\theta\cos^2\theta\,\Omega^6(\sin^2\phi + \cos^2\phi))\rangle$$

$$= -\frac{2}{5}\frac{G}{c^5}(I_1 - I_3)^2\Omega^6\sin^2\theta(16\sin^2\theta + \cos^2\theta). \qquad (16.6.14)$$

For small "wobble angle" θ, we get

$$\frac{dE}{dt} = -\frac{2}{5}\frac{G}{c^5}(I_1 - I_3)^2\Omega^6\theta^2. \qquad (16.6.15)$$

Note that for rotation about a principal axis [Eq. (16.6.7) or (16.6.9)], the frequency of the radiation is 2Ω as is clear from Eqs. (16.6.4) and (16.6.5). In the case of Eq. (16.6.15), however, the dominant radiation is at frequency Ω, since the $\cos^2\theta$ term in Eq. (16.6.14) comes from I_{xz} and I_{yz}.

It has been proposed to try to detect gravitational waves from a pulsar such as Crab or Vela by building a resonant detector carefully tuned to the right frequency, which is well known from radio observations. If the radiation is largely from a "mountain," with the rotation axis a principal axis, then this would be a good idea, with the resonant frequency chosen to be 2Ω. However, if the radiation is largely from a "wobble," then the frequency of a spot fixed on the pulsar is

$$\omega_z \approx \dot{\phi} + \dot{\psi}. \tag{16.6.16}$$

This would presumably be the frequency of the electromagnetic radiation, and it would differ from the gravitational wave frequency $\Omega = \dot{\phi}$ by the small unknown precession frequency $\dot{\psi}$.

In the general case of combined "mountain" and "wobble" radiation, dE/dt is given by the sum of Eqs. (16.6.7) and (16.6.15), provided the "mountain" and the "wobble" are small.[14]

For "mountain" radiation from a pulsar, the maximum value of the ellipticity defined in Eq. (16.6.10) will be of order the dimensionless breaking strain of the crust material. This is a very uncertain quantity. For laboratory materials, it is of order 10^{-2}–10^{-3} for perfect, pure crystals, but is $\leq 10^{-5}$ if impurities or defects are present. The starquake model (Section 10.11) is not able to pin down this quantity very well for pulsars, but it is generally believed to be of order 5×10^{-4} at most. A pulsar "mountain" is less than a few meters high!

For "wobble" radiation from pulsars,

$$I_3 - I_1 \simeq \varepsilon I_3,$$

where now ε means the oblateness parameter of Eq. (10.11.2). Values in the range 10^{-3}–10^{-4} are plausible (cf. Section 10.11). The wobble angle θ has been invoked to explain "microquakes" in the Crab and Vela pulsars. An upper bound for θ is around 10^{-1}; probably[15] $\theta \leq 10^{-2}$–10^{-3}.

Exercise   16.12   Estimate dE/dt separately for "mountain" and "wobble" radiation from Crab and Vela, and also estimate the wave amplitude h at the earth. Take the distances to be 2000 pc for Crab and 500 pc for Vela. What is the maximum value of h consistent with the pulsar's slowdown?

Exercise   16.13

(a) Show that

$$\frac{dE}{dt} = \Omega \frac{dJ}{dt}$$

for the case of "wobble" radiation. (Here $J = J_z$. You need not assume θ small.)

---

[14] See Zimmerman (1980) for details of the general case, including waveforms for the metric perturbation h.

[15] Pines and Shaham (1974).

(b)   Show that gravitational radiation causes θ to go to zero exponentially with time, while the angular momentum along the symmetry axis, $J \cos \theta$, remains constant.

16.7 Gravitational Waves from Collisions

Another potential source of gravitational radiation is collisions or close encounters between astrophysical objects.

Consider the idealized situation of two point masses m_1 and m_2 falling towards each other from rest at infinity. Assume that they collide head-on, so that the motion is completely along the x-axis. Choose the center of mass to be at rest at $x = 0$. Then

$$m_1 x_1 = -m_2 x_2 = \mu x, \tag{16.7.1}$$

where

$$x \equiv x_1 - x_2, \qquad \mu \equiv \frac{m_1 m_2}{M}, \qquad M \equiv m_1 + m_2. \tag{16.7.2}$$

Since

$$m_1 x_1^2 + m_2 x_2^2 = \mu x^2, \tag{16.7.3}$$

we get

$$\mathcal{I}_{xx} = \tfrac{2}{3}\mu x^2, \qquad \mathcal{I}_{yy} = \mathcal{I}_{zz} = -\tfrac{1}{3}\mu x^2. \tag{16.7.4}$$

Using the equation of motion

$$\ddot{x} = -\frac{GM}{x^2}, \tag{16.7.5}$$

or, equivalently,

$$\frac{\dot{x}^2}{2} = \frac{GM}{x}, \tag{16.7.6}$$

we find

$$\dddot{\mathcal{I}}_{xx} = -\frac{4}{3} G\mu M \frac{\dot{x}}{x^2}. \tag{16.7.7}$$

Thus

$$\frac{dE}{dt} = \frac{1}{5}\frac{G}{c^5}\langle \dddot{I}_{xx}^2 + \dddot{I}_{yy}^2 + \dddot{I}_{zz}^2 \rangle$$

$$= \frac{8}{15}\frac{G^3\mu^2 M^2}{c^5}\left\langle \frac{\dot{x}^2}{x^4} \right\rangle. \tag{16.7.8}$$

The total energy radiated during the collision is

$$\Delta E = \int \frac{dE}{dt} dt = \int \frac{dE}{dt}\frac{1}{\dot{x}}dx$$

$$= \frac{8}{15}\frac{G^3\mu^2 M^2}{c^5}(2GM)^{1/2}\int_{x_{\min}}^{\infty}\frac{dx}{x^{9/2}}. \tag{16.7.9}$$

The integral in Eq. (16.7.9) diverges as $x_{\min} \to 0$. However, our Newtonian treatment also clearly breaks down as $x_{\min} \to 0$. We will simply cut off the integral at the "horizon" $x_{\min} = 2GM/c^2$; results of some proper relativistic treatments are given below. Equation (16.7.9) gives, with $x_{\min} = 2GM/c^2$,

$$\Delta E = \frac{2}{105}\frac{\mu^2 c^2}{M}. \tag{16.7.10}$$

We can compare this result with the exact relativistic calculation by Davis et al. (1971) of the radiation from a test particle falling radially from rest at infinity into a Schwarzschild black hole. In this case the reduced mass μ is the mass of the test particle, while the total mass M is the mass of the black hole. They obtained

$$\Delta E = 0.0104\frac{\mu^2 c^2}{M}, \tag{16.7.11}$$

in reasonable agreement with Eq. (16.7.10) considering the arbitrariness in our cutoff.

Eppley and Smarr[16] have calculated the radiation from the head-on collision of two equal-mass Schwarzschild black holes by computer solution of the full Einstein equations for this situation. They obtained

$$\Delta E = 0.001Mc^2, \tag{16.7.12}$$

where the uncertainty in the numerical solution is probably less than a factor of 2.

---

[16] See Smarr (1979a).

Setting $m_1 = m_2$ in Eq. (16.7.10), we obtain

$$\Delta E = \tfrac{1}{840} Mc^2 = 0.0012\, Mc^2, \tag{16.7.13}$$

which is close to the numerical result. Presumably more radiation is emitted if the encounter is not exactly head-on.[17]

Exercise   16.14   Calculate the wave form $h_{jk}^{TT}(t)$ as seen by a distant observer at an angle θ from the x axis for the head-on collision analyzed in this section.

We can derive an important result for the low-frequency energy spectrum of gravitational waves from collapse events by considering our idealized collision. Define the Fourier transform $\tilde{\dddot{\mathfrak{I}}}_{jk}^{(3)}(\omega)$ of $\dddot{\mathfrak{I}}_{jk}$ by

$$\tilde{\dddot{\mathfrak{I}}}_{jk}^{(3)}(\omega) = \frac{1}{(2\pi)^{1/2}} \int_{-\infty}^{\infty} \dddot{\mathfrak{I}}_{jk}(t) e^{i\omega t}\, dt. \tag{16.7.14}$$

The inverse transform is

$$\dddot{\mathfrak{I}}_{jk}(t) = \frac{1}{(2\pi)^{1/2}} \int_{-\infty}^{\infty} \tilde{\dddot{\mathfrak{I}}}_{jk}^{(3)}(\omega) e^{-i\omega t}\, d\omega. \tag{16.7.15}$$

Since $\dddot{\mathfrak{I}}_{jk}(t)$ is real,

$$\tilde{\dddot{\mathfrak{I}}}_{jk}^{(3)*}(\omega) = \tilde{\dddot{\mathfrak{I}}}_{jk}^{(3)}(-\omega), \tag{16.7.16}$$

where * denotes complex conjugation.

We now derive Parseval's theorem for the energy spectrum, $dE/d\omega$. Since $\dddot{\mathfrak{I}}_{jk}^{(3)}(t)$ is real, we can write

$$\Delta E = \int_{-\infty}^{\infty} \frac{dE}{dt}\, dt = \frac{1}{5}\frac{G}{c^5} \int_{-\infty}^{\infty} \dddot{\mathfrak{I}}_{jk}(t) \dddot{\mathfrak{I}}_{jk}^{*}(t)\, dt$$

$$= \frac{1}{10\pi}\frac{G}{c^5} \int_{-\infty}^{\infty} dt \int_{-\infty}^{\infty} d\omega \int_{-\infty}^{\infty} d\omega'\, \tilde{\dddot{\mathfrak{I}}}_{jk}^{(3)}(\omega) \tilde{\dddot{\mathfrak{I}}}_{jk}^{(3)*}(\omega') e^{-i(\omega - \omega')t}.$$

$$\tag{16.7.17}$$

The integral over t gives a δ-function, and so

$$\Delta E = \frac{1}{5}\frac{G}{c^5} \int_{-\infty}^{\infty} d\omega \left| \tilde{\dddot{\mathfrak{I}}}_{jk}^{(3)}(\omega) \right|^2$$

$$= \frac{2}{5}\frac{G}{c^5} \int_{0}^{\infty} d\omega \left| \tilde{\dddot{\mathfrak{I}}}_{jk}^{(3)}(\omega) \right|^2, \tag{16.7.18}$$

---

[17]Compare Detweiler and Szedenits (1979).

where we have used Eq. (16.7.16). Thus

$$\Delta E = \int_0^\infty \frac{dE}{d\omega} d\omega, \qquad (16.7.19)$$

where

$$\frac{dE}{d\omega} = \frac{2}{5} \frac{G}{c^5} \left| \tilde{f}_{jk}^{(3)}(\omega) \right|^2. \qquad (16.7.20)$$

Now consider the low-frequency limit of Eq. (16.7.20). When $\omega \to 0$, Eq. (16.7.14) gives

$$\tilde{f}_{jk}^{(3)}(0) = \frac{1}{(2\pi)^{1/2}} \ddot{f}_{jk}(t) \Big|_{t_i}^{t_f}, \qquad (16.7.21)$$

where $\ddot{f}_{jk}(t)$ is nonzero only during the interval $t_i \leqslant t \leqslant t_f$. In the free-fall collision case described above, $t_i = -\infty$, $t_f = t_{\min}$. From Eqs. (16.7.20) and (16.7.21), we see that $dE/d\omega$ is nonzero at $\omega = 0$ if $\ddot{f}_{jk}(t)$ is nonzero at t_i or t_f [assuming $\ddot{f}_{jk}(t_i) \neq \ddot{f}_{jk}(t_f)$]. Thus coalescence following release from rest at a finite radius or from a hyperbolic encounter (finite velocity at infinity) leads to nonzero $dE/d\omega$ at $\omega = 0$. However, as we now show, a parabolic encounter (free-fall with zero velocity at infinity) gives $dE/d\omega \propto \omega^{4/3} \to 0$, as $\omega \to 0$.

If we substitute Eq. (16.7.4) directly in Eq. (16.7.14), we will encounter a divergent integral when we later take the limit $\omega \to 0$. We therefore first rewrite Eq. (16.7.14) by integrating by parts:

$$\tilde{f}_{jk}^{(3)}(\omega) = \frac{1}{(2\pi)^{1/2}} \left[\ddot{f}_{jk}(t) e^{i\omega t} \Big|_{-\infty}^{\infty} - i\omega \int_{-\infty}^{\infty} \ddot{f}_{jk}(t) e^{i\omega t} dt \right]. \qquad (16.7.22)$$

Now Eqs. (16.7.4)–(16.7.6) give

$$\ddot{f}_{xx}(t) = \frac{4}{3} \frac{GM\mu}{x}, \qquad (16.7.23)$$

where

$$x = \left(\tfrac{3}{2} \right)^{2/3} (2GM)^{1/3} (-t)^{2/3}, \qquad x \to \infty \text{ as } t \to -\infty. \qquad (16.7.24)$$

We cut off Eq. (16.7.24) at $x_{\min} = 2GM/c^2$, when $t = t_{\min}$. For $t > t_{\min}$, we take all derivatives of $f_{jk}(t)$ to be zero. Thus the first term on the right-hand side of

Eq. (16.7.22) is zero, and we have

$$\tilde{f}_{xx}^{(3)}(\omega) = \frac{-i\omega}{(2\pi)^{1/2}} \int_{-\infty}^{t_{\min}} \frac{2^{7/3}(GM)^{2/3}\mu}{3^{5/3}(-t)^{2/3}} e^{i\omega t} dt. \qquad (16.7.25)$$

Letting $y = -\omega t$, we get

$$\tilde{f}_{xx}^{(3)}(\omega) = \frac{-i\omega^{2/3}2^{7/3}(GM)^{2/3}\mu}{(2\pi)^{1/2}3^{5/3}} \int_{|\omega t_{\min}|}^{\infty} e^{-iy} \frac{dy}{y^{2/3}}. \qquad (16.7.26)$$

When $\omega \to 0$, we can set the lower limit of the integral to zero. The integral then has the value $\Gamma(\frac{1}{3})\exp(-i\pi/6)$. Substituting in Eq. (16.7.20) and using Eq. (16.7.4) for f_{yy} and f_{zz}, we find

$$\frac{dE}{d\omega} \to \frac{2^{11/3}\Gamma(\frac{1}{3})^2}{5\pi(3)^{7/3}} \frac{G}{c^5} (GM\omega)^{4/3}\mu^2, \qquad \omega \to 0. \qquad (16.7.27)$$

The $\omega^{4/3}$ power law in the limit $\omega \to 0$ comes from the behavior of $f_{jk}(t)$ as $t \to -\infty$. In this regime the source is Newtonian and the quadrupole formula used above should be an excellent approximation. Thus, even if a black hole is eventually formed by a collapse or collision process, the low-frequency gravitational wave spectrum is produced at early times and can be calculated using the quadrupole formula. In particular, free-fall behavior from rest at infinity quite generally leads to the power law $\omega^{4/3}$ as $\omega \to 0$.[18]

Gravitational radiation from the head-on collision of two equal-mass neutron stars is likely to be much greater than Eq. (16.7.12) predicts, if the neutron stars are sufficiently massive and compact. In this situation, the outward propagation of two recoil shock waves from the point of contact serves to decelerate the colliding matter, causing the quadrupole moment to change rather abruptly. Preliminary calculations[19] indicate radiation efficiencies of $\Delta E/Mc^2 \gtrsim 0.01$ for typical cases, with corresponding wave amplitudes of $h \sim 10^{-21}$ at distances of 10 Mpc (the Virgo Cluster). Such events, though not likely to be too common, might very well describe the final fate of the binary pulsars. It has been estimated[20] that there may be about one such a collision per year within about 40 Mpc of the Earth. To detect such a collision, therefore, detector sensitivies of $h \sim 10^{-22}$ may be required.

---

[18] Wagoner (1979).
[19] Clark and Eardley (1977); Smarr and Wilson as reported in Wilson (1979); and Shapiro (1980).
[20] Clark, Van den Heuvel, and Sutantyo (1979); Clark (1979a).

16.8   Gravitational Waves from Nonspherical Collapse

One of the most promising sources of gravitational radiation is gravitational collapse of a massive star to a neutron star or black hole. The dimensional estimates of Section 16.3 suggest that the star could radiate a substantial fraction of Mc^2 if the collapse were sufficiently nonspherical. Detailed calculations of such a collapse are only now becoming available. They require two- or three-dimensional hydrodynamical codes that are fully general relativistic. The equation of state of the collapsing matter must be accurately treated, and so must the spacetime geometry in order to compute the gravitational wave flux.[21]

Some insight into these complex calculations comes from simpler spheroidal or ellipsoidal model calculations.[22] Here one considers the Newtonian collapse of a homogeneous spheroid or ellipsoid. In order for the shape to remain ellipsoidal, the pressure law must be quadratic in the coordinates:

$$P = P_c(t)\left(1 - \frac{x^2}{a^2} - \frac{y^2}{b^2} - \frac{z^2}{c^2}\right), \qquad (16.8.1)$$

where a, b, and c are the semiaxes and P_c is the central pressure. An equation of state can be modeled by taking

$$P_c = P_c(\rho), \qquad (16.8.2)$$

where the uniform density ρ is given by

$$\rho = \frac{M}{4\pi abc/3}. \qquad (16.8.3)$$

The three-dimensional hydrodynamic equations of motion for such a configuration reduce to *ordinary* differential equations for a, b, and c as functions of time.

To obtain large fluxes of gravitational waves, one requires large deviations from spherical symmetry. Starting with an almost spherical configuration at large radius, one can induce ellipticity if the object has angular momentum. For spheroids with small angular momentum J, we have $J \propto e$, where the eccentricity is given by

$$e^2 = 1 - \frac{c^2}{a^2} \qquad (16.8.4)$$

(cf. Section 7.3).

[21]A discussion of this work can be found in Smarr (1979a).
[22]Saenz and Shapiro (1981) and references therein.

Exercise   16.15   Use the results of Section 7.3 to show that

$$J = \left(\frac{2M}{5}\right)^{3/2}(Ga)^{1/2}e \tag{16.8.5}$$

for a Maclaurin spheroid in the limit $e \to 0$.

Equation (16.8.5) suggests that during axisymmetric collapse with conserved J and M, e should grow. This is borne out by the detailed calculations. Angular momentum eventually provides centrifugal support for the equatorial a-axis, while the polar c-axis collapses unimpeded by rotation.

Exercise   16.16   Show that for spheroidal collapse

$$\frac{dE}{dt} = \frac{2}{375}\frac{GM^2}{c^5}\langle(\dddot{a^2} - \dddot{c^2})^2\rangle. \tag{16.8.6}$$

Equation (16.6.8) may be helpful.

From Eqs. (16.8.4)–(16.8.6), we see that for small J,

$$\frac{dE}{dt} \propto J^4 \quad (\text{small } J). \tag{16.8.7}$$

For large J, the configuration never collapses to small radii and high densities, because of the large centrifugal forces in the equatorial plane. Thus dE/dt goes to zero both at large and small J, and reaches a maximum at some intermediate value, J_c, which is typically about an order of magnitude larger than the sun's angular momentum for a solar mass object collapsing to a neutron star. When $J \sim J_c$, the rotational energy of the collapsing star becomes comparable to the gravitational potential energy and, hence, the eccentricity becomes large just above nuclear densities. Maximum radiation efficiencies, $\Delta E/Mc^2$, are at most 1% for extreme cases, but typical efficiencies could be several orders of magnitude lower.

One could also have efficient radiation if magnetic fields were strong enough to produce nonspherical distortion. Another possibility is to have a small initial eccentricity (from any source), which grows during collapse. If the configuration bounces several times, then during each successive collapse the eccentricity can grow, leading to a high efficiency. Such a scenario is depicted in Figure 16.2, based on spheroidal calculations by Saenz and Shapiro (1981). Which of these possibilities actually occurs in nature will be an open question until large-scale computer calculations are carried out or until gravitational wave detectors of sufficient sensitivity are built.

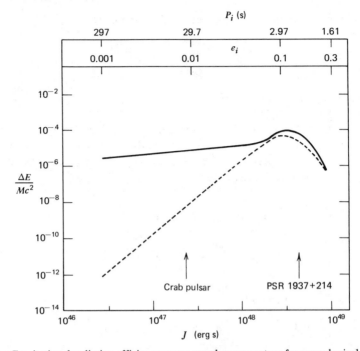

Figure 16.2 Gravitational radiation efficiency versus angular momentum for nonspherical stellar core collapse. Results are shown of numerical calculations employing homogeneous spheroids to model the collapse of a degenerate, homologous core of mass $M = 0.8M_\odot$. The initial density is $\rho_i = 4 \times 10^9$ g cm^{-3}, the eccentricity is e_i and the rotation period is P_i. Collapse starts from rest and is governed by a "realistic" hot, nuclear equation of state via Eq. (16.8.2) and by recoil "shocks" following bounce. (See Chapter 18.) The radiation efficiency resulting from the *initial* infall and rebound (at $\rho_{\text{bounce}} \sim 10^{15}$ g cm^{-3}) is shown by the *dashed* line [cf. Eq. (16.8.8)]. The *total* efficiency after all of the core oscillations have ceased is shown by the *solid* curve. The arrows indicate the estimated angular momentum of the Crab pulsar and PSR 1937 + 214, which have the shortest periods of the known pulsars (33 and 1.6 ms, respectively). [After Saenz and Shapiro (1981).]

Exercise 16.17

(a) Use dimensional analysis and Eq. (16.8.7) to show that collapse of a slowly rotating stellar core to a black hole releases an energy in the form of gravitational waves given by

$$\Delta E \sim \left(\frac{J}{GM^2/c} \right)^4 Mc^2. \tag{16.8.8}$$

(b) Estimate the efficiency $\Delta E/Mc^2$ numerically for the collapse of a stellar core with the same angular momentum as the Crab pulsar.

17
Supermassive Stars and Black Holes

17.1 Introduction

Supermassive stars are hypothetical equilibrium configurations with masses in the range 10^3–$10^8 M_\odot$. They can be quite compact, as we shall see, with surface potentials a small but nonnegligible fraction of c^2. Following quasistatic evolution, they can undergo catastrophic gravitational collapse. Consequently, supermassive stars are possible progenitors of supermassive black holes. Both objects—supermassive stars and supermassive black holes—are frequently invoked to explain the energy source responsible for the violent activity observed in quasars and active galactic nuclei. For these reasons, it is appropriate to discuss the properties of supermassive stars in a book on compact objects. As we shall see, the gross equilibrium and stability properties of supermassive stars can be readily understood by utilizing some mathematical machinery we have already developed for white dwarfs and neutron stars (e.g., the energy variational principle; cf. Chapter 6).

As one considers stars of larger and larger mass, more and more of the pressure support is supplied by radiation and less and less by the gas. We will demonstrate this fact below. According to theoretical calculations,[1] "normal" main sequence stars have a maximum mass around $60 M_\odot$. Above this mass, the combination of radiation pressure and nuclear energy generation in the core of the star leads to a violent pulsational instability. This theoretical prediction is supported by the lack of observed stars with masses greater then $60 M_\odot$.

In spite of this limit, it has long been speculated that supermassive stars with masses well above $1000 M_\odot$ may form in Nature. Such an object would be

---

[1]Schwarzschild and Härm (1959).

supported in hydrostatic equilibrium almost entirely by radiation pressure:

$$P = P_r = \tfrac{1}{3}aT^4,$$ (17.1.1)

$$\frac{dP}{dr} = -\frac{Gm(r)\rho}{r^2}.$$ (17.1.2)

In the surface layers, the energy flux is transported outward by the diffusion of photons. The luminosity is given by

$$L = -4\pi r^2 \frac{c}{3\kappa\rho}\frac{d}{dr}(aT^4),$$ (17.1.3)

where we take the opacity to be that due to Thomson scattering in an ionized hydrogen plasma:

$$\kappa = \frac{\sigma_T}{m_p} = 0.40 \text{ cm}^2 \text{ g}^{-1}.$$ (17.1.4)

Here σ_T is the Thomson cross section and m_p the proton mass [cf. Eqs. (I.31) and (I.41)]. Substituting Eqs. (17.1.1) and (17.1.2) in (17.1.3) and setting $m(r) = M$, we get

$$L = L_{Edd} = \frac{4\pi cGMm_p}{\sigma_T} = 1.3 \times 10^{38}\frac{M}{M_\odot} \text{ ergs}^{-1}.$$ (17.1.5)

The star thus radiates at the Eddington limit (cf. Section 13.7).

Hoyle and Fowler (1963a, b) were the first to propose supermassive *stars* as the energy sources for quasars and active galactic nuclei. A typical quasar luminosity of 10^{46} erg s^{-1} ($10^{13}L_\odot$!) leads naturally to the idea of radiation at the Eddington limit from an object with $M \sim 10^8 M_\odot$. Since 1963, there have been many studies of the possibility of releasing large amounts of nuclear and gravitational energy from such objects.[2]

Salpeter (1964) and Zel'dovich (1964) suggested, alternatively, that supermassive *black holes* may be the objects responsible for the energetic activity of quasars and galactic nuclei. Ever since, black hole models for the observed optical, radio, and X-ray emission from these sources have been the subject of intensive theoretical investigation.[3] Most of these models invoke gas accretion onto supermassive black holes to power the emission (cf. Chapter 14). As we have already

[2] See Wagoner (1969) for a review and references to early work; see Norgaard and Fricke (1976) for references to more recent developments.
[3] See Rees (1977b) and Blandford and Thorne (1979) for references.

remarked in Section 13.7, the maximum luminosity expected from accretion is likely to be comparable to L_{Edd}. The fact that supermassive stars *and* black holes can both radiate at the Eddington limit explains, in part, why they are both prominent candidates for quasar and active galactic nuclei models.

Recently, Thorne and Braginsky (1976) have proposed that the gravitational collapse of supermassive stars and the collision of supermassive black holes in quasars and galactic nuclei may give rise to bursts of long wavelength gravitational radiation, $\lambda \gtrsim 10^6 \, (M/10^6 M_\odot)$ km.

There is some direct observational evidence for nonluminous compact objects at the center of galaxies. For example, stars near the center of the giant elliptical galaxy M87 in the Virgo cluster have large velocities, consistent with motion in the gravitational field of a central black hole of mass $5 \times 10^9 M_\odot$.[4] In our own Galaxy, there is some evidence from the high velocities of gas clouds at the Galactic center for a compact mass of size $M \sim 10^6 M_\odot$ there.[5]

Exercise 17.1 The central object of the 30 Doradus Nebula in the Large Magellanic Cloud has been tentatively identified as a supermassive star (Cassinelli, Mathis, and Savage, 1981). The identification is still complicated by at least two major uncertainties: (i) contamination of the observed spectrum by other stars and the possibility that a *collection* of very hot, but otherwise "normal" stars produce the spectra, and (ii) appreciable and peculiar absorption of the light by dust. These uncertainties notwithstanding, observations from the International Ultraviolet Explorer (IUE) satellite indicate that the brightest, bluest component at the center of this giant region of ionized hydrogen is a peculiar hot object with a massive stellar wind. From the observed spectra the outflow speed of the wind V_w is estimated to be 3500 km s^{-1} and the color temperature T about 60,000 K. From the distance, flux (corrected for absorption) and color temperature, the luminosity L is estimated to be about $0.7 \times 10^8 L_\odot$.

(a) Assume that the source is a single, luminous star and use Eq. (17.1.5) to estimate its mass.

Answer: $M/M_\odot \approx 2000$

(b) Assume that the star radiates as a black body at temperature T and calculate its radius.

Answer: $R/R_\odot \approx 80$

(c) Is the supermassive star hypothesis consistent with a wind speed obeying the obvious constraint $V_w > V_{esc}$ where $V_{esc} = (2GM/R)^{1/2}$ is the escape velocity from the surface?

(d) Comment on the pulsational instability of a supermassive star in light of the appreciable observed stellar wind ($\dot{M}_w \approx 10^{-3.5} M_\odot$ yr^{-1}).

---

[4]Young et al. (1978); de Vaucouleurs and Nieto (1979); Sargent et al. (1978); for an interpretation that does not involve a black hole, see Duncan and Wheeler (1980).

[5]Lacy et al. (1979).

We shall study the properties of supermassive stars below, initially neglecting general relativity, nuclear burning and other complicating effects. Because radiation pressure implies an adiabatic index of $\frac{4}{3}$, a supermassive star is on the brink of dynamical instability. We thus next consider how general relativity influences the stability. Then we discuss additional effects due to nuclear reactions, rotation, and so on. Finally, we describe numerical calculations of the collapse of supermassive stars to supermassive black holes.

17.2   Basic Properties of Supermassive Stars

We consider the idealized case of a spherical cloud of completely ionized hydrogen. In zeroth order, we neglect the pressure of the hydrogen gas. The pressure and energy density of the radiation are

$$P_r = \tfrac{1}{3}aT^4, \qquad \varepsilon_r = aT^4. \tag{17.2.1}$$

Using the first law of thermodynamics, Eq. (2.1.3), we find for the photon entropy per baryon

$$s_r = \frac{4}{3}\frac{aT^3}{n}. \tag{17.2.2}$$

We assume that the entropy is constant throughout the star.[6] The density is

$$\rho = m_{\mathrm{H}}n, \tag{17.2.3}$$

where m_{H} is the mass of a hydrogen atom and we can neglect ε_r/c^2.

Equations (17.2.2) and (17.2.3) give

$$T = \left(\frac{3\rho s_r}{4m_{\mathrm{H}}a}\right)^{1/3}, \tag{17.2.4}$$

and so Eq. (17.2.1) gives

$$P = P_r = K\rho^{4/3}, \tag{17.2.5}$$

where

$$K = \frac{1}{3}a\left(\frac{3s_r}{4m_{\mathrm{H}}a}\right)^{4/3}. \tag{17.2.6}$$

Thus the structure of the star is that of an $n = 3$ polytrope.

---

[6] This is probably valid because of convection. See footnote 9 in Chapter 18 and Clayton (1968), Section 3.5 for a discussion of convection.

We can write the energy of such a configuration, to lowest order, in the form

$$E = E_{int} + E_{grav}$$

$$= k_1 KM\rho_c^{1/3} - k_2 GM^{5/3}\rho_c^{1/3}, \tag{17.2.7}$$

where $k_1 = 1.75579$, $k_2 = 0.639001$ [cf. Eqs. (6.10.9) and (6.10.10)]. Equilibrium occurs when $\partial E/\partial \rho_c = 0$, giving a relationship between entropy and mass [cf. Eq. (3.3.10)]:

$$\frac{s_r}{k} = \frac{4m_H a}{3k}\left(\frac{3k_2 G}{k_1 a}\right)^{3/4} M^{1/2} = 0.942\left(\frac{M}{M_\odot}\right)^{1/2}. \tag{17.2.8}$$

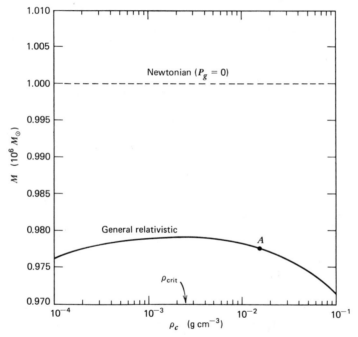

Figure 17.1 Equilibrium mass versus central density for a supermassive star of constant entropy. The entropy is specified by the value of β [see Eq. (17.3.5)]; here $\beta = 8.494 \times 10^{-3}$. The *dashed* curve shows the equilibrium mass of a Newtonian configuration supported exclusively by radiation pressure [cf. Eq. (17.2.8)]. The *solid* curve shows the equilibrium mass of a general relativistic star supported both by radiation and gas pressure. The *dashed* curve satisfies $dM/d\rho_c = 0$ everywhere, so all equilibrium configurations are marginally stable to radial oscillations. The *solid* curve has a maximum at $\rho_c = \rho_{crit}$, where $dM/d\rho_c = 0$. For $\rho_c \leqslant \rho_{crit}$ the stabilizing influence of gas pressure to radial perturbations dominates the destabilizing influence of nonlinear gravity, so these configurations are stable. The converse is true for $\rho_c > \rho_{crit}$. Accordingly, mass A lies on the unstable branch of the equilibrium curve. [After Shapiro and Teukolsky (1979).]

Here k is Boltzmann's constant. The temperature is given by Eq. (17.2.4):

$$T = 1.98 \times 10^7 \, \text{K} \left(\frac{\rho}{1 \, \text{g cm}^{-3}} \right)^{1/3} \left(\frac{M}{M_\odot} \right)^{1/6}. \tag{17.2.9}$$

To lowest order, therefore, the mass of a Newtonian equilibrium configuration depends only on the specific entropy and not on the central density. For a given value of s_r, then, the graph of M versus ρ_c is simply a horizontal line, as shown in Figure 17.1.

17.3 Effect of the Plasma

The hydrogen plasma supplies a gas pressure

$$P_g = 2nkT, \tag{17.3.1}$$

twice the pressure of the electrons or protons alone. The energy density is

$$\varepsilon_g = m_H nc^2 + 3nkT, \tag{17.3.2}$$

where we assume that the gas is nonrelativistic ($kT \ll m_e c^2$). The first law of thermodynamics thus gives for the entropy

$$s_g = 2k \ln\left(\frac{T^{3/2}}{n} \right) + \text{constant}. \tag{17.3.3}$$

The constant will turn out to be irrelevant for the rest of our treatment, but its numerical value is of interest. In our case, it is found to be [cf. Eq. (2.3.34)]

$$\frac{s_g}{k} = \ln\left(\frac{T^3}{\rho^2} \right) + \frac{s_0}{k}, \tag{17.3.4}$$

$$\frac{s_0}{k} \equiv 3 \ln\left(\frac{2\pi k}{h^2} \right) + \frac{3}{2} \ln m_e + \frac{7}{2} \ln m_p + 5 + 2 \ln 2$$

$$= -21.0,$$

where T is in Kelvin and ρ in g cm^{-3}.

Note that Eqs. (17.2.1), (17.2.2), and (17.3.1) give

$$\beta \equiv \frac{P_g}{P_r} = \frac{8}{s_r/k}. \tag{17.3.5}$$

Thus when $M \gg M_\odot$ so that $s_r/k \gg 1$ [Eq. (17.2.8)] and $\beta \ll 1$, the gas pressure is only a small perturbation.

Exercise   17.2   Compare s_r and s_g for $M = 10^6 M_\odot$, $\rho = 10^3$ g cm^{-3} and for $M = 10^8 M_\odot$, $\rho = 1$ g cm^{-3}. The temperature is given by Eq. (17.2.9).

   Answer: $s_g \ll s_r$.

Exercise   17.3

   (a)   Show that when both radiation *and* gas pressure are included, a supermassive star behaves like a Newtonian polytrope with $P = K\rho^\Gamma$, where K is given by Eq. (17.2.6) and Γ is given by

$$\Gamma \equiv 1 + \frac{P}{\varepsilon'} \simeq \frac{4}{3} + \frac{\beta}{6} + \mathcal{O}(\beta^2),$$

where ε' is the total internal energy density, excluding rest-mass energy.

   (b)   Use the results of part (a) to argue that, for *strictly* Newtonian gravitation, a supermassive star in equilibrium is *unconditionally* stable to adiabatic radial perturbations for *arbitrary* central densities.

   Hint: See Section 6.7.

   (c)   Fix the entropy by setting $\beta = 8.494 \times 10^{-3}$ and calculate the corresponding values of K and Γ.

   Answer: $K = 3.839 \times 10^{18}$ cgs, $\Gamma = 1.3347$.

   (d)   Plot on Figure 17.1 the equilibrium curve of M vs. ρ_c for the Newtonian polytropes corresponding to this value of entropy [you may approximate the required Lane–Emden quantities ξ_1 and $\xi_1^2|\theta'(\xi_1)|$ by employing Eq. (3.3.12)]. Compare your curve with the curve for a Newtonian polytrope supported purely by radiation pressure, and comment.

   Hint: What is the sign of $dM/d\rho_c$ for the two cases?

We wish to compute the correction to E_{int} due to the hydrogen plasma. The total internal energy per unit mass is

$$u = \frac{\varepsilon_r + \varepsilon_g}{\rho}$$

$$= \frac{aT^4}{\rho} + \frac{3kT}{m_{\text{H}}}, \tag{17.3.6}$$

where we have dropped the constant term arising from the rest-mass energy. We must eliminate T in terms of ρ and the total entropy

$$s = s_r + s_g$$

$$= \frac{4m_{\text{H}}aT^3}{3\rho} + k \ln \frac{T^3}{\rho^2} + s_0. \tag{17.3.7}$$

The zeroth-order expression for T in terms of s is obtained by neglecting s_g and is given by Eq. (17.2.4) with s_r replaced by s. Substitute this expression into the small term s_g, so that

$$\frac{4m_{\text{H}}aT^3}{3\rho} \simeq s\left(1 - \frac{s_0}{s} - \frac{k}{s}\ln\frac{3s}{4m_{\text{H}}a\rho}\right). \qquad (17.3.8)$$

Thus

$$T \simeq \left(\frac{3s\rho}{4m_{\text{H}}a}\right)^{1/3}\left(1 - \frac{s_0}{3s} - \frac{k}{3s}\ln\frac{3s}{4m_{\text{H}}a\rho}\right). \qquad (17.3.9)$$

For simplicity, we shall treat s_0/s as a small quantity, which is true for $M/M_\odot \gg 10^3$ [cf. Eqs. (17.2.8) and (17.3.4)] but not essential to the argument. Now substitute Eq. (17.3.9) into Eq. (17.3.6). In the small term $3kT/m_{\text{H}}$, it suffices to keep only the zero-order expression for T. We find

$$u = 3K\rho^{1/3} + \lambda\rho^{1/3} + \mu\rho^{1/3}\ln\rho, \qquad (17.3.10)$$

where

$$\lambda = -\frac{4as_0}{3s}\left(\frac{3s}{4m_{\text{H}}a}\right)^{4/3} + \frac{3k}{m}\left(\frac{3s}{4m_{\text{H}}a}\right)^{1/3}$$

$$-\frac{4ka}{3s}\left(\frac{3s}{4m_{\text{H}}a}\right)^{4/3}\ln\left(\frac{3s}{4m_{\text{H}}a}\right), \qquad (17.3.11)$$

$$\mu = \frac{4ka}{3s}\left(\frac{3s}{4m_{\text{H}}a}\right)^{4/3}. \qquad (17.3.12)$$

Now the total internal energy becomes

$$E_{\text{int}} + \Delta E_{\text{int}} = \int u\, dm, \qquad (17.3.13)$$

where we integrate over an $n = 3$ polytropic mass distribution. The first term in Eq. (17.3.10) gives E_{int} [cf. Eq. (17.2.7)]. Substituting $\rho = \rho_c\theta^3$ [Eq. (3.3.3)], we find

$$E_{\text{int}} + \Delta E_{\text{int}} = k_1\left(K + \frac{\lambda}{3} + \tau\right)M\rho_c^{1/3} + \tfrac{1}{3}k_1\mu M\rho_c^{1/3}\ln\rho_c, \qquad (17.3.14)$$

where

$$\tau = \frac{\mu}{Mk_1} \int \theta \ln \theta^3 \, dm. \qquad (17.3.15)$$

17.4   Stability of Supermassive Stars

Just as in the case of white dwarfs (Chapter 6), to investigate the stability of supermassive stars, we must include the effects of general relativity, Eq. (6.10.21). We thus write

$$E = E_{\text{int}} + E_{\text{grav}} + \Delta E_{\text{int}} + \Delta E_{\text{GTR}}$$

$$= AM\rho_c^{1/3} - BM^{5/3}\rho_c^{1/3} + CM\rho_c^{1/3}\ln \rho_c - DM^{7/3}\rho_c^{2/3}, \qquad (17.4.1)$$

where

$$A = k_1\left(K + \frac{\lambda}{3} + \tau\right), \qquad B = k_2 G,$$

$$C = \frac{k_1\mu}{3}, \qquad D = \frac{k_4 G^2}{c^2}, \qquad k_4 = 0.918294. \qquad (17.4.2)$$

Equilibrium requires $\partial E/\partial \rho_c = 0$, keeping M and s constant. This gives

$$0 = \tfrac{1}{3}\left(AM - BM^{5/3} + CM\ln \rho_c\right)\rho_c^{-2/3} + CM\rho_c^{-2/3} - \tfrac{2}{3}DM^{7/3}\rho_c^{-1/3}. \qquad (17.4.3)$$

The terms proportional to C and D give small corrections, depending on ρ_c, to the relation (17.2.8) between s and M for equilibrium. As a result, the equilibrium mass M now varies with ρ_c for fixed s, as shown in Figure 17.1.

The onset of instability occurs when $\partial^2 E/\partial \rho_c^2 = 0$, at which point $dM/d\rho_c = 0$ in Fig. 17.1. Differentiating Eq. (17.4.3) gives

$$0 = -\tfrac{2}{9}\left(AM - BM^{5/3} + CM\ln \rho_c\right)\rho_c^{-5/3} - \tfrac{1}{3}CM\rho_c^{-5/3} + \tfrac{2}{9}DM^{7/3}\rho_c^{-4/3}. \qquad (17.4.4)$$

Multiplying Eq. (17.4.4) by $\tfrac{3}{2}\rho_c$ and adding to Eq. (17.4.3) gives

$$0 = \tfrac{1}{2}CM\rho_c^{-2/3} - \tfrac{1}{3}DM^{7/3}\rho_c^{-1/3}, \qquad (17.4.5)$$

or, writing $\rho_c = \rho_{crit}$ at the onset of instability,

$$\rho_{crit} = \left(\frac{3C}{2D}\right)^3 \frac{1}{M^4}. \tag{17.4.6}$$

Substituting Eq. (17.4.2) for C and D, Eq. (17.3.12) for μ, and Eq. (17.2.8) for s, we find

$$\rho_{crit} = \left(\frac{k_1 kc^2}{2k_4 G^2 m_H}\right)^3 \left(\frac{3k_2 G}{k_1 a}\right)^{3/4} \frac{1}{M^{7/2}}$$

$$= 1.996 \times 10^{18} \left(\frac{M_\odot}{M}\right)^{7/2} \text{g cm}^{-3}. \tag{17.4.7}$$

Using Eq. (17.4.3) in Eq. (17.4.1), we find for the energy in equilibrium

$$E_{eq} = -3CM\rho_c^{1/3} + DM^{7/3}\rho_c^{2/3}. \tag{17.4.8}$$

The equilibrium energy at the onset of instability is, by Eqs. (17.4.5) and (17.4.7),

$$E_{crit} = -DM^{7/3}\rho_{crit}^{2/3} = -3.583 \times 10^{54} \text{ erg}, \tag{17.4.9}$$

independent of M.

Exercise 17.4

(a) Find the central temperature at the onset of instability.

(b) Find the redshift factor, GM/Rc^2, at the onset of instability.

Answer:

$$T_{crit} = (2.49 \times 10^{13} \text{ K}) \frac{M_\odot}{M}, \tag{17.4.10}$$

$$\left(\frac{GM}{Rc^2}\right)_{crit} = 0.6295 \left(\frac{M_\odot}{M}\right)^{1/2}. \tag{17.4.11}$$

Note from Eq. (17.4.11) that the maximum gravitational redshift for a stable supermassive star is very small, in the range 10^{-2}–10^{-4}. Nevertheless, as with white dwarfs, the small corrections due to general relativity are *crucial* in determining the stability of a seemingly Newtonian system.

17.5 Evolution of a Supermassive Star

Let us now attempt to trace the evolution of a supermassive star, starting with the contraction of a large, diffuse cloud of hydrogen gas. As we saw in Eq. (17.2.8), a spherical supermassive star requires an entropy per baryon in the range 10^2–10^4 to be in equilibrium. This is much higher than the entropy of hydrogen gas in typical astrophysical plasmas. [cf. Eq. (17.3.4)]. Thus the initial dynamical collapse must be dissipative (collisions of fragments, turbulence, shock waves) to lead to the equilibrium configuration we have discussed. Wagoner (1969) has considered the alternative possibility that before the entropy becomes large, centrifugal forces due to initial angular momentum become important, leading to supermassive disks.

We have assumed that the central temperature never becomes sufficiently high for nuclear burning to become important. A simple comparison of the nuclear energy generation rate with the photon luminosity, Eq. (17.1.5), shows[7] that this is a valid assumption for $M \gtrsim 6 \times 10^4 M_\odot$. The effect of electron-positron pairs is also negligible in this regime.

Once a supermassive star has formed, its evolution is a quasistatic progression through a sequence of equilibrium states of increasing central density. The star radiates at the Eddington limit, its mass remaining essentially constant while its entropy and energy decrease. When the energy has decreased from zero to E_{crit}, the density is high enough for the general relativistic instability to set in. The star undergoes catastrophic collapse.

The above scenario is valid provided the thermal evolution timescale is longer than the hydrodynamical timescale, so that the star can quickly readjust its structure to be always in equilibrium during the evolution. We have

$$t_{thermal} \sim \frac{|E_{crit}|}{L}$$

$$= 2.8 \times 10^{16} \left(\frac{M}{M_\odot}\right)^{-1} \text{s}, \tag{17.5.1}$$

using Eqs. (17.1.5) and (17.4.9). Also,

$$t_{hydro} \sim (G\rho)^{-1/2}$$

$$= 2.7 \times 10^{-6} \left(\frac{M}{M_\odot}\right)^{7/4} \text{s}, \tag{17.5.2}$$

where we have used Eq. (17.4.7). The two timescales are equal for $M \sim 10^8 M_\odot$; above this mass, there is no equilibrium phase of evolution for supermassive objects. Note that the equilibrium phase lasts only 10–10^5 yr for M in the range 10^8–$10^4 M_\odot$.

What is the fate of a supermassive star that has evolved to the critical density for collapse? One possible outcome is for nuclear reactions to become so rapid as the temperature and density increase that the star undergoes a violent explosion.

---

[7]Zel'dovich and Novikov (1971).

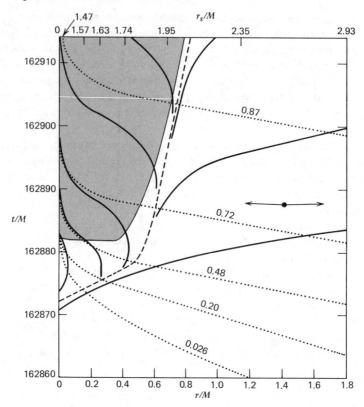

Figure 17.2 Spacetime diagram of the collapse of a $10^6 M_\odot$ supermassive star to a black hole. The initial configuration is the unstable, equilibrium configuration A shown in Figure 17.1. (See text for details.) [From Shapiro and Teukolsky (1979).]

This appears[8] to occur for stars with mass $M \lesssim 3 \times 10^5 M_\odot$. Above this mass, however, *nothing* can halt the collapse: a black hole is formed. Figure 17.2 shows the formation of such a black hole in a computer calculation of the collapse of a $10^6 M_\odot$ supermassive star by Shapiro and Teukolsky (1979). Shown is a spacetime diagram, time plotted vertically in units of light-travel time across the Schwarzschild radius GM/c^2, and radius plotted horizontally in units of the Schwarzschild radius. The radial coordinate adopted is the so-called "isotropic" radius, used to simplify the computer calculations. Shown at the top of the figure is the equivalent Schwarzschild radial coordinate r_s at the termination of the calculation. The coordinate r_s is defined so that the area of a spherical surface at r_s is $\mathcal{Q} = 4\pi r_s^2$. At large values of r/M the distinction between r, r_s, and the usual Newtonian radial coordinate vanishes.

[8]See, for example, Fricke (1973).

The dotted lines are the world lines of spherical shells of matter collapsing to smaller radius. Each world line is labeled with the fraction of mass interior to it. Solid lines denote *outgoing* radial light rays. The dashed line is the event horizon (surface of the black hole; see Chapter 12). It is determined by finding a pair of light rays emitted at different times from the same radius, one of which gets out to infinity and one of which does not. The shaded region inside the black hole is the region of "trapped surfaces." Here a spherical flash of light immediately decreases in area when it is emitted, because of the strong gravitational field. The straight lines with arrows show the world lines of radially ingoing and outgoing light rays in the absence of a gravitational field. The fact that the paths of outgoing photons emitted near and inside the black hole are not straight is thus a reflection of the strong gravitational field which is present there. Note how the horizon moves out to $r_s = 2GM/c^2$ as more and more matter falls into the black hole. The vertical asymptotes of particle and photon world lines at $r = 0$ are an artifact of the coordinate system.

A key feature of this and other calculations is that the collapse is *homologous*. The velocity is essentially linear with radius, and snapshots of the density profile at different times are self-similar in shape, though increasing in magnitude. Due to the homologous nature of the collapse, the entire mass moves inward coherently, crossing the event horizon in only a few light-travel times—that is, a few times GM/c^3. Such coherent motion on so short a timescale can lead to an appreciable burst of gravitational radiation, should the collapse be subjected to nonspherical perturbations (resulting, e.g., from rotation or magnetic fields). The quadrupole moment formula, Eq. (16.2.16), reveals just how sensitively the emitted radiation ΔE depends on the collapse timescale $\tau : \Delta E \propto \tau^{-5}$. The light-travel time across the Schwarzschild radius of a mass M represents the *shortest* possible timescale for significant changes in the bulk matter distribution of the mass. Only for homologous collapse can τ approach this short timescale. If instead the center of the star had collapsed first, followed by relatively slow accretion of the bulk of the mass onto the central black hole, τ would be much longer and much less gravitational radiation would be produced.

18
Stellar Collapse and Supernova Explosions

18.1 Introduction and Warning

No treatment of compact objects would be complete without some discussion of supernovae. As we have already described in Sections 9.1 and 10.1, it came as no surprise to some when the Crab and Vela pulsars were discovered in supernova remnants. Ever since Baade and Zwicky (1934) had shown that the gravitational energy released by the collapse of an evolved stellar core to a neutron star would be more than adequate to power observed supernova outbursts, many researchers regarded collapse and supernova explosions to be events intimately associated with neutron star (and, possibly, black hole) formation.

There is by now a wealth of "circumstantial" evidence linking the birth of neutron stars to collapse and supernova explosions. Evolutionary calculations of nonrotating stars with masses in the range $10 \leq M/M_\odot \leq 70$ all exhibit the development of an unstable core, containing roughly $1.5M_\odot$ of iron, which undergoes dynamical collapse.[1] Optical, radio, and X-ray observations of the expanding gas comprising "Type II" supernova remnants are nicely accounted for by assuming that these massive stars are disrupted following an explosion within the core.[2] For example, explosive nucleosynthesis in a massive star can explain the abundances of elements between oxygen and iron peak elements[3] as well as the neutron-rich ("r-process") elements. The element abundances in the

---

[1] Arnett (1977a); Barkat (1977); Weaver, Zimmerman, and Woosley (1978).

[2] Supernovae are divided into two types by observers. We shall only be concerned in this chapter with Type II supernovae, which are believed to result from collapse of massive stars $M \gtrsim 8M_\odot$. Type I supernovae are currently believed to result from the complete explosion of the degenerate core of a star of mass $4 \lesssim M \lesssim 8M_\odot$, leaving essentially no remnant.

[3] Arnett (1978); Weaver and Woosley (1980).

supernova remnants Cas A[4] and Puppis A[5] suggest that this material originates from the mantle of a massive star. The well-established Type II supernova light curves can be reliably modeled by an impulsive energy release of approximately 10^{51} ergs within the red-giant outer envelope of an evolved massive star.[6] This required energy is far less than the available gravitational binding energy released when the core collapses to a neutron star:

$$\Delta E_B \sim \frac{GM_{core}^2}{R} = 3 \times 10^{53} \left(\frac{M_{core}}{M_\odot}\right)^2 \left(\frac{R}{10 \text{ km}}\right)^{-1} \text{ erg.} \qquad (18.1.1)$$

And towering above everything else in significance are the pulsar identifications themselves, which clearly show neutron stars residing in young supernova remnants.

There is, however, one unsettling aspect to this otherwise tidy picture: no current numerical calculation of the gravitational collapse of an evolved stellar core shows an explosion! Over the many years in which this problem has been examined, several theoretical models have indeed generated explosions following core collapse.[7] But with improved understanding of (and even some consensus regarding) the complicated physics involved, recent simulations *fail* to produce *both* a gas remnant *and* a neutron star.

Now considering the great numerical and physical complexity of a collapsing stellar core, it is not surprising that determining the ultimate fate of a massive star is one of the most difficult problems of theoretical astrophysics. The difficulty results, in part, from the complicated global interplay between hydrodynamical and gravitational forces on the one hand, and neutrino transport on the other. In addition, there are still residual uncertainties in our understanding of the local microphysics of hot, dense matter. It is understandable why, to date, virtually all computer model calculations of core collapse and supernova explosion have some *known* fault that potentially alters the final outcome of the simulations.

Recalling the turbulent history and disparate predictions of collapse calculations over the past 15 or so years, it would be very premature to regard current "unsuccessful" models as definitive. To appreciate their tentative nature, we need only remind ourselves that these calculations deal with matter near nuclear densities and are, typically, confined to spherical, nonrotating, nonmagnetic configurations. For reasons of caution, as well as general space limitations, our discussion of supernova collapse will be rather brief. We will focus on several of the key physical ideas underlying the collapse of a stellar core rather than on

---

[4] Chevalier and Kirshner (1978).

[5] Canizares and Winkler (1981).

[6] Grassberg, Imshennik, and Nadyozhin (1971); Lasher (1975); Arnett and Falk (1976); Chevalier (1976, 1981b).

[7] For example, Colgate and White (1966); Arnett (1967); Schwartz (1967); Ivanova, Imshennik, and Nadyozhin (1969); Wilson (1971, 1980); Bruenn (1975).

specific models. Because even this discussion is likely to become outdated in the very near future, we urge the reader to consult the recent literature for a status report on the problem of collapse and supernova.[8]

18.2 The Onset of Collapse

Massive stars ($8-10 \leqslant M/M_\odot \lesssim 70$) are currently thought to evolve for $\sim 10^7$ yr via the thermonuclear burning of heavier and heavier nuclear fuels. The process begins with the fusion of hydrogen into helium (ignition temperature $T_{H, ig} \sim 2 \times 10^7$ K), proceeds at successively higher temperatures through the burning of ^{4}He, ^{12}C, ^{20}Ne, ^{16}O, and ^{28}Si (ignition temperature $T_{Si, ig} \sim 3 \times 10^9$ K), and terminates with the formation of ^{56}Fe, the nucleus with the maximum binding energy per nucleon. Virtually all calculations of this process show that evolved massive stars all develop central cores of mass $M_{core} \sim 1.5 M_\odot$ consisting primarily of iron group nuclei and supported primarily by electron degeneracy pressure.

Arnett (1979) has provided a simple explanation for the tendency of the physical processes during the evolution of massive stars to force all the model calculations to a similar final state ("core convergence"). Because of the large temperature gradient (i.e., entropy gradient), heat is transported through the core largely by *convection*. When convective motion is present, currents appear that tend to mix the gas and equalize the temperature.[9] The core is therefore relatively homogeneous in composition.

Focus now on the equation of state, which is dominated by the electrons:

$$\frac{P}{\rho} \approx \frac{Y_e kT}{m_B} + K_\Gamma Y_e^\Gamma \rho^{\Gamma - 1}. \tag{18.2.1}$$

Equation (18.2.1) represents an approximate interpolation between Maxwell–Boltzmann and degenerate behavior. Here $Y_e = n_e/n$ is the number of electrons per baryon and K_Γ is a constant in the nonrelativistic $\left(\Gamma = \frac{5}{3}\right)$ and extreme relativistic $\left(\Gamma = \frac{4}{3}\right)$ degeneracy limits (cf. Section 2.3). For a spherical configuration of mass M and radius R, the hydrostatic equilibrium equations (3.2.1) and

---

[8]A history of the early (pre-1970) models of core collapse and supernova explosions, with discussion of some of the neutrino processes they incorporated, is presented in Zel'dovich and Novikov (1971). A review of the more recent developments (pre-1983), with emphasis on the appropriate equation of state for hot, dense matter, is contained in Arnett (1979) and in Lattimer (1981). Recent discussions of gravitational radiation from stellar core collapse are contained in review articles in Smarr (1979b); see also Chapter 16.

[9]See Clayton (1968), Section 3.5 for a discussion of convection and a derivation of the Schwarzschild (1906) criterion for convection: $|(dT/dr)_{star}| > |(dT/dr)_{ad})|$, where $(dT/dr)_{star}$ is the actual stellar temperature gradient and $(dT/dr)_{ad} = (T/P)(1 - 1/\Gamma) \, dP/dr$ is the so-called adiabatic temperature gradient, where Γ is the adiabatic index. Clearly, if the temperature varies too rapidly with distance, convection occurs.

(3.2.2) require, crudely

$$\frac{P_c}{\rho_c} \approx \frac{GM}{R} \approx fGM^{2/3}\rho_c^{1/3},$$

(18.2.2)

where the subscript c denotes central values and f is a structure factor depending on the effective polytropic index or density profile. Combining Eqs. (18.2.1) and (18.2.2) yields

$$\frac{Y_e k T_c}{m_B} = fGM^{2/3}\rho_c^{1/3} - K_\Gamma Y_e^\Gamma \rho_c^{\Gamma-1}.$$

(18.2.3)

Consider now the maximum temperature a configuration of given mass can achieve. The result will indicate the mass required to ignite a given nuclear fuel. For large M, the first term on the right-hand side of Eq. (18.2.3) dominates so that $T_c \propto \rho_c^{1/3}$. Consequently, continued contraction leads to even higher temperature, so that any fuel will ignite eventually. However, for small M and $\Gamma > \frac{4}{3}$, T_c falls to zero when ρ increases to

$$\rho_{\text{crit}} \approx \left(\frac{fGM^{2/3}}{K_\Gamma Y_e^\Gamma}\right)^{1/(\Gamma-4/3)}.$$

(18.2.4)

For $\rho_c < \rho_{\text{crit}}$ we have $T_c \propto \rho_c^{1/3} M^{2/3}$, while larger ρ_c are unphysical ($T_c < 0$). Thus, for this case, the configuration will pass through a maximum T as it contracts, and then cool as the degenerate electrons provide the dominant support.

The two types of behavior are divided roughly by the Chandrasekhar limiting mass, for which $T_c = 0$ and $\Gamma = 4/3$ in Eq. (18.2.3):

$$M_{\text{Ch}} = \left(\frac{K_{4/3}}{fG}\right)^{3/2} Y_e^2 \approx 5.83 Y_e^2 M_\odot$$

(18.2.5)

[cf. Eq. (3.3.17)]. Now silicon burning, the final burning stage, is triggered at high temperatures, $kT \gtrsim 0.6 m_e c^2$. It is thus not surprising that to reach such high maximum temperatures and ignite silicon, masses with $M_{\text{Si, ig}} \approx M_{\text{Ch}}$ are required. If, after the exhaustion of the lighter nuclei in the core, $M_{\text{core}} < M_{\text{Si, ig}}$, the ignition of Si must wait until *shell* burning of lighter nuclei increases M_{core}. Since $Y_e \approx 0.42$ at the onset of Si burning [cf. Eq. (18.2.7)], the requirement that nuclear burning go to completion in the core is thus[10]

$$M_{\text{core}} \gtrsim M_{\text{Si, ig}} \approx M_{\text{Ch}}(Y_e \approx 0.42) \approx 1.2 M_\odot.$$

(18.2.6)

---

[10] For low-mass stars with *total* mass $M \lesssim 1.2 M_\odot$, nuclear burning can thus never go to completion because the required high ignition temperatures are never attained. Such stars form white dwarfs with interiors consisting of unburned ^{12}C and ^{16}O (cf. Chapter 3).

If instead, $M_{core} > M_{Si, ig}$ at some earlier epoch in the evolution of a massive star, hydrostatic contraction will proceed to rather high central temperatures. Between subsequent burning stages, neutrino cooling, which varies with a high power of T, will then become important near the center. Neutrino cooling reduces the entropy there, thereby generating a positive entropy gradient. The net effect will be to inhibit the growth of the convective core at each subsequent stage of nuclear burning, thereby shrinking the region over which chemical mixing occurs. The thermonuclear "ashes" will then encompass a *smaller* mass than contained in the original, unburned core.

The end result of either scenario—that is, M_{core}/M_{Ch} smaller or larger than unity prior to Si burning—is to drive M_{core} toward M_{Ch}. This fact is confirmed by all detailed evolution calculations. The core material (Si-burning "ashes") consists primarily of neutron-rich iron group nuclei (i.e., ^{56}Fe, ^{58}Fe, ^{60}Fe, ^{62}Ni, etc.).

Thus, a $15M_\odot$ star will have a core of about $1.5M_\odot$ at the onset of collapse with initial central temperature $T_{c,i}$, central density $\rho_{c,i}$, central lepton fraction $Y_{e,i}$ and entropy per baryon s_i approximately given by[11]:

$$\left.\begin{array}{l} T_{c,i} \approx 8.0 \times 10^9 \text{ K} = 0.69 \text{ MeV}/k \\ \rho_{c,i} \approx 3.7 \times 10^9 \text{ g cm}^{-3} \\ Y_{e,i} \approx 0.42 \\ s_i/k \approx 0.91 \end{array}\right\} M_{core} \approx 1.5 M_\odot \gtrsim M_{Ch}. \quad (18.2.7)$$

Following silicon burning, two different physical effects combine to drive the core dynamically unstable and lead to gravitational collapse. Initially, the collapse is triggered by the *partial dissociation of iron nuclei*. This partial dissociation uses up nuclear binding energy, and, as a result, lowers the pressure. As the collapse proceeds, the density rises, increasing the electron chemical potential. This leads to the *neutronization* of the core, as electrons are captured by protons in nuclei (cf. Sections 2.6 and 8.2). Both dissociation and neutronization lower the adiabatic index of the stellar material below $\frac{4}{3}$, thus leading to collapse.

18.3   Photodissociation

The partial photodissociation of nuclei into α-particles at high T is well understood.[12] At such high T and ρ, reactions involving the strong and electromagnetic

---

[11]Arnett (1977b); Weaver, Zimmerman, and Woosley (1978). Note that farther out in the core the density and temperature are lower, while the entropy is somewhat higher than at the center, but always of order unity throughout the core. Since s and Y_e do not vary much throughout the core, we drop the subscript c on these variables.

[12]Hoyle (1946); Burbidge et al. (1957).

interactions are sufficiently rapid that they proceed essentially in equilibrium with their inverses. The dissolution of $^{56}_{26}\mathrm{Fe}$ is typical:

$$\gamma + {}^{56}_{26}\mathrm{Fe} \rightleftarrows 13\alpha + 4n. \tag{18.3.1}$$

The energy required for this process is

$$Q = c^2(13m_\alpha + 4m_n - m_{\mathrm{Fe}}) = 124.4 \text{ MeV}. \tag{18.3.2}$$

In equilibrium Eq. (2.1.10) gives for reaction (18.3.1)

$$\mu_{\mathrm{Fe}} = 13\mu_\alpha + 4\mu_n, \tag{18.3.3}$$

where μ_i is the chemical potential of the ith species. For temperatures and densities of interest, the nuclei and nucleons are nondegenerate, so that Maxwell–Boltzmann statistics apply:

$$\frac{\mu_i - m_i c^2}{kT} = \ln\left[\frac{n_i}{g_i}\left(\frac{2\pi\hbar^2}{m_i kT}\right)^{3/2}\right] \tag{18.3.4}$$

[cf. Eq. (2.3.31)]. Recall that for a system with internal degrees of freedom, such as a nucleus that has excited states, the statistical weight g_i is the nuclear partition function (cf. Exercise 2.8)

$$g_i = \sum_r (2I_r + 1)e^{-E_r/kT}, \tag{18.3.5}$$

where I_r is the spin of the rth excited state and E_r is the energy above the ground state. For temperatures ≤ 1 MeV, one may set $g_\alpha = 1$ (ground state; $I = 0$), $g_n = 2$ (free fermion; $I = \frac{1}{2}$), and $g_{\mathrm{Fe}} \simeq 1.4$ (ground state with $I = 0$ plus lowest excited states).[13]

Substituting Eq. (18.3.4) in Eq. (18.3.3) yields the *Saha equation* for the equilibrium ratio of α particles and neutrons to iron nuclei:

$$\frac{n_\alpha^{13} n_n^4}{n_{\mathrm{Fe}}} = \frac{g_\alpha^{13} g_n^4}{g_{\mathrm{Fe}}}\left(\frac{kT}{2\pi\hbar^2}\right)^{24}\left(\frac{m_\alpha^{13} m_n^4}{m_{\mathrm{Fe}}}\right)^{3/2} e^{-Q/kT}. \tag{18.3.6}$$

To within an accuracy of better than one percent, we can replace the mass of a species with atomic weight A by Am_u, so that

$$\frac{n_\alpha^{13} n_n^4}{n_{\mathrm{Fe}}} = \frac{2^{43}}{(56)^{3/2}(1.4)}\left(\frac{m_u kT}{2\pi\hbar^2}\right)^{24} e^{-Q/kT}. \tag{18.3.7}$$

[13]Hoyle and Fowler (1960).

If we assume that ^{56}Fe is the most abundant heavy nucleus, then reaction (18.3.1) implies

$$n_n = \tfrac{4}{13} n_\alpha. \tag{18.3.8}$$

Equations (18.3.7) and (18.3.8) allow one to calculate the degree of dissociation at any value of ρ and T.

Exercise 18.1 Show that material consisting entirely of ^{56}Fe, α particles, and neutrons will be half ^{56}Fe by mass and half α particles and neutrons when

$$\log \rho = 11.62 + 1.5 \log T_9 - \frac{39.17}{T_9}, \tag{18.3.9}$$

where ρ is the density in g cm^{-3} and T_9 is the temperature in 10^9 degrees (Hoyle and Fowler, 1960). The logarithms are to base 10.

According to Eqs. (18.2.7) and (18.3.9), 50% dissociation at core densities near $\rho_{c,i}$ requires temperatures near $T_9 \approx 11$, which is not much higher than $T_{c,i}$. Hence it is not surprising that collapse sets in when the $1.5 M_\odot$ core enters this physical regime. At somewhat higher temperatures (for the same densities), the α particles photodissociate into nucleons via

$$\gamma + {}^4\text{He} \rightleftarrows 2p + 2n. \tag{18.3.10}$$

The energy required for this process is $Q' = 28.30$ MeV. Since the required energy per *new* particle created via reaction (18.3.10) ($\Delta N = 4 - 1 = 3$; $Q'/\Delta N \approx 9.5$ MeV) is larger than the energy per new particle created via reaction (18.3.1) ($\Delta N = 13 + 4 - 1 = 16$; $Q/\Delta N \approx 7.7$ MeV), higher temperatures are needed to dissociate ^{4}He than ^{56}Fe. Thus, there exists a temperature domain in which ^{56}Fe has dissociated into ^{4}He, but ^{4}He has not dissociated into free nucleons.

Exercise 18.2

(a) Show that the relevant Saha equation for the dissociation of α-particles into nucleons is

$$\frac{n_p^2 n_n^2}{n_\alpha} = 2 \left(\frac{m_u kT}{2\pi \hbar^2} \right)^{9/2} \exp\left(\frac{-Q'}{kT} \right). \tag{18.3.11}$$

(b) Calculate the temperature T^* for 50% dissociation of (i) ^{56}Fe into ^{4}He and free neutrons, and (ii) ^{4}He into nucleons when $\rho = 10^9$ g cm^{-3}, and when $\rho = \rho_{c,i}$.
Answer: $\rho = 10^9$ g cm^{-3}: $T_9^*(^{56}\text{Fe}) = 9.6$, $T_9^*(^4\text{He}) = 15.2$.

In any realistic calculation, one must treat the *entire* set of nuclei, stable and unstable, which exist at high temperature and density in nuclear statistical equilibrium ("NSE"). Typical reactions are of the form

$$(Z, A) + p \rightleftarrows (Z + 1, A + 1) + \gamma$$

$$(Z, A) + n \rightleftarrows (Z, A + 1) + \gamma, \tag{18.3.12}$$

as well as reactions such as (α, γ), (α, n), (α, p), (p, n), and so on. At the onset of collapse, the nuclei and nucleons are often approximated as an ideal nonrelativistic Maxwell–Boltzmann gas so that Eq. (18.3.4) applies for each species. For each reaction in the network, the equilibrium condition (2.1.10) applies. However, because all possible reactions of the form (18.3.12) are in equilibrium, we have only two independent chemical potentials, corresponding to conservation of baryon number and charge. Choosing these to be μ_p and μ_n, we have

$$\mu(Z, A) = Z\mu_p + (A - Z)\mu_n \tag{18.3.13}$$

as the condition of nuclear statistical equilibrium. Equation (18.3.4) then gives

$$n_i \equiv n(Z, A) = \frac{g(Z, A)A^{3/2}}{2^A \theta^{A-1}} n_p^Z n_n^{A-Z} \exp\left[\frac{Q(Z, A)}{kT}\right], \tag{18.3.14}$$

where

$$Q(Z, A) \equiv c^2 \left[Zm_p + (A - Z)m_n - M(Z, A) \right] \tag{18.3.15}$$

is the nuclear binding energy, $g(Z, A)$ is the nuclear partition function, Eq. (18.3.5), and

$$\theta \equiv \left(\frac{m_u kT}{2\pi\hbar^2}\right)^{3/2}. \tag{18.3.16}$$

To specify n_p and n_n we may invoke baryon and charge conservation:

$$\sum_i n_i A_i = \frac{\rho}{m_u} = n, \tag{18.3.17}$$

$$\sum_i n_i Z_i = nY_e, \tag{18.3.18}$$

where Y_e is the mean number of electrons per baryon. It is thus evident from Eqs. (18.3.14)–(18.3.18) that in NSE, the composition, equation of state, specific entropy, and so on, can be determined once the *three* quantities ρ, T, and Y_e are specified.[14] In applications to stellar evolution or collapse calculations, Y_e is

---

[14] Compare Epstein and Arnett (1975).

determined at any instant from the prior evolutionary history. In the early stages this history involves the thermonuclear synthesis of heavy elements from light elements via fusion, which influences Y_e. In the late stages following silicon burning, it is dominated by neutronization (see Section 18.4 below).

Exercise 18.3 In determining the composition for the "network" containing the *single* reaction given in Eq. (18.3.1), we did not specify Y_e, but, instead, imposed the constraint $n_n = \frac{4}{13} n_\alpha$. Show that this was equivalent to the condition that $Y_e = \frac{26}{56} = 0.464$.

Arnett (1979) has emphasized the complications that arise in solving the NSE network at the high densities achieved during core collapse. They include (i) uncertainties in the values of g_i and Q for the experimentally unknown nuclei found far from the domain of beta stability; (ii) the necessity of including nuclear matter corrections arising from the finite nuclear size and nuclear surface and Coulomb energy; and (iii) the need to include effects of nucleon interactions and neutron degeneracy at high density.[15] Recent work on the equation of state for hot, dense matter found during core collapse has focused precisely on these issues.[16]

18.4 Neutronization and Neutrino Emission

As the core density increases, the high electron Fermi energy drives electron capture onto nuclei and free protons ("neutronization"). This reduces Y_e and decreases the contribution of the degenerate electrons to the total pressure supporting the core against gravitational collapse. Eventually, the Chandrasekhar mass $M_{Ch}(Y_e)$, the maximum mass that can be supported by electron degeneracy pressure, falls below the core mass [cf. Eq. (18.2.5)]. From this moment on, core collapse proceeds in earnest: as the density increases further, more and more electron captures occur, the pressure $\left(\text{and the difference } \Gamma - \frac{4}{3} < 0\right)$ decreases further, and the collapse accelerates. In the neutronization process, neutrinos are emitted and, at least initially, many of them escape.

Neutrinos are generated during collapse both by *neutronization* and by *thermal emission*. Thermal mechanisms proceed via the annihilation of real and virtual $e^+ e^-$ pairs, forming $\nu\bar{\nu}$ pairs. If the neutrino pair escapes, the lepton number in the core is unchanged. The most significant *thermal emission* processes are:

1. Pair annihilation,

$$e^+ + e^- \xrightarrow{(W, Z)} \nu + \bar{\nu}, \qquad (18.4.1)$$

---

[15]See Chapters 2 and 8 for a discussion of these points for dense matter at $T = 0$.
[16]See Lamb et al. (1978); Bethe et al. (1979); Epstein and Pethick (1981).

2.   Plasmon decay,

$$(\text{plasma excitation}) \overset{(W,\,Z)}{\to} \nu + \bar{\nu}, \tag{18.4.2}$$

3.   Photoannihilation,

$$e^- + \gamma \overset{(W,\,Z)}{\to} e^- + \nu + \bar{\nu}, \tag{18.4.3}$$

4.   Bremsstrahlung,

$$e^- + (Z, A) \overset{(W,\,Z)}{\to} (Z, A) + e^- + \nu + \bar{\nu}. \tag{18.4.4}$$

All of the above processes can proceed both by charged currents via the exchange of $W^{\pm}$ vector bosons and by neutral currents through the exchange of Z vector mesons (cf. Chapter 11). In the case of $e^+ e^-$ annihilation (reaction 1), for example, an electron neutrino pair $\nu_e \bar{\nu}_e$ can be produced either by Z or by $W^{\pm}$ exchange. However, muon and tau neutrino pairs, $\nu_\mu \bar{\nu}_\mu$ and $\nu_\tau \bar{\nu}_\tau$ (or neutrino pairs associated with "heavy leptons," should they exist) are only produced by Z exchange. Rates for the above reactions have been calculated by Dicus (1972) and Dicus et al. (1976) using the Weinberg-Salam-Glashow theory of weak interactions. We point out that a "plasmon" is a quantized electromagnetic wave propagating in a dense, dielectric plasma. It behaves like a relativistic Bose particle with a rest mass $m_{\text{plasmon}} = \hbar \omega_p / c$, where ω_p is the plasma frequency. Unlike a free photon, a plasmon is energetically unstable to decaying into a neutrino–antineutrino pair by virtue of this rest-mass energy "excess."

Typically, thermal neutrino processes dominate in the core of a massive star until the collapse is fully underway. They are also important during the late stages of collapse, when they help carry thermal energy away from the shock-heated outer regions of the core. Neutrinos generated thermally have energies $E(\nu)$ near kT.

The most significant *neutronization* reactions are:

1.   Electron capture by nuclei,

$$e^- + (Z, A) \overset{(W)}{\to} \nu_e + (Z - 1, A), \tag{18.4.5}$$

2.   Electron capture by free protons,

$$e^- + p \overset{(W)}{\to} \nu_e + n. \tag{18.4.6}$$

Both of the above reactions proceed entirely via charged currents. The matrix element for reaction (18.4.6) above is well known: the process is closely related to free neutron decay, which we have already analyzed in Section 11.4 [cf. Eq. (11.4.9)]. During collapse, however, the nucleons and nuclei are nondegenerate and nonrelativistic Fermi gases, except at the highest densities when neutrinos no longer escape from the core. This situation contrasts with the later, extended cooling phase of a warm neutron star discussed in Chapter 11. This post-collapse cooling epoch is characterized by temperatures well below 1 MeV and nucleons that are nonrelativistic but highly degenerate, so that the above reactions are strongly suppressed.

During collapse, most of the nucleons remain bound in heavy nuclei, as we shall discuss in Section 18.6. The mass number A, proton number Z, and neutron number $A - Z$ all increase with increasing density and decreasing Y_e. Accordingly, electron capture onto free protons is limited by the small abundance of free protons. In addition, electron capture onto nuclei is restricted by the large threshold energies required as the nuclei become increasingly neutron-rich and because only "valence" protons are able to undergo transitions to "superallowed" states. Superallowed decays are so named because not only are they allowed in the sense that they do not change the orbital angular momenta of nucleons in nuclei, but the nuclear overlap integral [cf., e.g., Eq. (11.4.2) or (11.5.8)] is essentially unity.

Reliable shell model calculations of electron capture onto intermediate and heavy mass nuclei are only now being performed.[17] It has been shown by Fuller, Fowler, and Newman (1980) that electron capture onto nuclei is suppressed when the neutron number $N = A - Z$ exceeds 40. In this case the final-state neutron shell is full and the allowed Gamow–Teller transition to that state does not occur. The consequences of this neutron shell blocking phenomenon for core collapse have not been fully explored. However, preliminary analyses[18] suggest that electron capture rates on nuclei proceed rapidly at first ($\rho \lesssim 10^{11}$ g cm^{-3}) but are ultimately suppressed leading to the dominance of reaction (18.4.6) over reaction (18.4.5) and a decrease in the overall neutronization rate.

Once collapse gets underway, neutronization becomes the most important mechanism for generating neutrinos. Neutronization reduces Y_e, decreasing the electron pressure and influencing nuclear statistical equilibrium. The mean energy of neutrinos generated via electron capture is comparable to the electron Fermi energy:

$$\langle E(\nu) \rangle \sim \langle E_{e^-} \rangle \sim \mu_e = 51.6(Y_e \rho_{12})^{1/3} \text{ MeV}, \qquad (18.4.7)$$

---

[17]See, for example, Fuller, Fowler, and Newman (1980), (1982a, b).
[18]Fuller (1982); Van Riper and Lattimer (1981) and Zaringhalam (1982).

where ρ_{12} is the density in units of 10^{12} g cm^{-3}. [Equation (18.4.7) follows from Eqs. (2.3.1) and (2.3.15) when $x \gg 1$. Note that μ_e^{-1} in Eq. (2.3.15) is called Y_e here and does not refer to chemical potential.]

We now compute the electron capture rate onto free protons—that is, reaction (18.4.6). We perform this calculation because it is easier to do exactly than the case of capture onto nuclei and because it gives the correct order of magnitude of the total capture rate. We simply adopt the same perturbation approach and weak interaction matrix element used in Section 11.4 for the free neutron decay rate. We shall assume immediately that the nucleons are nonrelativistic and nondegenerate, which is valid for all but the final stages of hot core collapse. (In particular, this condition is satisfied during the early infall epoch in which the liberated neutrinos can freely escape from the collapsing core.) So we can again employ the Golden Rule in the nonrelativistic limit to obtain the capture rate for given initial proton and electron states:

$$d\Gamma = \frac{2\pi}{\hbar} \left(\frac{1}{2} \sum_{\text{spins}} |H_{fi}|^2 \right) (1 - f_\nu) \rho_\nu \, dE_\nu \, \delta(E_\nu + Q - E_e). \quad (18.4.8)$$

The various terms appearing in Eq. (18.4.8) have already been defined in Sections 11.4 and 11.5. In particular, the matrix element is given by Eq. (11.4.9) and, employing energy conservation via

$$E_\nu = E_e - Q, \quad (18.4.9)$$

the density of final neutrino states by Eq. (11.4.14). We are again employing a normalization to a box of unit volume. The quantity $1 - f_\nu$ is the neutrino "blocking factor" giving the fraction of unoccupied phase space for neutrinos. For both electrons and neutrinos we shall assume Fermi–Dirac equilibrium distributions,

$$f_j = \frac{1}{\exp\left[(E_j - \mu_j)/kT \right] + 1}. \quad (18.4.10)$$

Equation (18.4.10) will not be valid for the neutrinos until they become trapped in the core and their density builds up. However, even though neutrinos escape from the core easily prior to this stage and hence are not thermalized, we expect $f_\nu \ll 1$ so that the details of their distribution early on are not significant. Assembling the various factors gives for Eq. (18.4.8),

$$d\Gamma = \frac{2\pi}{\hbar} G_F^2 C_V^2 (1 + 3a^2)(1 - f_\nu) \frac{E_\nu^2}{2\pi^2 \hbar^3 c^3} \delta(E_\nu + Q - E_e) \, dE_\nu. \quad (18.4.11)$$

To find the total rate per proton, we integrate over all initial electron states and over dE_ν, using Eq. (11.4.11) for the density of electron states:

$$\Gamma = \frac{2\pi}{\hbar} \frac{G_F^2 C_V^2 (1 + 3a^2)}{(2\pi^2 \hbar^3 c^3)^2} \int E_\nu^2 \, dE_\nu \, SE_e \big(E_e^2 - m_e^2 c^4\big)^{1/2} \, dE_e \, \delta(E_\nu + Q - E_e),$$

(18.4.12)

where

$$S \equiv f_e(1 - f_\nu). \tag{18.4.13}$$

Using the δ-function to do the integral over dE_ν, we find

$$\Gamma = \frac{2\pi}{\hbar} \frac{G_F^2 C_V^2 (1 + 3a^2)}{(2\pi^2 \hbar^3 c^3)^2} I, \tag{18.4.14}$$

where

$$I \equiv \int_Q^\infty dE_e \, E_e \big(E_e^2 - m_e^2 c^4\big)^{1/2} (E_e - Q)^2 S. \tag{18.4.15}$$

Let us evaluate the integral I for the *initial* stages of collapse, when the electrons are extremely relativistic and degenerate ($E_e \gg m_e c^2$; $\mu_e \gg kT$) and the neutrinos escape freely from the core ($f_\nu \ll 1$). Since $f_e \simeq 0$ for $E_e > \mu_e$ in this limit, we get

$$I \simeq \int_Q^{\mu_e} dE_e \, E_e^2 (E_e - Q)^2 \simeq \int_Q^{\mu_e} dE_e \, E_e^4 = \frac{1}{5}\mu_e^5. \tag{18.4.16}$$

Exercise 18.4 Evaluate I in the low-temperature limit appropriate for the *late* stages of collapse when the neutrinos become trapped in the core. In this regime, $E_e \gg m_e c^2$, $\mu_e \gg kT$, and $\mu_\nu \gg kT$.

Answer:

$$I \simeq \int_{Q+\mu_\nu}^{\mu_e} dE_e \, (E_e - Q)^2 E_e^2$$

$$= \frac{1}{5}\big[\mu_e^5 - (Q + \mu_\nu)^5\big] - \frac{Q}{2}\big[\mu_e^4 - (Q + \mu_\nu)^4\big] + \frac{Q^2}{3}\big[\mu_e^3 - (Q + \mu_\nu)^3\big].$$

The rate per unit volume of electron captures onto free protons is found by multiplying Γ by n_p, the proton number density:

$$\frac{dn_e}{dt} = -\frac{dn_p}{dt} = -n_p\Gamma. \tag{18.4.17}$$

Dividing by n, we get from Eqs. (18.4.14) and (18.4.16),

$$\frac{dY_e}{dt} = -Y_p \frac{2\pi}{\hbar} \frac{G_F^2 C_V^2 (1 + 3a^2)}{(2\pi^2\hbar^3c^3)^2} \frac{\mu_e^5}{5}. \tag{18.4.18}$$

Exercise   18.5   What is the mean energy of the neutrino emitted following electron capture? [Assume conditions consistent with Eq. (18.4.16).]

   Hint: $\langle E_\nu \rangle = \int E_\nu \, d\Gamma / \Gamma.$

   Answer:

$$\langle E_\nu \rangle = \tfrac{5}{6}\mu_e = \tfrac{10}{9}\langle E_e \rangle, \tag{18.4.19}$$

where $\langle E_e \rangle$ is the mean electron energy in the relativistic Fermi sea.

Exercise   18.6

   (a)   Evaluate the electron capture rate (18.4.14) numerically for extremely relativistic, degenerate electrons and freely escaping neutrinos.

   (b)   Calculate the corresponding neutrino luminosity per proton.

   Answer:

$$\textbf{(a)}\quad \Gamma = 6.29 \times 10^{-4}\frac{1}{5}\left(\frac{\mu_e}{m_e c^2}\right)^5 \text{s}^{-1}\,\text{proton}^{-1}.$$

$$\textbf{(b)}\quad L_{\nu_e} = 5.15 \times 10^{-10}\frac{1}{6}\left(\frac{\mu_e}{m_e c^2}\right)^6 \text{erg s}^{-1}\,\text{proton}^{-1}.$$

18.5   Neutrino Opacity and Neutrino Trapping

Each neutrino emission process has an inverse process corresponding to absorption. Both *absorption* and *scattering* impede the free escape of neutrinos from a collapsing core. The most important processes are:

1.   Free nucleon scattering,

$$\nu + n \overset{(Z)}{\rightarrow} \nu + n,$$

$$\nu + p \overset{(Z)}{\rightarrow} \nu + p, \tag{18.5.1}$$

2.   "Coherent" scattering by heavy nuclei ($A > 1$),

$$\nu + (Z, A) \overset{(Z)}{\rightarrow} \nu + (Z, A), \tag{18.5.2}$$

3.   Nucleon absorption,

$$\nu_e + n \overset{(W)}{\to} p + e^-, \tag{18.5.3}$$

4.   Electron–neutrino scattering,

$$e^- + \nu \overset{(W,\,Z)}{\to} e^- + \nu. \tag{18.5.4}$$

Similar processes occur for antineutrinos. The first two opacity sources exist by virtue of neutral currents and were not considered seriously prior to the WSG theory of weak interactions. Their total cross sections, as measured in the matter rest frames, are given by[19]

$$\sigma_n \approx \frac{1}{4}\sigma_0 \left(\frac{E_\nu}{m_e c^2} \right)^2, \qquad E_\nu \ll m_n c^2, \tag{18.5.5}$$

for free neutron scattering, and

$$\sigma_A^{\text{coh}} \approx \frac{1}{16}\sigma_0 \left(\frac{E_\nu}{m_e c^2} \right)^2 A^2 \left[1 - \frac{Z}{A} + (4\sin^2\theta_w - 1)\frac{Z}{A} \right]^2,$$

$$E_\nu \ll 300A^{-1/3} \text{ MeV}, \tag{18.5.6}$$

for coherent scattering, where $\sigma_0 = 1.76 \times 10^{-44}$ cm^2 is given by Eq. (11.7.3) and where θ_w is the Weinberg angle with measured value given by Eq. (11.3.2). For low-energy neutrinos (i.e., $E_\nu \ll m_n c^2$), scattering from nucleons and nuclei is *elastic*: the initial and final neutrino energies are nearly equal. In addition, as first pointed out by Freedman (1974), neutrinos can also scatter *coherently* from nuclei so that the cross section varies as the atomic mass squared, A^2, as is seen from Eq. (18.5.6). The scattering is coherent in that the nucleus acts nonlinearly as a single particle and not simply as A separate nucleons. The persistence of heavy nuclei at high densities makes coherent scattering a dominant opacity source during "hot" core collapse.

The significance of neutral current reactions is that they increase the neutrino opacity without changing the emission rate appreciably. Thus, neutral current reactions inhibit neutrino transport in collapsing cores and enhance neutrino trapping.

In contrast to scattering from nucleons and nuclei, electron-neutrino scattering is *nonconservative* and changes the neutrino energy in the star rest frame. The inelastic nature of ν-e scattering results from the low rest-mass energy of the

---

[19]Equations (18.5.5) and (18.5.6), together with cross sections for the other dominant neutrino opacity sources, may be found in Tubbs and Schramm (1975).

electron so that, typically, $E_\nu > m_e c^2$. Collisions between neutrinos and degenerate electrons can lead to appreciable neutrino energy loss or "downscattering". High-energy neutrinos ($E_\nu \gg \mu_e$) can interact with degenerate electrons residing deep with the Fermi sea. For such interactions the total energy is roughly equal to E_ν and on the average this energy is shared equally by the electron and neutrino following scattering. For low-energy neutrinos ($E_\nu \ll \mu_e$) the effects of electron degeneracy are very important. One might expect that a neutrino would gain energy from such an interaction, leading to energy equilibration. However, an electron with energy $E_e < \mu_e$ cannot lose energy. As a result, in order for the interaction to proceed at all, a low-energy neutrino with $E_\nu \ll \mu_e$ can lose even more than half its energy in a scattering event. This process is thus quite important in *thermalizing* the neutrinos and helps drive them to local equilibrium once they are securely "trapped" in the core at high densities. The cross section for the scattering of a low-energy neutrino with degenerate electrons is given by Eq. (11.7.2).

Nucleon absorption—the inverse of neutronization—also contributes to the thermalization process. The cross section for the absorption of a neutrino by free, nonrelativistic neutrons, σ_a, can be calculated once again from the known charged current matrix element for free neutron decay, Eq. (11.4.9). The rate of reaction (18.5.3), per neutron, is given by

$$\sigma_a c = \int d\Gamma = \frac{2\pi}{\hbar} \big[G_F^2 C_V^2 (1 + 3a^2) \big] \int \rho_e \, dE_e (1 - f_e) \delta(E_\nu + Q - E_e),$$

$$(18.5.7)$$

where we again assume that the incident particles ν and n are confined to a box of unit volume. With the aid of Eqs. (11.4.11) and (18.4.10), Eq. (18.5.7) may be integrated to give

$$\sigma_a = \sigma_0 \frac{(1 + 3a^2)C_V^2}{4} \left(\frac{E_\nu}{m_e c^2} \right)^2 \left(1 + \frac{Q}{E_\nu} \right) \left[\left(1 + \frac{Q}{E_\nu} \right)^2 - \left(\frac{m_e c^2}{E_\nu} \right)^2 \right]^{1/2} I(E_\nu + Q),$$

$$(18.5.8)$$

where

$$I(E_\nu + Q) \equiv 1 - \frac{1}{\exp\big[(E_\nu + Q - \mu_e)/kT \big] + 1}. \qquad (18.5.9)$$

For extremely degenerate and ultrarelativistic electrons, the inhibition factor I goes to $\exp(-\mu_e/kT)$ for $E_\nu \ll \mu_e$ and to unity for $E_\nu \gtrsim \mu_e$, reflecting the important role of electron degeneracy in blocking low-energy absorptions.

Exercise 18.7 Argue that the cross section for antineutrino absorption via $\bar{\nu}_e + p \rightarrow n + e^+$ may be obtained by setting $I = 1$ and replacing Q by $-Q$ in Eq. (18.5.8). (Assume that the positrons are nondegenerate in a collapsing core, where $-\mu_{e^+} = \mu_{e^-} \gg kT$.)

As the density increases during core collapse and the opacity rises, the neutrinos experience greater difficulty escaping from the star before being dragged along with the matter.[20] At densities above

$$\rho_{\text{trap}} \sim 3 \times 10^{11} \text{ g cm}^{-3}, \tag{18.5.10}$$

the neutrinos are "trapped," comove with the matter, and build up a semidegenerate Fermi sea. By definition, at $\rho \sim \rho_{\text{trap}}$, the timescale for neutrinos to diffuse out of the core becomes comparable to the collapse timescale. This fact may be used to estimate ρ_{trap}. The hydrodynamical collapse timescale t_{coll} is in order of magnitude the free-fall timescale:

$$t_{\text{coll}} \sim \frac{1}{(G\rho)^{1/2}} \sim 4 \times 10^{-3} \rho_{12}^{-1/2} \text{ s}, \tag{18.5.11}$$

where $\rho \sim M/(4\pi R^3/3)$ is the mean density of the collapsing core and $M \sim M_\odot$. The diffusion timescale may be estimated by assuming that coherent scattering is the dominant opacity source, so that

$$t_{\text{diff}} \sim \frac{\lambda_A^{\text{coh}} N_{\text{scatt}}}{c}, \tag{18.5.12}$$

where λ_A^{coh} is the mean free path of a typical neutrino in the sea of heavy nuclei (A, Z) and $N_{\text{scatt}} \gg 1$ is the number of scatterings experienced by the neutrino prior to escape. Coherent scattering induces a random-walk trajectory for the neutrino as it zig-zags back and forth in the core without appreciably changing its energy before reaching the surface. N_{scatt} is thus determined by the random-walk relation[21]

$$\lambda_A^{\text{coh}} N_{\text{scatt}}^{1/2} \sim R. \tag{18.5.13}$$

The mean free path λ_A^{coh} may be estimated from Eq. (18.5.6),

$$\left(\lambda_A^{\text{coh}}\right)^{-1} = n_A \sigma_A^{\text{coh}} = \left(\frac{\rho}{Am_B}\right) \sigma_A^{\text{coh}}, \tag{18.5.14}$$

---

[20] Sato (1975a, b); Mazurek (1975, 1976); Lamb and Pethick (1976); Arnett (1977b).
[21] See Section 14.5(j) for an analogous discussion for photons undergoing elastic Thomson scatterings.

where we assume that all nucleons are heavy nuclei and take ^{56}Fe to be representative of these nuclei. In evaluating $\sigma_A^{\text{coh}}(E_\nu)$, we employ Eqs. (18.4.7) and (18.4.19), which give for the typical neutrino upon emission following electron capture onto a proton[22]

$$E_\nu \simeq \tfrac{5}{6}\mu_e \simeq 33\rho_{12}^{1/3} \text{ MeV}. \qquad (18.5.15)$$

Substituting Eq. (18.5.15) into Eq. (18.5.14) yields

$$\left(\lambda_A^{\text{coh}}\right)^{-1} \sim 3.9 \times 10^{-5} \text{ cm}^{-1}\rho_{12}^{5/3}. \qquad (18.5.16)$$

Solving Eq. (18.5.13) for N_{scatt} and substituting the result, together with Eq. (18.5.16), into Eq. (18.5.12) gives

$$t_{\text{diff}} \sim 0.08\rho_{12} \text{ s}. \qquad (18.5.17)$$

Equations (18.5.11) and (18.5.17) demonstrate that for sufficiently high densities $t_{\text{diff}} \gg t_{\text{coll}}$. The two timescales become comparable when

$$t_{\text{diff}} \sim t_{\text{coll}} \Rightarrow \rho \sim \rho_{\text{trap}} \sim 1.4 \times 10^{11} \text{ g cm}^{-3}, \qquad (18.5.18)$$

which is within a factor of 2 of the values found by more detailed hydrodynamical calculations[23] where allowance is made for central concentration, electron capture onto heavy nuclei, and neutrino downscattering.

Exercise   18.8   Estimate the density ρ_s at which neutrinos typically undergo a *single* scattering prior to escaping from the core [i.e., $N_{\text{scatt}}(\rho_s) \equiv 1$]. Verify that $\rho_s < \rho_{\text{trap}}$.

   Answer: $\rho_s \sim \rho_{\text{trap}}/10$

Neutrino trapping has enormous implications for core collapse. For $\rho \gtrsim \rho_{\text{trap}}$, most of the neutrinos from electron capture remain in the matter, and the lepton number per baryon, Y_l, does not change. The neutrino distribution approaches an equilibrium Fermi–Dirac form so the entire physical state of the system—baryons, leptons, and photons—can be uniquely specified by three quantities: T, ρ, and Y_e, say, or s, ρ, and Y_e.

Neutrino luminosities are greatly reduced due to trapping. Consider the luminosity once the center of the core achieves nuclear densities, $\rho_{\text{nuc}} = 2.8 \times 10^{14}$ g cm^{-3}. Above these densities, thermal pressure and nuclear forces cause the equation of state to stiffen, preventing further collapse (cf. Section 8.5). Now most of the gravitational binding energy of the core is ultimately released in the

---

[22] But see Exercise (18.14) for electron capture onto heavy nuclei.
[23] Eq. (18.5.10); Arnett (1977b); Wilson (1980).

form of neutrinos following collapse to nuclear densities. In the absence of neutrino trapping, the total binding energy would be completely emitted as neutrinos in a *collapse* timescale, the time for the core to contract from $2R_{nuc}$ to R_{nuc}, where $\rho_{nuc} \sim M/(4\pi R_{nuc}^3/3)$, giving $R_{nuc} \sim 12$ km for $M \sim M_\odot$. Accordingly, the neutrino luminosity would then achieve its maximum possible value,

$$L_{\nu,max} \sim \frac{GM^2/R_{nuc}}{t_{coll}} \sim 10^{57} \text{ erg s}^{-1}, \tag{18.5.19}$$

where we have used Eq. (18.5.11) to evaluate t_{coll}. In reality, neutrino trapping forces the liberated gravitational potential energy to be emitted on a much longer *diffusion* timescale, $t_{diff} \gg t_{coll}$ at $\rho \sim \rho_{nuc}$. As a result, the actual neutrino luminosity is closer to

$$L_\nu \sim \frac{GM^2/R_{nuc}}{t_{diff}} \sim 10^{52} \text{ erg s}^{-1}, \tag{18.5.20}$$

where we have employed Eq. (18.5.17) to evaluate t_{diff}.

Equation (18.5.20) is within an order of magnitude of more detailed model calculations,[24] which all exhibit the strong inequality $L_\nu \ll L_{\nu,max}$. This inequality demonstrates the inability of the neutrinos to stream freely out of the core during the advanced stages of collapse. The bulk of the liberated gravitational energy must therefore be converted into other forms of internal energy (i.e., thermal energy, energy of excited nuclear states, bounce kinetic energy, etc.) rather than being released immediately in the form of escaping neutrinos. On dynamical timescales, neutrino trapping thus causes the late stages of core collapse (i.e., $\rho \gtrsim \rho_{trap}$) to proceed *adiabatically* to high approximation.

Exercise 18.9 Assume that the neutrino luminosity given by Eq. (18.5.20) is emitted as thermal radiation from the surface of the core. Calculate the effective temperature of the emission, assuming that there are $N_\nu = 3$ massless neutrino types (ν_e, ν_μ, and ν_τ) and that neutrinos and antineutrinos are emitted in equal numbers. Compute the mean energy of an escaping neutrino.

Answer:

$$L_\nu = \left(\tfrac{7}{8}N_\nu\sigma T_{eff}^4\right)(4\pi R^2), \tag{18.5.21}$$

$$\left\langle E_{\nu(escape)}\right\rangle = 3.15kT_{eff}. \tag{18.5.22}$$

---

[24]See, for example, Arnett (1977b), who obtains a maximum infall luminosity of 8×10^{52} erg s^{-1} just prior to core bounce, and Wilson (1980).

18.6   Entropy and the Equation of State During Hot Collapse

In their recent analysis of the equation of state of hot, dense matter in stellar collapse, Bethe et al. (1979) (hereafter BBAL) have emphasized the significance of considering explicitly the entropy per baryon. The fact that the entropy is low initially ($s_i/k \sim 1$), increases only slightly prior to neutrino trapping ($\Delta s/k \lesssim 0.5$) and remains constant thereafter ($s_f/k \sim 1$–2) during the adiabatic infall epoch, has very important implications for the equation of state. In particular, the low entropy per baryon, together with the high lepton number due to neutrino trapping ($Y_l \sim 0.3$–0.4), preserves heavy nuclei right up to nuclear matter densities, at which point the nucleons are finally squeezed out of the nuclei. Thus, the partial dissolution of nuclei into α-particles and free neutrons at the onset of collapse is reversed at higher densities as the α-particles go back into the nuclei. The low value of s/k and high value of Y_l prevent the appearance of drip neutrons, which would add excessively to the net entropy. This situation in hot matter contrasts sharply with the behavior of cold matter at high densities (cf. Chapters 2 and 8), where $s/k \equiv 0$ and $Y_l \ll 1$. In cold matter neutron drip is already underway at $\rho = \rho_{\mathrm{drip}} \simeq 4.3 \times 10^{11}$ g cm^{-3} $\ll \rho_{\mathrm{nuc}}$ and free neutrons dominate the composition for $\rho \gtrsim \rho_{\mathrm{drip}}$.

Let us calculate the initial entropy at the onset of collapse near the center of the core, where $T_{c,i} \approx 8 \times 10^9$ K and $\rho_{c,i} \approx 3.7 \times 10^9$ g cm^{-3} [cf. Eq. (18.2.7)]. For a quick estimate, let us approximate the composition there by assuming that all the baryons are in the form of ideal, nondegenerate ^{56}Fe nuclei surrounded by an extremely degenerate, ultrarelativistic Fermi sea of electrons. Now the initial entropy *per nucleus* due to translational motion is given by Eq. (2.3.34),

$$\left(\frac{s_i}{k}\right)_{\text{per nucleus}} = \frac{5}{2} + \ln\left[\frac{1}{n_{\mathrm{Fe}}}\left(\frac{56 m_u k T_{c,i}}{2\pi \hbar^2}\right)^{3/2}\right] = 16.7, \qquad (18.6.1)$$

where $n_{\mathrm{Fe}} = \rho_{c,i}/(56 m_u)$ is the number density of iron nuclei. The initial entropy *per baryon* due to iron nuclei is then obtained by dividing Eq. (18.6.1) by 56:

$$\left(\frac{s_i}{k}\right)_{\text{per baryon}} = 0.30. \qquad (18.6.2)$$

For the electrons, the entropy may be easily obtained from Eq. (11.8.1) in the limit $x \equiv p_f/m_e c \gg 1$, which gives

$$\left(\frac{s_i}{k}\right)_{\text{per electron}} = \pi^2 \frac{k T_{c,i}}{\mu_e} = 1.10. \qquad (18.6.3)$$

Exercise 18.10   Derive the formula given in Eq. (18.6.3) from Eq. (11.8.1).
Hint: $c_v = T\, ds/dT$.

In evaluating Eq. (18.6.3) we have employed Eq. (18.4.7) with $Y_e = \frac{26}{56} = 0.464$, which gives a Fermi energy $\mu_e = 6.2$ MeV. Multiplying Eq. (18.6.3) by Y_e to give electron entropy per baryon and adding the result to Eq. (18.6.2) yields a total initial entropy per baryon of

$$\frac{s_i}{k} \simeq 0.30 + 0.51 = 0.81. \qquad (18.6.4)$$

BBAL perform a more careful calculation of the quantity s_i/k. They account for the partial dissociation of ^{56}Fe into α-particles and neutrons via Eq. (18.3.6). The α's and neutrons, treated as ideal, nondegenerate particles, increase the value of Eq. (18.6.1) by 3.6. BBAL also estimate the entropy contribution from excited nuclear states, which augments Eq. (18.6.1) by an additional 4.8.

Exercise 18.11 Use Eqs. (18.3.7) and (18.3.8) to estimate n_α, n_n, and n_{Fe} at $\rho_{c,i}$ and $T_{c,i}$. Hence find s_i/k.

Exercise 18.12 Treat the nucleus as a degenerate, nonrelativistic Fermi gas of A identical nucleons at density $\rho_{nuc} \simeq 2.8 \times 10^{14}$ g cm^{-3}. Each momentum state can contain *four* particles (two spin states for protons and neutrons).

(a) Show that the Fermi energy of the nucleus (excluding rest mass) is

$$E'_{F,nuc} = \frac{\left(3\pi^2\hbar^3\rho_{nuc}\right)^{2/3}}{\left(2m_B\right)^{5/3}} \simeq 39 \text{ MeV}.$$

(b) Use Eq. (11.8.1) in the nonrelativistic limit to show that the entropy of the nuclear excited states in this "finite temperature Fermi gas model" is

$$\left(\frac{s_i}{k}\right)_{exc,per\ nucleus} = A\frac{\pi^2}{2}\frac{kT_{c,i}}{E'_{F,nuc}} \simeq 4.9.$$

Correcting Eq. (18.6.3) to give the electron entropy for a realistic mixture of ironlike nuclei with $Y_{e,i} \simeq 0.42$ [cf. Eq. (18.2.7)] and adding the result to the baryon contribution, BBAL finally obtain for the total entropy per nucleon

$$\frac{s_i}{k} = \left(\frac{s_i}{k}\right)_{due\ to\ electrons} + \left(\frac{s_i}{k}\right)_{due\ to\ nucleons}$$

$$= 0.48 + 0.45 = 0.93 \quad (\text{total}), \qquad (18.6.5)$$

which is close to the crude estimate provided by Eq. (18.6.4).

Now that we know the initial entropy, let us consider how it changes during the early stages of collapse, prior to neutrino trapping. Entropy increase proceeds only through the weak interaction, since the matter is in equilibrium under the strong and electromagnetic interactions. According to the first law of thermodynamics [cf. Eqs. (2.1.1) and (2.1.6)], the rate of change of s is given by

$$\dot{q} = T\dot{s} + \sum_i \mu_i \dot{Y}_i, \tag{18.6.6}$$

where q is the heat absorbed per baryon, and i labels each species of nucleus, nucleon and lepton.

Consider the loss of heat due to the escape of neutrinos. Then we may write

$$\dot{q} \approx -\langle E_\nu \rangle_{\text{esc}} \frac{Y_\nu}{\tau_{\text{esc}}}, \tag{18.6.7}$$

where $\langle E_\nu \rangle_{\text{esc}}$ is the mean energy of the escaping neutrinos and τ_{esc} is their average escape time from the core. The quantity τ_{esc} may be estimated from

$$\tau_{\text{esc}} = \max\left(\frac{R}{c}, t_{\text{diff}} \right), \tag{18.6.8}$$

where t_{diff} is defined in Eq. (18.5.12) and gives τ_{esc} when $N_{\text{scatt}} > 1$.

We focus on electron neutrinos only since they dominate during the collapse. The total change in neutrino number is then determined by the difference between neutrino production via electron captures (the dominant emission mechanism during the initial collapse) and neutrino loss following escape from the core. This is described by the equation

$$\dot{Y}_\nu = -\dot{Y}_e - \frac{Y_\nu}{\tau_{\text{esc}}}. \tag{18.6.9}$$

The term $\sum_i \mu_i \dot{Y}_i$ in Eq. (18.6.6) can be written as

$$\sum_i \mu_i \dot{Y}_i = \mu_e \dot{Y}_e + \mu_p \dot{Y}_p + \mu_n \dot{Y}_n + \mu_\nu \dot{Y}_\nu + \sum_i \mu_{Z_i} \dot{Y}_{Z_i}, \tag{18.6.10}$$

where the last term describes the nuclei in reactions of the form

$$e^- + (Z_i + 1, A_i) \rightleftarrows (Z_i, A_i) + \nu. \tag{18.6.11}$$

Neutronization reactions conserve baryon number and charge, so

$$\sum_i \dot{Y}_{Z_i} A_i + \dot{Y}_p + \dot{Y}_n = 0,$$

$$\sum_i \dot{Y}_{Z_i} Z_i + \dot{Y}_p - \dot{Y}_e = 0. \tag{18.6.12}$$

Equation (18.3.13) for nuclear statistical equilibrium implies

$$\sum_i \mu_{Z_i} \dot{Y}_{Z_i} = \sum_i \left[Z_i \mu_p + (A_i - Z_i)\mu_n \right] \dot{Y}_{Z_i}. \qquad (18.6.13)$$

Equations (18.6.10), (18.6.12), and (18.6.13) give finally

$$\sum_i \mu_i \dot{Y}_i = (\mu_e + \mu_p - \mu_n)\dot{Y}_e + \mu_\nu \dot{Y}_\nu. \qquad (18.6.14)$$

Substituting Eqs. (18.6.7), (18.6.9), and (18.6.14) into Eq. (18.6.6) gives

$$T\dot{s} = -(\mu_e + \mu_p - \mu_n - \mu_\nu)\dot{Y}_e + (\mu_\nu - \langle E_\nu \rangle_{esc}) \frac{Y_\nu}{\tau_{esc}}. \qquad (18.6.15)$$

The first term in Eq. (18.6.15) represents the entropy change due to weak interaction processes that are out of β-equilibrium. The second term represents the entropy change due to loss of neutrinos.

Crucial to evaluating Eq. (18.6.15) for the entropy change is the associated equation for $\dot{Y}_e$, which gives ΔY_e due to neutronization. Naively employing Eq. (18.4.18) for this purpose, which gives the capture rate onto free protons, and setting $Y_p = Y_e$, yields

$$\Delta Y_e = \int \dot{Y}_e \, dt \sim \left[\dot{Y}_e(\rho) t_{coll}(\rho) \right]_{\rho = \rho_{trap}}, \qquad (18.6.16)$$

or, with the aid of Eqs. (18.4.7) and (18.5.11),

$$Y_{e,f} - Y_{e,i} \sim -(5.1 \times 10^3) Y_{e,f}^{8/3} \rho_{12, trap}^{7/6}. \qquad (18.6.17)$$

Equation (18.6.17) clearly shows that most of the captures prior to trapping occur at $\rho \sim \rho_{trap}$. Using Eq. (18.5.10) for ρ_{trap} yields $Y_{e,f} \sim 0.07$. A more careful analysis by BBAL, properly allowing for neutronization onto heavy nuclei instead of onto free protons, gives instead $Y_{e,f} \sim 0.31$ and $\Delta Y_e \sim -0.11$ for the characteristic core lepton fraction at neutrino trapping. Recent calculations by Fuller (1982), which account for the effect of a filled neutron shell in reducing the electron capture rate onto large nuclei, suggest an even larger core lepton fraction at trapping: $Y_{e,f} \sim 0.34$, and Brown et al. (1982) suggest $Y_{e,f} \sim 0.35$–0.36.

Given an equation for $\dot{Y}_e$ we can now calculate the entropy change from Eq. (18.6.15). Typically, the first term dominates whenever the matter is sufficiently cold so that there are few free protons and neutronization occurs primarily on nuclei.[25] In that case, the chemical potentials must then be sufficiently out of

---

[25] Epstein and Pethick (1981).

equilibrium to overcome the excitation energy of the daughter nucleus, Δ. Thus

$$\mu_e + \mu_p - \mu_n - \mu_\nu \gtrsim \Delta - \mathcal{O}(kT), \qquad (18.6.18)$$

where, typically, $\Delta \sim 3$ MeV and $\mu_\nu \sim 0$ prior to trapping. In this situation, the first term in Eq. (18.6.15) leads to an *increase* in the specific entropy. The second term can dominate at higher initial entropies, when there are more free protons and the free proton neutronization rate becomes important. In this case the neutrino energy is of order $\sim \mu_e$, which is much larger than $\mu_e + \mu_p - \mu_n$. Provided these neutrinos escape from the core without being downscattered in energy (cf. Section 18.5), the second term leads to an entropy *decrease*. In their pioneering calculation, BBAL find a final entropy (s_f/k) per baryon in the range 1–1.5. Thus, the entropy change is small prior to trapping and, hence, the entropy remains low thereafter.

The adiabatic nature of advanced core collapse can be seen from Eq. (18.6.15): following neutrino trapping, neutronization proceeds in β-equilibrium so the first term becomes small, while at the same time τ_{esc} becomes very large ($\tau_{esc} \gg \tau_{coll}$) and the second term becomes small as well. So entropy changes following trapping can be ignored during core collapse (Note: entropy generation via shock dissipation following core bounce at a central density near ρ_{nuc} *is* important eventually; cf. Section 18.7.)

Perhaps the most significant feature of recent core collapse calculations is that the properties of the core during the advanced stages of collapse—for example, $Y_{e,f}$ and s_f/k—appear to be rather insensitive to most variations of the assumptions about the underlying microphysics, global collapse rate, and initial conditions. In their parameter study, Epstein and Pethick (1981) found, for example, that the lepton number $Y_l = Y_e + Y_\nu$ was of the order $Y_l \sim 0.3$ and the change in specific entropy $\Delta s/k$ was less than ± 0.5, *independent* of wide variations in underlying physical and astronomical assumptions. This insensitivity of the state variables to details of the collapse scenario inspires confidence in the results and greatly simplifies the construction of an equation of state.

Since the entropy changes very little prior to neutrino trapping and remains constant thereafter, the collapse essentially proceeds along an adiabat of constant, low specific entropy. To lowest approximation the total entropy per baryon as a function of Y_e, ρ and T is [cf. Eqs. (18.6.1), (18.6.3), and Exercise 18.12]

$$\frac{s}{k} \approx \frac{1}{\bar{A}} \left\{ \frac{5}{2} + \ln \left[\frac{\bar{A} m_u}{\rho} \left(\frac{\bar{A} m_u kT}{2\pi \hbar^2} \right)^{3/2} \right] \right\} + \frac{\pi^2 kT Y_e}{\mu_e} + \frac{\pi^2 kT}{2 E'_{F, \, nuc}}$$

$$\approx \text{constant}, \qquad (18.6.19)$$

where we ignore the contribution from free baryons and α-particles and assume that all nucleons are bound in a nucleus of mean atomic weight $\bar{A}$.

Consider T vs. ρ during collapse. At the onset of collapse the first two terms in Eq. (18.6.19) dominate the entropy. Since the first term depends only logarithmically on ρ and T, we expect that, initially at least, the second term will control the T-ρ trajectory along an adiabat and, hence, will be roughly constant. Accordingly, recalling Eq. (18.4.7), we expect that the relation $T \propto \rho^{1/3}$ will apply during collapse, assuming $Y_e \sim$ constant. We thus have

$$T \approx T_{c,i}\left(\frac{\rho}{\rho_{c,i}}\right)^{1/3} \quad (s \approx \text{constant}; \ Y_e \approx \text{constant}), \qquad (18.6.20)$$

where $T_{c,i}$ and $\rho_{c,i}$ are given in Eq. (18.2.7). If the center of the core reaches nuclear density, $\rho_{\text{nuc}} \approx 2.8 \times 10^{14}$ g cm^{-3}, near the end of collapse, the temperature there will reach the value $T_{\text{nuc}} \approx 29$ MeV according to eq. (18.6.20). More careful consideration shows that T_{nuc} is likely to be somewhat smaller ($T_{\text{nuc}} \sim 10$ MeV) and that Eq. (18.6.20) predicts slightly too steep a rise in T (see Fig. 18.1). Modification of Eq. (18.6.20) is necessary, in part, due to the growing contribution of nuclear excited states to the entropy, as expressed by the third term in Eq. (18.6.19).

Computer Exercise   18.13   Use Eq. (18.6.19) with $\bar{A} = 56$ and $Y_e = 0.464$ to estimate T vs. ρ along constant entropy adiabats. Plot your results on a graph similar to Figure 18.1 for $s/k = 0.5$, 1.0, 1.5, 2, and 3. For each adiabat, determine T_{nuc} and the relative contribution of each term in Eq. (18.6.19) to the entropy at $\rho = \rho_{\text{nuc}}$.

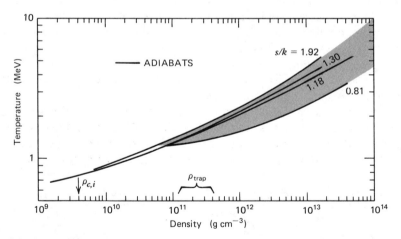

Figure 18.1   Range of low-entropy adiabats found during stellar core collapse. The collapse trajectories were obtained from hydrodynamical calculations of core collapse employing the hot, dense matter equation of state of Lamb et al. (1978, 1981). [After Van Riper and Lattimer (1981). Reprinted courtesy of the authors and *The Astrophysical Journal*, published by The University of Chicago Press; © 1981 The American Astronomical Society.]

Detailed calculations of the equation of state for hot, dense matter[26] assume that matter consists of a mixture of heavy nuclei, α-particles, neutrons, protons, electrons, and electron neutrinos. To the extent that s and Y_e can be treated as constants, the local equation of state calculation decouples from the global hydrodynamic calculation. It is this decoupling, brought about by neutrino trapping, that has made progress possible recently in developing a realistic hot equation of state for collapse. On the other hand, a coupled hydrodynamic-neutrino transport calculation is necessary to determine *changes* in s and Y_e due, for example, to neutronization, neutrino escape, shock dissipation, and so on. Of course, such a calculation is especially necessary to assess whether or not the gravitational potential energy released during the collapse can actually power a supernova explosion.

In existing hot equation of state calculations, the heavy nuclei are often characterized by a single species (A, Z) at each density. The mass-energy of the species is estimated from a semiempirical mass formula of the type considered in Chapter 2 for matter at $T = 0$. Such a formula tends to smear out pairing and shell structure effects, but this in fact may be a reasonable approximation at the finite temperatures encountered during core collapse ($\sim$ 1–10 MeV). Since temperatures remain low during collapse compared with the nucleon Fermi energy in nuclei ($\sim$ 40 MeV, cf. Exercise 18.12), finite temperature corrections to these nuclear energy expressions are often neglected in the determination of the equilibrium composition.

The density of each type of nucleon outside nuclei is determined from standard Fermi integrals (cf. Chapter 2), subject to the constraint that in nuclear equilibrium the chemical potentials inside and outside nuclei are equal. Finally, the conditions of charge neutrality and baryon conservation are employed to calculate the number of nucleons, nuclei and electrons [cf. Eqs. (18.3.17) and (18.3.18)].

Suppose we adopt the BBP "compressible liquid drop" model of the nuclei. Then the important quantity

$$\hat{\mu} \equiv \mu_n - \mu_p \tag{18.6.21}$$

appearing, for example, in Eq. (18.6.15) is given directly by equating the right-hand sides of Eqs. (8.2.8) and (8.2.11), yielding

$$\hat{\mu} = -\frac{\partial}{\partial Z}(W_N + W_L)_{A, n_N, V_N, n_n}. \tag{18.6.22}$$

BBAL find that the derivative in Eq. (18.6.22) may be approximated by

$$\hat{\mu}\ (\text{MeV}) \approx 207\left(0.45 - \frac{Z}{A}\right)\left(1.32 - \frac{Z}{A}\right) + Q, \qquad Q \equiv (m_n - m_p)c^2. \tag{18.6.23}$$

---

[26]For example, Lamb et al. (1978, 1981); Bethe et al. (1979) (BBAL); Epstein and Pethick (1981).

Ignoring neutron drip, we can replace Z/A by Y_e in the above equation. BBAL also find, using the BBP model, that the most likely atomic weight is given in the same limit by

$$\bar{A} \approx 194(1 - Y_e)^2 (1 - 0.236\rho_{12}^{1/3})^{-1}, \qquad (18.6.24)$$

and that $\sim 90\%$ of the nucleons remain bound into nuclei even up to nuclear densities.

Exercise   18.14   For electron capture onto heavy nuclei, the resulting neutrinos have an average energy

$$\langle E_\nu \rangle_{escape} = \tfrac{3}{5}(\mu_e - \hat{\mu} - \Delta)$$

(BBAL). For capture onto free protons, the neutrinos are more energetic,

$$\langle E_\nu \rangle_{escape} = \tfrac{5}{6}\mu_e$$

[cf. Eq. (18.4.19)]. Use the above to estimate numerically the total entropy change per baryon before trapping due to electron capture onto (a) heavy nuclei and (b) free protons, assuming that either (a) or (b) occurs exclusively. Assume that most captures occur just prior to trapping, at which point

$$\Delta Y_e \approx 0.35 - 0.42 \approx -0.07, \qquad kT \sim 1.5 \text{ MeV}, \qquad \mu_\nu \approx 0 \approx \dot{Y}_\nu, \quad \Delta \sim 3 \text{ MeV}.$$

Hint: Use Eqs. (18.4.7), (18.5.10), (18.6.9), (18.6.15), and (18.6.23).
Answer: (a) $T\Delta s = -\Delta Y_e[\tfrac{2}{3}(\mu_e - \hat{\mu}) + \tfrac{3}{5}\Delta] > 0$,
                  (b) $T\Delta s = -\Delta Y_e[\mu_e - \hat{\mu} - \tfrac{5}{6}\mu_e] < 0$.

The key finding that heavy nuclei persist at high densities is a consequence of the low entropy of collapse and of neutrino trapping. Trapping inhibits neutronization, since the presence of a Fermi sea of neutrinos drives reactions like

$$e^- + (Z, A) \rightleftarrows (Z - 1, A) + \nu \qquad (18.6.25)$$

toward the left. As a result, the electron number Y_e remains high during hot collapse, which, by charge neutrality, guarantees the presence of a high proton number. Since proton drip at low entropy is not very favorable energetically, the protons all remain in nuclei. But as the equilibrium ratio Z/A in nuclei is $\lesssim 0.3$–0.4 at high density, the nuclei must be very heavy to accommodate the required proton number.

Exercise   18.15   Evaluate $\bar{A}$ at ρ_{trap} from Eq. (18.6.24).

The resulting equation of state is rather simple. At all densities up to ρ_{nuc}, the pressure is dominated by relativistic degenerate electrons. Thus, the adiabatic index Γ during infall remains close to $\frac{4}{3}$, the slight decrease from $\frac{4}{3}$ due chiefly to electron capture. At the onset of collapse, the electron pressure is close to the hydrostatic equilibrium value throughout the core,

$$P_i \approx P_{eq} \propto \left(Y_{e,i}\rho \right)^{4/3}, \tag{18.6.26}$$

and the core has the density profile of an $n \approx 3$ polytrope. Following neutronization Y_e falls from ~ 0.42 to $\sim 0.34\text{–}0.36$, so the pressure at trapping falls below the equilibrium value according to

$$P_f \approx \left(\frac{Y_{e,f}}{Y_{e,i}} \right)^{4/3} P_{eq} \sim (0.7\text{–}0.8) P_{eq}. \tag{18.6.27}$$

Now recall from Chapter 6 that a $\Gamma = \frac{4}{3}$ configuration is neutrally stable to a homologous radial perturbation. The collapse thus proceeds homologously (as in the analogous situation for supermassive stars described in Chapter 17), and Eq. (18.6.26) is maintained (roughly) up to nuclear densities. At that point the equation of state stiffens and Γ increases sharply. At and above ρ_{nuc}, nuclei merge. Once this occurs, the matter may be treated as a uniform medium of individual neutrons and protons. Thus squeezed out of nuclei, these nucleons may be regarded as a degenerate Fermi gas with interactions. The results of $T = 0$ matter calculations at these densities may then be applied, with all of their associated uncertainties (cf. Chapter 8).

Soon after nuclear densities are reached at the center, the core decelerates rapidly and bounces in response to the increased nuclear matter pressure. In the next section we discuss the implications of the *core bounce*, which is currently believed to be at the heart of a supernova explosion.

18.7   Homologous Core Collapse, Bounce, and Then What?

Goldreich and Weber (1980) have shown that if the adiabatic index of a collapsing Newtonian gas sphere satisfies $\Gamma = \frac{4}{3}$, then the collapse of the inner fraction of the configuration will be *homologous*. That is, the position and velocity of a given mass point in this homologous "inner" core will vary as

$$r(t) = \alpha(t)r_0, \qquad \frac{\dot{r}}{r} = \frac{\dot{\alpha}}{\alpha}, \tag{18.7.1}$$

where r_0 is the initial position. The density, meanwhile, will evolve self-similarly according to

$$\rho(r(t), t) = \alpha^{-3}\rho_0(r_0), \tag{18.7.2}$$

thereby preserving the density profile during the collapse. Goldreich and Weber find that the mass of the homologous inner core can be determined from

$$M_{hc} = 1.0449d^{3/2}M, \tag{18.7.3}$$

where $d \equiv P/P_{eq} < 0.971$ is the pressure deficit and M is the mass of the initial equilibrium $n = 3$ polytrope[27].

We have already encountered homologous collapse behavior in Chapter 17, where we examined the collapse to a black hole of a supermassive star, with $\Gamma \approx \frac{4}{3}$ and $d \simeq 1$. We expect that the collapse of a stellar core will also be homologous in the inner regions, since the dominant contribution to the pressure is relativistic electron degeneracy pressure for which $\Gamma \approx \frac{4}{3}$. This expectation is corroborated by detailed hydrodynamical calculations of core collapse,[28] which suggest that Eq. (18.7.3) is satisfied to better than 20%. Thus Eqs. (18.2.5) and (18.2.7) imply $M \sim M_{Ch} \sim 5.83Y_{e,i}^2 M_\odot \sim 1M_\odot$ and so, by Eqs. (18.6.27) and (18.7.3) we estimate that the homologous inner core has a mass in the range[29]

$$M_{hc} \sim (0.6-0.8)M_\odot. \tag{18.7.4}$$

Once the homologous inner core reaches $\rho_{nuc} \sim 2.8 \times 10^{14}$ g cm^{-3}, the pressure increases rapidly as the nucleon component causes the equation of state to stiffen, and so $\Gamma > \frac{4}{3}$. At several times ρ_{nuc}, this pressure is sufficient to halt the collapse, causing the homologous core to "bounce" and then rebound somewhat, before it eventually settles down to hydrostatic equilibrium. The "outer" core, in the meantime, continues to fall towards the center at supersonic velocities. As a result, the rebounding inner core, acting as a piston, drives a *shock wave* into the infalling outer core.

The characteristic initial total energy imparted to the shock is[30] a few times 10^{51} ergs, corresponding to initial shock velocities $u_s \gtrsim 5 \times 10^9$ cm s^{-1}. This energy is more than adequate to disrupt the overlying massive star provided it can be deposited in the stellar mantle surrounding the iron core. In such a case, the shock would reverse the infall, causing an explosive outward motion as it propagated through the star. A supernova explosion would result, and a nascent neutron star would be left behind. In reality, however, energy is constantly being drained from the shock via neutrino dissipation and nuclear dissociation. It is a matter of current controversy whether or not this energy drain is adequately compensated by heat input from the kinetic energy of infalling matter and by neutrino diffusion up to the shock front.[31]

---

[27]See Yahil and Lattimer (1982) for further discussion.

[28]Arnett (1977b); Wilson (1980); Van Riper and Lattimer (1981).

[29]Typically, the initial core mass is somewhat larger, $M \sim 1.4M_\odot$, due to thermal enhancement of cold electron degeneracy pressure; Brown et al. (1982).

[30]Wilson (1980); Van Riper (1982).

[31]Brown et al. (1982); Lattimer (1981); Van Riper (1982).

Exercise 18.16 Estimate the binding energy of a $\sim 10 M_\odot$ star, with radius $R \sim 5 R_\odot$, surrounding a collapsing degenerate core, and compare with the initial shock energy.

Answer: $E_B \sim GM^2/R \sim 10^{50}$ erg following core bounce.

Exercise 18.17 Estimate the characteristic temperature of matter behind the shock front. What happens to the nuclei after passing through a shock?

Answer: $\frac{3}{2}kT \lesssim \frac{1}{2}m_B u_s^2 \sim 13$ MeV $>$ nuclear binding energy.

The outward expulsion of the outer layers of a massive star via the hydrodynamic "bounce" shock induced by the inner core remains the most promising mechanism for generating a supernova explosion from gravitational collapse. Curiously, this scenario is remarkably similar to one of the earliest supernova models, originally proposed by Colgate and Johnson (1960), which suggested that the infalling core could bounce sufficiently rapidly at high density (due to nuclear degeneracy) to generate a shock wave that deposits energy in the outlying matter.

Many different supernova models have been proposed over the years to explain how the gravitational energy released during collapse can be effectively transmitted to the outer layers of the star and expel them. Colgate and White (1966) suggested a neutrino "energy deposition" model in which neutrino-antineutrino pairs generated in the hot matter behind the shock deposit their energy in the overlying regions of lower density. The resulting heating then provides sufficient thermal pressure to blow off the outer parts of the star, leaving behind a warm neutron star. Unfortunately, detailed hydrodynamical calculations incorporating neutrino transport[32] showed that this mechanism was inefficient due to excessively high neutrino opacity; only black holes could form this way, without explosions.

Schramm and Arnett (1975) subsequently suggested that neutrinos flowing out of the core could transmit their outward momentum to the overlying matter and thereby cause an explosion. This neutrino "momentum deposition" model proved unlikely once careful numerical calculations incorporating weak neutral currents were performed.[33] The reason was that neutral currents lead to large neutrino opacities and low fluxes [cf. Eq. (18.5.20)]. Indeed, momentum deposition can be effective only if the neutrino luminosity exceeds the critical "Eddington neutrino luminosity," the value at which the outward force due to neutrino momentum deposition equals the inward force of gravity. By analogy with Eq. (13.7.6), which gives the Eddington photon luminosity, we may write

$$L_{\text{Edd},\nu} \approx \frac{4\pi GMc}{\kappa_\nu}, \qquad (18.7.5)$$

---

[32]Arnett (1967); Wilson (1971).
[33]For example, Bruenn (1975); Wilson et al. (1975).

where κ_ν is the dominant neutrino opacity and where we neglect any forward-back asymmetries in the scattering cross-section angles. Assuming that the opacity in the overlying mantle is dominated by coherent scattering off heavy nuclei and that the neutrinos have energies comparable to $3kT \lesssim 25$ MeV behind the outward propagating bounce shock [cf. Eq. (18.5.22) and Exercise 18.17], we find using Eq. (18.5.6)

$$\kappa_\nu \approx \frac{\sigma_A^{coh}}{A m_u}$$

$$\approx 2.2 \times 10^{-17} \text{ cm}^2 \text{ g}^{-1}, \tag{18.7.6}$$

where we set $Z \approx 26$, $A \approx 56$, and $E_\nu = 25$ MeV. Inserting Eq. (18.7.6) into Eq. (18.7.5) gives

$$L_{\text{Edd}, \nu} \approx 2 \times 10^{54} \text{ erg s}^{-1} > L_\nu, \tag{18.7.7}$$

where L_ν is estimated in Eq. (18.5.20). Inequality (18.7.7) demonstrates the ineffectiveness of momentum deposition as the sole mechanism responsible for ejection.

Although the bounce shock mechanism appears to be the most promising means of generating an explosion, recent detailed hydrodynamical-transport calculations have not shown it to be effective.[34] This situation might change, however, once needed refinements are incorporated in the numerical simulation of post-bounce behavior. Alternatively, other, more exotic, mechanisms may (single or jointly) play a role in driving the explosion. These may include convection,[35] Rayleigh–Taylor instabilities,[36] or departures from spherical symmetry induced by, for example, rotation or magnetic fields.[37]

One fact remains invariant: several neutron stars—for example, the Crab and Vela pulsars—are now sitting in supernova remnants! Undoubtedly gravitational collapse *can* generate a supernova explosion and produce a neutron star. The precise mechanism(s) will eventually be identified.

Exercise 18.18 Perform a literature search for the last 12 months, looking at abstracts dealing with gravitational core collapse and supernova explosions. What is the consensus, if any, regarding the likely mechanism for the explosion and the creation of a neutron star? Do detailed calculations support this opinion? Are you convinced?

---

[34]See, for example, Lattimer (1981) for a review.
[35]Epstein (1979).
[36]Colgate (1978); Livio et al. (1980); Smarr et al. (1981).
[37]LeBlanc and Wilson (1970); Müller and Hillebrandt (1979).

Appendix A
Astronomical Background

A.1 Parsecs and Magnitudes[1]

The standard astronomical unit of length is the *parsec*, which is the distance at which one second of arc is subtended by one Astronomical Unit (= mean Earth–Sun distance). The conversion factors are

$$1 \text{ pc} = 3.086 \times 10^{18} \text{ cm}$$

$$= 3.26 \text{ light years}. \tag{A.1}$$

In extragalactic astronomy the more convenient number is the megaparsec,

$$1 \text{ Mpc} = 3.086 \times 10^{24} \text{ cm}. \tag{A.2}$$

The observed brightness in the sky of a star or galaxy is expressed on a logarithmic scale. Let f_1 be the incident energy flux from an object in ergs cm^{-2} s^{-1} in some chosen wavelength band, and let f_2 be the energy flux from a second object in the same band. Then the *apparent* magnitudes of the two objects are related by the equation

$$m_2 - m_1 = 2.5 \log\left(\frac{f_1}{f_2}\right), \tag{A.3}$$

where the logarithm is to base 10. It will be noted that the dimmer the object the greater the apparent magnitude.

The incident flux f must be corrected for the effect of absorption and scattering within the Galaxy. Whenever we refer to m, it is understood that this correction has been carried out.

---

[1] The discussion in Section A.1 closely follows Peebles (1971), p. 279.

The *absolute* magnitude M of an object is a measure of its intrinsic luminosity—that is, power, radiated in a chosen wavelength band, on the same logarithmic scale as Eq. (A.3). The normalization of M is such that if the object were placed at the fiducial distance 10 pc from us its apparent magnitude would be equal to its absolute magnitude M. If the true distance is D, then by the inverse square law the incident flux is smaller than it would be for the fiducial distance by the factor $(D/10 \text{ pc})^2$, so by Eq. (A.3) its apparent magnitude at distance D satisfies

$$m - M = 5 \log\left(\frac{D}{10 \text{ pc}}\right). \tag{A.4}$$

The quantity $m - M$ is called the *distance modulus*.

Although it is difficult to measure a star's spectrum in detail, broad-band photometry, using a few filters of various colors, yields sufficient information to determine the approximate stellar surface temperature. The blue and visual (yellow) apparent magnitudes of a star are denoted by B and V, respectively, and their difference, $B - V$, is defined as the *color index* of the star. B is centered at $\lambda = 4400$ Å, while V is centered at $\lambda = 5500$ Å. A one-to-one correspondence exists between the color index of a star and the location of the spectral peak of its continuum emission, from which a *color temperature* may be defined. The color temperature is the temperature of a black body having the same relative intensity distribution, or color, as the star. Since stars do not radiate as black bodies, the color temperature varies with wavelength.

If the detector could respond to the entire radiant spectrum, the absolute magnitude measured would be the absolute *bolometric* magnitude M_b. Thus, the absolute visual magnitude is converted to the absolute bolometric magnitude of the star by correcting for the amount of radiant energy lying outside that portion received by the detecting instrument. If the *bolometric correction* is designated by BC (a negative quantity) and the absolute visual magnitude by M_V, then the absolute bolometric magnitude is defined by

$$M_b = M_V + BC. \tag{A.5}$$

A convenient reference point is that the absolute magnitude of the Sun is $M_V = 4.79$. Therefore, the luminosity of an object with absolute magnitude M_V in the visual band is

$$L(M_V) = 10^{0.4(4.79 - M_V)} L_\odot, \tag{A.6}$$

where

$$L_\odot = 3.90 \pm 0.04 \times 10^{33} \text{ erg s}^{-1}. \tag{A.7}$$

Since the bolometric correction for the sun is $BC = -0.07$, we may write for the *total* luminosity of any object

$$L = 10^{0.4(4.72 - M_b)} L_\odot. \qquad (A.8)$$

Note that the most luminous stars have the smallest magnitudes.

A.2 Stellar Types and the Hertzsprung–Russell Diagram

We will give here an elementary discussion of the Hertzsprung–Russell diagram, an essential tool in understanding the types and ages of stars. A more detailed discussion of stellar structure and evolution can be found in many standard texts.[2]

An early result of color photometry was the observed correlation between a star's color index and the strength of specific absorption lines. This observation led to a classification of stars into spectral types. In addition to other distinctive properties, each spectral type corresponds to a certain range in surface color temperature. The major spectral types and their temperature ranges are shown in Table A.1. The sun is a G star.

White dwarfs have colors ranging from O to M. From their spectra, they can be divided into two sequences, each of which spans the full range of colors. These are the "DA" sequence, which shows hydrogen lines, and the non-DA types (generally with helium-rich atmospheres). The basic spectral types, as defined by Greenstein (1960), are given in Table A.2. The letter "D" stands for "degenerate." More hybrid spectra besides DA, F are now known, and presumably a new classification scheme will soon be necessary.

In the early part of this century the Danish astronomer E. Hertzsprung, and slightly later the American astronomer H. N. Russell, made the remarkable discovery that stars populate only certain portions of the color–luminosity diagram. This finding is illustrated in Figure A.1.

Any graph that measures a quantity related to luminosity (e.g., luminosity, bolometric magnitude, visual magnitude, etc.) versus a quantity related to color (e.g., color index, color temperature, spectral type, etc.) is called a Hertzsprung–Russell diagram, or simply an H-R diagram.

When a large sample of all observed stars is plotted in an H-R diagram (Fig. A.1), it is found that more than 80% of the stars fall in a narrow diagonal band called the main sequence. The next largest class of stars—white dwarfs—represent about 10% of all stars. The primary significance of the H-R diagram is that it contains data on the evolutionary sequences of stars. From theoretical calculations of stellar evolution, it is found that stars move around in the H-R diagram

[2]See, for example, Clayton (1968).

Table A.1
The Main Sequence[a]

Spectral Type	Absolute Visual Magnitude M_v	Color Index $B - V$	Bolometric Correction	Effective Surface Temperature $T_e(K)$	Color Temperature $T_c(K)$	Absolute Bolometric Magnitude M_b	Logarithm of Luminosity $\log(L/L_\odot)$
O5	−6.0	−0.45	−4.6	35,000	70,000	−10.6	6.13
B0	−3.7	−0.31	−3.0	21,000	38,000	−6.7	4.56
B5	−0.9	−0.17	−1.6	13,500	23,000	−2.5	2.88
A0	0.7	0.00	−0.68	9,700	15,400	0.0	1.88
A5	2.0	0.16	−0.30	8,100	11,100	1.7	1.20
F0	2.8	0.30	−0.10	7,200	9,000	2.7	0.80
F5	3.8	0.45	−0.00	6,500	7,600	3.8	0.37
G0	4.6	0.57	−0.03	6,000	6,700	4.6	0.05
G5	5.2	0.70	−0.10	5,400	6,000	5.1	−0.15
K0	6.0	0.84	−0.20	4,700	5,400	5.8	−0.43
K5	7.4	1.11	−0.58	4,000	4,500	6.8	−0.83
M0	8.9	1.39	−1.20	3,300	3,800	7.6	−1.15
M5	12.0	1.61	−2.1	2,600	3,000	9.8	−2.03

[a]From Allen (1963). © C. W. Allen 1973. *Astrophysical Quantities* Third Edition, Published by the Athlone Press.

as they evolve and spend most of their time in the most populated areas of the diagram. The "track" of an isolated star is unique, given the star's initial chemical composition and mass. Comparison of the evolutionary calculations with the observed H-R diagram thus provides information on the initial parameters of stars, the duration of time they spend in each evolutionary phase, and their ages.

From theoretical calculations, the following simplified interpretation of the H-R diagram emerges: *Main sequence* stars are those stars that primarily convert

Table A.2
Spectral Classification of White Dwarfs

Name	Spectral Features
DA	H present, no He
DB	He I strong, no H
DO	He II strong, He I and/or H present
DF	Ca II present, no H
DG	Ca II, Fe I present, no H
DA, F	H present, weak Ca II, no He
DC	Continuous spectrum

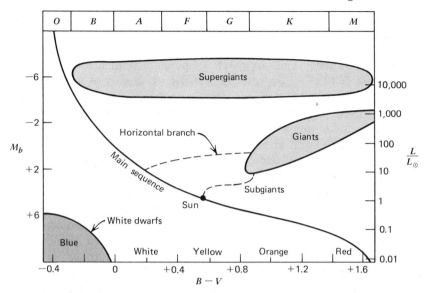

Figure A.1 A schematic representation of the heavily populated areas in the H-R diagram. A high percentage of stars lie near the main sequence. The next most populous groups are the white dwarfs and the giants. The subgiant and horizontal branches are conspicuous in those collections of stars having large numbers of giants, such as globular clusters. [From Clayton (1968).]

hydrogen into helium in their interiors by nuclear fusion; this activity constitutes the first and longest phase of a star's active life. The main sequence does not represent an evolutionary track. Rather, the position on the main sequence of a newborn star is determined by its mass and chemical composition. For a given composition more massive stars are hotter and brighter. The lifetime of a star on the main sequence is a steeply decreasing function of its initial mass M, corresponding to a steeply increasing dependence of luminosity on mass. The main sequence lifetime of a star is given approximately by Eq. (1.3.9). Thus, for example, stars with a mass near that of the sun remain on the main sequence for ~ 10 billion years, while more massive stars live only a few million years on the main sequence.

Typically, 70–80% of a star's mass is initially in the form of hydrogen, 20–30% is in the form of helium, and 0.1–3% is in the form of heavier elements. When 10–20% of the hydrogen in the central active nuclear burning region has been exhausted, nuclear energy generation ceases in the core but continues in a thin shell that moves outward toward the stellar surface. The core, no longer pressure supported, contracts somewhat under its self-gravitation and heats up. The liberated gravitational energy serves to expand the outer layers of the star while maintaining the luminosity. During this phase the star "evolves off the main sequence" and follows the *subgiant branch* of the H-R diagram upwards and to the right into the *red giant region* (see Fig. A.1). Stars redden tremendously during this evolutionary period because of their increasing surface area (the radius can increase by a factor of ~ 1000).

During the relatively rapid approach to the giant branch of a star with mass $M \lesssim 1 M_\odot$, the central region becomes sufficiently dense so that the electrons become degenerate (typical core densities and temperatures are 10^4–10^5 g cm^{-3} and $(3$–$7) \times 10^7$ K, respectively). When the degenerate core mass reaches about $0.5 M_\odot$, core temperatures are sufficiently high to ignite the triple alpha process, which converts helium into carbon in a rapid "helium flash." The released energy lifts the degeneracy of the electrons, the central region expands, the envelope contracts, and the star moves rather abruptly (e.g., $\sim 10^5$ yr for a $0.65 M_\odot$ star) to a position on the *horizontal branch*. The horizontal branch is not entirely an evolutionary track. The position of a star on the horizontal branch is again determined by the helium core and hydrogen envelope mass and the chemical composition of the precursor star at the upper tip of the red giant region. The situation is probably complicated by (ill-understood) mass loss processes experienced by a star during its evolution on the giant branch. A massive star with $M \gtrsim M_\odot$ does not become degenerate on the subgiant branch and ignites helium nonexplosively.

The subsequent evolution of stars on the horizontal branch is not entirely understood. A star may evolve back to the red giant tip of the H-R diagram several times, where it can ignite other sources of nuclear energy. Ultimately, stars exhaust their nuclear fuel and eventually cease to radiate. If the star is sufficiently massive it may explode as a supernova and/or collapse to form a black hole. Sufficiently small stellar remnants can contract to form either a white dwarf ($M \lesssim 1.4 M_\odot$, radius $R \sim 10^9$ cm), supported against collapse by electron degeneracy pressure and shown in the lower left region of the H-R diagram, or a neutron star ($M \lesssim 3 M_\odot$, radius $R \sim 10^6$ cm), supported by neutron degeneracy pressure and nuclear repulsive forces.

Appendix B
Proof That $V_2 < 0$ Implies Instability

The proof we give is based on the energy principle of Laval, Mercier, and Pellat (1965), and is valid for any symmetric, time-independent operator L_{ij}. Its strength is that it does not assume any properties (such as completeness) of the normal modes.

Suppose $V_2(t) < 0$ for some perturbation $\tilde{\xi}^i(\mathbf{x}, t)$ at a particular instant, t. Choose this t to be $t = 0$, say. Write $\tilde{\xi}^i$ at this instant as $\eta^i(\mathbf{x}, 0)$. Then by Eq. (6.6.8),

$$V_2(0) = -\frac{1}{2} \int \eta^i L_{ij} \eta^j \, d^3x$$

$$= -\frac{1}{2} \alpha^2 \int \rho (\eta^i)^2 \, d^3x \tag{B.1}$$

for an appropriate choice of $\alpha > 0$.

Now let us construct a *new* perturbation $\xi^i(\mathbf{x}, t)$ for which the kinetic energy, $T_2(t)$ grows without bound. (If we succeed, we will have explicitly demonstrated that the system is unstable.)

Consider the initial data

$$\xi^i(x, 0) = \eta(\mathbf{x}, 0), \qquad \partial_t \xi^i(\mathbf{x}, 0) = \alpha \eta^i(\mathbf{x}, 0), \tag{B.2}$$

which uniquely specifies the new perturbation. For this ξ^i, Eq. (6.6.7) gives

$$T_2(0) = \frac{1}{2} \alpha^2 \int \rho (\eta^i)^2 \, d^3x, \tag{B.3}$$

and so

$$E_2 = T_2 + V_2 = 0. \tag{B.4}$$

549

Since E_2 is conserved, $T_2 = -V_2$ for all t. Thus from Exercise 6.14,

$$\frac{1}{2}\frac{d^2I}{dt^2} = 2T_2.$$ (B.5)

Now the Schwarz inequality implies

$$\left[\int \rho(\partial_t\xi^i)\xi_i\,d^3x\right]^2 \leqslant \int \rho(\partial_t\xi^i)^2\,d^3x \int \rho(\xi^i)^2\,d^3x,$$ (B.6)

or

$$\left(\frac{dI}{dt}\right)^2 \leqslant 4IT_2.$$ (B.7)

Combining Eqs. (B.5) and (B.7) gives

$$\frac{d^2I}{dt^2} \geqslant \frac{1}{I}\left(\frac{dI}{dt}\right)^2,$$ (B.8)

or

$$\frac{d}{dt}\left(\frac{1}{I}\frac{dI}{dt}\right) \geqslant 0.$$ (B.9)

But from Eq. (B.2),

$$\left.\frac{1}{I}\frac{dI}{dt}\right|_{t=0} = 2\alpha,$$ (B.10)

and so

$$\frac{1}{I}\frac{dI}{dt} \geqslant 2\alpha \qquad \text{for all } t \geqslant 0.$$ (B.11)

Integrating, we get

$$I(t) \geqslant I(0)e^{2\alpha t}$$ (B.12)

and hence by Eq. (B.5)

$$T_2(t) \geqslant \alpha^2 I(0)e^{2\alpha t}.$$ (B.13)

Thus the kinetic energy grows without bound and the system is unstable.

Appendix C
Calculation of the Integral in Eq. (8.4.6)

As mentioned in the main text, we need only consider the case $b < 2R$. The integral can be evaluated by a tedious but straightforward geometrical evaluation of the region of integration. We present an alternative method, based on the representation of the step function as

$$H(x) = \frac{1}{2\pi i} \int_{-\infty}^{\infty} \frac{e^{ikx}}{k} dk, \qquad (C.1)$$

where the contour is indented infinitesimally below the origin near $k = 0$. Equation (C.1) can be verified by closing the contour in the upper half-plane for $x > 0$ and in the lower half-plane for $x < 0$.

Thus

$$\Omega^2 P(b, R) \equiv I = \frac{1}{2\pi i} \int_{-\infty}^{\infty} \frac{e^{ikb}}{k} dk \int \int d\mathbf{r}_1 \, d\mathbf{r}_2 e^{-ik|\mathbf{r}_1 - \mathbf{r}_2|}. \qquad (C.2)$$

We now use the expansion[1]

$$e^{-ik|\mathbf{r}_1 - \mathbf{r}_2|} = \frac{d}{dk} 4\pi k \sum_{l,\,m} j_l(kr_<) h_l^{(1)*}(kr_>) Y_{lm}^*(\Omega_1) Y_{lm}(\Omega_2). \qquad (C.3)$$

Only the $l = m = 0$ terms contribute to the integral (C.2), and so

$$I = \frac{(4\pi)^2}{2\pi i} \int_{-\infty}^{\infty} dk \, \frac{e^{ikb}}{k} \frac{d}{dk} k \int_0^R r_1^2 \, dr_1 \int_0^R r_2^2 \, dr_2 \, j_0(kr_<) h_0^{(1)*}(kr_>). \qquad (C.4)$$

Now write

$$\int_0^R dr_2 = \int_0^{r_1} dr_2 + \int_{r_1}^R dr_2. \qquad (C.5)$$

---

[1] This expansion follows from taking d/dk of the complex conjugate of Eq. (16.22) of Jackson (1975).

In the first integral, $r_< = r_2$ and $r_> = r_1$, while in the second integral these definitions are interchanged. Using the results

$$j_0(x) = \frac{\sin x}{x}, \qquad h_0^{(1)*}(x) = \frac{ie^{-ix}}{x}, \qquad \text{(C.6)}$$

we find

$$I = \frac{(4\pi)^2}{2\pi i} \int_{-\infty}^{\infty} dk \, \frac{e^{ikb}}{k} \frac{d}{dk} \int_0^R dr_1 \left[e^{-ikR} \left(\frac{i}{k^3} - \frac{R}{k^2} \right) r_1 \sin kr_1 - \frac{ir_1^2}{k^2} \right]$$

$$= \frac{(4\pi)^2}{2\pi i} \int_{-\infty}^{\infty} dk \left[e^{ikb} \left(\frac{2iR^3}{3k^4} - \frac{3R^2}{2k^5} - \frac{5}{2k^7} \right) \right.$$

$$\left. + e^{ik(b-2R)} \left(-\frac{iR^3}{k^4} - \frac{R^2}{2k^5} + \frac{5iR}{k^6} + \frac{5}{2k^7} \right) \right]. \qquad \text{(C.7)}$$

Since $b < 2R$, the term proportional to $\exp[ik(b - 2R)]$ gives no contribution, as can be seen by closing the contour in the lower half-plane. By the residue theorem,

$$\frac{1}{2\pi i} \int_{-\infty}^{\infty} dk \, \frac{e^{ikb}}{k^n} = \frac{(ib)^{n-1}}{(n-1)!}, \qquad \text{(C.8)}$$

and so

$$I = \left(\frac{4\pi R^3}{3} \right)^2 \frac{b^3}{R^3} \left(1 - \frac{9}{16} \frac{b}{R} + \frac{b^3}{32R^3} \right), \qquad \text{(C.9)}$$

which is the result quoted in the text.

The reader can verify that if $b > 2R$, the contribution of the second term in Eq. (C.7) added to the result (C.9) gives

$$I_{b>2R} = \left(\frac{4\pi R^3}{3} \right)^2. \qquad \text{(C.10)}$$

This result is trivially obtained by noting that for $b > 2R$, the step function in the definition of I is always unity.

Appendix D
Scalar and Vector Field Theories

The field theory the reader is probably most familiar with is electromagnetism, a vector theory since the basic field variable in the Lagrangian is the vector potential, A_α. The field equations follow from an action principle,

$$\delta S = 0, \qquad (D.1)$$

where the action is

$$S = \int \mathcal{L}\, d^4x, \qquad (D.2)$$

and $\mathcal{L}$ is the Lagrangian density. The quantity $\mathcal{L}$ must be a Lorentz scalar so that S is Lorentz invariant.

We write

$$S = S_{\text{em}} + S_{\text{p}} + S_{\text{int}}. \qquad (D.3)$$

Here S_{em} is the action for the free electromagnetic field, which depends only on A_α, S_{p} is the action for the particles, and S_{int} is the interaction term.

The free-field term is

$$\mathcal{L}_{\text{em}} = -\frac{1}{16\pi} F_{\alpha\beta} F^{\alpha\beta}, \qquad (D.4)$$

where the electromagnetic field tensor is defined to be

$$F_{\alpha\beta} \equiv A_{\beta,\,\alpha} - A_{\alpha,\,\beta}, \qquad (D.5)$$

and a comma denotes a partial derivative. The definition (D.5) automatically guarantees that half of Maxwell's equations are satisfied, namely,

$$F_{\alpha\beta,\gamma} + F_{\gamma\alpha,\beta} + F_{\beta\gamma,\alpha} = 0. \tag{D.6}$$

The remaining Maxwell equations follow from the variational principle. The Euler–Lagrange equations for the free field are

$$\frac{\delta \mathcal{L}}{\delta A_\alpha} \equiv \frac{\partial \mathcal{L}}{\partial A_\alpha} - \frac{\partial}{\partial x^\beta}\left(\frac{\partial \mathcal{L}}{\partial A_{\alpha,\beta}}\right) = 0. \tag{D.7}$$

The first term in Eq. (D.7) vanishes for $\mathcal{L}_{em}$, while the second term gives

$$-\frac{1}{4\pi} F^{\alpha\beta}{}_{,\beta} = 0. \tag{D.8}$$

The Lagrangian for a free particle was treated in Section 5.2. It is slightly more convenient to use the square of that Lagrangian, so that

$$S_p = \int L_p \, d\tau \tag{D.9}$$

$$= \int \mathcal{L}_p \, d^4x \tag{D.10}$$

where

$$L_p = \tfrac{1}{2} m \eta_{\alpha\beta} u^\alpha u^\beta, \tag{D.11}$$

$$\mathcal{L}_p = \tfrac{1}{2} m \int \eta_{\alpha\beta} u^\alpha u^\beta \delta^4[x^\gamma - z^\gamma(\tau)] \, d\tau. \tag{D.12}$$

Here $z^\alpha(\tau)$ is the world-line of the particle of mass m and $u^\alpha = dz^\alpha/d\tau$ is the 4-velocity. For a system of N particles, one has a sum of N terms each of the form (D.9).

If the particle has a charge e, than its 4-current is

$$J^\alpha = e \int u^\alpha \delta^4[x^\beta - z^\beta(\tau)] \, d\tau. \tag{D.13}$$

Exercise   D.1   By changing the variable of integration in Eq. (D.13) from τ to t, verify that the equation is equivalent to

$$J^0 = e \delta^3[\mathbf{x} - \mathbf{z}(t)], \tag{D.14}$$

$$\mathbf{J} = e \mathbf{v} \delta^3[\mathbf{x} - \mathbf{z}(t)]. \tag{D.15}$$

The interaction term is

$$S_{\text{int}} = \int L_{\text{int}} \, d\tau \qquad (\text{D}.16)$$

$$= \int \mathcal{L}_{\text{int}} \, d^4x, \qquad (\text{D}.17)$$

where

$$\mathcal{L}_{\text{int}} = J^\alpha A_\alpha, \qquad (\text{D}.18)$$

$$L_{\text{int}} = eA_\alpha u^\alpha. \qquad (\text{D}.19)$$

Thus Maxwell's equations including source terms are

$$\frac{\delta \mathcal{L}}{\delta A_\alpha} = \frac{\delta \mathcal{L}_{\text{em}}}{\delta A_\alpha} + \frac{\delta \mathcal{L}_{\text{int}}}{\delta A_\alpha} = 0, \qquad (\text{D}.20)$$

or, from Eqs. (D.8) and (D.18),

$$F^{\alpha\beta}{}_{,\beta} = 4\pi J^\alpha. \qquad (\text{D}.21)$$

The equation of motion for each particle is

$$\frac{\delta L}{\delta z^\alpha} \equiv \frac{\partial L}{\partial z^\alpha} - \frac{d}{d\tau}\left(\frac{\partial L}{\partial u^\alpha}\right) = 0, \qquad (\text{D}.22)$$

where $L = L_{\text{p}} + L_{\text{int}}$. From Eqs. (D.11) and (D.19), we have (since A_α is a function of z^α along the particle world-line),

$$0 = eA_{\beta,\alpha}u^\beta - \frac{d}{d\tau}\left(m\eta_{\alpha\beta}u^\beta + eA_\alpha\right). \qquad (\text{D}.23)$$

Thus

$$m\eta_{\alpha\beta}\frac{du^\beta}{d\tau} = e\left(A_{\beta,\alpha}u^\beta - A_{\alpha,\beta}u^\beta\right)$$

$$= eF_{\alpha\beta}u^\beta, \qquad (\text{D}.24)$$

which is the Lorentz force law.[1]

[1] For example, Jackson (1975), Eq. (11.144).

Given any field theory in Lagrangian form, there exist standard procedures for converting it to Hamiltonian form and for constructing the energy–momentum tensor of the field. The signs of the free-field terms S_{em} and S_p are fixed by the requirement that the corresponding free Hamiltonians (or, better, energy densities) be positive definite. The sign of S_{int} is arbitrary. While a sign change would change the sign of the field produced by a charge in Eq. (D.21), the resulting force produced on another charge would be the same because of a compensating sign change in Eq. (D.24).

Electromagnetism is a massless vector field theory. A massive vector field theory has a term

$$-\frac{1}{8\pi}\mu^2 A_\alpha A^\alpha \tag{D.25}$$

added to the field Lagrangian density (D.4). Equation (D.21) becomes

$$F^{\alpha\beta}{}_{,\beta} + \mu^2 A^\alpha = 4\pi J^\alpha. \tag{D.26}$$

Exercise   D.2   Show that charge conservation $J^\alpha{}_{,\alpha} = 0$ automatically guarantees that a massive vector field satisfies the Lorentz condition

$$A^\alpha{}_{,\alpha} = 0. \tag{D.27}$$

Substituting Eq. (D.5) in Eq. (D.26), we get

$$A^{\beta,\alpha}{}_\beta - A^{\alpha,\beta}{}_\beta + \mu^2 A^\alpha = 4\pi J^\alpha. \tag{D.28}$$

Interchanging the order of partial derivatives in the first term, we see that it vanishes by Eq. (D.27). The second term is simply minus the d'Alembertian of A^α, so

$$\Box A^\alpha - \mu^2 A^\alpha = -4\pi J^\alpha. \tag{D.29}$$

Exercise   D.3   Consider vacuum plane-wave solutions of Eq. (D.29) with $J^\alpha = 0$,

$$A^\alpha = a^\alpha e^{-i\omega t} e^{i\mathbf{k}\cdot\mathbf{x}}, \tag{D.30}$$

where a^α is constant. Show that

$$\frac{\omega^2}{c^2} - k^2 = \mu^2, \tag{D.31}$$

and hence conclude that the quanta of the corresponding quantum field have a mass $\hbar\mu/c$.

Let us now find the static interaction between two point "charges" at rest. A charge g at rest at the origin has [cf. Eqs. (D.14) and (D.15)]

$$J^0 = g\delta^3(\mathbf{x}), \qquad \mathbf{J} = 0. \tag{D.32}$$

In Eq. (D.29), we can therefore set

$$A^0 = \phi, \qquad \mathbf{A} = 0, \tag{D.33}$$

where

$$\nabla^2 \phi - \mu^2 \phi = -4\pi g \delta^3(\mathbf{x}). \tag{D.34}$$

Equation (D.34) can be solved, for example, by Fourier transforms. The result is

$$\phi = g \frac{e^{-\mu r}}{r}, \tag{D.35}$$

which is the *Yukawa potential*. Note from Eq. (D.24) that the force on another stationary charge g' is

$$m\mathbf{a} = -g'\nabla\phi, \tag{D.36}$$

so that *like charges repel*. The interaction energy between two equal charges is

$$V_{12} = g\phi = g^2 \frac{e^{-\mu r}}{r}, \tag{D.37}$$

and is *positive*.

Now consider a massive scalar field. The Lagrangian densities are

$$\mathcal{L}_{\text{field}} = -\frac{1}{8\pi}\left(\phi_{,\alpha}\phi^{,\alpha} + \mu^2\phi^2\right), \tag{D.38}$$

$$\mathcal{L}_{\text{int}} = -\rho\phi, \tag{D.39}$$

and $\mathcal{L}_p$ of course is still given by Eq. (D.12). Here ρ is the scalar charge density,

$$\rho = g\int d\tau\, \delta^4\left[x^\alpha - z^\alpha(\tau)\right]. \tag{D.40}$$

Varying ϕ, we obtain the field equation

$$\Box\phi - \mu^2\phi = 4\pi\rho, \tag{D.41}$$

while varying z^α gives the particle equation of motion,

$$m\eta_{\alpha\beta}\frac{du^\beta}{d\tau} = -g\frac{\partial\phi}{\partial z^\alpha}. \tag{D.42}$$

For a point charge at rest at the origin, we get

$$\nabla^2\phi - \mu^2\phi = 4\pi g\delta^3(\mathbf{x}),\tag{D.43}$$

so that

$$\phi = -g\frac{e^{-\mu r}}{r}.\tag{D.44}$$

Thus *like scalar charges attract*. The interaction energy

$$V_{12} = g\phi = -g^2\frac{e^{-\mu r}}{r}\tag{D.45}$$

is *negative*.

We conclude by noting that particles and antiparticles have charges of opposite sign so that particle–antiparticle forces are attractive for vector fields and repulsive for scalar fields.

Appendix E
Quarks

Particles are divided into two classes: *leptons*, which are not subject to the strong interaction and which show no signs of inner structure, and the strongly interacting *hadrons*. Three kinds of lepton are presently known: e, μ, and τ, each with its associated neutrino. (Experimental evidence for ν_τ is still circumstantial.)

Hadrons are divided into *baryons*, which ultimately decay into protons (i.e., carry baryon number 1), and *mesons* (baryon number 0). The quark model has been very successful in explaining many properties of hadrons. At present six "flavors" of quarks are believed to exist. These are called "up," "down," "strange," "charm," "top" and "bottom" (u, d, s, c, t, b). (There is no experimental confirmation of the t quark yet.) Quarks have spin $\frac{1}{2}$ and each comes in three "colors." Baryons are made out of three quarks, while mesons are made of a quark and an antiquark. The quarks u, c, and t all have charge $+\frac{2}{3}$ while d, s, and b have charge $-\frac{1}{3}$. All have strangeness zero, except s, which has strangeness -1.

The strong force between quarks is postulated to arise from the exchange of massless spin-1 particles, called "gluons," which carry color quantum numbers. This theory is aptly named "quantum chromodynamics" and is abbreviated QCD.

Quarks and leptons are grouped into three families:

$$
\begin{array}{cccccc}
e & \nu_e & \mu & \nu_\mu & \tau & \nu_\tau \\
d & u & s & c & b & t
\end{array}
$$

The weak interaction is responsible for transformations between different flavors of quarks and families of leptons. In the Weinberg–Salam model (which is incorporated into quantum chromodynamics), the weak interaction and electromagnetism are unified into a single theory. The electromagnetic force arises from the exchange of the massless spin-1 photon, while the weak force is produced by the exchange of very massive (~ 80 GeV) spin-1 particles, the W^+, W^-, and Z^0 (not yet seen directly in experiments). "Normal" matter consists

559

largely of the first family: a neutron is simply a *udd* triplet while a proton is a *uud* triplet.

Table E.1 contains a list of particle properties. Included are all the "stable" particles (lifetimes $> 10^{-20}$ s), plus some shorter-lived *resonances* of interest.

Table E.1
Particle Properties[a]

Particle	t	J^P	Mass (MeV)	Mean Life (s)
Leptons				
e		$\frac{1}{2}$	0.511003	Stable
μ		$\frac{1}{2}$	105.6594	2.19714×10^{-6}
τ		$\frac{1}{2}$	1784	5×10^{-13}
Nonstrange Baryons				
p	$\frac{1}{2}$	$\frac{1}{2}^+$	938.280	Stable
n	$\frac{1}{2}$	$\frac{1}{2}^+$	939.573	925
Δ	$\frac{3}{2}$	$\frac{3}{2}^+$	1232	6×10^{-24}
Strangeness -1 Baryons				
Λ	0	$\frac{1}{2}^+$	1115.60	2.63×10^{-10}
Σ^+	1	$\frac{1}{2}^+$	1189.36	8.00×10^{-11}
Σ^0	1	$\frac{1}{2}^+$	1192.46	6×10^{-20}
Σ^-	1	$\frac{1}{2}^+$	1197.34	1.48×10^{-10}
Strangeness -2 Baryons				
Ξ^0	$\frac{1}{2}$	$\frac{1}{2}^+$	1314.9	2.9×10^{-10}
Ξ^-	$\frac{1}{2}$	$\frac{1}{2}^+$	1321.3	1.64×10^{-10}
Strangeness -3 Baryon				
Ω^-	0	$\frac{3}{2}^+$	1672.5	8.2×10^{-11}
Nonstrange Charmed Baryon				
Λ_c^+	0	$\frac{1}{2}^+$	2282	1×10^{-13}
Nonstrange Mesons				
$\pi^\pm$	1	0^-	139.567	2.603×10^{-8}
π^0	1	0^-	134.963	8.3×10^{-17}
η	0	0^-	548.8	8×10^{-19}
ρ	1	1^-	769	4.3×10^{-24}
ω	0	1^-	782.6	6.6×10^{-23}
η'	0	0^-	957.6	2.4×10^{-21}
ϕ	0	1^-	1019.6	1.6×10^{-22}
J/Ψ	0	1^-	3096.9	1.0×10^{-20}
Υ		1^-	9456	1.6×10^{-20}

Table E.1

(*Continued*)

Particle	t	J^P	Mass (MeV)	Mean Life (s)
			Strangeness -1 *Mesons*	
$K^{\pm}$	$\frac{1}{2}$	0^-	493.67	1.237×10^{-8}
$K^0, \overline{K}^0$	$\frac{1}{2}$	0^-	497.7	$K_s: 8.92 \times 10^{-11}$
				$K_L: 5.18 \times 10^{-8}$
			Charmed Nonstrange Mesons	
$D^{\pm}$	$\frac{1}{2}$	0^-	1869.4	9×10^{-13}
$D^0, \overline{D}^0$	$\frac{1}{2}$	0^-	1864.7	5×10^{-13}
			Charmed Strange Meson	
$F^{\pm}$	0	0^-	2021	2×10^{-13}

[a]Taken from Particle Data Group (1982). The second column is the isospin t, while the next column is the spin and parity, J^P. Masses and lifetimes have generally been rounded; see the original reference for error bars and a complete listing of particle properties.

Appendix F
The Phase-Space
Factor, Eq. (11.5.16)

In this section we follow Bahcall and Wolf (1965) in evaluating the phase-space factor P of Eq. (11.5.14). Such integrals appear frequently in neutrino and condensed matter physics.[1] We make use of the approximation that kT is negligible compared with all the Fermi kinetic energies involved.

Using Eq. (11.5.2), rewrite P in the form

$$P = B \int \prod_{j=1}^{6} p_j^2 \, dp_j \, SE_{\bar{\nu}} \delta(E_f - E_i) A, \qquad (F.1)$$

where

$$B = (m_\pi c)^{-15} (2\pi)^{-18}, \qquad (F.2)$$

$$A = \hbar^{-3} \int \prod_{j=1}^{6} d\Omega_j \, \delta^3(\mathbf{k}_f - \mathbf{k}_i). \qquad (F.3)$$

In the angular integral A, it is sufficient to consider only that region of phase space in which particle energies are within a few kT of the Fermi energies. The corresponding Fermi momentum of the neutron is large compared with the Fermi momenta of the proton and the electron (cf. Exercise 11.1); the neutrino momentum kT/c is completely negligible. Recalling that

$$\mathbf{k}_i = \mathbf{k}_1 + \mathbf{k}_2, \qquad \mathbf{k}_f = \mathbf{k}_{1'} + \mathbf{k}_s, \qquad (F.4)$$

where

$$\mathbf{k}_s = \mathbf{k}_p + \mathbf{k}_e + \mathbf{k}_{\bar{\nu}}, \qquad (F.5)$$

[1] For example, the theory of ^{3}He.

we have

$$\hbar^3 A = \int d\Omega_1 \, d\Omega_2 \, d\Omega_{1'} \, d\Omega_p \, d\Omega_e \, d\Omega_{\bar{\nu}} \, \delta^3(\mathbf{k}_1 + \mathbf{k}_2 - \mathbf{k}_{1'} - \mathbf{k}_s). \qquad \text{(F.6)}$$

Do the $d\Omega_{1'}$ integral first, writing the δ-function as

$$\delta(k_{1'} - |\mathbf{k}_1 + \mathbf{k}_2 - \mathbf{k}_s|) \frac{\delta(\Omega_{1'} - \Omega_{1+2-s})}{k_{1'}^2}. \qquad \text{(F.7)}$$

The $d\Omega_{1'}$ integral gives unity, leaving

$$\hbar^3 A = \int d\Omega_1 \, d\Omega_2 \, d\Omega_p \, d\Omega_e \, d\Omega_{\bar{\nu}} \, \frac{\delta(k_{1'} - |\mathbf{k}_1 + \mathbf{k}_2 - \mathbf{k}_s|)}{k_{1'}^2}. \qquad \text{(F.8)}$$

Rewrite the remaining δ-function as

$$\delta\left[k_{1'} - \left(k_1^2 + |\mathbf{k}_2 - \mathbf{k}_s|^2 - 2k_1|\mathbf{k}_2 - \mathbf{k}_s|\cos\theta_1\right)^{1/2}\right]$$

$$= \frac{\delta\left[\cos\theta_1 - \left(k_{1'}^2 - k_1^2 - |\mathbf{k}_2 - \mathbf{k}_s|^2\right)/(2k_1|\mathbf{k}_2 - \mathbf{k}_s|)\right]}{k_1|\mathbf{k}_2 - \mathbf{k}_s|/k_{1'}}. \qquad \text{(F.9)}$$

Here we have chosen the z axis for $\mathbf{k}_1$ along $\mathbf{k}_2 - \mathbf{k}_s$, and we have used the identity $\delta[f(x)] = \delta(x - a)/|f'(a)|$, where $f(a) = 0$. Doing the $d\Omega_1$ integral, we get

$$\hbar^3 A = \frac{2\pi}{k_{1'}k_1} \int \frac{d\Omega_2 \, d\Omega_p \, d\Omega_e \, d\Omega_{\bar{\nu}}}{|\mathbf{k}_2 - \mathbf{k}_s|}. \qquad \text{(F.10)}$$

Recalling that $k_2 \gg k_s$, we find

$$\hbar^3 A = \frac{2\pi(4\pi)^4}{k_{1'}k_1 k_2}, \qquad \text{(F.11)}$$

or

$$A = (4\pi)^5 (2 p_1 p_2 p_{1'})^{-1}. \qquad \text{(F.12)}$$

The momentum differentials in Eq. (F.1) may be simplified by using

$$p_j \, dp_j = E_j \frac{dE_j}{c^2} \simeq m_j \, dE_j, \qquad j = n, p,$$

$$dp_e \simeq \frac{dE_e}{c}. \tag{F.13}$$

Then all the p_j (except $p_{\bar{\nu}}$) can be set equal to $p_F(j)$ and removed from the integral. This gives

$$P = 2^9 \pi^5 c^{-4} B m_n^3 m_p p_F(p) p_F^2(e) \int \prod_{j=1}^{6} dE_j \, E_{\bar{\nu}}^3 S \delta(E_f - E_i). \tag{F.14}$$

Now Eq. (11.5.5) for the statistical factor S can be written as

$$S = \prod_{j=1}^{5} (1 + e^{x_j})^{-1}, \tag{F.15}$$

where the nondimensional energies x_j are defined by

$$x_1 = \beta(E_1 - \mu_n),$$

$$x_2 = \beta(E_2 - \mu_n),$$

$$x_3 = -\beta(E_e - \mu_e),$$

$$x_4 = -\beta(E_{1'} - \mu_n),$$

$$x_5 = -\beta(E_p - \mu_p),$$

$$\beta \equiv \frac{1}{kT}. \tag{F.16}$$

Defining

$$y = \frac{E_{\bar{\nu}}}{kT}, \tag{F.17}$$

we get for Eq. (F.14)

$$P = 2^9 \pi^5 c^{-4} B m_n^3 m_p p_F(p) p_F^2(e)(kT)^8 I, \tag{F.18}$$

where

$$I \equiv \int_0^\infty dy\, y^3 J, \tag{F.19}$$

$$J \equiv \int \prod_{j=1}^5 dx_j (1 + e^{x_j})^{-1} \delta\left[\sum_{j=1}^5 x_j - y \right]. \tag{F.20}$$

The integral J does not extend over energies less than mc^2. However, we make errors of at most $\exp[-\beta E_F'(p)]$ by extending the range of integration from $-\infty$ to $+\infty$ for each x_j.

To evaluate J, start by introducing the representation

$$\delta(x) = \frac{1}{2\pi} \int_{-\infty}^\infty e^{izx}\, dz. \tag{F.21}$$

Then

$$J = \frac{1}{2\pi} \int_{-\infty}^\infty dz\, e^{-izy} [f(z)]^5, \tag{F.22}$$

where

$$f(z) \equiv \int_{-\infty}^\infty dx\, e^{izx} (e^x + 1)^{-1}. \tag{F.23}$$

In Eq. (F.23), z must have a small negative imaginary part for the integral to converge.

Evaluate $f(z)$ by integrating

$$K \equiv \oint dw\, e^{izw} (e^w + 1)^{-1} \tag{F.24}$$

around the contour shown in Figure F.1. The vertical segments give no contribution in the limit $R \to \infty$. The integral along the real axis gives $f(z)$, while the contribution along the line $\text{Im}(w) = 2\pi$ is $-\exp(-2\pi z)f(z)$. The only pole of the integrand in Eq. (F.24) enclosed by the contour is at $w = i\pi$; the residue there is $-\exp(-\pi z)$. Thus

$$K = f(z) - e^{-2\pi z} f(z) = -2\pi i e^{-\pi z}, \tag{F.25}$$

or

$$f(z) = \frac{\pi}{i \sinh \pi z}. \tag{F.26}$$

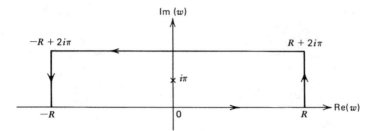

Figure F.1 Contour for Eq. (F.24), with $R \to \infty$.

Returning to Eq. (F.22), we have

$$J = \frac{1}{2\pi i} \int_{-\infty - i\varepsilon}^{\infty - i\varepsilon} dz \, e^{-izy} \left(\frac{\pi}{\sinh \pi z} \right)^5. \tag{F.27}$$

Here we have inserted $-i\varepsilon$ in the limits of integration as a reminder that z has a small negative imaginary part. We would like to evaluate J by finding a suitable closed contour. If we make the substitution

$$z = z' - i \tag{F.28}$$

in Eq. (F.27), we find

$$J = \frac{-e^{-y}}{2\pi i} \int_{-\infty - i\varepsilon + i}^{\infty - i\varepsilon + i} dz' \, e^{-iz'y} \left(\frac{\pi}{\sinh \pi z} \right)^5. \tag{F.29}$$

Equations (F.27) and (F.29) give

$$(1 + e^y) J = \frac{1}{2\pi i} \left[\int_{-\infty - i\varepsilon}^{\infty - i\varepsilon} + \int_{\infty - i\varepsilon + i}^{-\infty - i\varepsilon + i} \right] dz \, e^{-izy} \left(\frac{\pi}{\sinh \pi z} \right)^5$$

$$= \frac{1}{2\pi i} \oint dz \, d^{-izy} \left(\frac{\pi}{\sinh \pi z} \right)^5. \tag{F.30}$$

Here we have closed the contour as shown in Figure F.2, the vertical segments not contributing. The only pole enclosed is at $z = 0$. Thus

$$(1 + e^y) J = \text{Residue at } z = 0 \text{ of } \left[e^{-izy} \left(\frac{\pi}{\sinh \pi z} \right)^5 \right]. \tag{F.31}$$

The residue is found by making series expansions of the exponential and the sinh

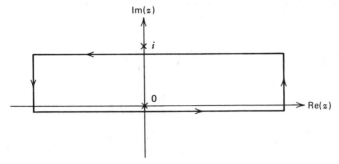

Figure F.2   Contour for Eq. (F.30).

about $z = 0$, and reading off the coefficient of $1/z$. This gives

$$(1 + e^y)J = \frac{3\pi^4}{8} + \frac{5\pi^2}{12}y^2 + \frac{1}{24}y^4. \tag{F.32}$$

Thus Eq. (F.19) gives

$$I = \int_0^\infty dy \left(\frac{3\pi^4}{8}y^3 + \frac{5\pi^2}{12}y^5 + \frac{1}{24}y^7 \right)(e^y + 1)^{-1}. \tag{F.33}$$

The integrand of Eq. (F.33) gives the energy spectrum of the antineutrinos. The integral can be found in Gradshteyn and Ryzhik (1965), Eq. (3.4114). The result is

$$I = \frac{11513\pi^8}{120960}. \tag{F.34}$$

Using the results of Exercise 11.1 in Eq. (F.18), we finally obtain

$$P \simeq 2.1 \times 10^{-30} \left(\frac{\rho}{\rho_{\text{nuc}}} \right)^2 T_9^8, \tag{F.35}$$

the result used in Eq. (11.5.16).

Note that throughout our treatment we have naively employed free particle masses m_j for the nucleons, rather than effective masses $m_j^* \lesssim m_j$. Effective masses take into account N-N interactions of the two-body nucleon system with the ambient many-body nucleon sea. Given the uncertainties in the effective masses, we have ignored this correction.[2]

---

[2] See Friman and Maxwell (1979) for further discussion of this point.

Appendix **G**
Spherical Accretion onto a Black Hole: The Relativistic Equations

In this appendix we discuss the general relativistic analogue of the Bondi (1952) equations for spherical, steady-state, adiabatic accretion onto a Schwarzschild black hole of mass M. We will show how the relativistic equations for black hole accretion demand a transition to supersonic flow in the solution.

The boundary conditions at infinity are the same as before; that is, the gas is at rest at baryon density n_∞, rest-mass density mn_∞, where m is the mean baryon mass, and *total* mass-energy density ρ_∞. As usual the connection between total and rest-mass energy density is

$$\rho = mn + \varepsilon', \tag{G.1}$$

where ε' is the internal energy density. (We set $c = G = 1$ in this appendix.) For adiabatic flow, we know that $P = P(n)$, which is all we need assume to prove that the flow is transonic. The key equations are baryon conservation,

$$(nu^\alpha)_{;\,\alpha} = 0, \tag{G.2}$$

momentum conservation (relativistic Euler equation),

$$(\rho + P)u_{\alpha;\,\beta}u^\beta = -P_{,\,\alpha} - u_\alpha P_{,\,\beta}u^\beta, \tag{G.3}$$

and mass-energy conservation [entropy equation, cf. Eq. (2.1.3)]

$$d\left(\frac{\rho}{n}\right) + P\,d\left(\frac{1}{n}\right) = T\,ds = 0. \tag{G.4}$$

The reader unfamiliar with tensor calculus will have to accept on faith that Eqs. (G.2) and (G.3) are the relativistic versions of Eqs. (6.1.1) and (6.1.2).[1] Here u^{α} denotes the components of the fluid 4-velocity. Equation (G.4) may be written as

$$\frac{d\rho}{dn} = \frac{\rho + P}{n}, \tag{G.5}$$

which will be useful below.

Equations (G.2) and (G.3) can be written in Schwarzschild coordinates as

$$\frac{n'}{n} + \frac{u'}{u} + \frac{2}{r} = 0, \tag{G.6}$$

and

$$uu' = -\frac{1}{\rho + P}\frac{dP}{dr}\left(1 + u^2 - \frac{2M}{r}\right) - \frac{M}{r^2}, \tag{G.7}$$

where prime ($'$) denotes differentiation with respect to r and u denotes the *inward* radial component $u \equiv |u^r|$. Define the sound speed according to

$$a^2 \equiv \frac{dP}{d\rho} = \frac{dP}{dn}\frac{n}{\rho + P}, \tag{G.8}$$

where we have employed Eq. (G.5). Then we can write Eq. (G.7) as

$$uu' + \frac{M}{r^2} + \left(1 - \frac{2M}{r} + u^2\right)a^2\frac{n'}{n} = 0, \tag{G.9}$$

where we have used $P' = (dP/dn)n'$. Equations (G.6) and (G.9) are immediately seen to be the relativistic generalizations of Eqs. (14.3.6) and (14.3.7).

Proceeding as in the Newtonian case to solve for u' and n', we obtain

$$u' = \frac{D_1}{D}, \qquad n' = \frac{-D_2}{D}, \tag{G.10}$$

where

$$D_1 = \frac{1}{n}\left[\left(1 - \frac{2M}{r} + u^2\right)\frac{2a^2}{r} - \frac{M}{r^2}\right], \tag{G.11}$$

$$D_2 = \frac{2u^2/r - M/r^2}{u}, \tag{G.12}$$

[1]See, for example, Misner, Thorne, and Wheeler (1973), Chapter 22.

and

$$D = \frac{u^2 - (1 - 2M/r + u^2)a^2}{un}.$$ (G.13)

Now we prove that, for any equation of state obeying the *causality constraint* $a^2 < 1$ (cf. Section 9.5), the flow must pass through a critical point outside the event horizon at $r = 2M$. For at large r, we know that the flow satisfies $u^2 \ll 1$ and is subsonic with $u^2 < a^2$ (e.g., as $r \to \infty$, $u \to 0$, $a \to a_\infty$), so that Eq. (G.13) implies

$$D \simeq \frac{u^2 - a^2}{un} < 0.$$ (G.14)

At $r = 2M$, however, we have

$$D = \frac{u}{n}(1 - a^2) > 0$$ (G.15)

since $a^2 < 1$. Thus D must pass through zero at *some* point $r = r_s$ outside $r = 2M$. To avoid singularities in the flow, Eq. (G.10) requires that

$$D_1 = D_2 = D = 0 \qquad \text{at } r = r_s.$$ (G.16)

From Eqs. (G.11)–(G.13), we find that at $r = r_s$,

$$u_s^2 = \frac{a_s^2}{1 + 3a_s^2} = \frac{M}{2r_s}.$$ (G.17)

Note that the proper velocity of the fluid $v^{\hat{f}}$ as measured by a local, stationary observer is related to u by

$$v^{\hat{f}} = \frac{u^{\hat{f}}}{u^{\hat{t}}} = \frac{\vec{u} \cdot \vec{e}_{\hat{f}}}{-\vec{u} \cdot \vec{e}_{\hat{t}}} = \frac{u^r}{u^t} \frac{1}{1 - 2M/r}$$ (G.18)

(cf. Sections 12.3 and 12.4). Using $-1 = \vec{u} \cdot \vec{u}$, which gives

$$-1 = -(u^t)^2\left(1 - \frac{2M}{r}\right) + \frac{(u^r)^2}{1 - 2M/r},$$ (G.19)

we find that Eq. (G.18) becomes

$$|v^{\hat{f}}| = \frac{u}{(1 - 2M/r + u^2)^{1/2}}.$$ (G.20)

So, according to Eq. (G.20), at large $r \gg 2M$, where $u \ll 1$, we have $|v^f| \simeq u$; as $r \to \infty$, the proper flow velocity $v^f \to 0$ and is subsonic. At $r = 2M$, $|v^f| \equiv 1 > a$ and the proper velocity, which equals the speed of light, is supersonic. It is significant that $|v^f| \equiv 1$ at $r = 2M$ *independent* of the magnitude of u. Thus, the requirement that the proper velocity equals the light velocity at the horizon is *not* sufficient by itself to guarantee that the flow pass through a critical point outside $2M$, as has been sometimes claimed.

To make further progress we recast Eqs. (G.6) and (G.9) into the form of conservation equations.[2] This gives

$$4\pi mnur^2 = \dot{M} = \text{constant} \quad (\text{independent of } r), \tag{G.21}$$

and

$$\left(\frac{\rho + P}{n}\right)^2 \left(1 - \frac{2M}{r} + u^2\right) = \text{constant}$$

$$= \left(\frac{\rho_\infty + P_\infty}{n_\infty}\right)^2. \tag{G.22}$$

Equation (G.21) gives the rest-mass accretion rate. Equation (G.22) is the relativistic Bernoulli equation.

To calculate an explicit value for $\dot{M}$, we need to adopt an equation of state. Following Bondi (1952), we assume a polytropic equation of state:

$$P = Kn^\Gamma, \quad K, \Gamma \text{ constant}. \tag{G.23}$$

Inserting Eq. (G.23) in Eq. (G.4) gives

$$d\left(\frac{\rho}{n}\right) = Kn^{\Gamma-2} \, dn. \tag{G.24}$$

Integrating gives

$$\rho = mn + \frac{Kn^\Gamma}{\Gamma - 1}, \tag{G.25}$$

where the constant of integration, m, is evaluated by comparison with Eq. (G.1). Hence

$$\frac{\rho + P}{n} = m + \frac{\Gamma}{\Gamma - 1} Kn^{\Gamma-1}, \tag{G.26}$$

[2] Michel (1972).

and by Eq. (G.8)

$$a^2 = \frac{\Gamma K n^{\Gamma - 1}}{m + \Gamma K n^{\Gamma - 1}/(\Gamma - 1)}, \qquad (G.27)$$

or

$$\Gamma K n^{\Gamma - 1} = \frac{a^2 m}{1 - a^2/(\Gamma - 1)}. \qquad (G.28)$$

Substituting Eqs. (G.26) and (G.28) in Eq. (G.22), we get

$$\left(1 - \frac{2M}{r} + u^2\right)\left(1 + \frac{a^2}{\Gamma - 1 - a^2}\right)^2 = \left(1 + \frac{a_\infty^2}{\Gamma - 1 - a_\infty^2}\right)^2. \qquad (G.29)$$

Inverting both sides of Eq. (G.29) and evaluating at the sonic point with the aid of Eq. (G.17), we get

$$\left(1 + 3a_s^2\right)\left(1 - \frac{a_s^2}{\Gamma - 1}\right)^2 = \left(1 - \frac{a_\infty^2}{\Gamma - 1}\right)^2. \qquad (G.30)$$

Now for large $r \geqslant r_s$, we expect the baryons to be nonrelativistic with $a \leqslant a_s \ll 1$ (i.e., $T \ll mc^2/k = 10^{13}$ K), provided they were nonrelativistic at infinity (i.e., $a_\infty \ll 1$). Expanding Eq. (G.30) to lowest nonvanishing order in a_s^2 and a_∞^2, we find

$$a_s^2 \approx \frac{2}{5 - 3\Gamma} a_\infty^2, \qquad \Gamma \neq \tfrac{5}{3},$$

$$\approx \tfrac{2}{3} a_\infty, \qquad \Gamma = \tfrac{5}{3}. \qquad (G.31)$$

Equation (G.31) justifies our expectation that $a_\infty \ll 1$ implies $a_s \ll 1$.

Exercise   G.1   Obtain the sonic radius in terms of M and a_∞ for $1 \leqslant \Gamma \leqslant \tfrac{5}{3}$ and compare with the Newtonian result, Eq. (14.3.14).
   Answer:

$$r_s \approx \frac{5 - 3\Gamma}{4} \frac{M}{a_\infty^2}, \qquad \Gamma \neq \frac{5}{3},$$

$$r_s \approx \frac{3}{4} \frac{M}{a_\infty}, \qquad \Gamma = \frac{5}{3}.$$

Now Eq. (G.28) gives, for $a^2/(\Gamma - 1) \ll 1$,

$$\frac{n_s}{n_\infty} \approx \left(\frac{a_s}{a_\infty}\right)^{2/(\Gamma - 1)}. \qquad (G.32)$$

Hence the accretion rate is

$$\dot{M} = 4\pi mn_s u_s r_s^2 \approx 4\pi \lambda_s M^2 mn_\infty a_\infty^{-3}, \tag{G.33}$$

where we have used Eqs. (G.21), (G.17), (G.31), and (G.32). In Eq. (G.33) the nondimensional accretion parameter λ_s is defined by Eq. (14.3.17) for $1 \leqslant \Gamma \leqslant \frac{5}{3}$. We thus find that to lowest order, the relativistic accretion rate is equal to the Bondi (1952) result for Newtonian, transonic flow. The equivalence of the Newtonian and general relativistic expressions is physically reasonable: the critical accretion rate is determined by conditions at the same sonic point $r = r_s$, which lies far outside the event horizon (i.e., $r_s \gg 2M$) and is uninfluenced by nonlinear gravity. On the other hand, the regularity condition imposed on the relativistic flow solution outside the horizon proved necessary to specify transonic flow in the first place!

To estimate the flow parameters at small radius $r \ll r_s$, we first note that, because the flow is transonic with $u > a$, Eq. (G.29) demands that for $\Gamma \neq \frac{5}{3}$

$$u^2 \approx \frac{2M}{r}, \quad r \ll r_s \quad \left(\Gamma \neq \frac{5}{3}\right). \tag{G.34}$$

From Eqs. (G.21), (G.33), and (G.34) we can calculate the gas compression,

$$\frac{n(r)}{n_\infty} \approx \frac{\lambda_s}{\sqrt{2}} \left(\frac{M}{a_\infty^2 r}\right)^{3/2}. \tag{G.35}$$

Assuming a Maxwell–Boltzmann gas, with $P = nkT$, where T is the temperature, we have from Eq. (G.23)

$$\frac{T(r)}{T_\infty} = \left(\frac{n(r)}{n_\infty}\right)^{\Gamma-1} \approx \left(\frac{\lambda_s}{\sqrt{2}}\right)^{\Gamma-1} \left(\frac{M}{a_\infty^2 r}\right)^{3/2(\Gamma-1)} \tag{G.36}$$

for the adiabatic temperature profile. At the event horizon (subscript h), we then find, independent of the black hole mass M,

$$u_h \approx 1 \quad \left(\Gamma \neq \frac{5}{3}\right),$$

$$\frac{n_h}{n_\infty} \approx \frac{\lambda_s}{4} \left(\frac{c}{a_\infty}\right)^3, \quad \frac{T_h}{T_\infty} \approx \left[\frac{\lambda_s}{4} \left(\frac{c}{a_\infty}\right)^3\right]^{\Gamma-1}, \tag{G.37}$$

where we have reinserted c, the speed of light.

The numerical coefficients in the above expressions are somewhat different for the special case $\Gamma = \frac{5}{3}$. For this case, a remains comparable to u inside the transonic radius. Since by Eq. (G.31) $a^2 \sim a_\infty$ for $r < r_s$, in Eq. (G.29) we can set the right-hand side to unity when $r = 2M$, giving

$$u_h\left(1 + \frac{a_h^2}{\frac{2}{3} - a_h^2}\right) \approx 1. \tag{G.38}$$

Using Eq. (G.27), this becomes

$$u_h\left(1 + \frac{5Kn_h^{2/3}}{2m}\right) \approx 1. \tag{G.39}$$

Now Eqs. (G.21) and (G.33) give, with $\lambda_s = \frac{1}{4}$,

$$n_h = \frac{n_\infty}{16a_\infty^3 u_h}. \tag{G.40}$$

Since by Eq. (G.27)

$$a_\infty^2 \approx \frac{5Kn_\infty^{2/3}}{3m}, \tag{G.41}$$

Eq. (G.39) becomes

$$u_h + \frac{3}{2^{11/3}}u_h^{1/3} - 1 \approx 0. \tag{G.42}$$

Solving Eq. (G.42) numerically yields

$$u_h \approx 0.782 \quad \left(\Gamma = \tfrac{5}{3}\right). \tag{G.43}$$

Equation (G.40) now gives

$$\frac{n_h}{n_\infty} \approx \frac{1}{16u_h}\left(\frac{c}{a_\infty}\right)^3, \qquad \frac{T_h}{T_\infty} \approx \left(\frac{1}{16u_h}\right)^{2/3}\left(\frac{c}{a_\infty}\right)^2. \tag{G.44}$$

Let us evaluate Eq. (G.44) for the case of nonrelativistic baryons accreting from the interstellar medium. The ambient sound speed is

$$a_\infty \approx \left(\frac{5}{3}\frac{kT_\infty}{m_B}\right)^{1/2} \approx 11.7\left(\frac{T_\infty}{10^4 \text{ K}}\right)^{1/2} \text{ km s}^{-1}. \tag{G.45}$$

Substituting Eqs. (G.43) and (G.45) into (G.44) gives

$$\frac{n_h}{n_\infty} \approx 1.33 \times 10^{12} \left(\frac{T_\infty}{10^4 \text{ K}} \right)^{-3/2},$$

$$T_h \approx \frac{3}{40} \left(\frac{2}{u_h^2} \right)^{1/3} \frac{m_B c^2}{k} \approx 1.21 \times 10^{12} \text{ K.} \qquad (G.46)$$

The temperature T_h is thus independent of T_∞. The reason is that for $\Gamma = \frac{5}{3}$, an appreciable fraction of the gravitational potential energy, which is comparable to the rest-mass energy at the horizon, is necessarily converted into thermal energy (both scale as r^{-1} inside r_s: $kT \sim GMm_B/r$). Adiabatic flow in the limit $\Gamma = \frac{5}{3}$ generates the *maximum* achievable gas temperatures at the horizon. Thermal energies are smaller for $\Gamma < \frac{5}{3}$.

Computer Exercise G.2 Solve Eqs. (G.21), (G.22), and (G.36) numerically for the accretion of nonrelativistic $\Gamma = \frac{5}{3}$ baryons with $T_\infty = 10^4$ K. Determine the nondimensional density profile $n(r)/n_\infty$, velocity profile $u(r)/c$, and temperature profile $T(r)/T_\infty$ as functions of r/M. Plot your results on log-log paper for the domain $2M \leqslant r \leqslant 10 r_a$, where $r_a \equiv GM/a_\infty^2$.

The stability of stationary, spherical accretion onto a Schwarzschild black hole has been examined by several groups. Most general is the analysis of Moncrief (1980).[3] Moncrief performs a general relativistic, normal mode analysis on the background, transonic flow solution. He finds that there are *no* unstable modes in either the subsonic or supersonic regions—the flow is stable.

[3]See Moncrief (1980) for references to earlier work.

Appendix H
Hydrodynamics of Viscous Fluid Flow

The equation of motion of a nonrelativistic viscous fluid is

$$\rho \frac{d\mathbf{v}}{dt} = -\nabla P + \nabla \cdot \underline{\mathbf{t}}, \tag{H.1}$$

where the viscous stress tensor $\underline{\mathbf{t}}$ has components

$$t_{ij} = t_{ji} = \eta \left(v_{i,j} + v_{j,i} - \tfrac{2}{3}\delta_{ij}v_{k,k} \right) \tag{H.2}$$

and is traceless. Here $\eta > 0$ is the dynamic or shear viscosity coefficient, and we are neglecting the bulk viscosity.[1] When $\underline{\mathbf{t}} = 0$, Eq. (H.1) is simply the Euler equation for a perfect fluid.

To find the rate at which the fluid is heated because of viscous dissipation, consider the rate of change of the energy of a unit mass of fluid (as measured comoving with the fluid):

$$\frac{d}{dt}\left(\frac{1}{2}v^2 + u \right) = \mathbf{v} \cdot \frac{d\mathbf{v}}{dt} + \frac{du}{dt}. \tag{H.3}$$

Here u is the internal energy per unit mass, and du/dt is given by the first law (6.1.8). Using Eq. (H.1) for $d\mathbf{v}/dt$ and the continuity Eq. (6.1.1) for $d\rho/dt$, we find

$$\frac{d}{dt}\left(\frac{1}{2}v^2 + u \right) = -\frac{1}{\rho}\mathbf{v} \cdot \nabla P + \frac{1}{\rho}\mathbf{v} \cdot (\nabla \cdot \underline{\mathbf{t}}) - \frac{P}{\rho}\nabla \cdot \mathbf{v} + T\frac{ds}{dt}$$

$$= -\frac{1}{\rho}\nabla \cdot (P\mathbf{v} - \underline{\mathbf{t}} \cdot \mathbf{v}) + T\frac{ds}{dt} - \frac{1}{\rho}t_{ij}v_{i,j}. \tag{H.4}$$

[1]See Landau and Lifshitz (1959), Ch. 2 for full details.

576

Now the divergence term on the right-hand side of Eq. (H.4) is the rate of doing work on a unit mass of fluid, for if we integrate over the mass of a fluid element in a volume $\mathcal{V}$ we get

$$\int \frac{1}{\rho} \nabla \cdot (P\mathbf{v} - \underset{\sim}{t} \cdot \mathbf{v}) \rho \, d\mathcal{V} = \int (P\mathbf{v} - \underset{\sim}{t} \cdot \mathbf{v}) \cdot d\mathbf{A}$$

$$= \int \mathbf{v} \cdot \mathbf{f} \, dA,$$

where

$$\mathbf{f} = P\mathbf{n} - \underset{\sim}{t} \cdot \mathbf{n} \tag{H.5}$$

is the external force per unit area acting on a surface with unit normal $\mathbf{n}$.

Thus the last two terms in Eq. (H.4), which are associated with internal heating, must exactly balance:

$$\rho T \frac{ds}{dt} = v_{i,j} t_{ij}$$

$$= \tfrac{1}{2}(v_{i,j} + v_{j,i}) t_{ij} \quad (\text{since } t_{ij} \text{ is symmetric})$$

$$= \frac{1}{2}\left(\frac{1}{\eta} t_{ij} + \frac{2}{3} \delta_{ij} v_{k,k}\right) t_{ij}$$

$$= \frac{1}{2\eta} t_{ij} t_{ij} \tag{H.6}$$

since $\mathrm{Trace}\, \underset{\sim}{t} = t_{ij} \delta_{ij} = 0$. Equation (H.6) shows that the entropy increase depends on the square of the departure from equilibrium (i.e., the square of the velocity gradient). This is as expected: entropy is a maximum in equilibrium.

We now specialize to a Keplerian accretion disk, where the velocity has essentially only a ϕ-component (i.e., $v_z \ll v_r \ll v_\phi$):

$$v_\phi = r\Omega = \left(\frac{GM}{r}\right)^{1/2}. \tag{H.7}$$

In cylindrical coordinates, the only non-negligible component of $\underset{\sim}{t}$ is[2]

$$t_{r\phi} = t_{\phi r} = \eta\left(\frac{1}{r} v_{r,\phi} + v_{\phi,r} - \frac{v_\phi}{r}\right). \tag{H.8}$$

[2]Landau and Lifshitz (1959), Eq. (15.15).

Using Eq. (H.7), we find

$$t_{r\phi} \simeq -\frac{3}{2}\eta\Omega = -\frac{3}{2}\eta\left(\frac{GM}{r^3}\right)^{1/2}.$$

(H.9)

The viscous force in the ϕ direction caused by the rubbing of adjacent fluid elements generates a torque that carries angular momentum outwards. The normal to the surface separating adjacent fluid elements in this case is in the radial direction, so the force is given by Eq. (H.5) to be

$$f_\phi = -t_{\phi r}.$$

(H.10)

Appendix I
Radiative Transport

In this section we assemble some of the basic facts regarding the transport of radiation in a gaseous medium. For a more thorough treatment, the reader is referred to the excellent textbooks by Chandrasekhar (1960) (which emphasizes the formal nature and solution of the transport equation), Mihalas (1978) (which emphasizes applications to stellar atmospheres), or Rybicki and Lightman (1979) (which emphasizes radiation processes). Also excellent are the discussions dealing with radiation transport found in the texts by Clayton (1968) (emphasizing applications to stellar interiors) and by Zel'dovich and Raizer (1966) (emphasizing applications to shock wave phenomena).

I.1 The Transport Equation

Recall that the fundamental quantity describing the thermodynamic state of an ensemble of particles—here, photons—is $f(\mathbf{r}, \mathbf{p}, t)$, the distribution function (cf. Chapter 2). The number of photons in a volume of phase space $d^3r\, d^3p$ about $\mathbf{r}$ and $\mathbf{p}$ at time t is related to f by

$$d\mathfrak{N} = \frac{2}{h^3} f\, d^3r\, d^3p = \frac{2}{c^3} f \nu^2\, d\nu\, d\Omega\, d^3r, \qquad (\mathrm{I}.1)$$

where h is Planck's constant, c is the speed of light, and ν is the frequency related to the momentum p by $p = h\nu/c$. The direction of propagation is specified by the element of solid angle $d\Omega$ about the unit vector $\mathbf{n}$. The energy carried by the photons in this volume is given by

$$dE \equiv h\nu\, d\mathfrak{N} \equiv \frac{I_\nu}{c} d\nu\, d\Omega\, d^3r. \qquad (\mathrm{I}.2)$$

Equation (I.2) thus defines the quantity $I_\nu(\mathbf{r}, \mathbf{n}, t)$, which is related to f by

$$I_\nu(\mathbf{r}, \mathbf{n}, t) \equiv \frac{2h\nu^3}{c^2} f(\nu, \mathbf{r}, \mathbf{n}, t). \tag{I.3}$$

The quantity I_ν, called the *specific intensity*, is constructed so that $I_\nu(\mathbf{r}, \mathbf{n}, t)\, d\nu\, d\Omega$ represents the photon energy in the interval ν to $\nu + d\nu$ passing per unit time through a unit area with normal $\mathbf{n}$ into the solid angle $d\Omega$ about $\mathbf{n}$. The radiation field is thus fully specified by either function, I_ν or f.

Associated with I_ν are its *moments*. Thus, the quantity

$$\varepsilon_\nu^r(\mathbf{r}, t) \equiv \frac{1}{c} \int_{4\pi} I_\nu\, d\Omega \tag{I.4}$$

gives the *specific energy density* at $\mathbf{r}$, while the vector function

$$\mathbf{F}_\nu(\mathbf{r}, t) \equiv \int_{4\pi} I_\nu \mathbf{n}\, d\Omega \tag{I.5}$$

measures the *specific flux* along the direction $\mathbf{n}$.[1] The specific flux through an area with a unit normal vector $\mathbf{m}$ is thus

$$F_\nu(\mathbf{r}, t, \mathbf{m}) \equiv \mathbf{F}_\nu \cdot \mathbf{m} = \int_{4\pi} I_\nu \cos\theta\, d\Omega, \tag{I.6}$$

where θ is the angle between the direction of motion of the photons $\mathbf{n}$ and the normal $\mathbf{m}$. Note that if the radiation field is isotropic—that is, I_ν is independent of $\mathbf{n}$—then by Eqs. (I.4) and (I.5)

$$\varepsilon_\nu^r = \frac{4\pi I_\nu}{c} \quad \text{(isotropic)}, \tag{I.7}$$

$$\mathbf{F}_\nu = 0.$$

The total intensity, energy density, and flux are obtained by integrating their monochromatic counterparts over frequency:

$$I = \int_0^\infty I_\nu\, d\nu, \quad \varepsilon_r = \int_0^\infty \varepsilon_\nu^r\, d\nu, \quad \mathbf{F} = \int_0^\infty \mathbf{F}_\nu\, d\nu. \tag{I.8}$$

For the special case in which *local thermodynamic equilibrium* (LTE) prevails at

---

[1] We use the term "specific" to denote monochromatic or "per unit frequency interval"; other names for these terms also exist in the literature.

temperature T, I_ν is given by the Planck function B_ν,

$$I_\nu = B_\nu = \frac{2h\nu^3}{c^2} \frac{1}{\exp(h\nu/kT) - 1} \quad \text{(LTE)}, \qquad (1.9)$$

while ε_ν^r and ε_r are given by

$$\varepsilon_\nu^r \equiv \varepsilon_\nu^P = \frac{4\pi B_\nu}{c} = \frac{8\pi h\nu^3}{c^3} \frac{1}{\exp(h\nu/kT) - 1} \quad \text{(LTE)},$$

$$\varepsilon_r \equiv \varepsilon_P = aT^4, \qquad (1.10)$$

where $a = 7.56 \times 10^{-15}$ erg cm^{-3} deg^{-4} is the usual radiation density constant and P denotes Planck.

In *strict* thermodynamic equilibrium, $\mathbf{F} = 0$ (no net flux in any direction). By contrast, in LTE, Eqs. (1.9) and (1.10) may apply to high precision even in the presence of a small nonzero flux $\mathbf{F}$ provided the gradient in the intensity (due, e.g., to a temperature gradient) is small.

Consider now the emission and absorption of photons in a gaseous medium. Define the emission coefficient, or *emissivity*, $j_\nu(\mathbf{r}, \mathbf{n}, t)$ so that $j_\nu\, d\nu\, d\Omega$ is the spontaneous energy emission rate per unit volume in the interval ν to $\nu + d\nu$ and in the solid angle $d\Omega$ about $\mathbf{n}$ at time t. Likewise, define the absorption coefficient or *opacity*, $\kappa_\nu(\mathbf{r}, \mathbf{n}, t)$ so that $\kappa_\nu \rho I_\nu\, d\nu\, d\Omega$ gives the corresponding energy absorbed per unit volume per unit time from a beam of given intensity I_ν (thus κ_ν is measured in cm^2 g^{-1}). Assuming that photons travel in straight lines in the medium, the total change in I_ν in a distance ds, measured along a light ray in the direction $\mathbf{n}$ equals

$$\frac{1}{c}\frac{\partial I_\nu}{\partial t} + \frac{\partial I_\nu}{\partial s} \equiv \frac{1}{c}\frac{\partial I_\nu}{\partial t} + \mathbf{n} \cdot \nabla I_\nu = j_\nu - \kappa_\nu \rho I_\nu, \qquad (1.11)$$

where we have allowed for local changes in time as well as for changes with distance. Equation (1.11) is the *equation of radiative transfer*. In typical applications, it is possible to treat the gas in LTE at every point in space and at each instant of time. In such a situation, the thermodynamic state of the gas can be described by two parameters, such as temperature and density. Accordingly, those functions that depend on the local state of matter—for example, the opacity κ_ν—will depend only on ρ and T.

For many circumstances, including LTE, j_ν and κ_ν assume the same relative values that they would have in *strict* thermodynamic equilibrium, when $I_\nu \equiv B_\nu \equiv$ constant. In this situation the left-hand side of Eq. (1.11) must vanish, giving

$$\frac{j_\nu}{\kappa_\nu \rho} = B_\nu(T), \qquad (1.12)$$

where T is the local matter temperature. Equation (I.12) is known as *Kirchhoff's law* and expresses the detailed balance that must prevail between absorption and emission in thermodynamic equilibrium. Note that stimulated emission can be considered to be "negative absorption," since it is also proportional to I_ν. Thus when we write κ_ν, we mean it to include the effects of stimulated emission. In Kirchhoff's law (I.12), κ_ν clearly does include stimulated emission, as required by detailed balance.

Inserting Eq. (I.12) into Eq. (I.11) yields

$$\frac{1}{c}\frac{\partial I_\nu}{\partial t} + \mathbf{n}\cdot\nabla I_\nu = \kappa_\nu\rho(B_\nu - I_\nu). \tag{I.13}$$

It is sometimes useful to consider the first moment of Eq. (I.13), obtained by integrating over $d\Omega$ and employing Eqs. (I.4), (I.5), and (I.10):

$$\frac{\partial \varepsilon_\nu^r}{\partial t} + \nabla\cdot\mathbf{F}_\nu = c\kappa_\nu\rho\left(\varepsilon_\nu^P - \varepsilon_\nu^r\right). \tag{I.14}$$

Here we have assumed that κ_ν is isotropic. Equation (I.14) is essentially an "equation of continuity" for radiation of a given frequency, including sources and sinks of radiation.

A formal solution to the equation of radiative transfer can be obtained by assuming that those functions depending upon the local thermodynamic state of the gas [e.g., $B_\nu(T)$, $\kappa_\nu(\rho, T)$] are *known* functions of space and time. For simplicity, we shall assume steady state and ignore time derivatives in Eq. (I.13). In this case Eq. (I.13) may be regarded as an ordinary linear differential equation for I_ν along the direction of propagation:

$$\frac{dI_\nu}{ds} = \kappa_\nu\rho(B_\nu - I_\nu). \tag{I.15}$$

Defining the *optical depth* τ_ν along the propagation direction according to

$$d\tau_\nu \equiv \kappa_\nu\rho\,ds, \tag{I.16}$$

we may integrate Eq. (I.15) to obtain the radiation received from a region of total optical depth τ_ν:

$$I_\nu = I_\nu(0)\exp(-\tau_\nu) + \int_0^{\tau_\nu} B_\nu(T)\exp\left[-(\tau_\nu - \tau_\nu')\right]d\tau_\nu'. \tag{I.17}$$

Here $I_\nu(0)$ is the intensity of radiation incident on the back side of the region, where τ_ν is defined to be zero. In general, the temperature may vary in the medium—that is, $T = T(\tau_\nu)$. For the special case in which T is constant, Eq.

(I.17) gives

$$I_\nu = I_\nu(0)\exp(-\tau_\nu) + B_\nu(T)[1 - \exp(-\tau_\nu)] \quad (T = \text{constant}). \quad (I.18)$$

Hence, for an "optically thick" region in which $\tau_\nu \gg 1$,

$$I_\nu \approx B_\nu(T), \qquad \tau_\nu \gg 1. \quad (I.19)$$

Equation (I.19) expresses the fact that the emission from an isothermal, optically thick slab is blackbody at the gas temperature. For an "optically thin" slab in which $\tau_\nu \ll 1$ with no incident radiation [i.e., $I_\nu(0) = 0$], the intensity is

$$I_\nu = \int_0^L j_\nu \, ds \quad (I.20)$$

$$\approx B_\nu(T)\tau_\nu \quad (T = \text{constant}),$$

where L is the total geometrical thickness of the slab along the light ray. Thus, for an optically thin slab, the reabsorption of the emitted radiation is unimportant. Equation (I.20) is valid whether or not T varies with position [cf. Eq. (I.11)]. Note that, depending upon the variation of the opacity with frequency, a region may be optically thick at some frequencies and optically thin at others.

I.2　The Diffusion Approximation

The equation of radiative transport simplifies considerably when the radiation field is only weakly anisotropic. Multiply Eq. (I.13) by $\mathbf{n}$ and integrate over all angles. Noting that $\kappa_\nu \rho B_\nu$ is independent of direction and so does not contribute to the integral, and recalling definition (I.5) for the flux, we obtain

$$\frac{1}{c}\frac{\partial}{\partial t}\mathbf{F}_\nu + \int \mathbf{n}(\mathbf{n} \cdot \nabla I_\nu) \, d\Omega = -\kappa_\nu \rho \mathbf{F}_\nu. \quad (I.21)$$

Assume next that the radiation may be treated as quasisteady at each instant of time, in which case Eq. (I.21) reduces to

$$\int \mathbf{n}(\mathbf{n} \cdot \nabla I_\nu) \, d\Omega = -\kappa_\nu \rho \mathbf{F}_\nu. \quad (I.22)$$

Equation (I.22) remains valid even if the radiation field changes with time, provided the changes are sufficiently slow that they may be described completely through the time variation of the gas temperature and density.[2]

---

[2]See Zel'dovich and Raizer (1966), Section II.6, for a discussion of this point.

Now recall that in a radiation field that is strictly isotropic, $\mathbf{F}_\nu$ vanishes. For a weakly anisotropic radiation field, we may evaluate the left-hand side of Eq. (I.22) by keeping only the leading-order, isotropic part of I_ν. The ith component is

$$\int n_i n_k \partial_k I_\nu \, d\Omega = (\partial_k I_\nu) \int n_i n_k \, d\Omega$$

$$= (\partial_k I_\nu) \frac{4\pi}{3} \delta_{ik}$$

$$= \frac{c}{3} \partial_i \varepsilon_\nu^r, \tag{I.23}$$

where we have used Eq. (I.7). Thus Eq. (I.22) becomes

$$\mathbf{F}_\nu = -\frac{c}{3\kappa_\nu \rho} \nabla \varepsilon_\nu^r. \tag{I.24}$$

Taken together with Eq. (I.14) in the quasisteady limit,

$$\nabla \cdot \mathbf{F}_\nu = c\kappa_\nu \rho \left(\varepsilon_\nu^P - \varepsilon_\nu^r \right), \tag{I.25}$$

Eqs. (I.24) and (I.25) constitute the *diffusion approximation* to the radiative transport equations. This approximation is valid whenever the radiation field is isotropic over distances comparable to or less than a radiation mean free path,

$$\lambda_\nu \equiv \frac{1}{\kappa_\nu \rho}. \tag{I.26}$$

Note that to be applicable the diffusion approximation does *not* require LTE. LTE is a sufficient, but not a necessary, condition for use of the diffusion approximation.

Let us now *assume* that LTE holds and consider the frequency-integrated version of Eq. (I.24). Given LTE we can set $\varepsilon_\nu^r = \varepsilon_\nu^P$, so that on integrating over ν we obtain

$$\mathbf{F} = -\frac{c}{3\rho} \int_0^\infty \frac{1}{\kappa_\nu} \nabla \varepsilon_\nu^P \, d\nu. \tag{I.27}$$

Defining an average opacity, the *Rosseland mean opacity*, by

$$\frac{1}{\kappa} \equiv \frac{\int_0^\infty (1/\kappa_\nu)\left(d\varepsilon_\nu^P/dT \right) d\nu}{\int_0^\infty \left(d\varepsilon_\nu^P/dT \right) d\nu}, \tag{I.28}$$

we may rewrite Eq. (I.27) as

$$\mathbf{F} = -\frac{c}{3\kappa\rho}\nabla\varepsilon_P = -\frac{c}{3\kappa\rho}\nabla(aT^4). \qquad (I.29)$$

We will occasionally put a bar over κ to emphasize that a Rosseland mean has been taken.

Exercise I.1 Show that the Rosseland weighting factor $d\varepsilon_\nu^P/dT$ has a maximum at $h\nu \simeq 3.8kT$, so that relatively high-energy photons play the major role in the transport process.

Although our entire discussion thus far has focused on *true absorption* opacity sources [via Eq. (I.12)] rather than on *scattering* sources, the results can be easily generalized to include scattering. Thus, whenever there is more than one source of opacity, Eq. (I.29) still applies, but with the Rosseland mean opacity defined by

$$\frac{1}{\kappa} = \left\langle \frac{1}{\kappa_1 + \kappa_2 + \cdots} \right\rangle. \qquad (I.30)$$

The average in Eq. (I.30) is taken with respect to the Rosseland weighting factor. Scattering opacities may be included in the sum appearing in Eq. (I.30), but since scattering does not obey Kirchhoff's law, there is no correction for stimulated emission for the scattering terms.

For transport in spherical stars, Eq. (I.29) is often written in the form

$$L(r) = 4\pi r^2 F = -\frac{16\pi r^2 ac T^3}{3\kappa\rho}\frac{dT}{dr}, \qquad (I.31)$$

where $L(r)$ is the luminosity (erg s^{-1}) at r. For such systems, the net flux of photons is driven by the temperature gradient in the interior.

The thermal structure of the *atmosphere* of a radiating region can be determined rather simply in the diffusion approximation. Suppose we treat the atmosphere as a semi-infinite, plane-parallel half-space in LTE. Imagine that throughout the atmosphere the flux is constant:

$$F = \text{constant}. \qquad (I.32)$$

Equation (I.32) will apply rather well to the outermost layers of an optically thick radiating region since most of the radiation is generated by the bulk of the gas far below the surface. Equation (I.32) also follows directly from Eq. (I.25) when LTE is assumed. In this case $\nabla \cdot \mathbf{F}_\nu = dF_\nu/dz = 0$, where z measures coordinate distance in the outward normal direction to the surface of the atmosphere. On replacing z by the mean optical thickness τ, where

$$d\tau \equiv -\kappa\rho\,dz \tag{I.33}$$

(the minus sign guarantees that τ, which is zero at the surface, *increases* as we move deeper into the atmosphere), we may re-write Eq. (I.29) in the form

$$F = \frac{c}{3}\frac{d\varepsilon_P}{d\tau} = \frac{c}{3}\frac{d(aT^4)}{d\tau}. \tag{I.34}$$

Integrating Eq. (I.34) with respect to τ gives

$$\varepsilon_P \equiv aT^4 = 3\tau\frac{F}{c} + \text{constant.} \tag{I.35}$$

The constant in Eq. (I.35) may be determined by imposing a boundary condition on the radiation field at $\tau = 0$. If we assume that the radiation field at the surface is isotropic along the outward directions but vanishes in the inward directions (i.e., no photons are incident on the surface from the vacuum side), then we easily obtain from Eqs. (I.4) and (I.5) the relation

$$F_\nu = \frac{c\varepsilon_\nu^r}{2}, \qquad \tau_\nu = 0, \tag{I.36}$$

or, in LTE,

$$F = \frac{c\varepsilon_P}{2}, \qquad \tau = 0. \tag{I.37}$$

Using Eq. (I.37) in (I.35), we obtain the matter temperature profile in terms of optical depth:

$$T^4 = T_0^4\left(1 + \frac{3\tau}{2}\right). \tag{I.38}$$

In Eq. (I.38) T_0 is the surface temperature—that is, $\varepsilon_P(\tau = 0) = aT_0^4$. It is thus related to the effective temperature, defined by $F = \sigma T_{\text{eff}}^4$, $\sigma \equiv ac/4$, by

$$T_{\text{eff}} = 2^{1/4}T_0. \tag{I.39}$$

   Equation (I.39) is often referred to as the "Eddington approximation" to the atmosphere profile. According to Eqs. (I.38) and (I.39) the effective temperature is the same as the matter temperature at $\tau = \frac{2}{3}$. If the opacity is dominated by true absorption processes, then the emergent photons originate near and above the layer at which $\tau \approx \frac{2}{3}$ [cf. Eq. (I.17)]. Thus, the characteristic photon spectral temperature (the "color temperature") will also be comparable to the effective temperature. If the opacity is dominated by scattering, Eq. (I.38) still describes the matter temperature profile (approximately). However, the emergent photons will originate, typically, deeper in, at say $\tau \equiv \tau_s \gg 1$. Following emission far below the surface at large optical depth τ_s, photons undergo many scatterings prior to emerging. If the scattering is elastic, then the photons will emerge with a color temperature comparable to the matter temperature at the point of generation, τ_s; hence[3] $T_{\text{color}} \approx T(\tau_s) \gg T_{\text{eff}}$. Thus elastic scattering tends to increase the characteristic energy of emerging photons *above* the value they would have if the emission were thermal blackbody radiation.

I.3   Opacity Sources

Radiative transport in astrophysical plasmas is governed by several different opacity sources. Which source dominates depends on the thermodynamic state of the gas—that is, ρ and T. For many applications discussed in this book (e.g., X-ray sources) the following nonrelativistic ($h\nu \ll m_e c^2$, $kT \ll m_e c^2$) sources of opacity can be important:

Scattering from Free Electrons

The cross section for the scattering of photons by free electrons is given by the Thomson formula,

$$\sigma_T = \frac{8\pi}{3}\left(\frac{e^2}{m_e c^2}\right)^2 = 0.665 \times 10^{-24} \text{ cm}^2. \tag{I.40}$$

This cross section is frequency independent and the scattering is elastic. The corresponding opacity is given by

$$\kappa_{\text{es}} = \frac{\sigma_T n_e}{\rho} = 0.40 f_e \text{ cm}^2 \text{ g}^{-1}, \tag{I.41}$$

where $n_e = f_e \rho / m_p$ is the free electron density and where f_e is the number of free electrons per baryon.[4] Because electron scattering is frequency independent, the Rosseland mean scattering opacity $\bar{\kappa}_{\text{es}}$ is also given by Eq. (I.41).

---

[3]See Section 14.5(j) for a more detailed discussion.
[4]We have previously used $Y_e \equiv f_e$ to denote the electron number per baryon for completely ionized matter.

Free-Free Absorption (Inverse Bremsstrahlung)

The opacity for free-free absorption (i.e., the absorption of photons by free electrons in the presence of positive ions of charge Z and atomic weight A) is given by[5]

$$\rho\kappa_\nu^{\rm ff} = \frac{4(2\pi)^{1/2}e^6Z^2n_en_ig_{\rm ff}}{(3m_e)^{3/2}c(kT)^{1/2}h\nu^3}\left[1 - \exp\left(\frac{-h\nu}{kT}\right)\right], \qquad (I.42)$$

where $n_i = f_i\rho/m_pA$ is the ion number density, f_i is the fraction of all atoms ionized to the state of interest, and $g_{\rm ff}$, which varies slowly with frequency ν, is the Gaunt factor for free-free transitions. The above opacity assumes that the electrons have a Maxwellian velocity distribution relative to the photons. Tables of Maxwell–Boltzmann-averaged Gaunt factors are given by Greene (1959) and Karzas and Latter (1961); for $h\nu/kT \le 1$ and $kT \le Z^2$ Ry, $g_{\rm ff} \approx 1$. The free-free opacity given by Eq. (I.42) can be re-written

$$\kappa_\nu^{\rm ff} = \left(1.50 \times 10^{25}\ {\rm cm^2\ g^{-1}}\right)\left(\frac{f_ef_iZ^2}{A}\right)\rho T^{-3.5}g_{\rm ff}\left(\frac{1 - \exp(-x)}{x^3}\right), \quad (I.43)$$

where $x = h\nu/kT$. The corresponding Rosseland mean opacity can be calculated from Eqs. (I.28) and (I.43) to yield

$$\bar\kappa_{\rm ff} = \left(0.645 \times 10^{23}\ {\rm cm^2\ g^{-1}}\right)\left(\frac{f_ef_iZ^2}{A}\right)\bar g_{\rm ff}\rho T^{-3.5}, \qquad (I.44)$$

where $\bar g_{\rm ff} \sim 1$ is the frequency-averaged Gaunt factor. Note that the mean opacity varies as $\rho T^{-3.5}$, as first demonstrated by Kramers. Any opacity that varies in this way is called a *Kramers opacity*.

The free-free emissivity can be calculated from Kirchhoff's law, Eq. (I.12), and Eq. (I.42). The result is

$$j_\nu^{\rm ff} = \frac{8}{3}\left(\frac{2\pi}{3}\right)^{1/2}\frac{e^6}{m_e^2c^3}\left(\frac{m_e}{kT}\right)^{1/2}g_{\rm ff}Z^2n_en_i\exp\left(\frac{-h\nu}{kT}\right)$$

$$= 5.44 \times 10^{-39}\frac{g_{\rm ff}Z^2n_en_i}{T^{1/2}}\exp\left(\frac{-h\nu}{kT}\right)\ {\rm erg\ cm^{-3}\ s^{-1}\ sr^{-1}\ Hz^{-1}}. \quad (I.45)$$

The total amount of energy emitted by free-free transitions per cm³ per sec is

[5]See, for example, Spitzer (1978); Novikov and Thorne (1973). Note that κ_ν defined by Spitzer is $\rho\kappa_\nu$ defined here.

obtained by integrating Eq. (I.45) over all frequency and multiplying by 4π:

$$\Lambda_{ff} = 1.42 \times 10^{-27} Z^2 n_e n_i T^{1/2} \bar{\bar{g}}_{ff} \text{ erg cm}^{-3} \text{ s}^{-1},\qquad(\text{I.46})$$

where the corresponding mean Gaunt factor $\bar{\bar{g}}_{ff}$ is again a slowly varying function near unity for conditions of interest.

Bound-Free Absorption (Photoionization)

Bound-free absorption occurs when an atom with bound electrons absorbs photons of sufficient energy to ionize the atom into a higher state. The reverse process is called "free-bound" radiation or "recombination" and results from the capture of a free electron in a bound state and the associated emission of a photon to conserve energy.

The cross section for the photoionization of a hydrogenic electron in a state with principal quantum number n may be written[6]

$$\sigma_{bf} = \frac{64\pi^4 m_e e^{10} Z^4}{3\sqrt{3} \, ch^6 n^5} \frac{g_{bf}}{\nu^3}$$

$$= (2.82 \times 10^{29} \text{ cm}^2) \frac{Z^4}{n^5 \nu^3} g_{bf}.\qquad(\text{I.47})$$

From Eq. (I.47) one may construct the bound-free opacity and its associated Rosseland mean. An approximate expression for the bound-free Rosseland mean opacity for a mixture of partially ionized atoms of different species is given by Schwarzschild (1958):

$$\bar{\kappa}_{bf} \approx (4.34 \times 10^{25} \text{ cm}^2 \text{ g}^{-1}) \bar{g}_{bf} Z (1 + X) \rho T^{-3.5},\qquad(\text{I.48})$$

where X and Z represent the mass fractions of hydrogen and heavy elements (i.e., elements other than H and He), respectively, in the mixture. Note that Eq. (I.48) also takes the form of a Kramers opacity law.

Although the dominant opacity source in any region depends in detail on ρ and T, the general behavior is as follows: at low T, where a large fraction of the atoms may be only partially ionized, bound-free absorption is likely to be important (at still lower T, the opacity resulting from "bound-bound" discrete transitions may dominate). At higher T, where the ionization nears completion, free-free absorption will dominate. Eventually, as T increases further, electron scattering will take over since $\bar{\kappa}_{ff}$ decreases with increasing T [cf. Eq. (I.44)]. New processes become important at relativistic photon energies ($h\nu \gtrsim m_e c^2$) and

[6] Clayton (1968), Eq. (3.151).

electron temperature ($kT \gtrsim m_e c^2$). In addition, the equation given above for bremsstrahlung and electron scattering must be modified.[7]

Comptonization

Comptonization is the process by which photons can gain or lose energy by scattering off thermal electrons. The process has proven to be significant for the generation of X-rays emitted from gas accreting onto compact objects. Accordingly, we summarize below some of its key features.[8]

Consider an infinite homogeneous Maxwellian gas of nonrelativistic electrons at temperature T and density n_e. The time evolution of the photon distribution function f due to repeated, nonrelativistic inverse Compton scattering off these electrons is governed by the *Kompaneets equation*.[9] The equation represents the Fokker–Planck approximation to the collisional Boltzmann equation for the photons. It is correct to second order in the energy exchange per collision, which is small (i.e., $\Delta E \ll kT$). The equation reads

$$\frac{\partial f}{\partial t} = \left[n_e \sigma_T c \left(\frac{kT}{m_e c^2} \right) \right] \frac{1}{x^2} \frac{\partial}{\partial x} \left\{ x^4 \left(\frac{\partial f}{\partial x} + f + f^2 \right) \right\}, \qquad (I.49)$$

where $x = h\nu/kT$. Not surprisingly, steady state is achieved when

$$f = \frac{1}{\exp[(E - \mu)/kT] - 1}, \qquad \left(\frac{\partial f}{\partial t} = 0 \right), \qquad (I.50)$$

where μ is a constant. When $\mu \leqslant 0$, Eq. (I.50) describes the general equilibrium distribution function for a Bose–Einstein gas—that is, equilibrium with a conserved particle number. Since Comptonization conserves photon *number*, Eq. (I.50) is quite reasonable. When, following repeated scatterings, the photon distribution achieves the Bose–Einstein form of Eq. (I.50), the process "saturates" and no further net energy exchange continues with the electrons. In this case, the spectrum assumes the familiar Wien shape when $f \ll 1$:

$$I_\nu \propto x^3 f \propto x^3 \exp(-x). \qquad (I.51)$$

The net rate of energy transfer from the electrons to the photons per unit volume is obtained by multiplying Eq. (I.49) by $h\nu$ and integrating over phase space.

---

[7] For details and references see, for example, Bethe and Salpeter (1957) or Rybicki and Lightman (1979).

[8] For a more thorough discussion, see references cited in Section 14.5(j).

[9] Kompaneets (1957).

Following integration by parts the result is[10]

$$\Lambda_{\text{Comp}} = \int h\nu \frac{\partial f}{\partial t} \frac{8\pi\nu^2 \, d\nu}{c^3}$$

$$= 4\sigma_T n_e c\varepsilon_r \left[\frac{kT}{m_e c^2} - \frac{2\pi(kT)^5}{(hc)^3 m_e c^2 \varepsilon_r} \int_0^\infty dx \, x^4 f(1+f) \right], \qquad (1.52)$$

where ε_r is the photon energy density given by

$$\varepsilon_r \equiv \frac{8\pi(kT)^4}{h^3 c^3} \int_0^\infty dx \, x^3 f. \qquad (1.53)$$

If we assume that the photons may be approximated by an equilibrium Planck distribution but at a temperature $T_{\text{ph}} \neq T$, Eq. (1.52) yields

$$\Lambda_{\text{Comp}} = 4\sigma_T n_e c\varepsilon_r \frac{k}{m_e c^2} (T - T_{\text{ph}}). \qquad (1.54)$$

Equation (1.54) shows that Comptonization causes the matter to cool when $T > T_{\text{ph}}$ and heat up when $T_{\text{ph}} < T$.

    Consider the modification of Eq. (1.49) for the case in which photons are generated in and escape from a medium of *finite* scattering depth τ_{es} (e.g., a homogeneous disk of half-thickness h, for which $\tau_{\text{es}} \approx \sigma_T n_e h$ measured from the disk midplane). Assume that photons are steadily produced at moderately low energies in the disk ($x \ll 1$), diffuse outward toward the surface via repeated scattering, and escape. The probability that a typical photon escapes per Compton scattering is then roughly equal to the reciprocal of the mean number of scatterings, N_s.

    Now

$$N_s \approx \max(\tau_{\text{es}}, \tau_{\text{es}}^2), \qquad (1.55)$$

since $\tau_{\text{es}} \lesssim 1$ leads to free escape while $\tau_{\text{es}} \gtrsim 1$ results in a random-walk through the medium prior to escape. In steady state, a "modified" Kompaneets Eq. (1.49) describing generation, Comptonization, and escape of photons from the medium may be written[11]

$$0 = \left(\frac{kT}{m_e c^2} \right) \frac{1}{x^2} \frac{\partial}{\partial x} \left[x^4 \left(\frac{\partial f}{\partial x} + f + f^2 \right) \right] + Q(x) - \frac{f}{\text{Max}(\tau_{\text{es}}, \tau_{\text{es}}^2)}, \qquad (1.56)$$

[10] Peebles (1971).
[11] Shapiro et al. (1976).

where $Q(x)$ describes the photon generation rate. Assume that photons are generated at some low frequency $x_s \ll 1$, above which Q is negligible ("soft photon source"). Focus attention on those emergent Comptonized photons with $x \gg x_s$, for which the terms $f^2 \ll 1$ and Q can be ignored in Eq. (I.56). Then the resulting equation yields the following approximate solution:

$$f \propto x^m, \qquad x_s \ll x \ll 1, \tag{I.57}$$

$$m = -\frac{3}{2} \pm \sqrt{\frac{9}{4} + \frac{4}{y}}, \qquad y \equiv \frac{4kT}{m_e c^2} N_s, \tag{I.58}$$

and

$$f \propto \exp(-x), \qquad x \gg 1. \tag{I.59}$$

The emergent intensity in the region between $x_s \ll x \ll 1$ has the power-law shape

$$I_\nu \sim I_{\nu_s} \left(\frac{\nu}{\nu_s} \right)^{3+m}. \tag{I.60}$$

For $y \geqslant 1$ or $m \geqslant -4$ a substantial increase in the "hard spectral" component above ν_s can thus emerge from the region due to Comptonization. This results from the repeated "upscattering" of soft photons by higher temperature electrons. An illustration of the emergent spectrum is shown in Figure I.1.

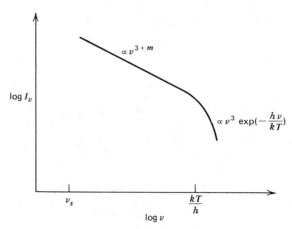

Figure I.1   Spectrum generated by the Comptonization of low energy photons by high-energy thermal electrons. [From Rybicki and Lightman (1979).]

References

Abramowicz, M. A., 1981. "Innermost Parts of Accretion Disks are Thermally and Secularly Stable," *Nature* **294**, 235.

Abramowicz, M. A. and R. V. Wagoner, 1976. "Variational Analysis of Rotating Neutron Stars," *Astrophys. J.* **204**, 896.

Abt, H. A., P. Hintzen and S. G. Levy, 1977. "A Search for a Third Star in the Cygnus X-1 System," *Astrophys. J.* **213**, 815.

Adams, W. S., 1915. "The Spectrum of the Companion of Sirius," *Pub. Astron. Soc. Pac.* **27**, 236.

_____ 1925. "The Relativity Displacement of the Spectral Lines in the Companion of Sirius," *Proc. Natl. Acad. Sci. USA* **11**, 382. Erratum: *Observatory* **49**, 88.

Aizu, K., 1973. "X-Ray Emission Region of a White Dwarf with Accretion," *Progr. Theor. Phys.* **49**, 1184.

Allen, C. W., 1963. *Astrophysical Quantities*, 2nd ed., University of London, London, England.

_____ 1973. *Astrophysical Quantities*, 3rd ed., Athlone, London, England.

Alloin, D., C. Cruz-González, and M. Peimbert, 1976. "On the Number of Planetary Nebulae in our Galaxy," *Astrophys J.* **205**, 74.

Alme, M. L. and J. R. Wilson, 1973. "X-Ray Emission from a Neutron Star Accreting Material," *Astrophys. J.* **186**, 1015.

Alpar, M. A., P. W. Anderson, D. Pines, and J. Shaham, 1981. "Giant Glitches and Pinned Vorticity in the Vela and Other Pulsars," *Astrophys. J. Lett.* **249**, L29.

Ambartsumyan, V. A. and G. S. Saakyan, 1960. "The Degenerate Superdense Gas of Elementary Particles," *Astron. Zhur.* **37**, 193 (English transl. in *Sov. Astron.-AJ* **4**, 187).

Angel, J. R. P., 1978. "Magnetic White Dwarfs," *Ann. Rev. Astron. Astrophys.* **16**, 487.

Applequist, T. and H. D. Politzer, 1975. "Heavy Quarks and e^+e^- Annihilation," *Phys. Rev. Lett.* **34**, 43.

Arnett, W. D., 1967. "Mass Dependence in Gravitational Collapse of Stellar Cores," *Can. J. Phys.* **45**, 1621.

_____ 1977a. "Advanced Evolution of Massive Stars: VII. Silicon Burning," *Astrophys. J. Suppl.* **35**, 145.

_____ 1977b. "Neutrino Trapping During Gravitational Collapse of Stars," *Astrophys. J.* **218**, 815.

_____ 1978. "On the Bulk Yields of Nucleosynthesis from Massive Stars," *Astrophys. J.* **219**, 1008.

_____ 1979. "Gravitational Collapse of Evolved Stars as a Problem in Physics," in *Sources of Gravitational Radiation*, L. L. Smarr, editor, Cambridge University Press, Cambridge, England.

Arnett, W. D. and R. L. Bowers, 1977. "A Microscopic Interpretation of Neutron Star Structure," *Astrophys. J. Suppl.* **33**, 415.

Arnett, W. D. and S. W. Falk, 1976. "Some Comparisons of Theoretical Supernova Light Curves with Supernova 1961 (Type II)," *Astrophys. J.* **210**, 733.

Arnett, W. D. and D. N. Schramm, 1973. "Origin of Cosmic Rays, Atomic Nuclei, and Pulsars in Explosions of Massive Stars," *Astrophys. J. Lett.* **184**, L47.

Arons, J. and S. M. Lea, 1976. "Accretion onto Magnetized Neutron Stars: Structure and Interchange Instability of a Model Magnetosphere," *Astrophys. J.* **207**, 914.

_____ 1980. "Accretion onto Magnetized Neutron Stars: The Fate of Sinking Filaments," *Astrophys. J.* **235**, 1016.

Au, C. K., 1976. "Equation of State for Pion Condensed Neutron Star Matter," *Phys. Lett. B* **61**, 300.

Audouze, J. and B. M. Tinsley, 1976. "Chemical Evolution of Galaxies," *Ann. Rev. Astron. Astrophys.* **14**, 43.

Auer, L. H. and D. Mihalas, 1972. "Non-LTE Model Atmospheres. VII. The Hydrogen and Helium Spectra of the O Stars," *Astrophys. J. Suppl.* **24**, 193.

Avni, Y. and J. N. Bahcall, 1975a. "Ellipsoidal Light Variations and Masses of X-Ray Binaries," *Astrophys. J.* **197**, 675.

_____ 1975b. "Masses for Vela X-1 and Other X-Ray Binaries," *Astrophys. J. Lett.* **202**, L131.

Baade, W., 1942. "The Crab Nebula," *Astrophys. J.* **96**, 188.

Baade, W. and F. Zwicky, 1934. "Supernovae and Cosmic Rays," *Phys. Rev.* **45**, 138.

Backer, D. C., S. R. Kulkarni, C. Heiles, M. M. Davis, and W. M. Goss, 1982. "A Millisecond Pulsar," *Nature* **300**, 615.

Bahcall, J. N., 1978a. "Masses of Neutron Stars and Black Holes in X-Ray Binaries," *Ann. Rev. Astron. Astrophys.* **16**, 241.

_____ 1978b. "Solar Neutrino Experiments," *Rev. Mod. Phys.* **50**, 881.

_____ 1978c. "Optical Properties of Binary X-Ray Sources," in *Physics and Astrophysics of Neutron Stars and Black Holes*, R. Giacconi and R. Ruffini, editors, North Holland, Amsterdam, Holland.

_____ 1981. "Solar Neutrinos: Rapporteur Talk," in *Neutrino-81*, R. J. Cence, E. Ma, and A. Roberts, editors, Department of Physics and Astronomy, University of Hawaii, Honolulu, Hawaii.

Bahcall, J. N. and T. J. Chester, 1977. "On the Mass Determination of Hercules X-1," *Astrophys. J. Lett.* **215**, L21.

Bahcall, J. N. and J. P. Ostriker, 1975. "Massive Black Holes in Globular Clusters," *Nature* **256**, 23.

Bahcall, J. N. and R. M. Soneira, 1980. "The Universe at Faint Magnitudes. I. Models for the Galaxy and the Predicted Star Counts," *Astrophys. J. Suppl.* **44**, 73.

_____ 1981. "The Distribution of Stars to $V = $ 16th Magnitude Near the North Galactic Pole: Normalization, Clustering Properties, and Counts in Various Bands," *Astrophys. J.* **246**, 122.

Bahcall, J. N. and R. A. Wolf, 1965. "Neutron Stars. II. Neutrino-Cooling and Observability," *Phys. Rev. B* **140**, 1452.

Bahcall, J. N., M. N. Rosenbluth, and R. M. Kulsrud, 1973. "Model for X-Ray Sources Based on Magnetic Field Twisting," *Nature Phys. Sci.* **243**, 27.

Bahcall, J. N., F. J. Dyson, J. I. Katz, and B. Paczynski, 1974. "Multiple Star Systems and X-Ray Sources," *Astrophys. J. Lett.* **189**, L17.

Bardeen, J. M., 1973. "Timelike and Null Geodesics in the Kerr Metric," in *Black Holes*, C. DeWitt and B. S. DeWitt, editors, Gordon and Breach, New York, New York.

Bardeen, J. M., W. H. Press, and S. A. Teukolsky, 1972. "Rotating Black Holes: Locally Nonrotating Frames, Energy Extraction, and Scalar Synchrotron Radiation," *Astrophys. J.* **178**, 347.

Bardeen, J. M., J. L. Friedman, B. F. Schutz, and R. Sorkin, 1977. "A New Criterion for Secular Instability of Rotating Stars," *Astrophys. J. Lett.* **217**, L49.

Barkat, Z., 1977. "Evolution of Supernova Progenitors," in *Supernovae*, D. N. Schramm, editor, Reidel, Dordrecht, Holland.

Bath, G. T. and J. E. Pringle, 1981. "The Evolution of Viscous Discs.—I. Mass Transfer Variations," *Mon. Not. Roy. Astron. Soc.* **194**, 967.

Baym, G. and S. A. Chin, 1976a. "Landau Theory of Relativistic Fermi Liquids," *Nucl. Phys. A* **262**, 527.

_____ 1976b. "Can a Neutron Star be a Giant MIT Bag?," *Phys. Lett. B* **62**, 241.

Baym, G. and C. Pethick, 1975. "Neutron Stars," *Ann. Rev. Nucl. Sci.* **25**, 27.

_____ 1979. "Physics of Neutron Stars," *Ann. Rev. Astron. Astrophys.* **17**, 415.

Baym, G. and D. Pines, 1971. "Neutron Starquakes and Pulsar Speedup," *Ann. Phys.* **66**, 816.

Baym, G., C. Pethick, D. Pines, and M. Ruderman, 1969. "Spin Up in Neutron Stars: The Future of the Vela Pulsar," *Nature* **224**, 872.

Baym, G., H. A. Bethe, and C. J. Pethick, 1971a (BBP). "Neutron Star Matter," *Nucl. Phys. A* **175**, 225.

Baym, G., C. Pethick, and P. Sutherland, 1971b (BPS). "The Ground State of Matter at High Densities: Equation of State and Stellar Models," *Astrophys. J.* **170**, 299.

Becker, R. H., J. H. Swank, E. A. Boldt, S. S. Holt, S. H. Pravdo, J. R. Saba, and P. J. Serlemitsos, 1977. "A1540-53, an Eclipsing X-Ray Binary Pulsator," *Astrophys. J. Lett.* **216**, L11.

Begelman, M. C., 1977. "Nearly Collisionless Spherical Accretion," *Mon. Not. Roy. Astron. Soc.* **181**, 347.

Bekenstein, J. D., 1973. "Black Holes and Entropy," *Phys. Rev. D* **7**, 2333.

_____ 1975. "Statistical Black-Hole Thermodynamics," *Phys. Rev. D* **12**, 3077.

Belian, R. D., J. P. Conner, and W. D. Evans, 1976. "The Discovery of X-Ray Bursts from a Region in the Constellation Norma," *Astrophys. J. Lett.* **206**, L135.

Bethe, H. A., 1953. "What Holds the Nucleus Together?," *Sci. Am.* **189**, No. 3, 58.

Bethe, H. A. and R. W. Jackiw, 1968. *Intermediate Quantum Mechanics*, 2nd ed., Benjamin, New York, New York.

Bethe, H. A. and M. B. Johnson, 1974. "Dense Baryon Matter Calculations with Realistic Potentials," *Nucl. Phys. A* **230**, 1.

Bethe, H. A. and E. E. Salpeter, 1957. *Quantum Mechanics of One- and Two-Electron Atoms*, Academic Press, New York, New York.

Bethe, H. A., G. E. Brown, J. Applegate, and J. M. Lattimer, 1979 (BBAL). "Equation of State in the Gravitational Collapse of Stars," *Nucl. Phys. A* **324**, 487.

Bisnovatyi-Kogan, G. S. and S. I. Blinnikov, 1977. "Disk Accretion onto a Black Hole at Subcritical Luminosity," *Astron. Astrophys.* **59**, 111.

Blaizot, J. P., D. Gogny, and B. Grammaticos, 1976. "Nuclear Compressibility and Monopole Resonances," *Nucl. Phys. A* **265**, 315.

Blandford, R. D., 1976. "Accretion Disc Electrodynamics—A Model for Double Radio Sources," *Mon. Not. Roy. Astron. Soc.* **176**, 465.

_____ 1977. "Spectrum of a Radio Pulse from an Exploding Black Hole," *Mon. Not. Roy. Astron. Soc.* **181**, 489.

Blandford, R. and S. A. Teukolsky, 1975. "On the Measurement of the Mass of PSR 1913 + 16," *Astrophys. J. Lett.* **198**, L27.

_____ 1976. "Arrival Time Analysis for a Pulsar in a Binary System," *Astrophys. J.* **205**, 580.

Blandford, R. D. and K. S. Thorne, 1979. "Black Hole Astrophysics," in *General Relativity. An Einstein Centenary Survey*, S. W. Hawking and W. Israel, editors, Cambridge University Press, Cambridge, England.

Blandford, R. D. and R. L. Znajek, 1977. "Electromagnetic Extraction of Energy from Kerr Black Holes," *Mon. Not. Roy. Astron. Soc.* **179**, 433.

Blatt, J. M. and V. F. Weisskopf, 1952. *Theoretical Nuclear Physics*, Wiley, New York, New York.

Bludman, S. A. and M. A. Ruderman, 1970. "Noncausality and Instability in Ultradense Matter," *Phys. Rev. D* **1**, 3243.

_____ 1975. "Bounds on Neutrino Burst Intensity Imposed by the Exclusion Principle and Causality," *Astrophys. J. Lett.* **195**, L19.

Bondi, H., 1952. "On Spherically Symmetrical Accretion," *Mon. Not. Roy. Astron. Soc.* **112**, 195.

_____ 1964. "Massive Spheres in General Relativity," *Proc. Roy. Soc. London Ser. A* **282**, 303.

Bondi, H. and F. Hoyle, 1944. "On the Mechanism of Accretion By Stars," *Mon. Not. Roy. Astron. Soc.* **104**, 273.

Boyer, R. H. and R. W. Lindquist, 1967. "Maximal Analytic Extension of the Kerr Metric," *J. Math. Phys.* **8**, 265.

Boynton, P. E., 1981. "Pulse Timing and Neutron Star Structure," in IAU Symposium No. 95 *Pulsars*, W. Sieber and R. Wielebinski, editors, Reidel, Dordrecht, Holland, p. 279.

Boynton, P. E., E. J. Groth, D. P. Hutchinson, G. P. Nanos, R. B. Partridge, and D. T. Wilkinson, 1972. "Optical Timing of the Crab Pulsar NP 0532," *Astrophys. J.* **175**, 217.

Bradt, H. V. D. and J. E. McClintock, 1983. "The Optical Counterparts of Compact Galactic X-Ray Sources," *Ann. Rev. Astron. Astrophys.* **21**, in press.

Branduardi, G., N. D. Kylafis, D. Q. Lamb, and K. O. Mason, 1980. "Evidence for the Degenerate Dwarf Nature of Cygnus X-2," *Astrophys. J. Lett.* **235**, L153.

Bregman, J., D. Butler, E. Kemper, A. Koski, R. P. Kraft, and R. P. S. Stone, 1973. "On the Distance to Cygnus X-1 (HDE 226868)," *Astrophys. J. Lett.* **185**, L117.

Brinkmann, W., 1980. "Adiabatic Accretion onto a Schwarzschild Black Hole," *Astron. Astrophys.* **85**, 146.

Brown, G. E., H. A. Bethe, and G. Baym, 1982. "Supernova Theory," *Nucl. Phys. A* **375**, 481.

Bruenn, S. W., 1975. "Neutrino Interactions and Supernovae," *Ann. N.Y. Acad. Sci.* **262**, 80 (Seventh Texas Symposium).

Brumberg, V. A., Ya. B. Zel'dovich, I. D. Novikov, and N. I. Shakura, 1975. "Component Masses and Inclination of Binary Systems Containing a Pulsar, Determined from Relativistic Effects," *Sov. Astron. Lett.* **1**, 2.

Brush, S. G., H. L. Sahlin, and E. Teller, 1966. "Monte Carlo Study of a One-Component Plasma. I.," *J. Chem. Phys.* **45**, 2102.

Burbidge, E. M., G. R. Burbidge, W. A. Fowler, and F. Hoyle, 1957. "Synthesis of the Elements in Stars," *Rev. Mod. Phys.* **29**, 547.

Burnell, S. J. B., 1977. "Petit Four," *Ann. N.Y. Acad. Sci.* **302**, 685 (Eighth Texas Symposium).

Burrows, A., 1980. "Beta Decay in Quark Stars," *Phys. Rev. Lett.* **44**, 1640.

Butterworth, E. M. and J. R. Ipser, 1975. "Rapidly Rotating Fluid Bodies in General Relativity," *Astrophys. J. Lett.* **200**, L103.

———— 1976. "On the Structure and Stability of Rapidly Rotating Fluid Bodies in General Relativity. I. The Numerical Method for Computing Structure and Its Application to Uniformly Rotating Homogeneous Bodies," *Astrophys. J.* **204**, 200.

Byram, E. T., T. A. Chubb, and H. Friedman, 1966. "Cosmic X-Ray Sources, Galactic and Extragalactic," *Science* **152**, 66.

Byrd, P. F. and M. D. Friedman, 1971. *Handbook of Elliptic Integrals for Engineers and Scientists*, 2nd ed., Springer-Verlag, New York, New York.

Cahn, J. H. and S. P. Wyatt, 1976. "The Birthrate of Planetary Nebulae," *Astrophys. J.* **210**, 508.

Cameron, A. G. W., 1959a. "Neutron Star Models," *Astrophys. J.* **130**, 884.

———— 1959b. "Pycnonuclear Reactions and Nova Explosions," *Astrophys. J.* **130**, 916.

Cameron, A. G. W. and M. Mock, 1967. "Stellar Accretion and X-Ray Emission," *Nature* **215**, 464.

Canizares, C. R. and P. F. Winkler, 1981. "Evidence for Elemental Enrichment of Puppis A by a Type II Supernova," *Astrophys. J. Lett.* **246**, L33.

Canuto, V., 1974. "Equation of State at Ultrahigh Densities, Part I," *Ann. Rev. Astron. Astrophys.* **12**, 167.

———— 1975. "Equation of State at Ultrahigh Densities, Part II," *Ann. Rev. Astron. Astrophys.* **13**, 335.

Carr, B. J., 1976. "Some Cosmological Consequences of Primordial Black-Hole Evaporation," *Astrophys. J.* **206**, 8.

Carr, W. J., 1961. "Energy, Specific Heat, and Magnetic Properties of the Low-Density Electron Gas," *Phys. Rev.* **122**, 1437.

Carter, B., 1968. "Global Structure of the Kerr Family of Gravitational Fields," *Phys. Rev.* **174**, 1559.

———— 1979. "The General Theory of the Mechanical, Electromagnetic and Thermodynamic Properties of Black Holes," in *General Relativity: An Einstein Centenary Survey*, S. W. Hawking and W. Israel, editors, Cambridge University Press, Cambridge, England.

Cassinelli, J. P., J. S. Mathis, and B. D. Savage, 1981. "Central Object of the 30 Doradus Nebula, a Supermassive Star," *Science* **212**, 1497.

598   **References**

Chandrasekhar, S., 1931a. "The Density of White Dwarf Stars," *Phil. Mag.* **11**, 592.

_____ 1931b. "The Maximum Mass of Ideal White Dwarfs," *Astrophys. J.* **74**, 81.

_____ 1934. "Stellar Configurations with Degenerate Cores," *Observatory*, **57**, 373.

_____ 1939. *An Introduction to the Study of Stellar Structure*, University of Chicago Press, Chicago, Illinois.

_____ 1960. *Radiative Transfer*, Dover, New York, New York.

_____ 1969. *Ellipsoidal Figures of Equilibrium*, Yale University Press, New Haven, Connecticut.

_____ 1970. "The Effect of Gravitational Radiation on the Secular Stability of the Maclaurin Spheroid," *Astrophys. J.* **161**, 561.

_____ 1980. "The Role of General Relativity in Astronomy: Retrospect and Prospect," in *Highlights of Astronomy*, Vol. 5, P. A. Wayman, editor, Reidel, Dordrecht, Holland.

Chandrasekhar, S. and R. F. Tooper, 1964. "The Dynamical Instability of the White-Dwarf Configurations Approaching the Limiting Mass," *Astrophys J.* **139**, 1396.

Chapline, G. F., 1975. "Cosmological Effects of Primordial Black Holes," *Nature* **253**, 251.

Chapline, G. F. and M. Nauenberg, 1976. "Phase Transition from Baryon to Quark Matter," *Nature* **264**, 235.

Chevalier, R. A., 1976. "The Hydrodynamics of Type II Supernovae," *Astrophys. J.* **207**, 872.

Chevalier, R. A., 1981a. "Exploding White Dwarf Models for Type I Supernovae," *Astrophys. J.* **246**, 267.

_____ 1981b. "The Interaction of the Radiation from a Type II Supernova with a Circumstellar Shell," *Astrophys. J.* **251**, 259.

Chevalier, R. A. and R. P. Kirshner, 1978. "Spectra of Cassiopeia A. II. Interpretation," *Astrophys. J.* **219**, 931.

Chiu, H.-Y., 1964. "Supernovae, Neutrinos, and Neutron Stars," *Ann. Phys.* **26**, 364.

Chiu, H.-Y. and E. E. Salpeter, 1964. "Surface X-Ray Emission from Neutron Stars," *Phys. Rev. Lett.* **12**, 413.

Chodos, A., R. L. Jaffe, K. Johnson, C. B. Thorn, and V. F. Weisskopf, 1974. "New Extended Model of Hadrons," *Phys. Rev. D* **9**, 3471.

Clark, C. B., 1958. "Coulomb Interactions in the Uniform-Background Lattice Model," *Phys. Rev.* **109**, 1133.

Clark, D. H. and J. L. Caswell, 1976. "A Study of Galactic Supernova Remnants, Based on Molongolo-Parkes Observational Data," *Mon. Not. Roy. Astron. Soc.* **174**, 267.

Clark, J. P. A., 1979a. "The Role of Binaries in Gravitational Wave Production," in *Sources of Gravitational Radiation*, L. Smarr, editor, Cambridge University Press, Cambridge, England, p. 447.

Clark, J. P. A. and D. M. Eardley, 1977. "Evolution of Close Neutron Star Binaries," *Astrophys. J.* **215**, 311.

Clark, J. P. A., E. P. J. Van den Heuvel, and W. Sutantyo, 1979. "Formation of Neutron Star Binaries and Their Importance for Gravitational Radiation," *Astron. Astrophys.* **72**, 120.

Clark, J. W., 1979b. "Variational Theory of Nuclear Matter," in *Progr. Part. Nucl. Phys.*, Vol. 2, Pergamon, Oxford, England, p. 89.

Clayton, D. B., 1968. *Principles of Stellar Evolution and Nucleosynthesis*, McGraw-Hill, New York, New York.

Cochran, S. G. and G. V. Chester, 1973. Unpublished work cited by Canuto (1975). See also Cochran, S. G., unpublished Ph.D. Thesis, Cornell University, and Ceperley, D. M., G. V. Chester, and M. H. Kalos, 1976. "Exact Calculations of the Ground State of Model Neutron Matter," *Phys. Rev. D* **13**, 3208.

Cocke, W. J., H. J. Disney, and D. J. Taylor, 1969. "Discovery of Optical Signals from Pulsar NP 0532," *Nature* **221**, 525.

Code, A. D., J. Davis, R. C. Bless, and R. Hanbury-Brown, 1976. "Empirical Effective Temperatures and Bolometric Corrections for Early-Type Stars," *Astrophys. J.* **203**, 417.

Cohen, J. M., A. Lapidus, and A. G. W. Cameron, 1969. "Treatment of Pulsating White Dwarfs Including General Relativistic Effects," *Astrophys. Space Sci.* **5**, 113.

Coldwell-Horsfall, R. A. and A. A. Maradudin, 1960. "Zero-Point Energy of an Electron Lattice," *J. Math. Phys.* **1**, 395.

Coleman, S., 1979. "The 1979 Nobel Prize in Physics," *Science* **206**, 1290.

Colgate, S. A., 1978. "Supernova Mass Ejection and Core Hydrodynamics," *Mem. Del. Soc. Astron. Ital.* **49**, 399.

Colgate, S. A. and M. H. Johnson, 1960. "Hydrodynamic Origin of Cosmic Rays," *Phys. Rev. Lett.* **5**, 235.

Colgate, S. A. and R. H. White, 1966. "The Hydrodynamic Behavior of Supernovae Explosions," *Astrophys. J.* **143**, 626.

Collins, J. C. and M. J. Perry, 1975. "Superdense Matter: Neutrons or Asymptotically Free Quarks?" *Phys. Rev. Lett.* **34**, 1353.

Cominsky, L., G. W. Clark, F. Li, W. Mayer, and S. Rappaport, 1978. "Discovery of 3.6-s X-Ray Pulsations from 4U 0115 + 63," *Nature* **273**, 367.

Conti, P. S., 1978. "Stellar Parameters of Five Early Type Companions of X-Ray Sources," *Astron. Astrophys.* **63**, 225.

Cordes, J. M., 1979. "Coherent Radio Emission from Pulsars," *Space Sci. Rev.* **24**, 567.

Cordes, J. M. and G. Greenstein, 1981. "Pulsar Timing. IV. Physical Models for Timing Noise Processes," *Astrophys. J.* **245**, 1060.

Cordes, J. M. and D. J. Helfand, 1980. "Pulsar Timing. III. Timing Noise of 50 Pulsars," *Astrophys. J.* **239**, 640.

Cowley, A. P. and D. Crampton, 1975. "The Spectroscopic Binary Scorpius X-1," *Astrophys. J. Lett.* **201**, L65.

Cox, J. P. and R. T. Giuli, 1968. *Principles of Stellar Structure*, Vols. I and II, Gordon and Breach, New York, New York.

Cunningham, C. T. and J. M. Bardeen, 1972. "The Optical Appearance of a Star Orbiting an Extreme Kerr Black Hole," *Astrophys. J. Lett.* **173**, L137.

Davidsen, A. and J. P. Henry, 1972. "A Physical Model of HZ Her," *Bull. Am. Astron. Soc.* **4**, 411.

Davidson, K. and J. P. Ostriker, 1973. "Neutron-Star Accretion in a Stellar Wind: Model for a Pulsed X-Ray Source," *Astrophys. J.* **179**, 585.

Davis, M., R. Ruffini, W. H. Press, and R. H. Price, 1971. "Gravitational Radiation from a Particle Falling Radially into a Schwarzschild Black Hole," *Phys. Rev. Lett.* **27**, 1466.

Davison, P. J. N., 1977. "A Regular Pulsation in the X-Ray Flux from A1540-53," *Mon. Not. Roy. Astron. Soc.* **179**, 35P.

Day, B. D., 1978. "Current State of Nuclear Matter Calculations," *Rev. Mod. Phys.* **50**, 495.

de Shalit, A. and H. Feshbach, 1974. *Theoretical Nuclear Physics*, Wiley, New York, New York.

Detweiler, S. L. and L. Lindblom, 1977. "On the Evolution of the Homogeneous Ellipsoidal Figures," *Astrophys. J.* **213**, 193.

Detweiler, S. L. and E. Szedenits, 1979. "Black Holes and Gravitational Waves. II. Trajectories Plunging into a Nonrotating Hole," *Astrophys. J.* **231**, 211.

de Vaucouleurs, G. and J.-L. Nieto, 1979. "Luminosity Distribution in the Central Regions of Messier 87: Isothermal Core, Point Source, or Black Hole?" *Astrophys. J.* **230**, 697.

Dicus, D. A., 1972. "Stellar Energy-Loss Rates in a Convergent Theory of Weak and Electromagnetic Interactions," *Phys. Rev. D* **6**, 941.

Dicus, D. A., E. W. Kolb, D. N. Schramm, and D. L. Tubbs, 1976. "Neutrino Pair Brehmsstrahlung Including Neutral Current Effects," *Astrophys. J.* **210**, 481.

Dirac, P. A. M., 1926. "On the Theory of Quantum Mechanics," *Proc. Roy. Soc. London Ser. A* **112**, 661.

Dolan, J. F., C. J. Crannell, B. R. Dennis, K. J. Frost, and L. E. Orwig, 1979. "High-Energy X-Ray Spectra of Cygnus XR-1 Observed from OSO 8," *Astrophys. J.* **230**, 551.

Dower, R. G., H. V. Bradt, and E. H. Morgan, 1982. "Circinus X-1: X-Ray Observations with SAS 3," *Astrophys. J.* **261**, 228.

Downs, G. S., 1981. "JPL Pulsar Timing Observations. I. The Vela Pulsar," *Astrophys. J.* **249**, 687.

Duncan, M. J. and J. C. Wheeler, 1980. "Anisotropic Velocity Distributions in M87: Is a Supermassive Black Hole Necessary?" *Astrophys. J. Lett.* **237**, L27.

Duncan, R. C., S. L. Shapiro, and I. Wasserman, 1983. "Equilibrium Composition and Neutrino Emissivity of Interacting Quark Matter in Neutron Stars," *Astrophys. J.* **267**, in press.

Durisen, R. H., 1973. "Viscous Effects in Rapidly Rotating Stars with Application to White-Dwarf Models. II. Numerical Results," *Astrophys. J.* **183**, 215.

_____ 1975. "Upper Mass Limits for Stable Rotating White Dwarfs," *Astrophys. J.* **199**, 179.

Durisen, R. H. and J. N. Imamura, 1981. "Improved Secular Stability Limits for Differentially Rotating Polytropes and Degenerate Dwarfs," *Astrophys. J.* **243**, 612.

Duyvendak, J. J. L., 1942. "Further Data Bearing on the Identification of the Crab Nebula with the Supernova of A.D. 1054, Part I: The Ancient Oriental Chronicles," *Proc. Astron. Soc. Pacific* **54**, 91.

Dyson, F. J., 1969. "Seismic Response of the Earth to a Gravitational Wave in the 1-Hz Band," *Astrophys. J.* **156**, 529.

Eardley, D. M. and A. P. Lightman, 1975. "Magnetic Viscosity in Relativistic Accretion Disks," *Astrophys. J.* **200**, 187.

Eardley, D. M., A. P. Lightman, D. G. Payne, and S. L. Shapiro, 1978. "Accretion Disks around Massive Black Holes: Persistent Emission Spectra," *Astrophys. J.* **224**, 53.

Eddington, A. S., 1922. *The Mathematical Theory of Relativity*, Cambridge University Press, Cambridge, England.

_____ 1926. *The Internal Constitution of the Stars*, Cambridge University Press, Cambridge, England.

_____ 1935. in Minutes of a Meeting of the Royal Astronomical Society, *Observatory* **58**, 37.

Elsner, R. F. and F. K. Lamb, 1976. "Accretion Flows in the Magnetospheres of Vela X-1, A0535 + 26 and Her X-1," *Nature* **262**, 356.

_____ 1977. "Accretion by Magnetic Neutron Stars. I. Magnetospheric Structure and Stability," *Astrophys. J.* **215**, 897.

Emden, R., 1907. *Gaskugeln*, B. G. Teubner, Leipzig, Germany.

Epstein, R., 1977. "The Binary Pulsar: Post-Newtonian Timing Effects," *Astrophys. J.* **216**, 92. Errata: **231**, 644.

Epstein, R. I., 1979. "Lepton-driven Convection in Supernovae," *Mon. Not. Roy. Astron. Soc.* **188**, 305.

Epstein, R. I. and W. D. Arnett, 1975. "Neutronization and Thermal Disintegration of Dense Stellar Matter," *Astrophys. J.* **201**, 202.

Epstein, R. I. and C. J. Pethick, 1981. "Lepton Loss and Entropy Generation in Stellar Collapse," *Astrophys. J.* **243**, 1003.

Ewart, G. M., R. A. Guyer and G. Greenstein, 1975. "Electrical Conductivity and Magnetic Field Decay in Neutron Stars," *Astrophys. J.* **202**, 238.

Fabian, A. C., J. E. Pringle, and J. A. J. Whelan, 1974. "Is Cyg X-1 a Neutron Star?" *Nature* **247**, 351.

Fabian, A. C., J. E. Pringle, and M. J. Rees, 1976. "X-Ray Emission from Accretion onto White Dwarfs," *Mon. Not. Roy. Astron. Soc.* **175**, 43.

Fahlman, G. G. and P. C. Gregory, 1981. "An X-Ray Pulsar in SNR G109.1-1.0," *Nature* **293**, 202.

Fechner, W. B. and P. C. Joss, 1978. "Quark Stars with 'Realistic' Equations of State," *Nature* **274**, 347.

Feinberg, G., 1969. "Pulsar Test of a Variation of the Speed of Light with Frequency," *Science* **166**, 879.

Felten, J. E. and M. J. Rees, 1972. "Continuum Radiative Transfer in a Hot Plasma with Application to Scorpius X-1," *Astron. Astrophys.* **17**, 226.

Feynman, R. P., 1972. *Statistical Mechanics: A Set of Lectures*, Benjamin, Reading, Massachusetts.

Feynman, R. P., N. Metropolis, and E. Teller, 1949. "Equations of State of Elements Based on the Generalized Fermi–Thomas Theory," *Phys. Rev.* **75**, 1561.

Fichtel, C. E., R. C. Hartman, D. A. Kniffen, D. J. Thompson, G. F. Bignami, H. Ögelman, M. E. Özel, and T. Tümer, 1975. "High-Energy Gamma-Ray Results from the Second Small Astronomy Satellite," *Astrophys. J.* **198**, 163.

Flowers, E. G., P. G. Sutherland, and J. R. Bond, 1975. "Neutrino Pair Bremsstrahlung by Nucleons in Neutron-Star Matter," *Phys. Rev. D* **12**, 315.

Fontaine, G. and H. M. Van Horn, 1976. "Convective White-Dwarf Envelope Model Grids for H-, He-, and C-Rich Compositions," *Astrophys. J. Suppl.* **31**, 467.

Forman, W., C. Jones, L. Cominsky, P. Julien, S. Murray, G. Peters, H. Tananbaum, and R. Giacconi, 1978. "The Fourth Uhuru Catalog of X-Ray Sources," *Astrophys. J. Suppl.* **38**, 357.

Forward, R. L., 1978. "Wideband Laser-Interferometer Gravitational-Radiation Experiment," *Phys. Rev. D* **17**, 379.

Fowler, R. H., 1926. "Dense Matter," *Mon. Not. Roy. Astron. Soc.* **87**, 114.

Fowler, W. A., G. R. Caughlan, and B. A. Zimmerman, 1975. "Thermonuclear Reaction Rates, II," *Ann. Rev. Astron. Astrophys.* **13**, 69.

Frautschi, S., J. N. Bahcall, G. Steigman and J. C. Wheeler, 1971. "Ultradense Matter," *Comments Astrophys. Space Phys.* **3**, 121.

Freedman, D. Z., 1974. "Coherent Effects of a Weak Neutral Current," *Phys. Rev. D* **9**, 1389.

Fricke, K. J., 1973. "Dynamical Phases of Supermassive Stars," *Astrophys. J.* **183**, 941.

Friedman, B. and V. R. Pandharipande, 1981. "Hot and Cold, Nuclear and Neutron Matter," *Nucl. Phys. A* **361**, 502.

Friedman, J. L. and B. F. Schutz, 1975. "Gravitational Radiation Instability in Rotating Stars," *Astrophys. J. Lett.* **199**, L157. Erratum: **221**, L91 (1978).

_____ 1978a. "Lagrangian Perturbation Theory of Nonrelativistic Fluids," *Astrophys. J.* **221**, 937.

_____ 1978b. "Secular Instability of Rotating Newtonian Stars," *Astrophys. J.* **222**, 281.

Friman, B. L. and O. V. Maxwell, 1979. "Neutrino Emissivities of Neutron Stars," *Astrophys. J.* **232**, 541.

Fuller, G. M., 1982. "Neutron Shell Blocking of Electron Capture During Gravitational Collapse," *Astrophys. J.* **252**, 741.

Fuller, G. M., W. A. Fowler and M. J. Newman, 1980. "Stellar Weak-Interaction Rates for sd-Shell Nuclei. I. Nuclear Matrix Element Systematics with Application to ^{26}Al and Selected Nuclei of Importance to the Supernova Problem," *Astrophys. J. Suppl.* **42**, 447.

_____ 1982a. "Stellar Weak Interaction Rates for Intermediate-Mass Nuclei. II. $A = 21$ to $A = 60$," *Astrophys. J.* **252**, 715.

_____ 1982b. "Stellar Weak Interaction Rates for Intermediate Mass Nuclei. III. Rate Tables for the Free Nucleons and Nuclei with $A = 21$ to $A = 60$," *Astrophys. J. Suppl.* **48**, 279.

Gamow, G., 1937. *Structure of Atomic Nuclei and Nuclear Transformations*, Clarendon, Oxford, England.

Gatewood, G. D. and C. V. Gatewood, 1978. "A Study of Sirius," *Astrophys. J.* **225**, 191.

Gatewood, G. and J. Russell, 1974. "Astrometric Determination of the Gravitational Redshift of Van Maanen 2 (EG 5)," *Astron. J.* **79**, 815.

Ghosh, P. and F. K. Lamb, 1978. "Disk Accretion by Magnetic Neutron Stars," *Astrophys. J. Lett.* **223**, L83.

_____ 1979a. "Accretion by Rotating Magnetic Neutron Stars. II. Radial and Vertical Structure of the Transition Zone in Disk Accretion," *Astrophys. J.* **232**, 259.

_____ 1979b. "Accretion by Rotating Magnetic Neutron Stars. III. Accretion Torques and Period Changes in Pulsating X-Ray Sources," *Astrophys. J.* **234**, 296.

Giacconi, R. and R. Ruffini, 1978, eds. *Physics and Astrophysics of Neutron Stars and Black Holes*, North Holland, Amsterdam, Holland.

Giacconi, R., H. Gursky, F. R. Paolini, and B. B. Rossi, 1962. "Evidence for X-Rays from Sources outside the Solar System," *Phys. Rev. Lett.* **9**, 439.

Giacconi, R., H. Gursky, E. Kellogg, E. Schreier, and H. Tananbaum, 1971. "Discovery of Periodic X-Ray Pulsations in Centaurus X-3 from UHURU," *Astrophys. J. Lett.* **167**, L67.

Gies, D. R. and C. T. Bolton, 1982. "The Optical Spectrum of HDE 226868 = Cygnus X-1. I. Radial Velocities and Orbital Elements," *Astrophys. J.* **260**, 240.

Glen, G. and P. Sutherland, 1980. "On the Cooling of Neutron Stars," *Astrophys. J.* **239**, 671.

Gold, T., 1968. "Rotating Neutron Stars as the Origin of the Pulsating Radio Sources," *Nature* **218**, 731.

———— 1969. "Rotating Neutron Stars and the Nature of Pulsars," *Nature* **221**, 25.

Goldman, I., 1979. "General Relativistic Effects and the Radius and Mass of X-Ray Bursters," *Astron. Astrophys.* **78**, L15.

Goldreich, P. and W. H. Julian, 1969. "Pulsar Electrodynamics," *Astrophys. J.* **157**, 869.

Goldreich, P. and S. V. Weber, 1980. "Homologously Collapsing Stellar Cores," *Astrophys. J.* **238**, 991.

Goldstein, H., 1981. *Classical Mechanics*, 2nd ed., Addison-Wesley, Reading, Massachusetts.

Gottlieb, E. W., E. L. Wright, and W. Liller, 1975. "Optical Studies of Uhuru Sources. XI. A Probable Period for Scorpius X-1 = V818 Scorpii," *Astrophys. J. Lett.* **195**, L33.

Gradshteyn, I. S. and I. W. Ryzhik, 1965. *Table of Integrals, Series, and Products*, 4th ed., Academic Press, New York, New York.

Grassberg, E. K., V. S. Imshennik, and D. K. Nadyozhin, 1971. "On the Theory of the Light Curves of Supernovae," *Astrophys. Space Sci.* **10**, 28.

Green, A. E. S., 1955. *Nuclear Physics*, McGraw-Hill, New York, New York.

Green, A. M. and P. Haapakoski, 1974. "The Effect of the $\Delta(1236)$ in the Two-Nucleon Problem and in Neutron Matter," *Nucl. Phys. A* **221**, 429.

Green, R. F., 1977. Unpublished Ph.D. Thesis, California Institute of Technology, Pasadena, California.

———— 1980. "The Luminosity Function of Hot White Dwarfs," *Astrophys. J.* **238**, 685.

Greene, J., 1959. "Bremsstrahlung from a Maxwellian Gas," *Astrophys. J.* **130**, 693.

Greenstein, G., 1979. "Pulsar Timing Observations, X-Ray Transients, and the Thermal/Timing Instability in Neutron Stars," *Astrophys. J.* **231**, 880.

Greenstein, J. L., 1960. "Spectra of Stars Below the Main Sequence," in *Stars and Stellar Systems*, Vol. 6, J. L. Greenstein, editor, University of Chicago Press, Chicago, Illinois, p. 676.

Greenstein, J. L. and V. L. Trimble, 1967. "The Einstein Redshift in White Dwarfs," *Astrophys. J.* **149**, 283.

Greenstein, J. L., J. B. Oke, and H. L. Shipman, 1971. "Effective Temperature, Radius, and Gravitational Redshift of Sirius B," *Astrophys. J.* **169**, 563.

Greenstein, J. L., A. Boksenberg, R. Carswell, and K. Shortridge, 1977. "The Rotation and Gravitational Redshift of White Dwarfs," *Astrophys. J.* **212**, 186.

Grindlay, J. E., 1981. "X-Ray Sources in Globular Clusters," in *X-Ray Astronomy with the Einstein Satellite*, R. Giacconi, editor, Reidel, Dordrecht, Holland.

Grindlay, J., H. Gursky, H. Schnopper, D. R. Parsignault, J. Heise, A. C. Brinkman, and J. Schrijver, 1976. "Discovery of Intense X-Ray Bursts from the Globular Cluster NGC 6624," *Astrophys. J. Lett.* **205**, L127.

Groth, E. J., 1975a. "Observational Properties of Pulsars," in *Neutron Stars, Black Holes and Binary X-Ray Sources*, H. Gursky and R. Ruffini, editors, Reidel, Dordrecht, Holland.

———— 1975b. "Timing of the Crab Pulsar. III. The Slowing Down and the Nature of the Random Process," *Astrophys. J. Suppl.* **29**, 453.

Gullahorn, G. E., R. Isaacman, J. M. Rankin, and R. R. Payne, 1977. "The Crab Nebula Pulsar: Six Years of Radio-Frequency Arrival Times," *Astron. J.* **82**, 309.

Gunn, J. E. and J. P. Ostriker, 1969. "Magnetic Dipole Radiation from Pulsars," *Nature* **221**, 454.

_____ 1970. "On the Nature of Pulsars. III. Analysis of Observations," *Astrophys. J.* **160**, 979.

Gursky, H. and R. Ruffini, 1975, eds. *Neutron Stars, Black Holes and Binary X-Ray Sources*, Reidel, Dordrecht, Holland.

Hagedorn, R., 1970. "Remarks on the Thermodynamical Model of Strong Interactions," *Nucl. Phys. B* **24**, 231.

Hamada, T. and E. E. Salpeter, 1961. "Models for Zero-Temperature Stars," *Astrophys. J.* **134**, 683.

Harnden, F. R., B. Buehler, R. Giacconi, J. Grindlay, P. Hertz, E. Schreier, F. Seward, H. Tananbaum, and L. Van Speybroeck, 1979a. "X-Ray Observations of the Crab Nebula with the Einstein Observatory," *Bull. Am. Astron. Soc.* **11**, 789.

Harnden, F. R., P. Hertz, P. Gorenstein, J. Grindlay, E. Schreier, and F. Seward, 1979b. "Observations of the Vela Pulsar from the Einstein Observatory," *Bull. Am. Astron. Soc.* **11**, 424.

Harris, D. L., 1963. "The Stellar Temperature Scale and Bolometric Corrections," in *Basic Astronomical Data*, Vol. III, K. A. Strand, editor, University of Chicago Press, Chicago, Illinois.

Harrison, B. K. and J. A. Wheeler, 1958 (HW). See Harrison et al. (1958, 1965).

Harrison, B. K., M. Wakano, and J. A. Wheeler, 1958. "Matter-Energy at High Density; End Point of Thermonuclear Evolution," in *La Structure et l'évolution de l'univers*, Onzième Conseil de Physique Solvay, Stoops, Brussels, Belgium, p. 124.

Harrison, B. K., K. S. Thorne, M. Wakano, and J. A. Wheeler, 1965 (HTWW). *Gravitation Theory and Gravitational Collapse*, University of Chicago Press, Chicago, Illinois.

Hartle, J. B., 1967. "Slowly Rotating Relativistic Stars, I: Equations of Structure," *Astrophys. J.* **150**, 1005.

Hartle, J. B. and A. G. Sabbadini, 1977. "The Equation of State and Bounds on the Mass of Nonrotating Neutron Stars," *Astrophys. J.* **213**, 831.

Hartle, J. B. and K. S. Thorne, 1968. "Slowly Rotating Relativistic Stars, II: Models for Neutron Stars and Supermassive Stars," *Astrophys. J.* **153**, 807.

Hawking, S. W., 1971. "Gravitationally Collapsed Objects of Very Low Mass," *Mon. Not. Roy. Astron. Soc.* **152**, 75.

_____ 1974. "Black Hole Explosions?" *Nature* **248**, 30.

_____ 1975. "Particle Creation by Black Holes," *Commun. Math. Phys.* **43**, 199.

Hawking, S. W. and G. F. R. Ellis, 1973. *The Large Scale Structure of Space-time*, Cambridge University Press, Cambridge, England.

Hayakawa, S. and M. Matsuoka, 1964. "Origin of Cosmic X-Rays," *Progr. Theor. Phys. Suppl.*, No. 30, **204**.

Heintz, W. D., 1974. "Astrometric Study of Four Visual Binaries," *Astron. J.* **79**, 819.

Helfand, D. J., 1981. "Unpulsed X-Rays from Pulsars," in IAU Symposium 95, *Pulsars*, W. Sieber and R. Wielebinski, editors, Reidel, Dordrecht, Holland.

Helfand, D. J., G. A. Chanan, and R. Novick, 1980. "Thermal X-Ray Emission from Neutron Stars," *Nature* **283**, 337.

Hershey, J., 1978. "Astrometric Study of the Sproul Plate Series on Van Maanen's Star, Including Gravitational Redshift," *Astron. J.* **83**, 197.

Hewish, A., S. J. Bell, J. D. H. Pilkington, P. F. Scott, and R. A. Collins, 1968. "Observation of a Rapidly Pulsating Radio Source," *Nature* **217**, 709.

Hjellming, R. M., 1973. "Radio Variability of HDE 226868 (Cygnus X-1)," *Astrophys. J. Lett.* **182**, L29.

Hoffman, J. A., H. L. Marshall, and W. H. G. Lewin, 1978. "Dual Character of the Rapid Burster and a Classification of X-Ray Bursts," *Nature* **271**, 630.

Holt, S. S. and R. McCray, 1982. "Spectra of Cosmic X-Ray Sources," *Ann. Rev. Astron. Astrophys.* **20**, 323.

Holt, S. S., E. A. Boldt, P. J. Serlemitsos, and L. J. Kaluzienski, 1976. "New Results from Long-Term Observations of Cygnus X-1," *Astrophys. J. Lett.* **203**, L63.

Hoshi, R., 1973. "X-Ray Emission from White Dwarfs in Close Binary Systems," *Progr. Theor. Phys.* **49**, 776.

Hoyle, F., 1946. "The Synthesis of the Elements from Hydrogen," *Mon. Not. Roy. Astron. Soc.* **106**, 343.

Hoyle, F. and W. A. Fowler, 1960. "Nucleosynthesis in Supernovae," *Astrophys. J.* **132**, 565.

———— 1963a. "On the Nature of Strong Radio Sources," *Mon. Not. Roy. Astron. Soc.* **125**, 169.

———— 1963b. "Nature of Strong Radio Sources," *Nature* **197**, 533.

Hoyle, F. and R. A. Lyttleton, 1939. "Evolution of Stars," *Proc. Camb. Phil. Soc.* **35**, 592.

Hoyle, F., J. V. Narlikar, and J. A. Wheeler, 1964. "Electromagnetic Waves from Very Dense Stars," *Nature* **203**, 914.

Huckle, H. E., K. O. Mason, N. E. White, P. W. Sanford, L. Maraschi, M. Tarenghi, and S. Tapia, 1977. "Discovery of Two Periodic X-Ray Pulsators," *Mon. Not. Roy. Astron. Soc.* **180**, 21P.

Hulse, R. A. and J. H. Taylor, 1975. "Discovery of a Pulsar in a Binary System," *Astrophys. J. Lett.* **195**, L51.

Hunt, R., 1971, "A Fluid Dynamical Study of the Accretion Process," *Mon. Not. Roy. Astron. Soc.* **154**, 141.

Hunter, C., 1977. "On Secular Stability, Secular Instability and Points of Bifurcation of Rotating Gaseous Masses," *Astrophys. J.* **213**, 497.

Iben, I., 1974. "Post Main Sequence Evolution of Single Stars," *Ann. Rev. Astron. Astrophys.* **12**, 215.

Illarionov, A. F. and R. A. Sunyaev, 1972. "Compton Scattering by Thermal Electrons in X-Ray Sources," *Astron. Zhur.* **49**, 58 (English trans. in *Sov. Astron. -AJ* **16**, 45).

———— 1975. "Why the Number of Galactic X-Ray Stars is so Small?" *Astron. Astrophys.* **39**, 185.

Ipser, J. R. and R. H. Price, 1977. "Accretion onto Pregalactic Black Holes," *Astrophys. J.* **216**, 578.

Ivanova, L. N., V. S. Imshennik, and D. K. Nadyozhin, 1969. *Nauchyiye Informatsii Astron. Sovieta USSR Acad. Sci.* **13**, 3 (in Russian).

Ives, J. C., P. W. Sanford, and S. J. Bell Burnell, 1975. "Observations of a Transient X-Ray Source with Regular Periodicity of 6.75 min," *Nature* **254**, 578.

Iwamoto, N., 1980. "Quark Beta Decay and the Cooling of Neutron Stars," *Phys. Rev. Lett.* **44**, 1637.

Jackson, J. D., 1975. *Classical Electrodynamics*, 2nd ed., Wiley, New York, New York.

James, R. A., 1964. "The Structure and Stability of Rotating Gas Masses," *Astrophys. J.* **140**, 552.

Jeans, J. H., 1919. *Problems of Cosmogony and Stellar Dynamics*, Cambridge University Press, Cambridge, England.

Jones, C. and W. Forman, 1976. "UHURU Observations of Hercules X-1 During the Low State of the 35-Day Cycle," *Astrophys. J. Lett.* **209**, L131.

Joss, P. C., 1977. "X-Ray Bursts and Neutron-Star Thermonuclear Flashes," *Nature* **270**, 310.

Joss, P. C. and S. Rappaport, 1976. "Observational Constraints on the Masses of Neutron Stars," *Nature* **264**, 219.

Kaplan, S. A., 1949, "Superdense Stars," *Naukovy Zapiski (Sci. Notes Univ. Lwow)* **15**, 109.

Karzas, W. J. and R. Latter, 1961. "Electron Radiative Transitions in a Coulomb Field," *Astrophys. J. Suppl.* **6**, 167.

Katz, J. I., 1973. "Thirty-Five-Day Periodicity in Her X-1," *Nature Phys. Sci.* **246**, 87.

―――― 1975. "Two Kinds of Stellar Collapse," *Nature* **253**, 698.

―――― 1977. "X-Rays from Spherical Accretion onto Degenerate Dwarfs," *Astrophys. J.* **215**, 265.

Keister, B. D. and L. S. Kisslinger, 1976. "Free-Quark Phases in Dense Stars," *Phys. Lett.* B **64**, 117.

Kelley, R. L., K. M. V. Apparao, R. E. Doxsey, J. G. Jernigan, S. Naranan, and S. Rappaport, 1981. "Discovery of X-Ray Pulsations from 2S 1417-624," *Astrophys. J.* **243**, 251.

Kelley, R. L., J. G. Jernigan, A. Levine, L. D. Petro, and S. Rappaport, 1983. "Discovery of 13.5 s X-Ray Pulsations from LMC X-4 and an Orbital Determination," *Astrophys. J.*, **264**, 568.

Kerr, R. P., 1963, "Gravitational Field of a Spinning Mass as an Example of Algebraically Special Metrics," *Phys. Rev. Lett.* **11**, 237.

Kippenhahn, R., 1969. "Mass Exchange in a Massive Close Binary System," *Astron. Astrophys.* **3**, 83.

Klebesadel, R. W., I. B. Strong, and R. A. Olson, 1973. "Observations of Gamma-Ray Bursts of Cosmic Origin," *Astrophys. J. Lett.* **182**, L85.

Klebesadel, R. W., E. E. Fenimore, J. G. Laros, and J. Terrell, 1982. "Gamma-Ray Burst Systematics," in *Gamma-Ray Transients and Related Astrophysical Phenomena*, R. E. Lingenfelter, H. S. Hudson, and D. M. Worrall, editors, American Institute of Physics, New York, New York.

Klein, R. I., H. S. Stockman, and R. A. Chevalier, 1980. "Supercritical Time-Dependent Accretion onto Compact Objects. I. Neutron Stars," *Astrophys. J.* **237**, 912.

Kobayashi, S., T. Matsukuma, S. Nagai, and K. Umeda, 1955. "Accurate Value of the Initial Slope of the Ordinary TF Function," *J. Phys. Soc. Jap.* **10**, 759.

Koester, D., H. Schulz, and V. Weidemann, 1979. "Atmospheric Parameters and Mass Distribution of DA White Dwarfs," *Astron. Astrophys.* **76**, 262.

Kompaneets, A. S., 1957. "The Establishment of Thermal Equilibrium between Quanta and Electrons," *Zh. Eksp. Teor. Fiz.* **31**, 876 (1956) (English trans. in *Sov. Phys.-JETP* **4**, 730).

Krisciunas, K., 1977. "Toward the Resolution of the Local Missing Mass Problem," *Astron. J.* **82**, 195.

Kruskal, M. D., 1960. "Maximal Extension of Schwarzschild Metric," *Phys. Rev.* **119**, 1743.

Kylafis, N. D. and D. Q. Lamb, 1979. "X-Ray and UV Radiation from Accreting Nonmagnetic Degenerate Dwarfs," *Astrophys. J. Lett.* **228**, L105.

Kylafis, N. D., D. Q. Lamb, A. R. Masters, and G. J. Weast, 1980. "X-Ray Spectra of Accreting Degenerate Stars," *Ann. N.Y. Acad. Sci.* **336**, 520 (Ninth Texas Symposium).

Lacy, J. H., F. Baas, C. H. Townes, and T. R. Geballe, 1979. "Observations of the Motion and Distribution of the Ionized Gas in the Central Parsec of the Galaxy," *Astrophys. J. Lett.* **227**, L17.

Lamb, F. K., 1979a. "Structure of the Magnetospheres of Accreting Neutron Stars," in *Proceedings of the Sydney Chapman Conference on Magnetospheric Boundary Layers*, edited by B. Battrick (ESA SP Series).

Lamb, D. Q., 1979b. "Degenerate Dwarf X-Ray Sources," in *Compact Galactic X-Ray Sources: Current Status and Future Prospects*, F. K. Lamb and D. Pines, editors, Physics Dept., University of Illinois, Urbana, Illinois.

―――― 1982. "Surface and Magnetospheric Physics of Neutron Stars and Gamma Ray Bursts," in *Gamma Ray Transients and Related Astrophysical Phenomena*, R. E. Lingenfelter, H. S. Hudson, and D. M. Worrall, editors, American Institute of Physics, New York, New York.

Lamb, D. Q. and F. K. Lamb, 1978. "Nuclear Burning in Accreting Neutron Stars and X-Ray Bursts," *Astrophys. J.* **220**, 291.

Lamb, D. Q. and A. R. M. Masters, 1979. "Radiation from Accreting Magnetic Degenerate Dwarfs," *Astrophys. J. Lett.* **234**, L117.

Lamb, D. Q. and C. J. Pethick, 1976. "Effects of Neutrino Degeneracy in Supernovae Models," *Astrophys. J. Lett.* **209**, L77.

Lamb, F. K. and D. Pines, 1979. eds. *Compact Galactic X-Ray Sources: Current Status and Future Prospects*, Physics Dept., University of Illinois, Urbana, Illinois.

Lamb, D. Q. and H. M. Van Horn, 1975. "Evolution of Crystallizing Pure ^{12}C White Dwarfs," *Astrophys. J.* **200**, 306.

Lamb, F. K., C. J. Pethick, and D. Pines, 1973. "A Model for Compact X-Ray Sources: Accretion by Rotating Magnetic Stars," *Astrophys. J.* **184**, 271.

Lamb, D. Q., F. K. Lamb, D. Pines, and J. Shaham, 1975. "Neutron Star Wobble in Binary X-Ray Sources," *Astrophys J. Lett.* **198**, L21.

Lamb, D. Q., J. M. Lattimer, C. J. Pethick, and D. G. Ravenhall, 1978. "Hot Dense Matter and Stellar Collapse," *Phys. Rev. Lett.* **41**, 1623.

―――― 1981. "Physical Properties of Hot, Dense Matter: The Bulk Equilibrium Approximation," *Nucl. Phys. A* **360**, 459.

Lamb, R. C., T. H. Markert, R. C. Hartman, D. J. Thompson, and G. F. Bignami, 1980. "Two X-Ray Pulsars: 2S 1145-619 and 1E 1145.1-6141," *Astrophys. J.* **239**, 651.

Landau, L. D., 1932. "On the Theory of Stars," *Phys. Z. Sowjetunion* **1**, 285.

―――― 1938. "Origin of Stellar Energy," *Nature* **141**, 333.

Landau, L. D. and E. M. Lifshitz, 1959. *Fluid Mechanics*, Pergamon, Elmsford, New York.

―――― 1969. *Statistical Physics*, 2nd ed., Addison-Wesley, Reading, Massachusetts.

―――― 1975. *The Classical Theory of Fields*, 4th rev. ed., Pergamon, Elmsford, New York.

_____ 1977. *Quantum Mechanics: Non-Relativistic Theory*, 3rd ed., Pergamon, Elmsford, New York.

Landstreet, J. D., 1979. "The Magnetic Fields of Single White Dwarfs," in *White Dwarfs and Variable Degenerate Stars*, IAU Colloq. No. 53, H. M. Van Horn and V. Weidemann, editors, University of Rochester Press, Rochester, New York.

Laplace, P. S., 1795. *Le Système du Monde*, Vol. II, Paris (English edition: *The System of the World*, W. Flint, London, 1809).

Large, M. I., A. E. Vaughan, and B. Y. Mills, 1968. "A Pulsar Supernova Association?" *Nature* **220**, 340.

Lasher, G., 1975. "A Simple Model for the Early Part of Type I Supernova Light Curves," *Astrophys. J.* **201**, 194.

Lattimer, J. M., 1981. "The Equation of State of Hot Dense Matter and Supernovae," *Ann. Rev. Nucl. Part. Sci.* **31**, 337.

Laval, G., C. Mercier, and R. Pellat, 1965. "Necessity of the Energy Principles for Magnetostatic Stability," *Nuclear Fusion* **5**, 156.

LeBlanc, J. M. and J. R. Wilson, 1970. "A Numerical Example of the Collapse of a Rotating Magnetized Star," *Astrophys. J.* **161**, 541.

Lee, B. W. and J. Zinn-Justin, 1972. "Spontaneously Broken Gauge Symmetries. I. Preliminaries," *Phys. Rev. D* **5**, 3121.

Legaris, I. E. and V. R. Pandharipande, 1981. "Variational Calculations of Realistic Models of Nuclear Matter," *Nucl. Phys. A* **359**, 349.

Leung, Y. C. and C. G. Wang, 1973. "Equation of State of Matter at Supernuclear Densities," *Astrophys. J.* **181**, 895.

Lewin, W. H. G. and G. W. Clark, 1980. "Galactic Bulge Sources, What Are They?," *Ann. N.Y. Acad. Sci.* **336**, 451 (Ninth Texas Symposium).

Lewin, W. H. G. and P. C. Joss, 1977. "X-Ray Burst Sources," *Nature* **270**, 211.

_____ 1981. "X-Ray Bursters and the X-Ray Sources of the Galactic Bulge," *Space Sci. Rev.* **28**, 3.

Lewin, W. H. G., G. R. Ricker, and J. E. McClintock, 1971. "X-Rays from a New Variable Source GX 1 + 4," *Astrophys. J. Lett.* **169**, L17.

Liang, E. P. T., 1980. "Test of the Inverse Compton Model for Cygnus X-1," *Nature* **283**, 642.

Liang, E. P. T. and R. H. Price, 1977. "Accretion Disk Coronae and Cygnus X-1," *Astrophys. J.* **218**, 247.

Liang, E. P. T. and K. A. Thompson, 1980. "Transonic Disk Accretion onto Black Holes," *Astrophys. J.* **240**, 271.

Liebert, J., 1980. "White Dwarf Stars," *Ann. Rev. Astron. Astrophys.* **18**, 363.

Liebert, J., C. C. Dahn, M. Gresham, and P. A. Strittmatter, 1979. "New Results from a Survey of Faint Proper-Motion Stars: A Probable Deficiency of Very Low Luminosity Degenerates," *Astrophys. J.* **233**, 226.

Lightman, A. P. and D. M. Eardley, 1974. "Black Holes in Binary Systems: Instability of Disk Accretion," *Astrophys. J. Lett.* **187**, L1.

Lightman, A. P. and G. B. Rybicki, 1979. "X-Rays from Active Galactic Nuclei: Inverse Compton Reflection," *Astrophys. J. Lett.* **229**, L15.

Lightman, A. P. and S. L. Shapiro, 1975. "Spectrum and Polarization of X-Rays from Accretion Disks around Black Holes," *Astrophys. J. Lett.* **198**, L73.

_____ 1978. "The Dynamical Evolution of Globular Clusters," *Rev. Mod. Phys.* **50**, 437.

Lightman, A. P., W. H. Press, R. H. Price, and S. A. Teukolsky, 1975. *Problem Book in Relativity and Gravitation*, Princeton University Press, Princeton, New Jersey.

Lightman, A. P., S. L. Shapiro, and M. J. Rees, 1978. "Accretion onto Compact Objects," in *Physics and Astrophysics of Neutron Stars and Black Holes*, R. Giacconi and R. Ruffini, editors, North Holland, Amsterdam, Holland.

Lightman, A. P., P. Hertz, and J. E. Grindlay, 1980. "A New Statistical Test with Application to Globular Cluster X-Ray Source Masses," *Astrophys. J.* **241**, 367.

Lindemann, F. A., 1910. "Molecular Frequencies," *Phys. Z.* **11**, 609.

Livio, M., J. R. Buchler, and S. A. Colgate, 1980. "Rayleigh–Taylor Driven Supernova Explosions: A Two-Dimensional Numerical Study," *Astrophys. J. Lett.* **238**, L139.

Löhsen, E., 1975. "Third Speed-up of the Crab Pulsar," *Nature* **258**, 688.

Lovelace, R. V. E., 1976. "Dynamo Model of Double Radio Sources," *Nature* **262**, 649.

Lucke, R., D. Yentis, H. Friedman, G. Fritz, and S. Shulman, 1976. "Discovery of X-Ray Pulsations in SMC X-1," *Astrophys. J. Lett.* **206**, L25.

Lui, Y. W., P. Bogucki, J. D. Bronson, U. Garg, C. M. Rozsa, and D. H. Youngblood, 1980. "Observation of the Giant Monopole Resonance in ^{64,66}Zn," *Phys. Lett. B* **93**, 31.

Lynden-Bell, D., 1969. "Galactic Nuclei as Collapsed Old Quasars," *Nature* **223**, 690.

Lynden-Bell, D. and J. P. Ostriker, 1967. "On the Stability of Differentially Rotating Bodies," *Mon. Not. Roy. Astron. Soc.* **136**, 293.

Lynden-Bell, D. and J. E. Pringle, 1974. "The Evolution of Viscous Discs and the Origin of the Nebular Variables," *Mon. Not. Roy. Astron. Soc.* **168**, 603.

Macy, W. W., 1974. "Pulsar Magnetic Axis Alignment and Counteralignment," *Astrophys. J.* **190**, 153.

Malone, R. C., 1974. Unpublished Ph.D. Thesis, Cornell University, Ithaca, New York.

Manchester, R. N. and J. H. Taylor, 1977. *Pulsars*, Freeman, San Francisco, California.

Manchester, R. N., L. M. Newton, W. M. Goss, and P. A. Hamilton, 1978. "Detection of a Large Period Discontinuity in the Longest Period Pulsar PSR 1641–45," *Mon. Not. Roy. Astron. Soc.* **184**, 35P.

Manchester, R. N., I. R. Tuohy, and N. D'Amico, 1982. "Discovery of Radio Pulsations from the X-ray Pulsar in the Supernova Remnant G32.4–1.2," *Astrophys. J. Lett.* **262**, L31.

Mandrou, P., M. Niel, G. Vedrenne, A. Dupont, and K. Hurley, 1978. "Observation of Cygnus X-1 in the Energy Range 100 keV–3 MeV," *Astrophys. J.* **219**, 288.

Maraschi, L. and A. Cavaliere, 1977. "X-Ray Bursts of Nuclear Origin?" in *Highlights of Astronomy*, Vol. 4, Part 1, p. 127, E. A. Müller, editor, Reidel, Dordrecht, Holland.

Margon, B. and J. P. Ostriker, 1973. "The Luminosity Function of Galactic X-Ray Sources: A Cutoff and a 'Standard Candle'?" *Astrophys. J.* **186**, 91.

Margon, B., S. Bowyer, and R. P. S. Stone, 1973. "On the Distance to Cygnus X-1," *Astrophys. J. Lett.* **185**, L113.

Mauder, H., 1973. "On the Mass Limit of the X-Ray Source in Cygnus X-1," *Astron. Astrophys.* **28**, 473.

Maxon, S., 1972. "Bremsstrahlung Rate and Spectra from a Hot Gas ($Z = 1$)," *Phys. Rev. A* **5**, 1630.

Maxwell, O. V., 1979. "Neutron Star Cooling," *Astrophys. J.* **231**, 201.

Maxwell, O., G. E. Brown, D. K. Campbell, R. F. Dashen, and J. T. Manassah, 1977. "Beta Decay of Pion Condensates as a Cooling Mechanism for Neutron Stars," *Astrophys. J.* **216**, 77.

Mayall, N. U. and J. H. Oort, 1942. "Further Data Bearing on the Identification of the Crab Nebula with the Supernova of 1054 A.D., Part II: The Astronomical Aspects," *Proc. Astron. Soc. Pacific* **54**, 95.

Mazurek, T. J., 1975. "Chemical Potential Effects on Neutrino Diffusion in Supernovae," *Astrophys. Space Sci.* **35**, 117.

Mazurek, T. J., 1976. "Pauli Constriction of the Low-Energy Window in Neutrino Supernova Models," *Astrophys. J. Lett.* **207**, L87.

McClintock, J. E. and L. D. Petro, 1981. IAU Circular No. 3615.

McClintock, J. E., S. Rappaport, P. C. Joss, H. Bradt, J. Buff, G. W. Clark, D. Hearn, W. H. G. Lewin, T. Matilsky, W. Mayer, and F. Primini, 1976. "Discovery of a 283-Second Periodic Variation in the X-Ray Source 3U 0900-40," *Astrophys. J. Lett.* **206**, L99.

McClintock, J. E., S. A. Rappaport, J. J. Nugent, and F. K. Li, 1977. "Discovery of a 272 Second Periodic Variation in the X-Ray Source GX 304-1," *Astrophys. J. Lett.* **216**, L15.

McCulloch, P. M., P. A. Hamilton, G. W. R. Royle, and R. N. Manchester, 1981. IAU Circular No. 3644.

McCuskey, S. W., 1966. "The Stellar Luminosity Function," in *Vistas in Astronomy*, Vol. 7, A. Beer, editor, Pergamon, Oxford, England, p. 141.

Merzbacher, E., 1970. *Quantum Mechanics*, Wiley, New York, New York.

Mestel, L., 1952. "On the Theory of White Dwarf Stars I. The Energy Sources of White Dwarfs," *Mon. Not. Roy. Astron. Soc.* **112**, 583.

Mestel, L. and M. A. Ruderman, 1967. "The Energy Content of a White Dwarf and its Rate of Cooling," *Mon. Not. Roy. Astron. Soc.* **136**, 27.

Meszaros, P., 1975. "Radiation from Spherical Accretion onto Black Holes," *Astron. Astrophys.* **44**, 59.

Michel, F. C., 1972. "Accretion of Matter by Condensed Objects," *Astrophys. Space Sci.* **15**, 153.

_____ 1982. "Theory of Pulsar Magnetospheres," *Rev. Mod. Phys.* **54**, 1.

Migdal, A. B., 1978. "Pion Fields in Nuclear Matter," *Rev. Mod. Phys.* **50**, 107.

Mihalas, D., 1978. *Stellar Atmospheres*, 2nd ed., Freeman, San Francisco, California.

Milgrom, M., 1978. "On the Nature of the Galactic Bulge X-Ray Sources," *Astron. Astrophys.* **67**, L25.

Miller, G. E. and J. M. Scalo, 1979. "The Initial Mass Function and Stellar Birthrate in the Solar Neighborhood," *Astrophys. J. Suppl.* **41**, 513.

Milne, E. A., 1927. "The Total Energy of Binding of a Heavy Atom," *Proc. Camb. Phil. Soc.* **23**, 794.

Minkowski, R., 1942. "The Crab Nebula," *Astrophys. J.* **96**, 199.

Misner, C. W., K. S. Thorne, and J. A. Wheeler, 1973. *Gravitation*, Freeman, San Francisco, California.

Moncrief, V., 1980. "Stability of Stationary, Spherical Accretion onto a Schwarzschild Black Hole," *Astrophys. J.* **235**, 1038.

Morrison, P., 1967. "Extrasolar X-Ray Sources," *Ann. Rev. Astron. Astrophys.* **5**, 325.

Morse, P. M. and H. Feshbach, 1953. *Methods of Theoretical Physics*, Vol. 1, McGraw-Hill, New York, New York.

Müller, E. and W. Hillebrandt, 1979. "A Magnetohydrodynamical Supernova Model," *Astron. Astrophys.* **80**, 147.

Murray, S. S., G. Fabbiano, A. C. Fabian, A. Epstein, and R. Giacconi, 1979. "High-Resolution X-Ray Observations of the Cassiopeia A Supernova Remnant with the Einstein Observatory," *Astrophys. J. Lett.* **234**, L69.

Myers, W. D. and W. J. Swiatecki, 1966. "Nuclear Masses and Deformations," *Nucl. Phys.* **81**, 1.

Nauenberg, M. and G. Chapline, 1973. "Determination of Properties of Cold Stars in General Relativity by a Variational Method," *Astrophys. J.* **179**, 277.

Newman, E. T., E. Couch, K. Chinnapared, A. Exton, A. Prakash, and R. Torrence, 1965. "Metric of a Rotating, Charged Mass," *J. Math. Phys.* **6**, 918.

Nomoto, K. and S. Tsuruta, 1981. "Cooling of Young Neutron Stars and the Einstein X-Ray Observations," *Astrophys. J. Lett.* **250**, L19.

Norgaard, H. and K. J. Fricke, 1976. "^{7}Li Production in Bouncing Supermassive Stars," *Astron. Astrophys.* **49**, 337.

Novikov, I. D. and K. S. Thorne, 1973. "Black Hole Astrophysics," in *Black Holes*, C. DeWitt and B. DeWitt, editors, Gordon and Breach, New York, New York.

Oda, M. 1977. "Cyg X-1/ A Candidate of the Black Hole," *Space Sci. Rev.* **20**, 757.

Oort, J. H., 1960. "Note on the Determination of K_z and on the Mass Density Near the Sun," *Bull. Astron. Inst. Neth.* **15**, 45.

Oppenheimer, J. R. and R. Serber, 1938. "On the Stability of Stellar Neutron Cores," *Phys. Rev.* **54**, 540.

Oppenheimer, J. R. and H. Snyder, 1939. "On Continued Gravitational Contraction," *Phys. Rev.* **56**, 455.

Oppenheimer, J. R. and G. M. Volkoff, 1939 (OV). "On Massive Neutron Cores," *Phys. Rev.* **55**, 374.

Osterbrock, D. E., 1973. "The Origin and Evolution of Planetary Nebulae," *Mem. Soc. Roy. Sci. Liège* **5**, 391.

Ostriker, J. P., 1976. As cited in E. P. T. Liang and R. H. Price, "Convective Accretion Disks and X-Ray Bursters," *Astrophys. J.* **218**, 243.

Ostriker, J. P. and P. Bodenheimer, 1968. "Rapidly Rotating Stars. II. Massive White Dwarfs," *Astrophys. J.* **151**, 1089.

―――― 1973. "On the Oscillations and Stability of Rapidly Rotating Stellar Models. III. Zero-Viscosity Polytropic Sequences," *Astrophys. J.* **180** 171. Erratum: **182**, 1037.

Ostriker, J. P. and J. E. Gunn, 1969. "On the Nature of Pulsars. I. Theory," *Astrophys. J.* **157**, 1395.

Ostriker, J. P. and F. D. A. Hartwick, 1968. "Rapidly Rotating Stars. IV. Magnetic White Dwarfs," *Astrophys. J.* **153**, 797.

Ostriker, J. P. and J. L. Tassoul, 1969. "On the Oscillations and Stability of Rotating Stellar Models. II. Rapidly Rotating White Dwarfs," *Astrophys. J.* **155**, 987.

Ostriker, J. P., D. O. Richstone, and T. X. Thuan, 1974. "On the Numbers, Birthrates, and Final States of Moderate- and High-Mass Stars," *Astrophys. J. Lett.* **188**, L87.

Overbeck, J. W. and H. D. Tananbaum, 1968. "Twofold Increase of the High-Energy X-Ray Flux from Cygnus XR-1," *Phys. Rev. Lett.* **20**, 24.

Pacini, F., 1967. "Energy Emission from a Neutron Star," *Nature* **216**, 567.

_____ 1968. "Rotating Neutron Stars, Pulsars and Supernova Remnants," *Nature* **219**, 145.

Paczyński, B., 1971. "Evolutionary Processes in Close Binary Systems," *Ann. Rev. Astron. Astrophys.* **9**, 183.

_____ 1974. "Mass of Cygnus X-1," *Astron. Astrophys.* **34**, 161.

Page, D. N., 1976. "Particle Emission Rates from a Black Hole: Massless Particles from an Uncharged, Nonrotating Hole," *Phys. Rev. D* **13**, 198.

Page, D. N. and S. W. Hawking, 1976. "Gamma Rays from Primordial Black Holes," *Astrophys. J.* **206**, 1.

Pandharipande, V. R., 1971. "Hyperonic Matter," *Nucl. Phys. A* **178**, 123.

Pandharipande, V. R. and R. A. Smith, 1975a. "A Model Neutron Solid with π° Condensate," *Nucl. Phys. A* **237**, 507.

_____ 1975b. "Nuclear Matter Calculations with Mean Scalar Fields," *Phys. Lett. B* **59**, 15.

Pandharipande, V. R., D. Pines, and R. A. Smith, 1976. "Neutron Star Structure: Theory, Observation and Speculation," *Astrophys. J.* **208**, 550.

Parker, E. N., 1965. "Dynamical Properties of Stellar Coronas and Stellar Winds. IV. The Separate Existence of Subsonic and Supersonic Solutions," *Astrophys. J.* **141**, 1463.

Particle Data Group, 1982. "Review of Particle Properties," *Phys. Lett. B* **111**, 1.

Payne, D. G., 1980. "Time-dependent Comptonization: X-Ray Reverberations," *Astrophys. J.* **237**, 951.

Payne, D. G. and D. M. Eardley, 1977. "X-Ray Spectrum from Disk Accretion onto Massive Black Holes," *Astrophys. Lett.* **19**, 39.

Pedersen, H., J. van Paradijs, and W. H. G. Lewin, 1981. "Evidence for a Four Hour Orbital Period of 4U/MXB 1636–53," *Nature* **294**, 725.

Peebles, P. J. E., 1971. *Physical Cosmology*, Princeton University Press, Princeton, New Jersey.

Penrose, R., 1969. "Gravitational Collapse: The Role of General Relativity," *Riv. Nuovo Cim.* **1**, 252.

Perkins, D. H., 1972. *Introduction to High-Energy Physics*, Addison-Wesley, Reading, Massachusetts.

Peters, P. C., 1964. "Gravitational Radiation and the Motion of Two Point Masses," *Phys. Rev.* **136**, 1224.

Petterson, J. A., 1975. "Hercules X-1: A Neutron Star with a Twisted Accretion Disk?" *Astrophys. J. Lett.* **201**, L61.

Pines, D., 1963. *Elementary Excitations in Solids*, Benjamin, New York, New York.

_____ 1980a. "Pulsars and Compact X-Ray Sources: Cosmic Laboratories for the Study of Neutron Stars and Hadron Matter," *J. Phys. Colloq.* **41**, p. C2/111.

_____ 1980b. "Accreting Neutron Stars, Black Holes, and Degenerate Dwarf Stars," *Science* **207**, 597.

Pines, D. and J. Shaham, 1974. "Free Precession of Neutron Stars: Some Plain Truths, Cautionary Remarks, and Assorted Speculations," *Comments Astrophys. Space Phys.* **6**, 37.

Pines, D., J. Shaham, and M. Ruderman, 1972. "Corequakes and the Vela Pulsar," *Nature Phys. Sci.* **237**, 83.

Pines, D., J. Shaham, and M. A. Ruderman, 1974. "Neutron Star Structure from Pulsar Observations," in IAU Symposium No. 53, *Physics of Dense Matter*, C. J. Hansen, editor, Reidel, Dordrecht, Holland.

Polidan, R. S., G. S. G. Pollard, P. W. Sanford, and M. C. Locke, 1978. "X-Ray Emission from the Companion to V861Sco," *Nature* **275**, 296.

Prendergast, K. H. and G. R. Burbidge, 1968. "On the Nature of Some Galactic X-Ray Sources," *Astrophys. J. Lett.* **151**, L83.

Press, W. H. and S. A. Teukolsky, 1972. "Floating Orbits, Superradiant Scattering and the Black Hole Bomb," *Nature* **238**, 211.

_____ 1973. "Perturbations of a Rotating Black Hole. II. Dynamical Stability of the Kerr Metric," *Astrophys. J.* **185**, 649.

Primini, F., S. Rappaport, P. C. Joss, G. W. Clark, W. Lewin, F. Li, W. Mayer, and J. McClintock, 1976. "Orbital Elements and Masses for the SMC X-1/Sanduleak 160 Binary System," *Astrophys. J. Lett.* **210**, L71.

Pringle, J. E., 1976. "Thermal Instabilities in Accretion Discs," *Mon. Not. Roy. Astron. Soc.* **177**, 65.

_____ 1981. "Accretion Discs in Astrophysics," *Ann. Rev. Astron. Astrophys.* **19**, 137.

Pringle, J. E. and M. J. Rees, 1972. "Accretion Disc Models for Compact X-Ray Sources," *Astron. Astrophys.* **21**, 1.

Pringle, J. E., M. J. Rees, and A. G. Pacholczyk, 1973. "Accretion onto Massive Black Holes," *Astron. Astrophys.* **29**, 179.

Pye, J. P., K. A. Pounds, D. P. Rolf, F. D. Seward, A. Smith, and R. Willingale, 1981. "An X-Ray Map of SN 1006 from the Einstein Observatory," *Mon. Not. Roy. Astron. Soc.* **194**, 569.

Rappaport, S. and P. C. Joss, 1981. "Binary X-Ray Pulsars," in *X-Ray Astronomy with the Einstein Satellite*, R. Giacconi, editor, Reidel, Dordrecht, Holland.

_____ 1983. "X-Ray Pulsars in Massive Binary Systems," in *Accretion Driven Stellar X-Ray Sources*, W. H. G. Lewin and E. P. J. Van den Heuvel, editors, Cambridge University Press, Cambridge, England, in press.

Rappaport, S., T. Markert, F. K. Li, G. W. Clark, J. G. Jernigan and J. E. McClintock, 1977. "Discovery of a 7.68 Second X-Ray Periodicity in 3U 1626–67," *Astrophys. J. Lett.* **217**, L29.

Rees, M. J., 1977a. "A Better Way of Searching for Black Hole Explosions?" *Nature* **266**, 333.

_____ 1977b. "Sources of Gravitational Waves at Low Frequencies," in *Proc. Internat. Symp. Experimental Gravitation*, Academia Nazionale dei Lincei, Rome, Italy, p. 423.

Reid, R. V., 1968. "Local Phenomenological Nucleon-Nucleon Potentials," *Ann. of Phys.* **50**, 411.

Reif, F., 1965. *Fundamentals of Statistical and Thermal Physics*, McGraw-Hill, New York, New York.

Rhoades, C. E. and R. Ruffini, 1974. "Maximum Mass of a Neutron Star," *Phys. Rev. Lett.* **32**, 324.

Richards, D. W. and J. W. Comella, 1969. "The Period of Pulsar NP 0532," *Nature* **222**, 551.

Roberts, D. H. and P. A. Sturrock, 1972. "The Structure of Pulsar Magnetospheres," *Astrophys. J. Lett.*, **173**, L33.

Roberts, W. J., 1974. "A Slaved Disk Model for Hercules X-1," *Astrophys. J.* **187**, 575.

Robertson, H. P., 1938. "Note on the Preceding Paper: The Two-Body Problem in General Relativity," *Ann. of Math.* **39**, 101.

Romanishin, W. and J. R. P. Angel, 1980. "Determination of the Upper Mass Limit for Stars Producing White-Dwarf Remnants," *Astrophys. J.* **235**, 992.

Rosenberg, F. D., C. J. Eyles, G. K. Skinner, and A. P. Willmore, 1975. "Observations of a Transient X-Ray Source with a Period of 104 s," *Nature* **256**, 628.

Rosenfeld, L., 1974. In *Astrophysics and Gravitation*, Proc. 16th Solvay Conference on Physics, Editions de l'Université de Bruxelles, Brussels, Belgium, p. 174.

Rothschild, R. E., E. A. Boldt, S. S. Holt, and P. J. Serlemitsos, 1974. "Millisecond Temporal Structure in Cygnus X-1," *Astrophys. J. Lett.* **189**, L13.

Ruderman, M., 1969. "Neutron Starquakes and Pulsar Periods," *Nature* **223**, 597.

_____ 1972. "Pulsars: Structure and Dynamics," *Ann. Rev. Astron. Astrophys.* **10**, 427.

_____ 1975. "Theories of Gamma Ray Bursts," *Ann. N.Y. Acad. Sci.* **262**, 164 (Seventh Texas Symposium).

_____ 1980. "Pulsar Radiation Mechanisms," *Ann. N.Y. Acad. Sci.* **336**, 409 (Ninth Texas Symposium).

Rybicki, G. B. and A. P. Lightman, 1979. *Radiative Processes in Astrophysics*, Wiley, New York, New York.

Saenz, R. A. and S. L. Shapiro, 1981. "Gravitational Radiation from Stellar Core Collapse. III. Damped Ellipsoidal Oscillations," *Astrophys. J.* **244**, 1033.

Salpeter, E. E., 1955. "The Luminosity Function and Stellar Evolution," *Astrophys. J.* **121**, 161.

_____ 1961. "Energy and Pressure of a Zero-Temperature Plasma," *Astrophys. J.* **134**, 669.

_____ 1964. "Accretion of Interstellar Matter by Massive Objects," *Astrophys. J.* **140**, 796.

_____ 1965. "Superdense Equilibrium Stars," in *Quasi-Stellar Sources and Gravitational Collapse*, I. Robinson, A. Schild, and E. L. Schucking, editors, University of Chicago Press, Chicago, Illinois.

Salpeter, E. E. and H. M. Van Horn, 1969. "Nuclear Reaction Rates at High Densities," *Astrophys. J.* **155**, 183.

Salpeter, E. E. and H. S. Zapolsky, 1967. "Theoretical High-Pressure Equations of State Including Correlation Energy," *Phys. Rev.* **158**, 876.

Salter, M. J., A. G. Lyne, and B. Anderson, 1979. "Measurements of the Trigonometric Parallax of Pulsars," *Nature* **280**, 477.

Sandage, A., P. Osmer, R. Giacconi, P. Gorenstein, H. Gursky, J. Waters, H. Bradt, G. Garmire, B. V. Sreekantan, M. Oda, K. Osawa, and J. Jugaku, 1966. "On the Optical Identification of Sco X-1," *Astrophys. J.* **146**, 316.

Sargent, W. L. W., P. J. Young, A. Boksenberg, K. Shortridge, C. R. Lynds, and F. D. A. Hartwick, 1978. "Dynamical Evidence for a Central Mass Concentration in the Galaxy M87," *Astrophys. J.* **221**, 731.

Sato, K., 1975a. "Neutrino Degeneracy in Supernova Cores and Neutral Current of Weak Interaction," *Progr. Theor. Phys.* **53**, 595.

_____ 1975b. "Supernova Explosion and Neutral Currents of Weak Interaction," *Progr, Theor. Phys.* **54**, 1325.

Sawyer, R. F., 1972. "Energy Shifts of Excited Nucleons in Neutron-Star Matter," *Astrophys. J.* **176**, 205. Erratum: **178**, 279.

Scargle, J. D. and F. Pacini, 1971. "On the Mechanism of the Glitches in the Crab Nebula Pulsar," *Nature Phys. Sci.* **232**, 144.

Schatzman, E., 1956. "Influence of the Nucleon-Electron Equilibrium on the Internal Structure of White Dwarfs," *Astron. Zhur.* **33**, 800.

_____ 1958a. *White Dwarfs*, Interscience, New York, New York.

_____ 1958b. "Theory of White Dwarfs," in *Hdb. d. Phys.*, Vol. 51, S. Flügge, editor, Springer-Verlag, Berlin, Germany.

Schramm, D. N. and W. D. Arnett, 1975. "The Weak Interaction and Gravitational Collapse," *Astrophys. J.* **198**, 629.

Schreier, E., R. Levinson, H. Gursky, E. Kellogg, H. Tananbaum, and R. Giacconi, 1972. "Evidence for the Binary Nature of Centaurus X-3 from Uhuru X-Ray Observations," *Astrophys. J. Lett.* **172**, L79. Errata: **173** L51.

Schwartz, R. A., 1967. "Gravitational Collapse, Neutrinos and Supernovae," *Ann. Phys.* **43**, 42.

Schwarzschild, K., 1906. "Equilibrium of the Sun's Atmosphere," *Nachr. Ges. Wiss. Göttingen Math. Phys. Kl.* **1**, 41 (in German).

_____ 1916. "On the Gravitational Field of a Point Mass in Einstein's Theory," *Sitzungsber. Dtsch. Akad. Wiss. Berlin, Kl. Math. Phys. Tech.*, p. 189 (in German).

Schwarzschild, M., 1958. *Structure and Evolution of the Stars*, Princeton University Press, Princeton, New Jersey.

Schwarzschild, M. and R. Härm, 1959. "On the Maximum Mass of Stable Stars," *Astrophys. J.* **129**, 637.

Seaton, M. J., 1979. "Extinction of NGC 7027," *Mon. Not. Roy. Astron. Soc.* **187**, 785.

Service, A. T., 1977. "Concise Approximation Formulae for the Lane–Emden Functions," *Astrophys. J.* **211**, 908.

Seward, F. D. and F. R. Harnden, 1982. "A New, Fast X-Ray Pulsar in the Supernova Remnant MSH 15–52," *Astrophys. J. Lett.* **256**, L45.

Shakura, N. I., 1972. "Disk Model of Gas Accretion on a Relativistic Star in a Close Binary System," *Astron. Zhur.* **49**, 921 (English trans. in *Sov. Astron.-AJ* **16**, 756).

Shakura, N. I. and R. A. Sunyaev, 1973. "Black Holes in Binary Systems. Observational Appearance," *Astron. Astrophys.* **24**, 337.

_____ 1976. "A Theory of the Instability of Disk Accretion on to Black Holes and the Variability of Binary X-Ray Sources, Galactic Nuclei and Quasars," *Mon. Not. Roy. Astron. Soc.* **175**, 613.

Shapiro, I. I., 1964. "Fourth Test of General Relativity," *Phys. Rev. Lett.* **13**, 789.

Shapiro, S. L., 1973a. "Accretion onto Black Holes: The Emergent Radiation Spectrum," *Astrophys. J.* **180**, 531.

_____ 1973b. "Accretion onto Black Holes: The Emergent Radiation Spectrum. II. Magnetic Effects," *Astrophys. J.* **185**, 69.

_____ 1974. "Accretion onto Black Holes: The Emergent Radiation Spectrum. III. Rotating (Kerr) Black Holes," *Astrophys. J.* **189**, 343.

_____ 1980. "Gravitational Radiation from Colliding, Compact Stars: Hydrodynamic Calculations in One Dimension," *Astrophys. J.* **240**, 246.

Shapiro, S. L. and A. P. Lightman, 1976a. "Black Holes in X-Ray Binaries: Marginal Existence and Rotation Reversals of Accretion Disks," *Astrophys. J.* **204**, 555.

_____ 1976b. "Rapidly Rotating, Post-Newtonian Neutron Stars," *Astrophys. J.* **207**, 263.

Shapiro, S. L. and E. E. Salpeter, 1975. "Accretion onto Neutron Stars under Adiabatic Shock Conditions," *Astrophys. J.* **198**, 671.

Shapiro, S. L. and S. A. Teukolsky, 1976. "On the Maximum Gravitational Redshift of White Dwarfs," *Astrophys. J.* **203**, 697.

_____ 1979. "Gravitational Collapse of Supermassive Stars to Black Holes: Numerical Solution of the Einstein Equations," *Astrophys. J. Lett.* **234**, L177.

Shapiro, S. L., A. P. Lightman, and D. M. Eardley, 1976. "A Two-Temperature Accretion Disk Model for Cygnus X-1: Structure and Spectrum," *Astrophys. J.* **204**, 187.

Shaviv, G. and A. Kovetz, 1976. "The Cooling of Carbon–Oxygen White Dwarfs," *Astron. Astrophys.* **51**, 383.

Shipman, H. L., 1979. "Masses and Radii of White-Dwarf Stars. III. Results for 110 Hydrogen-Rich and 28 Helium-Rich Stars," *Astrophys. J.* **228**, 240.

Shipman, H. L. and R. F. Green, 1980. "Revised Stellar Birthrates and the Genesis of Pulsars," *Astrophys. J. Lett.* **239**, L111.

Shklovskii, I. S., 1967. "The Nature of the X-Ray Source Sco X-1," *Astron. Zhur.* **44**, 930 (English trans. in *Sov. Astron.-AJ* **11**, 749).

Shvartsman, V. F., 1971. "Halos Around Black Holes," *Astron. Zhur.* **48**, 479 (English trans. in *Sov. Astron.-AJ* **15**, 377).

Sieber, W. and R. Wielebinski, 1981. eds. IAU Symposium No. 95, *Pulsars*, Reidel, Dordrecht, Holland.

Silk, J. and J. Arons, 1975. "On the Nature of the Globular Cluster X-Ray Sources," *Astrophys. J. Lett.* **200**, L131.

Sion, E. M., 1979. "Statistical Investigations of the Luminosity Distribution of the Spectroscopic White Dwarf Sample," in *White Dwarfs and Variable Degenerate Stars*, IAU Colloq. No. 53, H. M. Van Horn and V. Weidemann, editors, University of Rochester Press, Rochester, New York.

Slattery, W. L., G. D. Doolen, and H. E. DeWitt, 1980. "Improved Equation of State for the Classical One-Component Plasma," *Phys. Rev. A* **21**, 2087.

Smarr, L. L., 1979a. "Gauge Conditions, Radiation Formulae and the Two Black Hole Collision," in *Sources of Gravitational Radiation*, L. L. Smarr, editor, Cambridge University Press, Cambridge, England.

_____ 1979b. ed. *Sources of Gravitational Radiation*, Cambridge University Press, Cambridge, England.

Smarr, L., J. R. Wilson, R. T. Barton, and R. L. Bowers, 1981. "Rayleigh–Taylor Overturn in Supernova Core Collapse," *Astrophys. J.* **246**, 515.

Spitzer, L., 1962. *Physics of Fully Ionized Gases*, 2nd ed., Interscience, New York, New York.

_____ 1975. "Dynamical Theory of Spherical Stellar Systems with Large N," in IAU Symposium No. 69, *Dynamics of Stellar Systems*, A. Hayli, editor, Reidel, Dordrecht, Holland.

_____ 1978. *Physical Processes in the Interstellar Medium*, Wiley, New York, New York.

Staelin, D. H. and E. C. Reifenstein, 1968. "Pulsating Radio Sources near the Crab Nebula," *Science* **162**, 1481.

Stoeger, W. R., 1980. "Boundary-Layer Behavior of the Flow at the Inner Edge of Black Hole Accretion Disks," *Astrophys. J.* **235**, 216.

Strand, K. Aa., 1977. "Triple System Stein 2051 (G175–34)," *Astron. J.* **82**, 745.

Swank, J. H., R. H. Becker, E. A. Boldt, S. S. Holt, S. H. Pravdo and P. J. Serlemitsos, 1977. "Spectral Evolution of a Long X-Ray Burst," *Astrophys. J. Lett.* **212**, L73.

Sweeney, M. A., 1976. "Cooling Times, Luminosity Functions and Progenitor Masses of Degenerate Dwarfs," *Astron. Astrophys.* **49**, 375.

Szekeres, G., 1960. "On the Singularities of a Riemannian Manifold," *Publ. Mat. Debrecen* **7**, 285.

Tammann, G., 1978. "Supernova Rates," in *Supernovae and Supernova Remnants*, edited by J. Danziger and A. Renzini, *Mem. Soc. Astron. Italiana*, **49**, 299.

Tananbaum, H., H. Gursky, E. M. Kellogg, R. Levinson, E. Schreier, and R. Giacconi, 1972a. "Discovery of a Periodic Pulsating Binary X-Ray Source in Hercules from Uhuru," *Astrophys. J. Lett.* **174**, L143.

Tananbaum, H., H. Gursky, E. Kellogg, R. Giacconi, and C. Jones, 1972b. "Observation of a Correlated X-Ray–Radio Transition in Cygnus X-1," *Astrophys. J. Lett.* **177**, L5.

Tassoul, J. L., 1978. *Theory of Rotating Stars*, Princeton University Press, Princeton, New Jersey.

Tassoul, J. L. and J. P. Ostriker, 1968. "On the Oscillations and Stability of Rotating Stellar Models. I. Mathematical Techniques," *Astrophys. J.* **154**, 613.

Taylor, J. H. and R. N. Manchester, 1977. "Galactic Distribution and Evolution of Pulsars," *Astrophys. J.* **215**, 885.

Taylor, J. H. and J. M. Weisberg, 1982. "A New Test of General Relativity: Gravitational Radiation and the Binary Pulsar PSR 1913 + 16," *Astrophys. J.* **253**, 908.

Teukolsky, S. A. and W. H. Press, 1974. "Perturbations of a Rotating Black Hole. III. Interaction of the Hole with Electromagnetic and Gravitational Radiation," *Astrophys. J.* **193**, 443.

't Hooft, G., 1971. "Renormalizable Lagrangians for Massive Yang–Mills Fields," *Nucl. Phys. B* **35**, 167.

't Hooft, G. and M. Veltman, 1972a. "Regularization and Renormalization of Gauge Fields," *Nucl. Phys. B* **44**, 189.

_____ 1972b. "Combinatorics of Gauge Fields," *Nucl. Phys. B* **50**, 318.

Thorne, K. S., 1967. "The General Relativistic Theory of Stellar Structure and Dynamics," in *High-Energy Astrophysics*, Vol. III, C. DeWitt, E. Schatzman, and P. Véron, editors, Gordon and Breach, New York, New York.

_____ 1980. "Gravitational-Wave Research: Current Status and Future Prospects," *Rev. Mod. Phys.* **52**, 285.

Thorne, K. S. and V. B. Braginsky, 1976. "Gravitational-Wave Bursts from the Nuclei of Distant Galaxies and Quasars: Proposal for Detection Using Doppler Tracking of Interplanetary Spacecraft," *Astrophys. J. Lett.* **204**, L1.

Thorne, K. S. and R. H. Price, 1975. "Cygnus X-1: An Interpretation of the Spectrum and Its Variability," *Astrophys. J. Lett.* **195**, L101.

Toor, A. and F. D. Seward, 1977. "Observation of X-Rays from the Crab Pulsar," *Astrophys. J.* **216**, 560.

Trimble, V. L. and K. S. Thorne, 1969. "Spectroscopic Binaries and Collapsed Stars," *Astrophys. J.* **156**, 1013.

Trimble, V., W. K. Rose, and J. Weber, 1973. "A Low-Mass Primary for Cygnus X-1?" *Mon. Not. Roy. Astron. Soc.* **162**, 1P.

Trümper, J., W. Pietsch, C. Reppin, W. Voges, R. Staubert, and E. Kendziorra, 1978. "Evidence for Strong Cyclotron Line Emission in the Hard X-Ray Spectrum of Hercules X-1," *Astrophys. J. Lett.* **219**, L105.

Tsuruta, S., 1974. "Cooling of Dense Stars," in IAU Symposium 53, *Physics of Dense Matter*, C. J. Hansen, editor, Reidel, Dordrecht, Holland, p. 456.

_____ 1979. "Thermal Properties and Detectability of Neutron Stars—I. Cooling and Heating of Neutron Stars," *Phys. Rept.* **56**, 237.

Tsuruta, S. and A. G. W. Cameron, 1965. "Cooling of Neutron Stars," *Nature* **207**, 364.

_____ 1966. "Rotation of Neutron Stars," *Nature* **211**, 356.

Tubbs, D. L. and D. N. Schramm, 1975. "Neutrino Opacities at High Temperatures and Densities," *Astrophys. J.* **201**, 467.

Tuchman, Y. and R. Z. Yahel, 1977. "Accretion onto Neutron Star: The X-Ray Spectra and Luminosity," *Astrophys. Space Sci.* **50**, 473.

Tuohy, I. and G. Garmire, 1980. "Discovery of a Compact X-Ray Source at the Center of the Supernova Remnant RCW 103," *Astrophys. J. Lett.* **239**, L107.

Van den Bergh, S., 1978. "Comment on Supernova Rates," in *Supernovae and Supernova Remnants*, edited by J. Danziger and A. Renzini, *Mem. Soc. Astron. Italiana*, **49**, 299.

Van den Heuvel, E. P. J., 1974. "Modes of Mass Transfer and Classes of Binary X-Ray Sources," *Astrophys. J. Lett.* **198**, L109.

Van den Heuvel, E. P. J. and J. Heise, 1972. "Centaurus, X-3, Possible Reactivation of an Old Neutron Star by Mass Exchange in a Close Binary," *Nature Phys. Sci.* **239**, 67.

Van Horn, H. M., 1968. "Crystallization of White Dwarfs," *Astrophys. J.* **151**, 227.

Van Paradijs, J., 1978. "Average Properties of X-Ray Burst Sources," *Nature* **274**, 650.

———— 1979. "Possible Observational Constraints on the Mass-Radius Relation of Neutron Stars," *Astrophys. J.* **234**, 609.

———— 1980. "On the Maximum Luminosity in X-Ray Bursts," *Astron. Astrophys.* **101**, 174.

Van Riper, K. A., 1982. "Stellar Core Collapse. II. Inner Core Bounce and Shock Propagation," *Astrophys. J.* **257**, 793.

Van Riper, K. A. and D. Q. Lamb, 1981. "Neutron Star Evolution and Results from the Einstein X-Ray Observatory," *Astrophys. J. Lett.* **244**, L13.

Van Riper, K. A. and J. M. Lattimer, 1981. "Stellar Core Collapse. I. Infall Epoch," *Astrophys. J.* **249**, 270.

Wagoner, R. V., 1969. "Physics of Massive Objects," *Ann. Rev. Astron. Astrophys.* **7**, 553.

———— 1979. "Low-Frequency Gravitational Radiation from Collapsing Systems," *Phys. Rev. D* **19**, 2897.

Walborn, N. R., 1973. "The Spectrum of HDE 226868 (Cygnus X-1)," *Astrophys. J. Lett.* **179**, L123.

Walecka, J. D., 1974. "A Theory of Highly Condensed Matter," *Ann. Phys.* **83**, 491.

Walter, F. M., N. E. White, and J. H. Swank, 1981. IAU Circular No. 3611.

Walter, F. M., S. Bowyer, K. O. Mason, J. J. Clarke, J. P. Henry, J. Halpern and J. E. Grindlay, 1982. "Discovery of a 50 Minute Binary Period and a Likely 22 Magnitude Optical Counterpart for the X-Ray Burster 4U 1915–05," *Astrophys. J. Lett.* **253**, L67.

Wapstra, A. H. and K. Bos, 1976. "A 1975 Midstream Atomic Mass Evaluation," *Atomic Data and Nuclear Data Tables* **17**, 474.

———— 1977. "The 1977 Atomic Mass Evaluation," *Atomic Data and Nuclear Data Tables* **19**, 175.

Weaver, T. A. and S. E. Woosley, 1980. "Evolution and Explosion of Massive Stars," *Ann. N.Y. Acad. Sci.* **336**, 335 (Ninth Texas Symposium).

Weaver, T. A., G. B. Zimmerman, and S. E. Woosley, 1978. "Presupernova Evolution of Massive Stars," *Astrophys. J.* **225**, 1021.

Weber, J., 1960. "Detection and Generation of Gravitational Waves," *Phys. Rev.* **117**, 306.

Wegner, G., 1980. "A New Gravitational Redshift for the White Dwarf σ^2 Eri B," *Astron. J.* **85**, 1255.

Weidemann, V., 1967. "Luminosity Function and Space Density of White Dwarfs," *Z. Astrophys.* **67**, 286 (in German).

———— 1977. "Mass Loss towards the White Dwarf Stage," *Astron. Astrophys.* **59**, 411.

Weinberg, S., 1972. *Gravitation and Cosmology*, Wiley, New York, New York.

———— 1979. "Cosmological Production of Baryons," *Phys. Rev. Lett.* **42**, 850.

Weisskopf, V. F., 1981. "The Formation of Cooper Pairs and the Nature of Superconducting Currents," *Contemp. Phys.* **22**, 375.

Wheaton, W. A., J. P. Doty, F. A. Primini, B. A. Cooke, C. A. Dobson, A. Goldman, M. Hecht, J. A. Hoffman, S. K. Howe, A. Scheepmaker, E. Y. Tsiang, W. H. G. Lewin, J. L. Matteson, D. E. Gruber, W. A. Baity, R. Rothschild, F. K. Knight, P. Nolan, and L. E. Peterson, 1979. "An Absorption Feature in the Spectrum of the Pulsed Hard X-Ray Flux from 4U 0115 + 63," *Nature*, **282**, 240.

Wheeler, J. A., 1966. "Superdense Stars," *Ann. Rev. Astron. Astrophys.* **4**, 393.

_____ 1968. "Our Universe: The Known and the Unknown," *American Scientist* **56**, 1.

White, N. E. and S. H. Pravdo, 1979. "The Discovery of 38.22 Second X-Ray Pulsations from the Vicinity of OAO 1653–40," *Astrophys. J. Lett.* **233**, L121.

White, N. E. and J. H. Swank, 1982. "The Discovery of 50 Minute Periodic Absorption Events from 4U 1915–05," *Astrophys. J. Lett.* **253**, L61.

White, N. E., K. O. Mason, H. E. Huckle, P. A. Charles, and P. W. Sanford, 1976a. "Periodic Modulation of Three Galactic X-Ray Sources," *Astrophys. J. Lett.* **209**, L119.

White, N. E., K. O. Mason and P. W. Sanford, 1976b. "The X-Ray Behaviour of 3U 0352 + 30 (X Per)," *Mon. Not. Roy. Astron. Soc.* **176**, 201.

White, N. E., G. E. Parkes, P. W. Sanford, K. O. Mason, and P. G. Murdin, 1978. "Two X-Ray Periodicities from the Vicinity of 4U 1145–61," *Nature* **274**, 665.

Wilson, J. R., 1971. "A Numerical Study of Gravitational Stellar Collapse," *Astrophys. J.* **163**, 209.

_____ 1979. "A Numerical Method for Relativistic Hydrodynamics," in *Sources of Gravitational Radiation*, L. L. Smarr, editor, Cambridge University Press, Cambridge, England p. 423.

_____ 1980. "Neutrino Flow and Stellar Core Collapse," *Ann. N.Y. Acad. Sci.* **336**, 358 (Ninth Texas Symposium).

Wilson, J. R., R. Couch, S. Cochran, J. LeBlanc, and Z. Barkat, 1975. "Neutrino Flow and the Collapse of Stellar Cores," *Ann. N.Y. Acad. Sci.* **262**, 54 (Seventh Texas Symposium).

Withbroe, G. L., 1971. "The Chemical Composition of the Photosphere and the Corona," in *The Menzel Symposium* (NBS Special Pub. 353), K. B. Gebbie, editor, Government Printing Office, Washington, D.C.

Wolff, R. S., H. L. Kestenbaum, W. Ku, and R. Novick, 1975. "Search for Continuous X-Ray Emission from NP 0532," *Astrophys. J. Lett.* **202**, L77.

Woltjer, L., 1964. "X-Rays from Type I Supernova Remnants," *Astrophys. J.* **140**, 1309.

_____ 1972. "Supernova Remnants," *Ann. Rev. Astron. Astrophys.* **10**, 129.

Yahil, A. and J. M. Lattimer, 1982. "Supernovae for Pedestrians," in *Supernovae: A Survey of Current Research*, M. J. Rees and R. J. Stoneham, editors, Reidel, Dordrecht, Holland.

Young, P. J., J. A. Westphal, J. Kristian, C. P. Wilson, and F. P. Landauer, 1978. "Evidence for a Supermassive Object in the Nucleus of the Galaxy M87 from SIT and CCD Area Photometry," *Astrophys. J.* **221**, 721.

Youngblood, D. H., C. M. Rosza, J. M. Moss, D. R. Brown, and J. D. Bronson, 1977. "Isoscalar Breathing-Mode State in ^{144}Sm and ^{208}Pb," *Phys. Rev. Lett.* **39**, 1188.

_____ 1978. "The Breathing Mode State in ^{144}Sm and ^{208}Pb," in *Proceedings of the International Conference on Nuclear Structure*, *J. Phys. Soc. Japan (Suppl.)* **44**, 197.

Zapolsky, H. S. and E. E. Salpeter, 1969. "Mass-Radius Relation for Cold Spheres of Low Mass," *Astrophys. J.* **158**, 809.

Zaringhalam, A., 1982. Cited in Brown, Bethe, and Baym (1982).

Zel'dovich, Ya. B., 1958. "Nuclear Reactions in Super-Dense Cold Hydrogen," *Sov. Phys.-JETP* **6**, 760.

____ 1962. "The Equation of State at Ultrahigh Densities and Its Relativistic Limitations," *J. Exp. Theor. Phys.* (*U.S.S.R.*) **41**, 1609 (English trans. in *Sov. Phys.-JETP* **14**, 1143).

____ 1964. "The Fate of a Star and the Evolution of Gravitational Energy upon Accretion," *Dokl. Akad. Nauk SSSR* **155**, 67 (English Trans. in *Sov. Phys.-Dokl.* **9**, 195).

Zel'dovich. Ya. B. and O. H. Guseynov, 1965. "Collapsed Stars in Binaries," *Astrophys. J.* **144**, 840.

Zel'dovich. Ya. B. and I. D. Novikov, 1966. "The Hypothesis of Cores Retarded during Expansion and the Hot Cosmological Model," *Astron. Zhur.* **43**, 758 (English trans. in *Sov. Astron.-AJ* **10**, 602).

____ 1971. *Relativistic Astrophysics*, Vol. 1, University of Chicago Press, Chicago, Illinois.

Zel'dovich, Ya. B. and Yu. P. Raizer, 1966. *Physics of Shock Waves and High-Temperature Hydrodynamic Phenomena*, Academic Press, New York, New York.

Zel'dovich, Ya. B. and N. I. Shakura, 1969. "X-Ray Emission, Accompanying the Accretion of Gas by a Neutron Star," *Astron. Zhur.* **46**, 225 (English trans. in *Sov. Astron.-AJ* **13**, 175).

Zimmerman, M., 1980. "Gravitational Waves from Rotating and Precessing Rigid Bodies. II. General Solutions and Computationally Useful Formulas," *Phys. Rev. D* **21**, 891.

Author Index

Page numbers in *italics* are for co-authors designated by et al. in the cited page but listed in full in the References.

Subject Index

Accretion:
  complexity of calculation discussed, 403
  as source of X-rays from compact objects,
    371, 394
  in X-ray binary pulsars, 243
  in X-ray bursters, 393
  see also Accretion onto black holes; Accretion
    onto neutron stars; Accretion onto white
    dwarfs; Bondi solution for spherical
    accretion; Disk accretion; Spherical
    accretion; Spherical accretion onto black
    holes; Spherical accretion onto neutron
    stars; Spherical accretion onto white
    dwarfs
Accretion onto black holes:
  Cygnus X-1, 389, 447
  in extenso, 403ff
  overview, 403
  as quasar model, 371, 403, 500
  relativistic treatment, 568
  see also Accretion; Cygnus X-1
Accretion onto neutron stars:
  disk model, 453
  efficiency, 394, 403
  in extenso, 450ff
  Ghosh-Lamb disk model, 453
  luminosity from in general relativity, 430
  magnetosphere, 450
  radiation spectrum, 460
  simple spherical model, 394
  spin-up, 456, 459, 460
  in X-ray binary pulsars, 377
  see also Accretion
Accretion onto white dwarfs:
  gravitational versus nuclear energy release,
    464
  magnetic versus nonmagnetic, 464

radiation spectrum, 465
simple spherical model, 395
see also Accretion
Accretion radius, defined, 409
Accretion rate, in X-ray binaries, 395, 401
Action:
  sign of, 556
  in variational principles, 553
Adiabatic, *see* Adiabatic index; Fluids, adiabatic
  flow
Adiabatic index, 58, 68:
  for Baym-Bethe-Pethick (BBP) equation of
    state, 49, 196
  for Bethe-Pethick-Sutherland (BPS) equation
    of state, 49
  critical value for radial stability, 140, 145,
    146, 149, 151
  critical value for radial stability in general
    relativity, 152, 160
  critical value for radial stabililty in rotating
    stars, 185
  critical value for stability of uniform density
    relativistic stars, 260
  during supernova formation, 520, 539
  governing perturbations, 133
  for Harrison-Wheeler (HW) equation of state,
    49
  for ideal cold n-p-e gas, 49, 233
  for ideal cold n-p-e gas with pions, 233
  for ideal degenerate Fermi gas, 49
  modification for relativistic electrons,
    424
  near neutron drip, 196, 252
  near nuclear density, 221
  in spherical accretion, 412
Affine parameter, 114
Alfvén radius, 451, 456